Biodegradable Polymers in the Circular Plastics Economy

Biodegradable Polymers in the Circular Plastics Economy

Edited by Michiel Dusselier and Jean-Paul Lange

WILEY-VCH

Editors

Prof. Dr. ir. Michiel Dusselier
KU Leuven
Faculty of Bioscience Engineering
Center for Sustainable Catalysis &
Engineering
Celestijnenlaan 200F
3001 Heverlee
Belgium

Prof. Dr. ir. Jean-Paul Lange
Shell Global Solutions Int. B.V.
Grasweg 31
1031 HW Amsterdam
The Netherlands

Cover Image: Margo Togni

Library of Congress Card No.: applied for

British Library Cataloguing-in-Publication Data
A catalogue record for this book is available from the British Library.

Bibliographic information published by the Deutsche Nationalbibliothek
The Deutsche Nationalbibliothek lists this publication in the Deutsche Nationalbibliografie; detailed bibliographic data are available on the Internet at <http://dnb.d-nb.de>.

Print ISBN: 978-3-527-34761-2
ePDF ISBN: 978-3-527-82756-5
ePub ISBN: 978-3-527-82757-2
oBook ISBN: 978-3-527-82758-9

Cover Design: Adam Design, Weinheim, Germany
Typesetting Straive, Chennai, India

Printed and bound by CPI Group (UK) Ltd, Croydon, CR0 4
C9783527347612_310724

Contents

Preface

Since the start of the mass production of plastics in the 1950s, synthetic polymers have profoundly improved human comfort through their pallet of properties, their ease of shaping, and their low cost. Yet, these plastics are affecting more than just humans.

The presence of plastic litter in the environment and its impact on wildlife has hit all forms of media since the past five years. Due to their very slow – practically nonexistent – degradation they are accumulating everywhere, even in the remotest of places, in an unsustainable manner. For a large part, this problem stems from irresponsible human behavior and should be tackled at that level by public authorities, e.g. through public sensibilization and effective waste collection infrastructure. Once collected, the plastic waste becomes a valuable resource for recycling to polymers, monomers, or feedstock. Hence, much effort in academia and industry is devoted to developing and deploying the necessary technologies. But

- Will recycling fully solve the litter problem?
- Should humanity replace "persistent" products for non-persistent/degradable ones?
- Should we keep some persistent materials for specific applications?
- Do we foresee value for degradation beyond the litter challenge?
- Do we understand the degradation mechanisms? And can we steer or trigger them?
- What do we really mean by (bio)degradable, compostable, bio-, or biobased plastics?
- Will there be a need to compromise in properties for allowing non-persistence?
- How can we introduce new polymers, non-persistent ones, without jeopardizing the sorting and recycling of the whole polymer mix?

In summary, how can we balance non-persistence with durable, functional, reusable, and/or recyclable while designing plastics?

This book tries to offer the groundwork to enable a balanced discussion toward answering these questions. Short of answering all *key questions* posed above, it will hopefully offer an insight into the wonderful chemistry and engineering of degradable and biobased plastics that have come to the fore through a comprehensive approach. It starts by setting the scene and clarifying the terminology. It then discusses the mechanisms of biodegradation and the environmental impact of persistent polymers. A second section provides a short tutorial on polymers and then covers, in depth, the various families of degradable/hydrolysable polymers, namely polyesters, polysaccharides, and lignin as well as covalent adaptable polymer networks (also known as vitrimers). Finally, a third section places (bio)degradation in the bigger context of plastic waste and recycling, analyzes the life cycle of plastics, and introduces the commercialization barriers that industry (and society) needs to pass to deploy a truly circular plastic economy.

The world needs a lot of different scientific angles to design the new plastics of tomorrow. This book thus hopes to inspire chemists, engineers, and biologists, but also lawmakers and consumers (as we all are). We hope you'll enjoy this book and its chapters, and appreciate the nuanced vision on a new plastics economy we tried to convey.

We wish to thank the Wiley team for their invitation and support, Margo Togni for designing the cover, and Dr. Stefaan De Wildeman for his help in initiating this project. Finally, we are grateful to all our co-authors for teaching us so much and for accommodating all our requests during the editing process.

January 2022

Prof. Dr. ir. Michiel Dusselier
Prof. Dr. ir. Jean-Paul Lange
KU Leuven
Pellenberg

1

Biodegradable Polymers – A Tutorial for a Circular Plastics Economy

Jean-Paul Lange[1,2], Michiel Dusselier[3], and Stefaan De Wildeman[4]

[1] *Shell Global Solutions International B.V., Grasweg 31, Amsterdam 1031 HW, Netherlands*
[2] *University of Twente, Sustainable Process Technology, Faculty of Science and Technology, Drienerlolaan 5, P.O. Box 217, Enschede 7522 NB, Netherlands*
[3] *KU Leuven, Center for Sustainable Catalysis and Engineering (CSCE), Faculty of Bioscience Engineering, Celestijnenlaan 200F, 3001 Leuven, Belgium*
[4] *B4Plastics R&D, IQ-Parklaan 2A, Dilsen-Stokkem 3650, Belgium*

1.1 Context

From the appearance of the first bakelite object in the early twentieth century to the ban on certain single-use plastics in 2021 in the EU, the modern world has changed drastically due to the advent of synthetic plastics. Although the mass production of most plastics – notably polypropylene and polyethylene – only truly started in the 1950s, the cumulative volumes produced since have grown rapidly [1–3]. This growth comes from their utility on our comfort, convenience, and health that plastics have delivered so far and, arguably, will deliver in the future – the age of 3D printing at home has just begun. However, these benefits also come at a price. There is a growing accumulation of plastic litter on land, in rivers and seas all over the world, and even inside living organisms along the whole food chain, from mussels to birds and mammals. Best exemplifying the scale of the issue are the plastic soups, i.e. concentrated floating plastic debris in all major subtropical ocean gyres. Sources estimate that more than 67% of the plastic marine load originates from 20 major rivers [3].

This pollution problem has two main root causes. The first and most important one is us, the consumers, that are discarding waste plastic without consideration for the consequences and/or, by lack of proper waste collection infrastructure. Corporations also cause spillage and ill-disposal. Plastics were never meant to end up in the environment, although little consideration has been given to their end-of-life in the first 50 years since their mass production. As a matter of fact, nowadays 30% of the production is littered, notwithstanding an additional 40% that ends up in landfill and waste dumps [1]. The second root cause is the high chemical stability of the materials: plastics should indeed be considered as persistent pollutants that were never present together with fauna and flora during their almost four billion years of evolution.

Biodegradable Polymers in the Circular Plastics Economy, First Edition.
Edited by Michiel Dusselier and Jean-Paul Lange.

The first root cause would be subject of a book on social and political science and will be kept out of scope here. But the second is a consequence of technology, leaning close to the purpose of this book. Is there a reason to run the plastics economy further with almost completely inert materials, or do we have alternatives that lead us to a new plastics economy, as advocated by the Ellen McArthur Foundation? [1]. Of course, as the challenges of the current plastics industry are rapidly urging us to take actions, we foresee interwoven debate and regulations to align both root causes.

Diving to the heart of technological issue, the book aims at presenting a comprehensive view on biodegradable plastics, encompassing a range of aspects from synthesis, e.g. based on renewable resources, to properties and applications, recycling, biodegradation, and environmental impacts as well as business challenges. The following sections will summarize and analyze for you the 13 chapters of the book with a critical look, extract their major learnings, and formulate an outlook for the future of biodegradable plastics. The structure of the book is visually outlined in Figure 1.1.

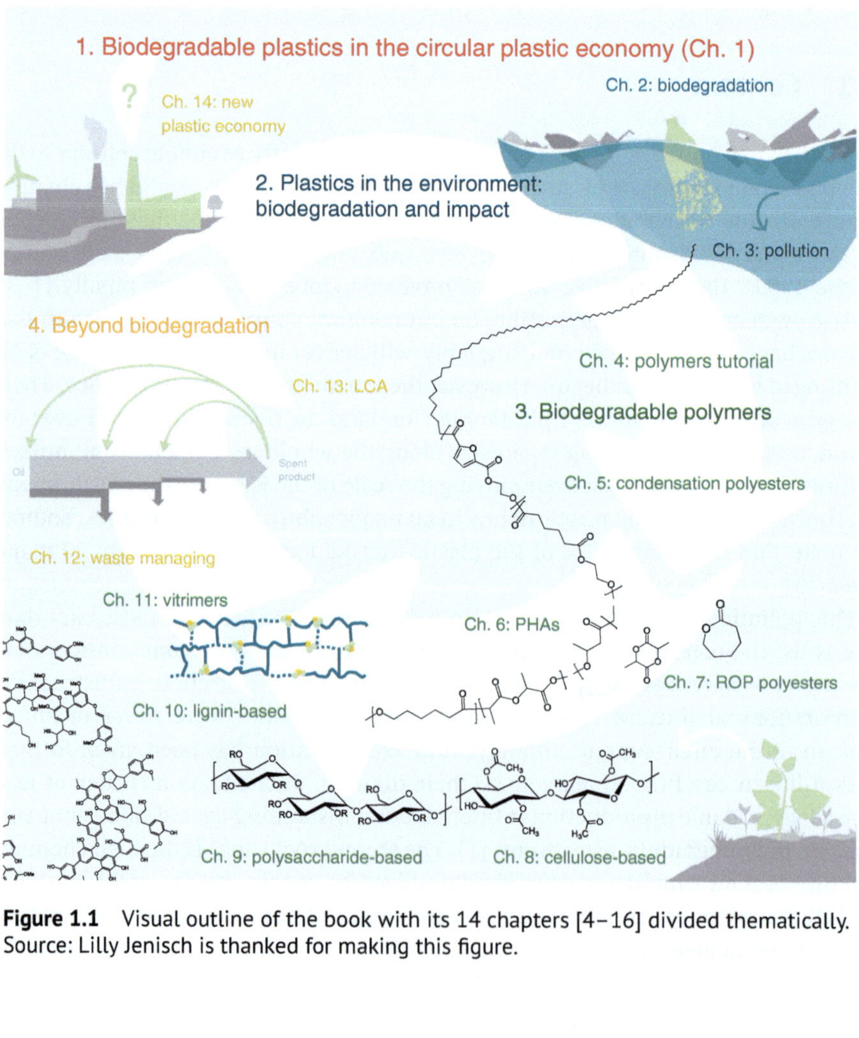

Figure 1.1 Visual outline of the book with its 14 chapters [4–16] divided thematically. Source: Lilly Jenisch is thanked for making this figure.

But the plastic litter is not the sole problem to solve to develop a truly circular economy. The chemical industry also needs to disconnect from fossil hydrocarbons as feedstock and energy sources and minimize all its emissions, particularly eliminating those greenhouse gas emissions that contribute to climate change [2]. Accordingly, the industry must eventually switch to recycled and renewable carbon as feedstock for chemicals and plastics [2]. A major role is therefore laid down for the use of atmospheric CO_2, via sustainable biomass or direct air capture, to produce sustainable plastics. These sustainable plastics represent the nexus of plastic litter and climate change. They will therefore be an inevitable subject of discussion in this book on biodegradable plastics.

The book kicks off by sketching the generation of plastic litter and its environmental impact (Chapter 3, [4]) and analyzing the mechanisms of biodegradation (Chapter 2, [5]) as potential remediation to the problem (Chapter 3). Before diving into the different types of (semi-)degradable plastics out there, the book reviews today's polymers, both fossil- and bio-based (Chapter 4, [6]), their manufacture, properties, and applications. Among the various families of potentially biodegradable polymers, polyesters (Chapters 5–7, [7–9]) and polysaccharide-based plastics (Chapters 8 and 9 [10, 11]) feature prominently, but lignin-derived materials (Chapter 10, [12]) and more specialized recyclable thermosets – vitrimers – (Chapter 11, [13]) are also covered. Going beyond these technical options, degradable plastics are discussed in relation to plastic waste (end-of-life) management (Chapter 12, [14]) and from a life-cycle (analytical) perspective (Chapter 13, [15]). Finally, business challenges and marketing strategies for launching new polymers are outlined (Chapter 14, [16]).

Before diving into the heart of the subject, we wish to clarify a few points of confusion: the difference between biodegradable, compostable, bio-based, and bioplastics. These labels are often assumed to cover the same plastics and to oppose them to the traditional fossil-based plastics. But bio-based origin and biodegradable function are in fact unrelated; the former refers to its source while the latter to its end-of-life [17]. Bio-based plastics are literally derived from biological feedstocks, being from animal, vegetal, or other living organisms (e.g. fungi and bacteria). Many of them show significant biodegradability, but others are equally environmentally persistent as today's fossil plastics. For instance, biological feedstocks are sometimes the source for molecular structures that are identical to or resemble those derived from fossil feedstock and, therefore, show a similar lack of biodegradability. An obvious example is the inert bio-based polyethylene that is produced by converting sugars to ethanol, ethylene, and finally polyethylene. Along the same line, however, a limited number of fossil-based plastics show chemical structures that are readily biodegradable, as illustrated by oil-based polycaprolactone (PCL) (Chapter 7). We also need to differentiate compostable from biodegradable plastics. Compostable plastics are a sub-family of biodegradable plastics; they degrade most swiftly to make industrial composting affordable, as discussed below. Finally, we would recommend to avoid using the label "bioplastics" that is often confusingly used to cover all bio-based plastics (biodegradable and non-degradable) as well as the fossil-based biodegradable plastics. One could reserve the label of "bioplastics" to "bio-based and bio-made plastics," such as polyhydroxyalkanoates (PHAs) or biopolymers such as silk, cellulose, or starch (see Chapter 4). Such distinction should be limited to the scientific arena however, for it adds little value for the larger public, beyond confusion on source and end-of-life. An appendix in the tutorial (Chapter 4) further clarifies the spectrum of these alternative plastics.

1.2 Plastics in the Environment – Biodegradation and Impact of Litter

The biodegradation of polymers is a complex process that depends on many factors. Some are determined by the polymer itself, but others are determined by the environment such as T, pH, air, light, and density in microorganisms [1]. The suitability of conditions ranges from industrial composting as most favorable environment to home composting, soil and sludges, to fresh water and, finally, sea water and landfill as least favorable ones.

Rao and coworkers (Chapter 2, [5]) elegantly showed that biodegradation proceeds in three major steps: (i) formation of a biofilm – a colony of various wild microorganisms – on the plastic, (ii) depolymerization to low-M_w components after excretion of various enzymes onto the plastic, and (iii) ingestion and conversion of the low-M_w components as "feed" for the microorganisms. The polyolefins and aromatics polymers that form most today's polymers are generally recalcitrant to the first two steps of biodegradation: their high hydrophobicity makes them inhospitable for microorganism to form biofilms and their continuous C–C chains are difficult the break by excreted enzymes. In contrast, condensation polymers such as polyesters, polysaccharides, and polyamides are generally more prone to biodegradation as their hydrophilicity allows biofilm formation and their weaker C–O or C–N bridges in the main chain are prone to enzymatic hydrolysis and cleavage. In this latter case, the rate of biodegradation can further be tuned by numerous parameters. For instance, one can tune the accessibility of the main chain e.g. by tuning its crystallinity. But one can also tune the hydrolytic reactivity of the weak links by selecting a reactive hemiketal bond (in polysaccharides), a reactive aliphatic ester bond (in aliphatic polyesters), or a more recalcitrant aromatic ester (in polyethylene terephthalate [PET]), or amide one. The intrinsic recalcitrance of polyolefins can also be tamed by pretreatment or adding oxidation or photooxidation catalysts. Such tunability will come handy to adjust the biodegradability of plastic to their use and end-of-life. Interestingly, for some biopolymers such as the natural polyamide silk, the degradation pathways and microorganisms have not yet been elucidated. This is perhaps a good proof that research into biodegradation mechanisms for plastics is still developing and that many open questions remain.

Tian et al. (Chapter 3, [4]) quantitatively describe the magnitude of the problem of plastic litter and its environmental impact, which is striking, both with very visible and truly hidden components. They also provide estimates on the extremely long lifetime of plastic items, with over 200 years for a plastic straw. Their chapter focuses on the scale of the issue and the different sources of macroplastics as well as micro- and nanoplastics. Microplastics are generated through various routes, e.g. mismanagement of waste, accidental release, microplastics additives to products, the degradation of outdoor objects or natural wear of products, e.g. from rolling tires and washing textiles. While the ecological impact is visible, e.g. in birds strangled in waste plastics, the economic and human health impact, especially of microplastics, is less well documented. In the context of this chapter, the importance of biodegradability is introduced: even though some degradable plastics are not biodegradable according to home or industrial composting standards, they still degrade at significant rate to be considered as "non-persistent" chemicals. Polylactide (PLA), for instance, would fail home composting in under a year, but it is expected to degrade within a few years in ambient conditions, while PET would require 100+ years. Time will tell what rates of degradation would lead to a manageable steady state vis-a-vis leakage and littering. Finally, Tian et al. present a list of solutions and mitigations to the plastic conundrum with, among others, a focus on material design and on waste management.

Building on these two chapters, it is important to stress that biodegradability should be discussed in a more nuanced way than the binary "biodegradable and non-biodegradable" that is presently applied. Today's strict biodegradation label is meant to make industrial biodigesters affordable by ensuring a limited reactor size and residence time for a given load of generally known plastics. But this definition should really be limited to compostable materials, while the broader label "biodegradable" should be used to also cover materials that slowly degrade in nature [17] and can, thereby, qualify as non-persistent pollutants, which is arguably the truly critical property to consider for protecting our environment. The difficulty is to assess (and certify) the level of non-persistence in natural conditions or thus to establish what an acceptable rate of biodegradation would be.

1.3 Biodegradable Polymers

The scene being set, we will now dive deeper into the world of polymers, with their raw materials, synthesis, structure, performance, and end-of-life options. We will start with the fossil-based polymers that are dominating the market today and then, present some bio-based alternative materials, some old ones and some very novel.

Today's polymers are mostly derived from fossil hydrocarbons, mainly crude oil, as discussed by Gruter and Lange (Chapter 4, [6]). The largest fraction is made of vinylic monomers $H_2C{=}C(R){-}X$ that can be polymerized in a variety of vinyl polymers with continuous carbon chains such as polyolefins, polystyrene, polyvinylchloride (PVC), or polyacrylate. But a smaller fraction of polymers contains heteroatoms in their main carbon chain, which come from polymerizing monomers through ester bonds (—C(O)O—), amide bonds (—C(O)—NH—), acetal bonds (—C(OR)$_2$—), urethane or carbamate bonds (—C(NH)—O—), or carbonate bonds (—O—C(O)—O—). These can simplistically be referred to as condensation polymers,[1] whereas the former ones are referred to as addition polymers. The majority of the polymers consist of thermoplastics that are shaped and reshaped at will by melting, shaping (e.g. to desired bottle, film, or fiber), and finally cooling for use. Occasionally, some of these polymer chains can be bound or cross-linked to one another to form rigid thermosets for specific application such as tires, mattresses, or coating. While most polymers are still derived from fossil feedstock today, a small, though growing fraction is made of renewable feedstock such as sugars or vegetable oils.

Polymers have become very popular today because they combine several valuable properties (Chapter 4): lightweight, tunable strength and elasticity, easy shaping and coloring, tunable transparency, thermal/electric insulation, and many more. These properties are rewardingly exploited for a variety of applications. Some include rigid containers and castings, flexible films for packaging, rigid sheets for finished two-dimensional articles, fibers for textiles and ropes, foams for cushions and mattresses or all kinds of surface coverings, shock-absorbing joints or assembling agents such as coatings, adhesives, sealants, and elastomers. These various applications require totally different combinations of properties. Particularly demanding applications can further be reached by reinforcing polymers with strong materials such as glass or carbon fibers and, thereby forming composites.

The chemical composition of these polymers has a significant impact on the options that are available to responsibly manage plastic waste. For instance, most vinyl polymers are persistent pollutants when released into the environment. When collected, however, they are generally good candidates for chemical recycling to hydrocarbon feedstock, eventually via cracking or other stringent chemical modifications. Condensation polymers are more readily susceptible to chemical depolymerization: their constituting monomers can directly be recuperated after solvolysis with for instance alcohols or even water. Moreover, these polymers are often also more prone to biodegradation, depending on the type of heteroatom linkages and of the concerned monomers. For instance, PET with aromatic flanked ester bonds degrades several orders of magnitude slower in natural conditions (and could be called persistent) than linear aliphatic polyesters such as polybutylenesuccinate (PBS). Condensation polymers are essential in all biodegradable plastic developments. They thus receive more attention in this book.

1.3.1 Polyesters

Polyesters are very promising materials because they offer a wide variety of properties, from rigids to flexible films and from hydrophobic to hydrophilic materials [18–22]. Moreover, they are readily depolymerized to their monomers by solvolysis, such as hydrolysis or alcoholysis, which is valuable for chemical recycling

1 Condensation polymers are mostly known in this context as broad term. From a polymer chemistry point of view, condensation polymerization equates removing water from the monomer during its polymerization. PLA can for instance be made via condensation polymerization or via ring opening from the anhydrous lactide. Chapter 5 [7] deals with true condensation polyesters.

or biodegradation. These depolymerization technologies are technically well developed but not yet deployed at scale by lack of economic rivers.

Stouten and Bernaerts (Chapter 5, [7]) present the synthesis and properties of a variety of aliphatic polyesters that can be made via dehydration (condensation polymerization). Some are based on aliphatic diols and diacids (e.g. 1,4-butanediol and succinic acid) or on aliphatic hydroxyacids (e.g. lactic acid), while others may consist of aromatic monomers such as terephthalic acid. Their properties can be derived from the expanding set of known polyesters tested by many academic and industrial groups. For instance, aliphatic polyesters built on short-chain monomers tend to be rather stiff, whereas those built on long-chain monomers are more flexible and show a lower T_g. But polyesters can also be made stronger and stiffer by using more rigid monomers such as terephthalic acid and its emerging incumbent furanic dicarboxylic acid (FDCA) but also cyclic aliphatic diols such as isosorbides. Aliphatic polyesters are notoriously easier to depolymerize by hydrolysis or alcoholysis than their aromatic counterparts. Hence, they are also widely used in fast(er) degrading single-use applications e.g. in medical articles, in agriculture (e.g. for covering fields) and, increasingly, in packaging or single-use goods (Chapter 5, [7]).

Koller and Mukherjee (Chapter 6, [8]) dive further into a specific type of aliphatic polyesters, the polyhydroxyalkanoates (PHA), and discuss their unique manufacture and tunability. In contrast to most polymers discussed so far, PHA are synthesized *in vivo* by various microorganisms and stored as energy-rich granules to be metabolized when the microorganism falls short of carbon source; PHA are stored in microorganisms much like starch is stored in cereal and tubers for later plant/seed growth. This means that the PHA are designed to be biodegradable. These microorganisms can produce PHA from a variety of feedstocks, e.g. sugars, glycerol, or fatty acids but also synthesis gas ($CO + H_2$) or CO_2. They can also produce a large variety of PHA. The most common is based on 3-hydroxybutyric acid. But over 150 monomers have been identified. This versatility allows the microorganisms to produce PHAs with a large variety of properties. Koller et al. discuss the conditions required to produce various PHA and the challenges encountered to recover the PHA from the organisms. The PHA is often recovered from the microorganism by solvent extraction, occasionally assisted with cell disruption. This has for long remained a critical step in the production cost and has, therefore, hampered the deployment of PHA. Recent advances in the field help in reducing the production cost to promote their long-awaited growth in the market.

Narmon et al. (Chapter 7, [9]) deliver a detailed tutorial on ring-opening polymerization (ROP) as a sustainable and scalable strategy for high-molecular-weight polymers. The chapter dives into (mostly aliphatic) polyesters made from lactones. The most famous and commercially available polyesters from ROP are PLA and PCL. A primer to the kinetics and thermodynamics of ROP is given, where ring strains are key, as well as routes to functionalized ROP-based polyesters. Functionalization can be a valuable strategy to lower crystallinity and increase hydrophilicity, which are crucial in determining the polymer's biodegradation potential. Functionalization can be introduced either through ROP of substituted lactone monomers or through modification and grafting of the polymer. ROP-based polyesters can be categorized based on the ring sizes (mostly 5–12 membered) of their lactone and dilactone monomers, which largely determine polymer properties and sometimes also their relation to degradability. An interesting example is the four-membered ring β-butyrolactone, which offers an alternative route (via ROP) to the most popular PHA, poly(3-hydroxybutyrate). Although a lot of functional and structurally different lactones (and their polyesters) have been reported, only few clear structure–degradation relations have been developed thoroughly, most notably for co-polymers of lactide (PLA-based co-polymers). Assessment methods for hydrolytic degradability are not uniform and thus briefly but critically discussed.

1.3.2 Polysaccharides

Bio-based and biodegradable polymers are not limited to polyesters, however. The most traditional and abundant biodegradable polymers are not polyesters but polysaccharides [23]. Regenerated celluloses are traditional material for textile fibers (rayon) or films (cellophane), while thermoplastic (i.e. plasticized) starches are used in packaging and plastic bags. The originally stiff polysaccharide materials are softened by capping the free hydroxyl groups, e.g. in ester (–O–(O)C–) or xanthate groups (–O–(S)C–S–) or by inserting low-M_w plasticizers such as glycerol [24, 25]. But a variety of novel polysaccharide materials are also being developed, e.g. based on microbial polysaccharides.

In Chapter 8, [10] Molenveld and Slaghek discuss the class of cellulose-based plastics in detail. Being among the first thermoplastic polymers, they have found their way to applications based on decent mechanical and optical properties. Key to rendering the non-plastic cellulose into a thermoplast is chemical modification, e.g. by esterification of 1–3 of the three OH groups available (per mole of glucose entity) to tune the solubility and thermal properties and therefore its application. The chapter reports new developments on the esterification process, e.g. using new solvents or mixed esters, as well as on plasticizing and blending studies. Critically, the cost and sustainability of the cellulose plastics is to a large extent determined by the post-processing or derivatization steps. Finally, even though these plastics have been around since the nineteenth century, there is still little fundamental insight into the biodegradation mechanisms of the ester derivatives.

In Chapter 9, [11] Iwata and coworkers review the development of novel materials based on starch alternatives that are naturally produced by microorganisms instead of plants. These more exotic microbial polysaccharides connect glucose in different ways than do plants in cellulose and starch, namely through 1,3-, 1,4- or 1,6-glucosidic bonds in equatorial (α) or axial (β) configuration. Inspired by the development of modified cellulose and starches discussed in Chapter 8 and elsewhere [8, 9], the free hydroxyls of these microbial polysaccharides can be capped by various groups, e.g. by various ester groups, to tune the final properties of the material. This allows the production of adhesives, polyolefin-like plastics, or high-performance engineering plastics, occasionally with unique optical transparency. Such modifications are also affecting the biodegradability of the polysaccharides by affecting their affinity for biofilm formation and affecting the accessibility of the hydrolysable hemiketal/glycosidic bond. Lee et al. thus briefly review the microbial synthesis and modification (esterification). As in the case of PHA discussed above, the extraction of the polysaccharide is likely to be a critical part in the industrial production of such materials.

1.3.3 Lignin

If polysaccharides are the most abundant biopolymers on earth, lignin is the second most abundant one. It is therefore not surprising to see much research devoted to upgrading lignin to materials [26–28]. Thermosets have historically been prime targets. Their synthesis generally consists of fixing functional groups to the lignin to allow cross-linking, or by using lignin as partial replacement of phenol in phenol–formaldehyde wood glue. But much more can be done with lignin, as discussed by Sarkanen and Chen in Chapter 10 [12].

Sarkanen and Chen (Chapter 10, [12]) first shatter the myth that process lignin such as kraft lignin would be highly cross-linked material. Instead, lignin is a linear phenolic polymer of ~2 kDa that reversibly associates to form complexes of ~20 kDa in alkaline water and of 2000 kDa in polar organic solvents after derivatization. This tendency to associate can be exploited to produce materials with tensile strength comparable to polystyrene (PS) by derivatizing various lignins and by adding minor amounts of monomeric or polymeric components. Interestingly, the lignin is biodegradable by organisms with strong oxidation capabilities and radical-generating enzymes such as laccases found in white-rot fungi. This opens the path toward fully biodegradable lignin-rich thermoplastics, when blended with biodegradable additives. Apart from that, the randomly substituted aromatic moieties in lignin have attracted the attention of chemical experts that want to link them to BTX (benzene–toluene–xylene) – the classical precursors for many high-performance materials, including many polyesters and polyamides – or to renewable bisphenol alternatives – monomer to polyesters, epoxy-resins, and polycarbonates [29].

1.3.4 Vitrimers – Recyclable Thermosets

So far, the discussion has been focused on thermoplastic materials and on the options that we have to make them bio-based and biodegradable. But one also needs to pay attention to the smaller, though important family of thermoset resins that are valuable (or required) in specific applications, e.g. for durable rigid structures, for elastomers, or for foams. These materials generally present challenges for responsible end-of-life management because they are difficult to recycle and equally difficult to biodegrade, as discussed in Chapter 11 [13].

An emerging family of polymers is attempting to mitigate these challenges, namely the covalent adaptive networks (CANs) and, particularly, the vitrimers (associative CAN). As demonstrated by Pitet and coworkers (Chapter 11, [13]), these materials consist of polymers that are cross-linked by reversible covalent bonds and are designed such that the cross-link can open and close reversibly and flexibly upon heating. Hence, these materials combine the typical creep resistance of thermosets at application temperature with thermoplasticity at elevated temperature that allow mechanical recycling by remolding/reshaping. Pitet et al. nicely present the various types of adaptable cross-links that have been developed so far, discuss the resulting thermal behavior and the impact on self-healing and recyclability.

Vitrimers are intrinsically condensation polymers, with heteroatoms in the main chain and with cross-linkers to build the network. This offers the potential (though not the promise) to also recycle them chemically by depolymerization to the monomer. Similarly, this makes them candidates for non-persistent materials and, possibly, even biodegradable materials. The complex chemistry required to produce the monomers may however limit their use to high-value niche applications.

1.4 Beyond Biodegradation

1.4.1 Recycling and End-of-Life

Coming back to the root causes of the accumulation of plastic litter in the environment, this book has now discussed the second root cause: the persistence of today's plastics and the potential to design novel non-persistent plastics. However, it has left the solution of the first root cause, the human behavior and lack of infrastructure, to the social and political sciences. Strong governmental leadership can minimize the first root cause by sensibilizing the consumer and providing efficient collection infrastructure for plastic waste, as found in numerous high-income countries. Once that is reached, the challenge comes back to the technology: what do we do with this collected plastic?

The industry has been struggling to (mechanically) recycle collected waste plastic in an economically viable manner for a few decades already. But this seems to be changing, now that consumers and society at large are pushing for solving this problem (Lange, Chapter 12, [14]). Much progress has already been achieved in mechanical recycling of the highest quality materials. The industry is now developing or resuscitating complementary technologies that can recycle spent polymers of lesser quality, i.e. by depolymerizing condensation polymers back to their monomers and by cracking addition polymers back to their general hydrocarbon feedstock. The residual waste fractions of the poorest quality can still be valorized by gasification, pyrolysis, or liquefaction technologies to produce chemical feedstock or liquid hydrocarbon fuels. But the residual waste, which is often called RDF or residue-derived fuel, can also be incinerated for energy recovery. Incineration can proceed either directly or indirectly after digestion of the organic fraction to biogas. Biodegradation eventually comes at the bottom of this valorization cascade, as it destroys the waste without producing any valuable product, save for some low-value compost. But biodegradation becomes invaluable for the fraction of materials that eventually ends up in the environment. Also, material recycling can't proceed eternally, for it suffers from material losses at all stage. Full closure of the carbon cycle requires to compensate such carbon losses by virgin intake of atmospheric CO_2 via natural or artificial photosynthesis (Lange, Chapter 12). Hence, bio-based plastics and, preferably, non-persistent ones are a necessary complement to recycling and an essential element of a circular chemical economy.

1.4.2 LCA

In view of all facets of plastics discussed above, there is a growing need to understand what makes (sustainable) sense and what not. Will proper end-of-life management allow the plastic economy to become environment friendly? Should society switch to novel and "greener" plastics, e.g. derived from biomass? Or should it return to more traditional materials such as wood/paper, metal, and glass/ceramic? These are very challenging questions to answer, for they require analysis of the whole life cycle of each option and consideration of many factors to properly assess the impact on land, water, air, and life and even external effects such as land-use change.

Li has devoted Chapter 13 [15] to highlight some of the challenges of such life-cycle analysis (LCA) and illustrating them by a study that tried to assess the environmental impact of replacing conventional plastics by bio-based ones in well-selected applications. Li reports important benefits for such substitutions. But she also shows that the benefits reached in some environmental aspects are often accompanied by penalties in others. This trade-off varies highly between cases and could change in the future by improving practices. This makes it difficult to draw general conclusions.

But LCA can also be applied to assess the environmental impact of recycling. Lange in Chapter 12 mentions a few studies that seem to converge in prioritizing mechanical recycling, over chemical recycling, over energy recovery and, finally, biodegradation.

1.4.3 Implementing the "New Plastics Economy"

This book has tried to address a few important technical aspects of the "New Plastics Economy." But there is clearly more than technology in the overall toolbox to achieve this New Economy. Beyond *why* we need it, and *which technical solutions* are at hand to create it, there is one pervasive question that needs attention: "How will we navigate there?" While announcements for new plastics have been manifold over the last decades, their successes have been rare. The hurdles to overcome have been numerous: too long time to market, too much focus (no pivot possible), too little focus (no stable path found), too costly in capital expenditure (CAPEX) or operational expenditure (OPEX), and many more "too's." For mitigating all these excesses, some business experiences and guidelines are discussed by Nevejans and De Wildeman (Chapter 14, [16]). For instance, they warn for trying to copy the mature industry and discuss the Technology Readiness levels (TRLs) in the context of these new plastics. The chapter concludes with five useful innovation rules to keep in mind when companies or individuals want to participate in the production chain of the alternative plastics economy, three of which are worth mentioning here: start with small-volume high-value and applications (#1), take to time to develop it right (#2), and develop truly "green" product (#5).

1.5 Conclusions and Outlook

Overall, plastic biodegradability or, more accurately, non-persistence must become a design parameter in plastic products. But the discussion is generally oversimplified and mixed or confused with other properties such as bio- versus fossil-based. This has resulted in the general perception that bio-based plastics are biodegradable, while fossil-based ones are not. Additional confusion has been born from mixing strict composting requirements with the broader biodegradation requirements that make littered material non-persistent in nature. Environmental non-persistence should indeed be considered over a wide timescale, from weeks to decennia (maybe more), depending on the application of the polymer, on its expected end-of-life, and on the risk of it ending up in the environment as unintentional litter or as microplastics generated during normal use (e.g. tires and textiles). On one end of the scale, fast biodegradation is very valuable in medical applications such as sewing wires. It would also be valuable for single-use products that get contaminated heavily with food, e.g. throw-away cutlery and beakers or packaging of wet food and that could be composted along with the food waste. On the other end of the scale, durable products, e.g. those used in construction, particularly outdoor, should resist biodegradation for decennia. Many other applications fall between these extremes (medical degradation vs. durable multi-decade long functionality). It is therefore tempting to design biodegradation as the last step of the waste management hierarchy, after "reduce, reuse, and recycle." Biodegradation should indeed neither hinder normal use and longevity of the product nor hinder its efficient recycling after use.

The chapters of this book have discussed opportunities in non-persistent materials. For instance, aliphatic polyesters such as PHA and PBS generally degrade rapidly in natural environments, whereas aromatic polyesters such as PET survive for much longer times, and PLA would be somewhere in between. Similarly, the modification of polysaccharides based, e.g. on starch or cellulose, allows to tune the rate of degradation besides tuning the mechanical properties. Sometimes, aligning the thermal and mechanical properties of such plastics with the application in mind can be a challenge. Even, lignin-based materials promise to be non-persistent through oxidative biodegradation by fungi. But other polymeric materials should also be considered, e.g. polyamides, polyethers, and polycarbonates. They all contain heteroatoms in their main polymer chain that offer anchor points to biofilms and biological attack. Selecting or chemically tailoring a condensation polymer (bio-based or not) for an application with substantial leakage to the environment – thus where the persistence has to be on the order of less than a few years – is in practice not an easy task. The degradation rate has to be somewhat in line with the leakage rate and what is deemed an acceptable accumulation in the environment. Multidisciplinary collaboration from the early stage of the research to high TRL levels may be necessary here, as the input of chemists (feedstock conversion and polymer design), engineers (processing, thermomechanical product properties), and biologists (biodegradation, toxicity, persistence) is needed almost simultaneously.

Interestingly, numerous non-persistent polymers are well suited for manufacturing from renewable carbon, being atmospheric CO_2 or biomass, e.g. sugars.

They are indeed rich in oxygen and/or nitrogen and contain, thereby, partially oxidized carbons, like these renewable resources. If today's polymers are dominated by oxygen-free polyolefins, it's largely due to the fact that these materials were well suited for production from the fossil oxygen-free hydrocarbon feedstock that the industry has been relying on throughout the twentieth century up to today. This represents an additional opportunity for deploying a new, sustainable, and circular plastic economy.

Looking beyond biodegradable plastics, a new circular plastic economy needs to start by recycling the spent plastic back to new materials [1]. Combining mechanical and chemicals recycling is not expected to fully close the carbon cycle, however. Carbon will inevitably be lost at all stages of the material cycle. These losses need to be compensated for by input of fresh carbon, preferably, fresh renewable carbon. Non-persistent bio-based plastic are well positioned to fulfill this task and simultaneously address the problems of carbon sourcing and carbon disposal in a circular and sustainable way [1, 2].

The transition from fossil to recycled and renewable carbon will take time, likely a few decades, as observed during the emergence of the petrochemical industry some 50 years ago. Technologies are indeed expected to start at modest scale and modest efficiency and, consequently, at high cost. The new polymers will likely grow slowly by proving themselves in high-value and small-volume applications at first instance. As technologies and feedstock supply mature and develop in scale, the production costs will decrease, which opens access to larger and more diverse markets, eventually, up to commodity applications. A simple, though ambitious deployment model suggests indeed that the use of fossil carbon may peak at twice today's level by 2040 (Figure 1.2), to be overtaken by recycled and renewable carbon after 2050 and drop to 10% of the chemical diet only by 2070 [2].

To proceed smoothly and cost effectively, this transition must be supported by numerous measures, many of non-technical nature. Starting with the non-technical ones, society must stimulate and reward producers, distributors, and consumers to switch to sustainable products, even if these are less convenient or (initially) more expensive. Stimulation may start with ethics, by promoting the harmony between humans themselves and with their environment. But more will be needed. Society must adjust the regulation and certification to help consumers recognize truly circular and sustainable products and nudge them to choose these products. Regulators must also support sustainable solutions financially, e.g. by taxing resource utilization and waste production at higher rate while taxing labor at lower one. They must also harmonize their support for sustainably energy and materials to allow both sectors to compete fairly for the same feedstock. This is presently not the case in many regions that heavily support subsidize or mandate low-carbon energy but not low-carbon chemicals.

But society must also stimulate technology innovation, while watching out for undesired technology lock-in. Society must stimulate the sorting and recycling of today's fossil plastics while preparing the recycling of tomorrow's renewable plastics. The same warning applies to the "design for recycling" that will naturally attempt to restrict the number of materials but should nevertheless stimulate the emergence of

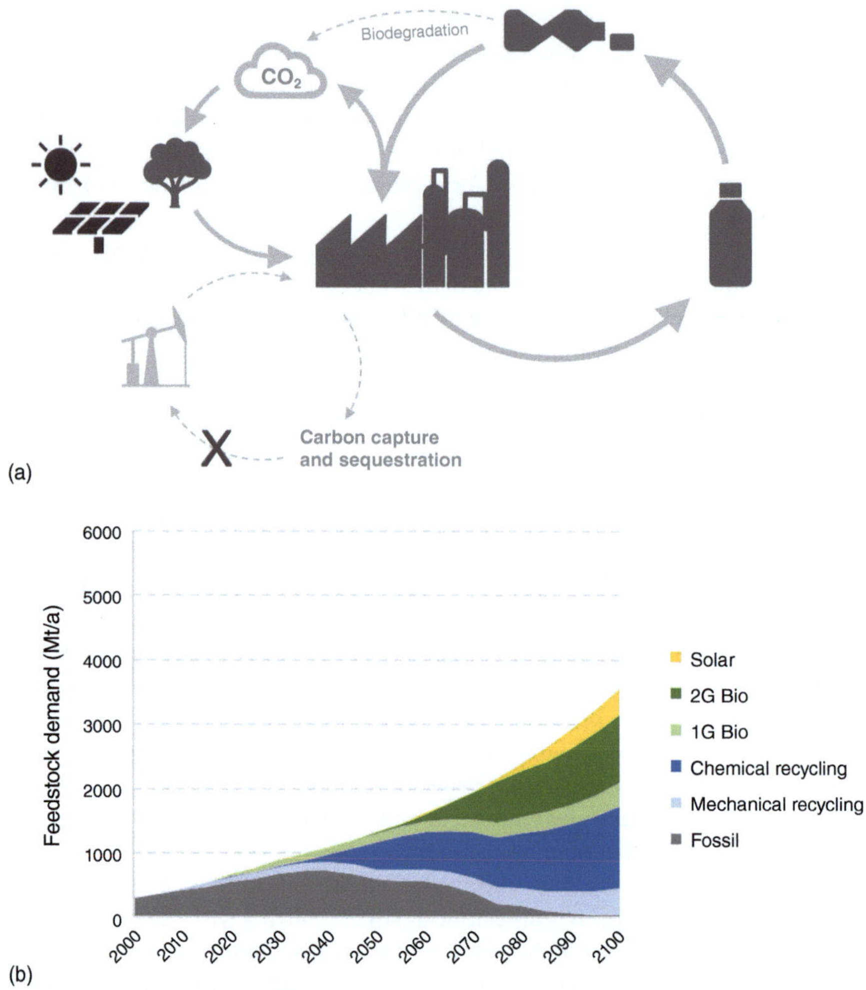

Figure 1.2 Artist's impression of future sustainable carbo-chemical cycles with fossil carbon being gradually displaced by recycled and renewable carbon (via natural/artificial photosynthesis) (a) and possible deployment rate following an aggressive scenario (b; technologies are deployed with an annual growth of 30%; global demand growth at 2% after 2025). Source: Lange [2].

tomorrow's renewable and non-persistent plastics. Finally, regulators and financial instances must support pioneers to try out sometimes far-stretched ideas, for some could form a cornerstone of the New Plastics Economy one day.

The definition and development of a new circular plastic economy presents enormous challenges. But it also offers unique opportunities for young scientists to contribute to the design of their future. For all of us, a mantra (or a cascade of measures) that frames the role for biodegradable plastics in the overall rethink of the plastics economy, but equally holds for other material economies, is found in Box 1.1.

Box 1.1 4R mantra for rethinking material economies, managing natural resources, and safeguarding the environment

- **Reduce** the consumption of goods.
- **Reuse** spent goods in all possible ways.
- **Recycle** the materials into new goods.
- **Regenerate** the environment with the carbon that we can't use or that escaped our control (biodegradation in case of leaked plastics)

Note in proof: *While finalizing this book we recognized some shortcoming in our scope. For instance, we could have presented a better coverage of the physical and mechanical properties of the emerging plastics and of the challenges these, as well as in situ degradation, may present in the processing to final goods. We also failed to recognize the emergence of design for "triggered" (bio)degradation that aims at controlling the onset of biodegradation (mostly after use) by triggers like radiation, heat, humidity, or chemical stimuli. This could be of use in niche applications, but it might conflict with a design for recycle. The reader is invited to explore these fields to complete the picture developed in this book.*

References

1 World Economic Forum, Ellen MacArthur Foundation and McKinsey & Company, The New Plastics Economy — Rethinking the future of plastics (2016, http://www.ellenmacarthurfoundation.org/publications).

2 Lange, J.-P. (2021). *Energy and Environmental Science* 14: 4358–4376.

3 Ritchie, H. and Roser, M. (2018). Plastic pollution. Published online at http://OurWorldInData.org. https://ourworldindata.org/plastic-pollution.

4 Tian, L., van Putten, R.-J., and Gruter, G.J.M. Chapter 3 Plastic pollution. The role of (bio)degradable plastics and other solutions. In: *Biodegradable Polymers in the Circular Plastics Economy.*

5 Joseph, E., Tohidifar, P., Sarver, C.T. et al. Chapter 2 Fundamentals of polymer biodegradation mechanisms. In: *Biodegradable Polymers in the Circular Plastics Economy.*

6 Chapter 4 Gruter, G.J.M. and Lange, J.-P. Tutorial on polymers – manufacture, properties and applications. In: *Biodegradable Polymers in the Circular Plastics Economy.*

7 Chapter 5 Stouten, J. and Bernaerts, K.V. Condensation polyesters. In: *Biodegradable Polymers in the Circular Plastics Economy.*

8 Chapter 6 Koller, M. and Mukherjee, A. Polyhydroxyalkanoates (PHA) – production, properties and biodegradation. In: *Biodegradable Polymers in the Circular Plastics Economy.*

9 Chapter 7 Narmon, A.S., Jenisch, L.M., Pitet, L.M., and Dusselier, M. Ring-opening polymerization strategies for degradable polyesters. In: *Biodegradable Polymers in the Circular Plastics Economy.*

10 Chapter 8 Molenveld, K. and Slaghek, T.M. Recent developments in biodegradable cellulose-based plastics. In: *Biodegradable Polymers in the Circular Plastics Economy.*

11 Chapter 9 Lee, H., Gan, H., Togo, A. et al. Ester derivatives of microbial synthetic polysaccharides. In: *Biodegradable Polymers in the Circular Plastics Economy.*

12 Chapter 10 Chen, Y.r. and Sarkanen, S. Biodegradable lignin-based plastics. In: *Biodegradable Polymers in the Circular Plastics Economy.*

13 Chapter 11 Monnery, B.D., Karanastasis, A., and Pitet, L.M. Design of recyclable thermosets. In: *Biodegradable Polymers in the Circular Plastics Economy.*

14 Chapter 12Lange, J.-P. Managing plastic waste. In: *Biodegradable Polymers in the Circular Plastics Economy.*

15 Chapter 13 Shen, L. Life cycle assessment of bio-based plastics: concepts, findings and pitfalls. In: *Biodegradable Polymers in the Circular Plastics Economy.*

16 Chapter 14 Nevejans, S. and De Wildeman, S. How to create a "new plastics economy"? Marketing strategies and hurdles – finding application niches. In: *Biodegradable Polymers in the Circular Plastics Economy.*

17 European Environmental Agency, Biodegradable and compostable plastics — challenges and opportunities; 2020, https://www.eea.europa.eu/publications/biodegradable-and-compostable-plastics.

18 Noordover, B.A.J. (2012). *Renewable Polymers, Synthesis, Processing and Technology* (ed. V. Mittal), 305–354. Scrivener/Wiley.

19 Kabasci, S. and Bretz, I. (2012). *Renewable Polymers, Synthesis, Processing and Technology* (ed. V. Mittal), 354–380. Scrivener/Wiley.

20 Koller, M., Salermo, A., and Braunegg, G. (2014). *Bio-based Plastics, Materials and Applications* (ed. S. Kabasci), 137–170. Wiley.

21 Tsuji, H. (2014). *Bio-based Plastics, Materials and Applications* (ed. S. Kabasci), 171–239. Wiley.

22 van Es, D.S., van der Klis, F., Knoop, R.J.I. et al. (2014). *Bio-based Plastics, Materials and Applications* (ed. S. Kabasci), 241–274. Wiley.

23 Rajesh, N., Uma, N., and Valluru, R. (2012). *Renewable Polymers, Synthesis, Processing and Technology* (ed. V. Mittal), 429–472. Scrivener/Wiley.

24 Bastioli, C., Magistrali, P., and Gesti Garcia, S. (2014). *Bio-based Plastics, Materials and Applications* (ed. S. Kabasci), 9–33. Wiley.

25 Ganster, J. and Fink, H.-P. (2014). *Bio-based Plastics, Materials and Applications* (ed. S. Kabasci), 35–62. Wiley.

26 Glasser, W.G. and Sarkanen, S. 1989 Joint Meeting of the 195th National Meeting of the American Chemical Society and the 3rd Chemical Congress of North America, Toronto, Canada (5–11 June 1988).

27 Hu, T.Q. (2002). *Chemical Modification, Properties, and Usage of Lignin.* Springer.

28 Laurichesse, S. and Avérous, L. (2014). *Progress in Polymer Science* 39: 1266–1290.

29 Trullemans, L., Koelewijn, S.-F., Scodeller, I. et al. (2021). *Polymer Chemistry* 12: 5870–5901.

2

Fundamentals of Polymer Biodegradation Mechanisms

Ebin Joseph[1], Payman Tohidifar[1], Cara T. Sarver[1], Roderick I. Mackie[2], and Christopher V. Rao[1]

[1] *University of Illinois at Urbana-Champaign, Department of Chemical and Biomolecular Engineering, 600 South Mathews Avenue, Urbana, IL, 61801, USA*
[2] *University of Illinois at Urbana-Champaign, Department of Animal Sciences, 1207 West Gregory Drive, Urbana, IL, 61801, USA*

2.1 Introduction

Plastics wastes can be removed from circulation mainly via mechanical recycling, thermal recycling, or chemical recycling. Each one of these recycling methods alone can face inherent limitations that may severely impact large-scale feasibility. Therefore, selection of one or multiple proper recycling methods largely depends on the chemical structure, purity, and the application of the specific plastic polymer. For example, thermomechanical recycling often results in a loss of the mechanical properties of the recycled polymer that cannot compare to those of the virgin polymer [1]. Mechanical recycling (e.g. grinding and microionization) is further complicated by the presence of additives and plasticizers within the plastic, and heating and remolding during the process cause chain scission to the degree that most plastics cannot be used beyond three recycled stages. Consequently, de novo polymer synthesis is preferred, leading to increasing accumulation of plastic polymers.

Thermal recycling typically includes incineration, gasification, pyrolysis, and plasma arc. Incineration can help produce energy from municipal waste, but it usually results in incomplete combustion, releasing toxic compounds into the atmosphere where it becomes a public health hazard [2]. Although new technologies have recently been introduced to clean up toxic exhausts, their full practical implementation is still far from a viable solution, partly due to political incompetency. In the past two decades, gasification of municipal waste has also been considered for waste disposal and energy recovery in the form of syngas, which is eventually converted into methanol, ethanol, or diesel. Although gasification appears to be a promising alternative for disposing municipal waste compared to

Ebin Joseph, Payman Tohidifar, and Cara T. Sarver contributed equally to this chapter.

Biodegradable Polymers in the Circular Plastics Economy, First Edition.
Edited by Michiel Dusselier and Jean-Paul Lange.

other approaches such as landfill burial or incineration, its success in the long run depends on appropriate governmental policies. Despite the advantages mentioned, gasification cannot be applied to the disposal of plastic waste and microplastics that have already contaminated the oceans. Moreover, this technology cannot rescue us from the need for de novo synthesis of plastics from oil, which eventually adds more carbon on earth every year.

Chemical recycling can be achieved via abiotic and biotic methods. It is typically used to convert plastic wastes back to oil and, in some cases, can achieve full depolymerization, where the resulting monomers can be reused to resynthesize new polymers [3]. For example, polymers with carbon chain structures such as hydrocarbon polymers can be reverted back to oil, while hydrolysable polymers with heterologous chain structures including polyesters can be converted into raw monomers and oligomers via physiochemical processes such as glycolysis, amine decomposition, pyrolysis, supercritical decomposition, and enzyme-based hydrolysis [4]. Although abiotic methods are currently considered as the predominant approach for chemical recycling of plastic polymers, they can pose limitations in a few situations depending on the type of plastic waste. For example, these methods require nearly pure polymer feedstock, which is realistically impossible to obtain from plastics synthesized with additives or plasticizers and/or compounded by the contamination of plastics sourced from landfills or municipal waste. Moreover, every year, an increasing amount of plastic waste including polyester-based microplastics present in laundry products, for example, enters our oceans through sewage systems. In most cases, collection and disposal of these microplastics are not practical. Therefore, their degradation and assimilation by microbes is highly considered. Given these limitations, chemical recycling based on biodegradation methods can offer extra advantages for polymer waste disposal.

By its very nature, material degradation by microorganisms is a dirty process and benefits from mixed microbial communities. Therefore, biodegradable polymers exposed to the environment, compost, or landfill automatically benefit from exposure to as many microorganisms and weathering conditions as possible. Some polymers can be assimilated into microbial biomass and converted into energy (e.g. biogas) or other value-added bioproducts. These microbial processes usually require no or little energy inputs outside of the typical metabolism. The use of additives and plasticizers can sometimes inhibit microbial degradation but can also serve as an energy source for secondary metabolizers or even attract primary colonizers instead of the polymer itself. Finally, the scale of plastic accumulation in landfills and the environment is substantially high that recollection of all such plastics would be monumental. These facts suggest that widespread biodegradable plastic use in the future would at least enable plastic waste degradation on a wide scale while non-biodegradable plastics in general may require a more deliberate disposal process. It is important to note that biodegradation is a last resort.

This chapter will explore degradation of several types of common polymers grouped by the bonds that hold their monomer units together in polymer chains. More specifically, degradation of polysaccharides, polyamides, polyesters, hydrocarbon polymers, halogenated polymers, and polyethers will be discussed. Each section will begin with a brief introduction to the chemical structure of a specific polymer, followed by a discussion about available abiotic and biotic degradation mechanisms. Next, secondary characteristics such as the polymer's crystallinity, melting and glass transition temperatures, the presence of plasticizers or additives, and conditions under which these polymers reach the end of their lifecycle will be explained. Moreover, we will explore how a combination of mechanical pretreatments and chemical recycling using engineered enzymes would enable depolymerization and repolymerization of the most abundant polyester, poly(ethylene terephthalate) (PET), in controlled industrial environment, introducing a path toward a full circular plastic economy. We will conclude the chapter by presenting applications of biodegradation and the current challenges and possible future directions in the management of plastic waste disposal.

2.2 Overall Scheme of Polymer Degradation

Polymer degradation can occur through two separate mechanisms: abiotic and biotic. Abiotic degradation results in fragmentation. This fragmentation usually comes in the form of pretreatments. Ultraviolet (UV) and gamma irradiation often produce radicals, which propagate throughout the polymer chain, thereby increasing the reactivity. The sun readily provides UV exposure from which photo-oxidation occurs by function of the intensity of electromagnetic radiation. This often occurs simultaneously with oxidation since oxygen is available during these pretreatments. Peroxides and hydroperoxides tend toward homolytic cleavages in their bonds from UV radiation in the 240–400 nm range [5]. Thermal pretreatments can also introduce hydrophilic functional groups allowing for more microbial contact, facilitating further biotic degradation. In addition to forming radicals, which can be generated through radiation, oxidative chemicals can also impact the polymer through depolymerization and inducing bulk physical changes, resulting in increased permeability and tensile strength [6]. However, the downsides to these reactions include the long time frames (up to two years) for any discernable effect to take place for buried samples. The most promising results, thus far, come through *in vitro* experiments on polyesters via gamma radiation [7].

Biotic degradation is assisted through microbial action. A more comprehensive degradation occurs through bioassimilation, in which polymer fragments are consumed by microorganisms and converted to products such as water, carbon dioxide, methane, and biomass depending on the presence or the absence of oxygen. The extent of this process depends on, but is not limited to, terrestrial or marine environments, the microorganism itself, the presence of additives in the polymer, and even the polymer's shape. Biotic degradation generally necessitates the formation of biofilms. A biofilm is a consortium of microorganisms that produce a network of extracellular polymeric substances that stabilizes the microorganisms and adheres them onto the target surface [8]. After this adherence, extracellular enzymes are secreted, which then hydrolyze the surface of the polymer, which have been primed via abiotic degradation. The released oligomers, dimers, and monomers, as the hydrolysis products, are then used by microbes as nutrients [7, 8]. This process can become a function of surface area in that the biodegradability increases proportionally with biofilm colonization. Besides hydrolytic enzymes, microbes can release biosurfactants, which would reduce the surface tension at the interface between the biofilm and plastic. They are also amphiphilic compounds, which can increase the bioavailability and biodegradability of the polymer chain [6].

2.3 Biodegradation of Polysaccharides

2.3.1 Cellulose

As a naturally produced polymer, cellulose is a glucose molecule connected via β-1,4 linkages [9]. These covalent bonds are referred to as glycosidic bonds. There are two stable conformers of natural cellulose (cellulose I), one of which contains a bifurcated bond with the oxygen on the C-6 hydroxyl and the ring oxygen with the C-3 hydroxyl on the adjacent chain. The other variant consists of the ring oxygen bonding with the C-3 hydroxyl on the next residue (Figure 2.1a) [10]. Cellulose may contain one variant in surplus, while both variants found together in disordered configurations are typical of amorphous regions. Further polymorphs (cellulose II,

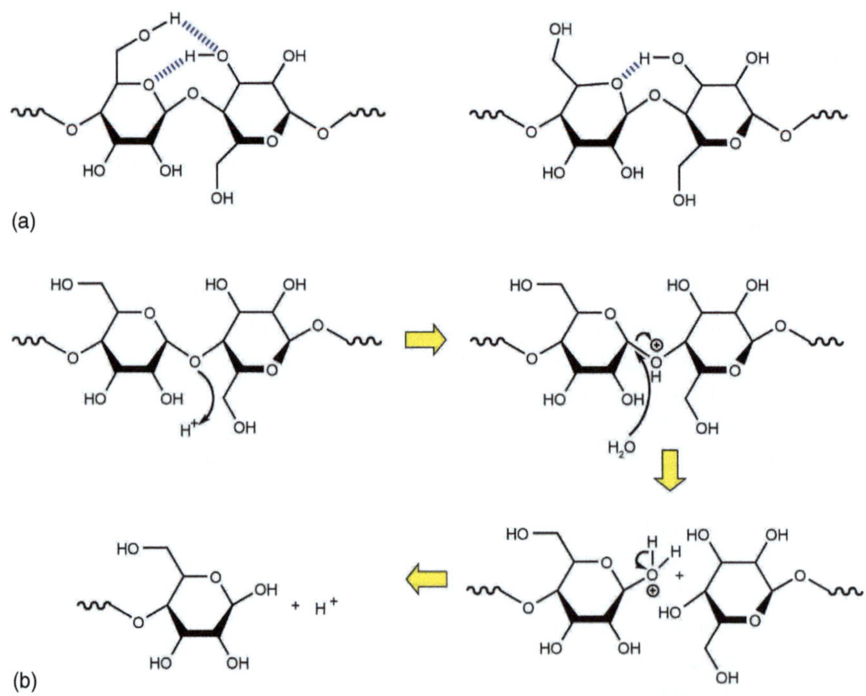

Figure 2.1 Biodegradation of cellulose. (a) Conformers of cellulose. (b) Cellulose hydrolysis mechanism.

III_I, III_{II}, IV_I, and IV_{II}) can be generated through various treatments such as with liquid ammonia [11]. Depending on the type of the microorganism, biodegradation often occurs in the amorphous parts of the polymer [12]. Depending on the chain length of the polymer, also known as the degree of polymerization, the bonds can be very strong due to London dispersion forces that increase proportionally with chain length. These chains are in the form of fibers that are further embedded in a hemicellulose and lignin-composed matrix. Because of this crystalline and inaccessible structure, penetration of enzymes to the fragment cellulose would be impaired [13].

As noted above, cellulose contains enough crystallinity to deter enzymatic hydrolysis. This barrier can only be overcome through a medley of enzymes that work in unison to degrade this natural polymer, which also requires up to a 100-fold enzyme concentration than that of starch [14]. Studies have shown that cellulose-hydrolyzing enzymes, cellulases, in particular, have larger catalytic active sites that can accommodate 5 to 6 units of glucose. These enzymes have considerable diversity but belong to a common class called cellulase, which contain endoglucanases, cellobiohydrolases, and cell biases [15]. Despite these differences, these enzymes can also contain a variety of similarities relating to folding patterns, catalytic residues, and reaction mechanisms. Other enzymes are involved in auxiliary functions such as assisting in degradative product recovery, focusing hydrolysis efforts, and securing enzymes along the surface of the polymer [16].

During cellulose hydrolysis, enzymes attack the β-1,4 glycosidic bonds. This process promotes fragmentation, resulting in glucose, cellobiose (a double-glucose molecule), and cellodextrins (cyclic oligomers of cellulose) (Figure 2.1b). Endo- and exolytic enzymes work synergistically to unzip various layers of cellulose by moving along the surface. For example, cellobiohydrolase acts to pull away chains from the strata of the substrate, thereby undoing the hydrogen bonds holding the lattice together [9]. β-Glucosidases wrap up hydrolysis by cleaving cellobiose units and extracting the glucose monomer from the non-reducing end – a group that has no free aldehyde or ketone to be oxidized [9, 17].

In the next step, cellulose fragments are consumed by microorganisms. It can be attacked by a myriad of microorganisms as its polymer constituents are natural metabolites. The high nutrient utilization by microbes highlights the cellulose's biodegradability. These microorganisms have cellulosomes (complexes on the cell surface) that adhere onto cellulose. The cellulosomes facilitate multiple cuts resulting in cellobiose, which is then bioassimilated [18]. *Clostridium thermocellum* and *Thermomonospora curvata* are examples of microbes that have this capability [19]. In fact, *C. thermocellum* (an anaerobe) can reach up to 93% utilization of cellulose at a rate of 2.9 g/l/h. during fermentation [20]. Although this organism shows the highest growth on cellulosic substrates, it does not produce tremendous amounts of enzymes [21]. Comparing burial and exposure methods shows that the consortium of microorganisms present in soil have a greater effect than the oxidative benefit from sunlight. In fact, soil burial boasted 65.7% molecular weight decrease in nine weeks, whereas cellulose hung outside merely lost 38.9% [22]. Many species from commonly recognized genera degrade cellulose such as *Aspergillus*, *Penicillium*, and

Trichoderma [19]. Overall, the entire process should result in carbon dioxide and water as by-products. The same by-products are present in anaerobic biodegradation in addition to methane, hydrogen sulfide, hydrogen gas, and ammonia [7].

2.3.2 Starch

Starch is another biodegradable polysaccharide. It consists of amylose and amylopectin. The former is a linear polymer, which contains α-1,4 linkages of glucose units. The latter is a branched polymer connected by α-1,6 bonds (Figure 2.2a). At least 10–20% of the material is composed of the linear amylose, while the remainder is branched amylopectin, which is responsible for the polymer's crystallinity [23]. Each iteration of the glycosidic bond in amylose proffers a natural twist, which is reminiscent of a helix. In the case of cellulose, however, the successive glycosidic bonds have a 180° rotation per monomer. Starch residues contain the hydroxyl groups on the external side of the helix, allowing for abundant hydrogen bonding with adjacent chains, other similar functional groups, and water molecules [24], which varies physical properties depending on extent of network formation. Little is known about the structure of amylopectin, but it is estimated to have a variety of chain lengths.

Starch can undergo hydrolysis. It comes in the form of granules, which are naturally formed by many plants to be used for storage. These granules are initially

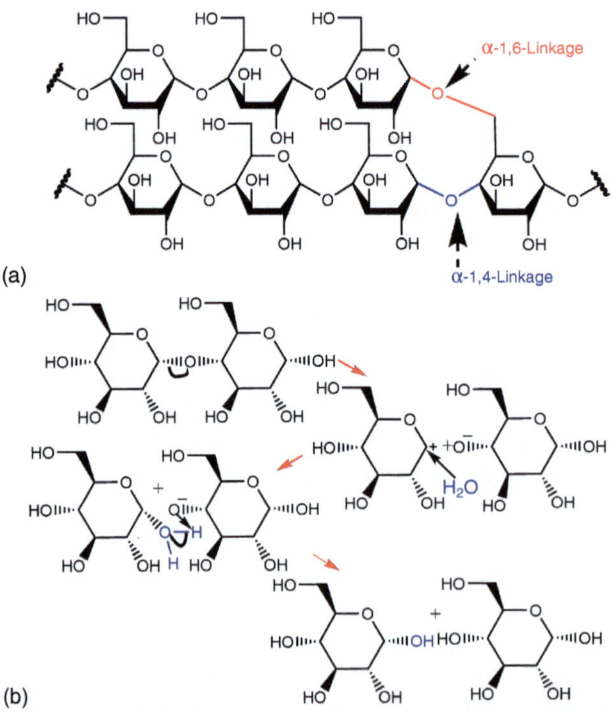

Figure 2.2 Hydrolysis of starch. (a) Two types of starch linkages. (b) Hydrolysis mechanism of starch.

degraded on the surface. Some starches have specific zones susceptible to corrosion from within. These holes or pits grow larger and form channels permeating through each stratum of starch [24]. Amylases are the main enzymes that degrade starch. Amylases were shown to disentangle the helix structure of amylose. As with other polymers, starch greatly benefits from pretreatment strategies, particularly gelatinization, in which heating a starch suspension allows for swelling within the amorphous regions of the polymer and opening the crystal structure. This process greatly improves the reactivity to hydrolytic enzymes. Exo- and endolytic enzymes play a role in starch degradation as well. For example, β-amylase and glucoamylase catalyze the non-reducing end of the polymer chain, which is the location of the anomeric carbon. This is where glycosidic bond cleavages occur whose product leaves a newly formed hydroxyl group in the C1 position. The ketal bond in the glucose ring is especially susceptible to breakage. The overall rate of hydrolysis is initially quick but slowly tapers off [25]. Glycosidic hydrolysis is very sensitive to ring shape and size. Formation of the enzyme–substrate complex allows for ring distortion, thereby decreasing the dihedral angle. This transition state is more thermodynamically favored toward hydrolysis [26]. Research has shown that *Aspergillus niger* can degrade starch completely in 20 days in both solid and liquid cultures without light and heat, which is faster than the case for cellulose [27].

Starch is often used in other polymers to promote biodegradation. Pure starch is brittle and moisture sensitive, enough to disintegrate in the presence of water [28]. Because of its sensitivity to humidity and temperature, starch is typically added to other polymers as a degradation enhancer, filler, and cross-linking agent [7]. Because of the availability and relative inexpensiveness of starch, there is a push to incorporate it into more traditional polymers. However, there are inherent limitations to this process because of starch's subpar mechanical properties [29]. Regardless, studies show that starch-infused plastics are more heat resistant with the added benefit of attracting more microbes for biodegradation and consequently improving the overall biodegradability [7]. This biodegradation comes through initial hydrolysis (Figure 2.2b).

Use of starch as a biodegradation inducer of other polymers has grown in the compost market because of its pairing with low-density polyethylene (LDPE) and linear low-density polyethylene (LLDPE) even though these blends are a long way from being considered fully biodegradable [29]. However, starch–PE blends have been shown to degrade quickly – initial starch degradation forms pores and holes in the composite film and facilitates high oxygen saturation and therefore oxidation and biodegradation of PE [7, 30]. There is a balance to achieve in that higher starch content does not yield higher biodegradability. In 10 weeks, 40% starch–LDPE samples had a weight loss of 5.24%, which is higher than that of 50% starch composites that had 2.75% in the compost [31]. This shows that there is an upper limit to starch blends that minimizes the amount of recalcitrant PE used in the composite polymer, which may be a better option compared to a pure hydrocarbon polymer. Although there is degradation of PE, this may also harm terrestrial and aquatic environments in that PE fragments may be ingested by animals and propagate through the food chain in the form of microplastics [32].

2.4 Biodegradation of Polyamides

Polyamides are also an important polymer coming in both natural and synthetic forms. The amide linkages combine various monomers of different types composed of aliphatic, semiaromatic, or aromatic molecules. It is known for its high crystallinity and strong interchain interactions [33]. The most common synthetic polyamide applications are in nylon and Kevlar. Natural polyamides such as silk are prevalent; however, scientists have not yet specified which microorganisms biodegrade the polymer [34]. There are bacteria that can degrade lower weight oligomers such as 6-aminohexanoic acids (cyclic and linear) typically found in waste streams in nylon production facilities [35]. The *Arthrobacter* sp. have been found using several hydrolases and aminotransferases that aid in the degradation and eventual bioassimilation [34]. The hydrolysis of amide bonds leads to the metabolism of 6-aminohexanoate to adipate semialdehyde. This is further catalyzed by adipate semialdehyde dehydrogenase to form adipate through the use of a NADP+ cofactor [36]. Some researchers have shown *Bacillus cereus*, *Bacillus sphaericus*, *Vibrio furnissii*, and *Brevundimonas vesicularis* to degrade nylon, although it has not demonstrated that these bacteria are not just degrading the polymer additives, which is a common issue in measuring biodegradability [37]. Other researchers have reported *Pseudomonas* sp. to utilize only 6-aminohexanoate dimers as its carbon and nitrogen source [38]. No enzymes or pathways are described for nylon degradation by bacteria but one has been identified as a manganese peroxidase isolated from a white-rot fungus. Alas, the evidence for its biodegrading capabilities is scant at best [39]. However, polyamide 4 (PA4), a common type of nylon, have been found to be degraded in activated sludge [40]. In addition to activated sludge, soil and seawater tests have been performed with seawater tests showing biodegradation tapering off at 50% in 12 days [41]. PA4 has not shown any discernable degradation *in vitro* when incubated at 37 °C in phosphate buffers [42].

2.5 Biodegradation of Polyesters

Polyesters are classified into three groups: aliphatic, aliphatic-co-aromatic, and aromatic. Because of their good material performance, there is a great interest of aliphatic-co-aromatic polyesters such as PET in various industries, which has led to accumulation of large amount of post-consumer polyesters in the environment. Therefore, recycling these polymers is absolutely necessary. Although aliphatic polyesters are generally susceptible to biodegradation, aliphatic-co-aromatic and aromatic polyesters are extremely resistant to enzymatic attack largely due to the presence of aromatic groups. Although a number of native and engineered enzymes have been shown to degrade these polyesters, no enzyme is still known to fully degrade high-crystalline variants of these polymers. Here, we first review the fundamentals of biodegradation of common aliphatic polyesters such as polylactic acid (PLA), poly(ε-caprolactone) (PCL), and polyhydroxyalkanoates (PHAs). Then, we present a more comprehensive discussion surrounding the biodegradation of PET, which has been investigated in greater detail by the scientific community compared to other polymers.

2.5.1 Polylactic Acid

Polylactic acid or polylactide (PLA) is an aliphatic polyester synthesized from lactic acid and can be produced from entirely renewable resources, such as corn and sugar beets [43, 44]. PLA can be degraded back into its biologically benign monomers and is currently one of the most widely produced polymers in the biodegradable polymer market. The degradability of PLA has been mostly measured following chemical hydrolysis [45]. Much of PLA biodegradation has been studied under aerobic conditions such as composting [46]. For example, industrial composting at 58 °C can achieve 90% biodegradation within a maximum of six months [47]. Based on these studies, PLA can be considered sustainable under industrial composting conditions. However, only certain types of PLAs are anaerobically degradable under thermophilic (52 °C) and mesophilic (37 °C) conditions [48]. For example, anaerobic incubation of amorphous PLA at 35 °C for 180 days [49] and PLA cup at 58 °C for 56 days [50] resulted in 56% and 40% PLA degradation, respectively. These numbers are calculated based on the amount of methane released from 1 g of PLA during anaerobic digestion compared to that of theoretical yield (467 ml of CH_4/g of PLA) [49]. Finally, biodegradation of PLA in landfills is still not proven [47].

PLA polymer exists in three stereoforms: poly(L-lactide) (L-PLA), poly(D-lactide) (D-PLA), and poly(DL-lactide) (DL-PLA) [51]. The synthesis of pure L-PLA or pure D-PLA facilitates partial stacking of the polymer chains, yielding semicrystalline polymers. However, random polymerization of the two monomers, depending on the amount of D and L monomers, does not facilitate the formation of hydrogen bonds and therefore yields more amorphous polymer and less crystallinity overall [51]. Although pure L-PLA and D-PLA appear to have similar chemical hydrolysis rates, DL-PLA degrades much faster because of more amorphous structure. The blends of the two polymers (stereocomplexed) by contrast appear to facilitate the formation of crystal structures and therefore hydrolyze more slowly [52]. It is thought that increases in polymer crystallinity can inhibit water molecules from accessing the surface of the polymer chain, thereby retarding chemical hydrolysis overall. Similar to chemical hydrolysis, PLA biodegradability is also highly dependent on its level of crystallinity [53–55].

Similar to other aliphatic polyesters, PLA is susceptible to microbial degradation. However, the number of PLA-degrading microorganisms in the environment is limited compared to those involved in the degradation of PHA and PCL. PLA-degrading microorganisms mostly belong to actinomycetes, particularly *Pseudonocardiaceae* family, and are also found in bacteria and fungi. The PLA-degrading bacteria mainly belong to Firmicutes phylum. Fungal degradation of PLA has been reported for *Tritirachium album* (ATCC 22563), which was able to degrade L-PLA in liquid culture after 14 days in the presence of 0.1% (w/v) gelatin [56]. Gelatin, silk fibroin, elastin, a number of peptides, and amino acids are known to induce L-PLA degrading activity in several microorganisms by stimulating the production of PLA-degrading enzymes [57–59]. Most of these inducers are composed of the L-alanine unit, whose stereochemical chiral carbon is similar to that of the L-lactic acid unit of PLA.

Microbial degradation of PLA begins with secretion of extracellular PLA depolymerase by the PLA-degrading microbe, where production of extracellular depolymerase is often stimulated by some inducers, as noted above. The enzyme is then absorbed on the surface of the polymer through its surface-binding domain and, in the next step, attacks intramolecular ester bonds in PLA. This process results in production of oligomers, dimers, and monomers, where the low molecular weight compounds are assimilated by the microbe and eventually converted into carbon dioxide, water, or methane by intercellular enzymes [60–62].

PLA-degrading enzymes are mainly proteases, lipases (esterases), and cutinases. Protease-type enzymes, such as proteinase K, show preference toward L-PLA, while lipase/cutinase-type enzymes show specificity toward D-PLA [63]. The activity of PLA-degrading enzymes is also affected by local pH, temperature, oxygen or water availability, and polymer crystallinity. For example, enzymatic hydrolysis of PLA amorphous regions is faster than its crystalline regions [63, 64].

Proteases and lipases/cutinases belong to serine protease and α/β hydrolase protein families, respectively. Although these enzymes share same amino acid residues (e.g. serine, histidine, and aspartate) in their catalytic active sites, the positioning of these residues in serine proteases is the mirror image of that in α/β hydrolases [65]. This structural difference may explain why these enzymes show different preferences toward L-PLA and D-PLA substrates [63].

A typical catalytic mechanism of proteinase K against ester bonds in L-PLA generally involves four steps: (i) binding of the enzyme to polymer surface, (ii) nucleophilic attack by the serine residue, (iii) protonation of substrate enzyme–substrate complex, and (iv) hydrolysis of ester bond [63, 66]. Briefly, the enzyme first interacts with L-PLA side chain through its substrate recognition site. Next, in the acylation half of the reaction, the catalytic serine residue attacks the carbonyl of the L-PLA substrate, which is mediated by the catalytic histidine acting as a general base. This reaction yields a tetrahedral intermediate and the protonated catalytic histidine (His-H^+). Here, the oxyanion of tetrahedral intermediate is stabilized by the main chain NHs of the oxyanion hole and His-H^+ is stabilized by the hydrogen bond to the catalytic aspartate. His-H^+ then mediates the collapse of tetrahedral intermediate by donating a proton, yielding the acyl enzyme intermediate and cleaved L-PLA^+. In the deacylation half of the reaction, water attacks the acyl enzyme, assisted by the catalytic histidine, yielding a second tetrahedral intermediate. Finally, the resulting intermediate collapses and forms the second product L-PLA^- with the carboxyl group and regenerates the catalytic serine (Figure 2.3).

Studies exploring the effect of UV irradiation on PLA degradation confirm that polymer chains are cleaved following exposure, although only to a partial extent [68]. Other studies reveal that UV irradiation, followed by exposure to PLA-degrading microorganisms can enhance overall polymer degradation rates to completion. Under environmental conditions, such as in landfill, the polymer is susceptible to chemical hydrolysis, UV exposure, and microbial action and can take a few weeks to a few years to degrade depending on the polymer's crystallinity [68]. Composting of PLA results in a faster degradation: elevated temperatures and humidity facilitate early chemical hydrolysis while enabling the growth of thermophilic organisms effective at degrading PLA [69]. The only environment in which PLA does not readily degrade is in marine waters [70]. Collectively, it appears that isolation and screening of new microbial strains from

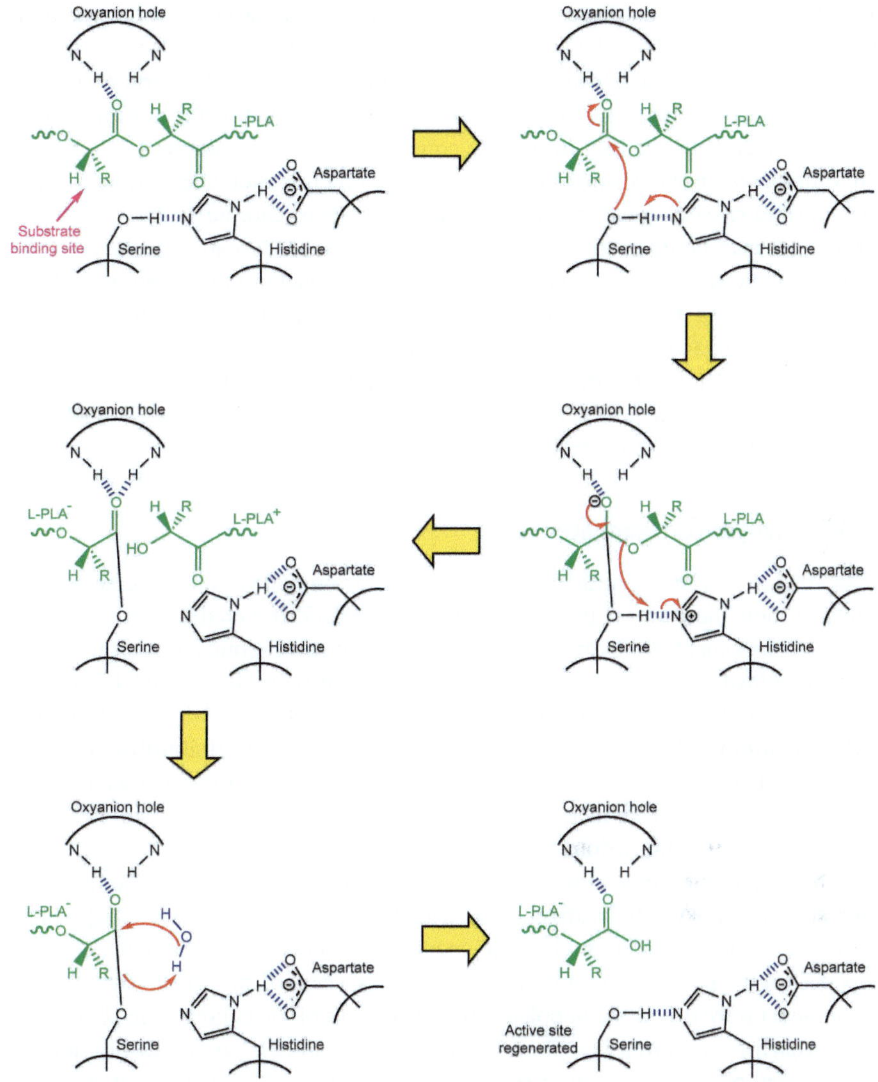

Figure 2.3 Hydrolysis mechanism of protease K against L-PLA. Source: The mechanism is adapted from Ref. [67].

the environment and identification of degrading enzymes are necessary to improve the overall degradation PLA in the industrial level.

2.5.2 Poly(ε-caprolactone)

PCL is polymerized from synthetic monomers but nevertheless degrades readily. Degradation typically occurs via chemical hydrolysis or enzymatic action, both of which target the polymer's ester bonds, and degradation can take a few weeks up to a year in the environment. Relatively less has been studied regarding polymer–enzyme interactions in PCL degradation. Most enzymes that degrade PCL are lipases, cutinases, and esterases [71–73]. One molecular dynamics study suggests the presence of a typical α/β hydrolase fold and the classic serine, histidine, and aspartate catalytic triad. Similar to enzymatic hydrolysis of PLA (see Figure 2.3), the nucleophilic serine attacks the PCL substrate, followed by a concerted proton transfer and C—O bond formation between the serine and the substrate, which is then attacked by a water molecule and releases the degradation product 6-hydroxycaproic acid.

In the environments such as compost, sea, or pond, PCL can degrade over the course of a few weeks up to a year [74]. A variety of bacteria and fungi have been isolated degrading the polymer [71–73]), and the typical degradation product, 6-hydroxycaproic acid, can easily be absorbed by cells and incorporated into their metabolisms [75]. Degradation is enhanced under compost conditions, where elevated temperatures and humidity facilitate early hydrolysis of the polymer backbone and make it more accessible to biological attack [76].

2.5.3 Polyhydroxyalkanoates

PHAs are a family of aliphatic polyesters derived from hydroxyalkanoic acids. The variety of their mechanical properties comes from differing chain lengths and the isomeric positions of the hydroxyl groups present in the chemical structure. PHAs are thermoplastic and are produced via a single-step fermentation derived from sugars, starches, fatty acids, or other renewable resources [28]. Poly(3-hydroxybutyrate) (PHB), a common PHA variant, is produced by methanotrophs that utilize methane *in vivo* [77]. PHAs, in general, are made via microorganisms that use it as a carbon and energy source [78]. In fact, up to 80% of the microorganisms' dry cell mass can be accounted for by the various PHAs [79].

The biodegradation mechanism begins with hydrolytic cleavages at ester linkages that result in carboxyl and hydroxyl groups leading to smaller fragments [80]. This is a surface area operation, and while the enzymes are released superficially, the bulk properties of the polymer remain unaffected. It has a synergistic effect with physical or chemical pretreatments that bolster the subsequent biotic degradation [81]. However, in marine environments, plastics such as PLA and PHA are often submerged, where typical physical conditions such as UV exposure and heat are limited, which prevent further abiotic degradation. So far, the current biodegradation of a PHA film takes up to 20 days in laboratory compost for full disintegration. As biodegradation occurs, the formation of pores and holes on the polymer surface increases its surface area and promotes overall bulk degradation [82].

The entire process for polyesters can be broken down to three important steps: biofilm formation, hydrolysis, and bioassimilation. Biofilms adhere onto surfaces and release extracellular enzymes that oxidize and fragment the substrate, which can then be used as food. The surface increases in porosity, which allows for further degradation that compromises the overall properties of the polymer [83]. PHAs can be degraded by many bacteria and fungi that are present in most environments such as in compost and ocean surfaces which degrade the polymer both aerobically and anaerobically [84].

PHAs are denser than water and tend to sink in marine environments. In addition, they are more likely to settle in sediment, thus limiting UV exposure and, consequently, photo-oxidation rates [85]. PHAs present in deeper sediment layers are privy to a wider consortium of bacteria but a lower dissolved oxygen content. Those that are closer to the shore experience warmer temperatures and sunlight that can permeate shallow waters with more active bacterial populations. PHA biodegradation is therefore more effective along shallow shorelines than the open ocean [8].

2.5.4 Polyethylene Terephthalate

PET belongs to aliphatic-co-aromatic polyesters, which is formed from terephthalic acid (TPA) and ethylene glycol (EG). PET is a semicrystalline polymer with a glass transition temperature (T_g) around 70–80 °C (assumed lower in water) [86]. In its purest form, PET is an amorphous polymer. However, it develops crystallinity upon addition of modifying additives or heat treatment of the polymer melt [87]. The crystallinity rate varies among PET-based products. For example, PET bottles and textiles have 30–40% crystallinity, while PET used for packaging has 8% crystallinity [88].

Unlike aliphatic polyesters, which can be hydrolyzed at ambient temperatures by various ester bond hydrolases, the presence of repeating aromatic terephthalate units in the PET backbone results in slow hydrolysis of PET. Several other factors are also reported to hamper enzymatic hydrolysis of PET surface and its inner building blocks. For example, high surface hydrophobicity and low surface area can significantly reduce enzyme access to the polymer, resulting in lower degradation overall [89]. In this case, generating hydrophilic carboxylic acid or hydroxy residues on PET surface using PET surface-modifying enzymes and increasing the polymer surface area through mechanical micrionization can greatly improve the subsequent enzymatic hydrolysis [90]. High crystallinity also poses additional inhibition to PET biodegradation, likely due to limited mobility and fluctuation of molecular chains in crystalline regions of PET, which reduces the accessibility of enzymes to the polymer inner blocks. Using high reaction temperatures (e.g. 65–70 °C) can significantly improve hydrolysis of crystalline PET in aqueous solutions. The likely reason is that the polymer has higher chain mobility and flexibility at temperatures near PET's T_g, which consequently exposes crystalline regions of PET to hydrolases. In this regard, thermophilic PET hydrolases with high melting temperatures ($T_{m,e} > 70$ °C) are absolutely required to achieve noticeable PET degradation.

Various aromatic polyesterases from fungal and bacterial species have been reported to degrade PET to different levels (see Figure 2.A.1 and Table 2.A.1 in Appendix 2.A for species and enzymes). Significant PET degradation has been mostly observed with hydrolases, which belong to cutinase family (EC 3.1.1.74). First discovered in phytopathogenic fungi, cutinases are generally known to initiate fungal invasion of plants by degrading the waxy polyester cutin of plant surface cuticle layer [109]. In the case of PET hydrolysis, for example, notable degradation was observed when a commercialized thermophilic cutinase (HiC), purified from the thermophilic fungus *Humicola insolens*, was incubated with amorphous PET (PET-GF) at 70 °C. Several other thermophilic cutinases capable

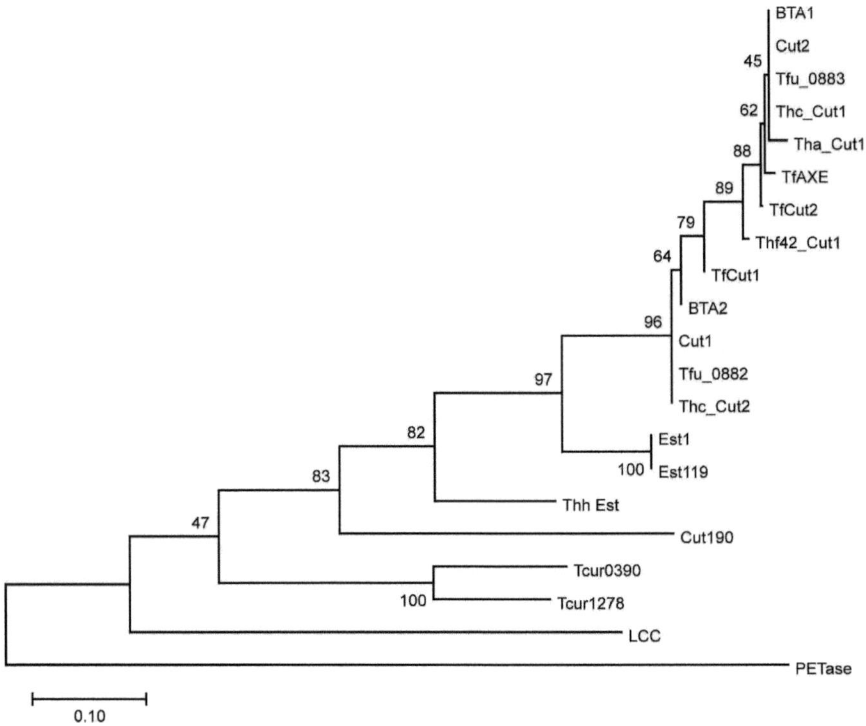

Figure 2.A.1 Phylogenetic tree for amino acid sequences of PET hydrolases reported in Table 2.A.1. The amino acid alignments were performed using MUSCLE (v-3.8.31) [108]. The unrooted tree was generated using the maximum likelihood method. Numbers on the nodes are the bootstrap values calculated from 1000 simulations and represent the reliability of the nodes on the tree.

of PET hydrolysis have also been identified from thermophilic actinomycetes, namely, BTA-1 (TfH) from *Thermobifida fusca* DSM43793 [91] and TfCut2 from *T. fusca* KW3 [110]. In addition, leaf-branch compost cutinase (LCC) identified using metagenomic approaches [102] and a cutinase-type polyesterase (Cut190) from *Saccharomonospora viridis* AHK190 [97] have been shown to degrade PET extensively at high temperatures (e.g. 50–70 °C).

In a recent study, a mesophilic bacterium *Ideonella sakaiensis* was isolated, which can grow on a low-crystallinity (1.9%) PET film [88] and possesses a mesophilic PET hydrolase (IsPETase). In this study, IsPETase could hydrolyze the amorphous PET film after 18 hours of incubation at 30 °C. However, it appeared to hydrolyze only the PET film surface, leaving the polymer inner blocks nearly intact. Additionally, the total PET degradation level induced by IsPETase was significantly lower than those of cutinases (1% at 30 °C with IsPETase versus 40% at 70 °C with LCC). IsPETase's PET hydrolysis activity also dropped dramatically over a relatively long time, and the enzyme lost almost its entire hydrolytic activity after 24 hours at 37 °C. These behaviors are likely attributed to IsPETase's low thermostability ($T_{m,e} \sim 40\,°C$) even though its crystal structure is highly similar to those of actinomycetes thermophilic

Table 2.A.1 PET hydrolase homologs.

Name	NCBI accession number	Source organism	References
BTA-1 (TfH)	AJ810119.1	*Thermobifida fusca* DSM43793	[91]
BTA-2	AJ810119.1	*Thermobifida fusca* DSM43793	[92]
Tfu_0882	AAZ54920.1	*Thermobifida fusca* YX	[93]
Tfu_0883	AAZ54921.1	*Thermobifida fusca* WSH03-11	[94]
TfCut1	CBY05529.1	*Thermobifida fusca* KW3	[95]
TfCut2	CBY05530.1	*Thermobifida fusca* KW3	[95]
Est1	BAI99230.2	*Thermobifida alba*	[96]
Est119	BAK48590.1	*Thermobifida alba* AHK119	[97, 98]
Thc_Cut1	ADV92526.1	*Thermobifida cellulosilytica* DSM44535	[95]
Thc_Cut2	ADV92527.1	*Thermobifida cellulosilytica* DSM44535	[95]
Thf42_Cut1	ADV92528.1	*Thermobifida fusca* DSM44342	[95]
Tha_Cut1	ADV92525.1	*Thermobifida alba* DSM43185	[99]
Thh_Est	AFA45122.1	*Thermobifida halotolerans* DSM44931	[100]
TfAXE	ADM47605.1	*Thermobifida fusca* NTU22	[101]
LC-cutinase	AEV21261.1	Metagenome from leaf-branch compost	[102]
Cut1	JN129499.1	*Thermobifida fusca* NRRL B-8184	[103]
Cut2	JN129500.1	*Thermobifida fusca* NRRL B-8184	[103]
Tcur1278	ACY96861.1	*Thermomonospora curvata* DSM43183	[104]
Tcur0390	ACY95991.1	*Thermomonospora curvata* DSM43183	[105]
Cut190	AB728484	*Saccharomonospora viridis* AHK190	[106]
PETase	GAP38373.1	*Ideonella Sakaiensis* 201-F6	[88]
HiC	4OYY_A	*Humicola insolens*	[107]

cutinases, such as TfH and LCC. In addition to IsPETase, there have been reports of several other bacterial PET-degrading enzymes, such as esterases from *Bacillus subtilis* [111] and *Clostridium botulinum* [112] and a cutinase from *Pseudomonas putida* [113]. Again, these enzymes are only capable of hydrolyzing the PET surface.

Taken these observations together, the PET-degrading enzymes can be classified into two groups of PET hydrolases and PET surface-modifying enzymes [4]. PET hydrolases must be able to hydrolyze the main body of PET, preferably at temperatures above the polymer T_g (60–70 °C in aqueous solution), resulting in PET surface erosion visible in SEM and releasing millimolar level (>10% of initial PET) of degradation products. As of now, only five cutinases – TfH from *T. fusca* DSM43793 [91], a variant of TfCut2 from *T. fusca* KW3 [114], a variant of Cut190 from *S. viridis* AHK190 [106], LCC from uncultured bacteria [115], and HiC from the fungus *Humicola insolens* [107] – meet these criteria and are qualified as PET hydrolases [4]. On the other hand, PET surface-modifying enzymes are expected to hydrolyze only the cyclic trimer of mono-(2-hydoxyethyl)terephthalic acid (MHET) present inside and outside of PET, releasing micromolar level (<1–2% of initial PET) of degradation products. IsPETase and most mesophilic bacterial polyesterases belong to this group.

Although hydrolytic activity against PET varies immensely among PET-degrading enzymes, they share highly similar crystal structures. In particular, they all belong to α/β hydrolase superfamily with nine β strands surrounded by seven to eight α helices depending on the enzyme [116]. Fungal cutinases such as HiC generally possess smaller structures compared to those from actinomycetes and bacteria (Figure 2.4a) [113, 117]. Notably, three amino acid residues constituting the catalytic site are almost identical among all PET-degrading cutinases (Figure 2.4b), implying that a similar mechanism is likely employed by these enzymes to cleave the PET scissile ester bonds (Figure 2.4c). It is noteworthy that the mechanism for enzymatic cleavage of the ester bonds is highly similar among different polyesters including PLA, PHA, and PET (see Figure 2.3). In the case of PET, however, the presence of aromatic group seems to slow this process, causing PET hydrolysis more challenging compared to aliphatic polyesters (see Appendix 2.A).

The activities and thermal stabilities of the cutinases were shown to increase in the presence of Mg^{2+} and Ca^{2+}. Molecular dynamics simulations indicate that binding of Ca^{2+} to the hydrolase decreases fluctuation in its structure, leading to higher thermal stability [118, 119]. For example, the $T_{m,e}$ values of *T. fusca* cutinases were increased by 10–14 °C in the presence of 10 mM Ca^{2+} or Mg^{2+}, which consequently led to significant degradation of PET-GF (supplied by Goodfellow Cambridge, Ltd) at 65 °C [98]. Similar to catalytic sites, the Ca^{2+} binding sites on PET hydrolases are also highly conserved [4], highlighting their important role in improving thermal stability and PET hydrolysis activity of the enzymes. Interestingly, introduction of new disulfide bonds via cysteine residues at the Ca^{2+} binding sites of TfCut2, Cut190, and LCC, for example, resulted in higher $T_{m,e}$ values and improved durability of the enzymes at high reaction temperatures [106, 120]. This method indeed would eliminate high costs associated with purification of Ca^{2+} salts following large-scale PET hydrolysis reaction [120] and would enable direct use of microbes for the treatment of PET waste where high salt concentrations pose toxicity to cells.

Finally, unlike enzymes such as cellulase, PET hydrolases lack a specific substrate-binding site. Docking calculations supported by mutagenesis analyses suggest that the PET surface initially binds a long, shallow hydrophobic cleft on the surface of the PET hydrolases. These analyses further demonstrate that the binding cleft should be relatively wide to accommodate bulky aromatic residues of PET molecular chains [121]. Remarkably, all the four main PET hydrolases have wider binding cleft compared to those of PET surface-modifying enzymes, namely, IsPETase [4].

Although studies of PET biodegradation have mainly focused on the characterization of efficient PET hydrolases, not much is known about assimilation of PET by microorganisms. As noted earlier, one group recently isolated *I. sakaiensis* bacterium, which can degrade the amorphous PET film and metabolize the degradation products near ambient temperature [88]. This discovery, since then, has encouraged

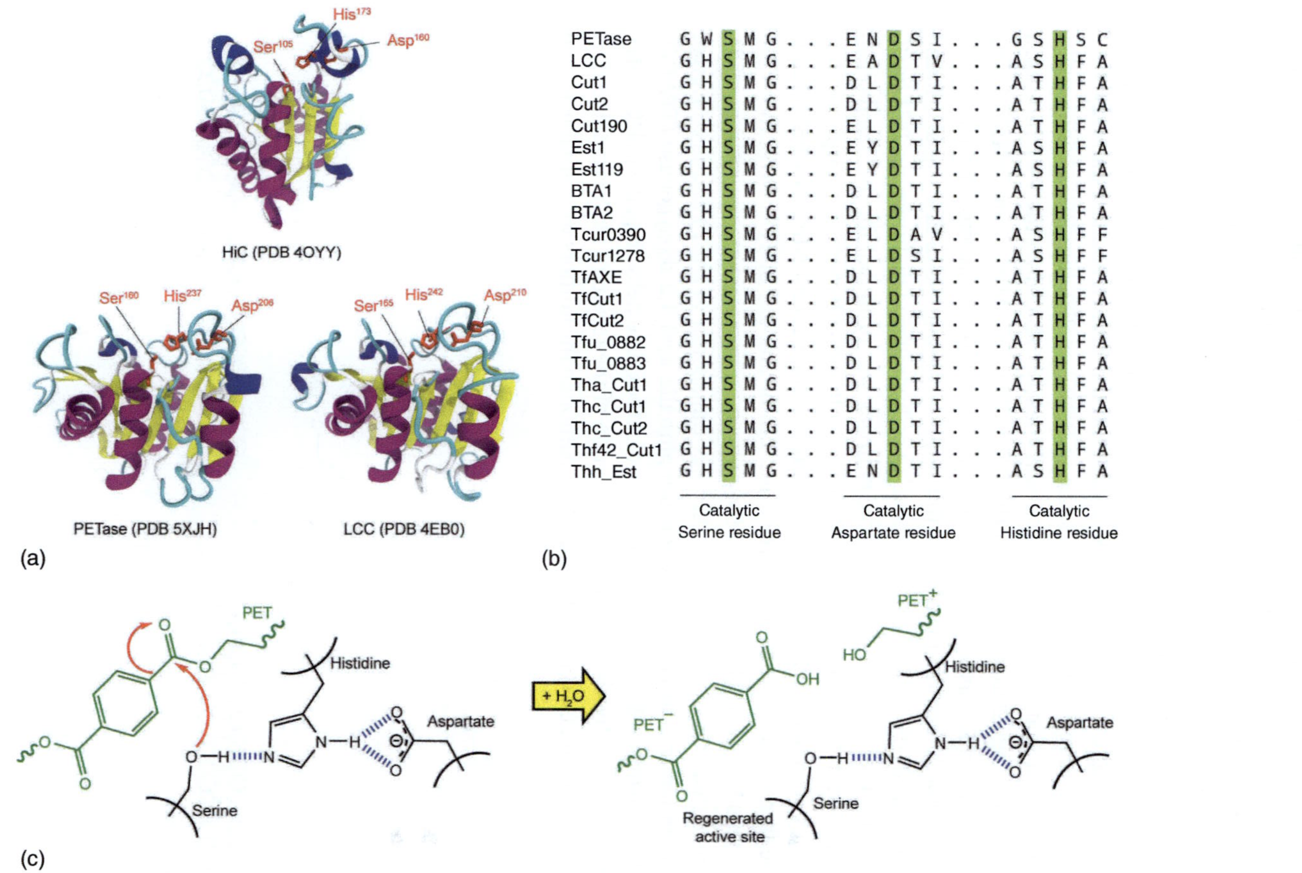

Figure 2.4 **PET hydrolases**. (a) Crystal structures of PET hydrolases; HiC: PDB 4OYY [114], PETase: PDB 5XJH [117], and LCC: PDB 4EBO (to be published). (b) Conservation of catalytic triad residues on different PET hydrolases. (c) Overall mechanism of PET hydrolysis.

the scientific community to attempt to isolate new microbes from nature or engineer mutants, which can efficiently degrade and metabolize PET waste.

To illustrate how PET chains can be metabolized by bacteria, we briefly describe the assimilation pathway of PET film products in *I. sakaiensis*. Using transcriptomic analysis of the cells cultured on maltose, sodium terephthalate BHET, or PET film, Yoshida et al. found that genes encoding PETase, a MHET hydrolase (later termed IsMHETase), and a cluster of genes involved in the TPA catabolism pathway were significantly upregulated in the presence of BHET or PET film [88]. Additionally, they suggested that PETase expression was mainly induced by the PET film itself and/or some degradation products other than TPA, EG, MHET, and BHET. Using bioinformatics analysis, they next predicted that PETase is a secretory protein, while MHETase is a lipoprotein localized in the outer membrane. MHETase was also found to hydrolyze MHET to TPA and EG, showing no hydrolytic activity against PET itself.

Based on these results, the following pathway was proposed for the metabolism of the PET film by *I. sakaiensis*. Briefly, when cells are cultured on amorphous PET film, they secrete out PETase to the medium. PETase then hydrolyzes PET chains on the film surface, producing MHET and minor amount of TPA. MHETase then hydrolyzes MHET to TPA and EG at the outer membrane, where the resultant TPA is transported into cytosol through the TPA transporter (TPATP) and is converted to pyroglutamic acid (PCA) through terephthalic acid 1,2 dioxygenase (TPADO) and 1,2-dihydroxy-3,5-cyclohexadiene-1,4-dicarboxylate dehydrogenase (DCDDH), respectively. Finally, the ring opening reaction of the resultant PCA is catalyzed by PCA 3,4-dioxygenase (Pca34) (Figure 2.5). Although the authors were unable to find

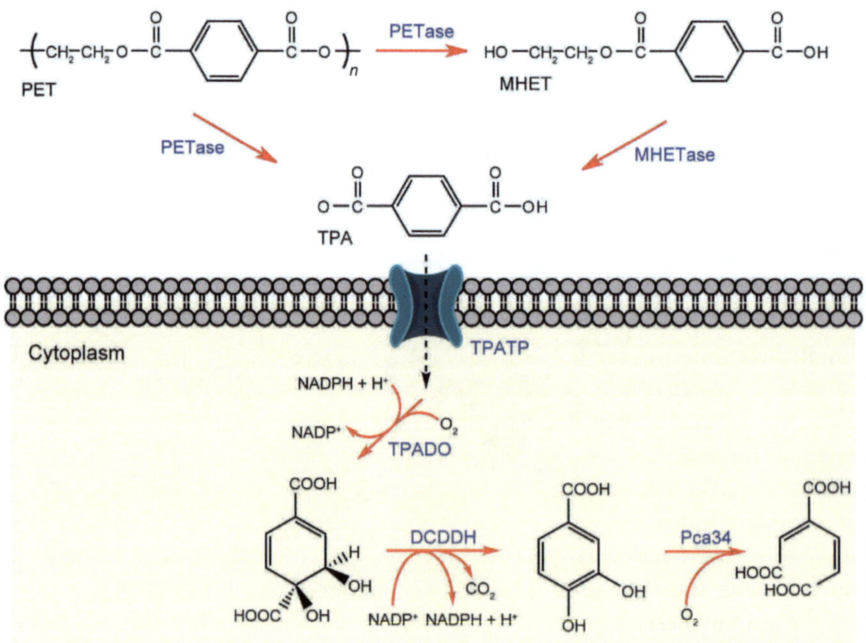

Figure 2.5 PET metabolism pathway by *I. sakaiensis*. Source: Adapted from Ref. [83].

the complete set of genes involved in PET metabolism using fully sequence genomes at the time, they found that of 92 microorganisms possess MHETase homologs and 33 had homologs for both TPA and PCA dioxygenases. These observations suggest that many microorganisms are capable of metabolizing MHET and its analogs.

Efficient degradation of PET-based plastics, particularly high-crystallinity PET film and bottles, requires high reaction temperatures, ideally near PET glass temperature (e.g. 65–75 °C). Unfortunately, the thermal stability of the PET-degrading enzymes is generally low and is known as a major limiting factor in PET degradation. The most obvious example is mesophilic IsPETase with $T_{m,e}$ around 46–49 °C, which is reported to lose its hydrolytic activity against amorphous PET after 24 hours at 37 °C, likely because the enzyme denatures over time. Even thermophilic cutinases with $T_{m,e}$ ranging from 56 to 85 °C show reduced hydrolytic activity against PET over time, resulting in partial depolymerization [120]. Over the past few years, researchers have employed rational and computational based enzyme engineering techniques to create variants of hydrolases with improved thermal stability and enhanced PET hydrolysis activity [98, 120]. For example, using a computational-based protein engineering approach, Cui et al. were able to create an IsPETase variant, which exhibited an increased $T_{m,e}$ by 31 °C and an improved degradation performance by more than 100-fold against semicrystalline (23%) PET films at mild temperatures [122].

Moreover, using enzyme engineering based on *in silico* docking calculations, Tournier et al. were able to generate a LCC variant F243I/D238C/S283C/Y127G (ICCG) with an improved $T_{m,e}$ of +9.3 °C and a hydrolytic activity comparable to that of wild type [120]. Remarkably, when tested against pretreated (extruded and micronized) post-consumer colored flake PET waste (PcW-PEt) with 14.6% crystallinity level in optimal bioreactor conditions, ICCG variant achieved 90% depolymerization conversion in only 10.5 hour with a mean productivity of 16.7 g_{TA}/l/h. Later molecular calculations suggested that the significant improvement in the hydrolytic activity of ICCG variant was likely due to catalytically favorable binding of the mutant enzyme to the substrate and improved enzyme–substrate affinity compared with the parental LCC. It is important to note that the ICCG variant currently outperforms all PET hydrolases reported so far, and its productivity rate is considerably higher than that of starch degradation (4 g/l/h), which is a common biodegradable polymer [120, 123].

Finally, using optimized ICCG variant in a 150-l-pilot scale, the same authors were able to produce terephthalate acid monomers at 99.8% purity, which was subsequently reused to synthesize virgin PET. The newly made PET was ultimately blown into bottles where their physical and mechanical properties were comparable to those synthesized using petrochemical TPA. Overall, this work illustrates how computer-aided enzyme engineering can enable efficient conversion of post-consumer PET plastics into their building monomers and back to commercial plastics, paving the way toward circular economy in the plastic industry. It is noteworthy that these results are relatively comparable to those obtained from chemical hydrolysis. In general, chemical hydrolysis of PET can be achieved by acidic, neutral, and alkaline hydrolysis [124]. Acidic hydrolysis typically produces high yields of TPA monomer, but high amounts of acid would decrease the purity of EG monomer, increase cost of the process, and raise environmental concerns [125]. On the other hand, neutral hydrolysis is eco-friendly, but all mechanical impurities present in the polymer remain in the TPA, resulting in lower TPA purity, overall [126]. Alkaline hydrolysis appears to be a better alternative. One study showed that the two-step aqueous alkaline hydrolysis of PET resulted in the highest yield (~95%) at 80 °C with the initial PET particle size lower than 500 μm in a solution containing 60 : 40 vol% EtOH:H_2O and 5 wt% NaOH in 20 minutes [127]. Moreover, among the real post-consumer PET waste samples including multilayer and monolayer PET trays and films and bottles tested in the same study, the highest PET conversion (~90%) was obtained with monolayer PET films at the smallest particle size (<500 μm). The conversion rate was diminished for monolayer PET trays, bottles, and multilayer PET samples. Especially with PET bottles and multilayer trays, the degradation yield did not exceed 70% [127].

2.6 Biodegradation of Hydrocarbons

2.6.1 Polyethylene

PE is the number one produced plastic, in volume, in the world. It comes in a variety of grades, for example, LLDPE, LDPE, and HDPE (high-density polyethylene). Ultrahigh molecular weight polyethylene (UHMWPE) is another PE variant with a high degree of polymerization (36000+) [128]. Each grade has its own unique applications, mainly because of the type and degrees of branching and degree of polymerization. Higher branching with longer side chains leads to less molecular packing because of higher steric hindrance within the polymer chains [129]. Similar to having high molecular weight, high branching also impairs biodegradation.

A number of studies have reported on biodegradation of PE grades, where oxidases, peroxidases, monooxygenases, etc., produced by aerobic microflora could cleave polymer chains to lower weight fragments [130, 131]. These resulting fragments were also shown to be taken up by bacteria and yeasts and converted into energy through β-oxidation pathways [130, 132]. For example, *Arthrobacter parraffineus* was shown to degrade LDPE after three years via β-oxidation [133]. Despite this, biodegradation can still take years because bioassimilation often being the rate-limiting step.

Biodegradation of PE can be enhanced through various pretreatment strategies such as thermal and UV exposure. UV oxidation produces free radicals from (C–H) and (C–C) homolytic bond cleavages. These radicals in the presence of oxygen form peroxy radicals, which abstract a hydrogen from the polymer backbone leaving behind another radical in its wake [134]. This process is continuous and leads to many hydrophilic functional groups, including ketones, carboxylic acids, esters, and alcohols, which are favored by many hydrolases. Indeed, an increase of UV exposure correlates with an increase of PE biodegradation [128]. For example, UV exposure via lamps led to a 25% increase in the biodegradation of LDPE. Thermal

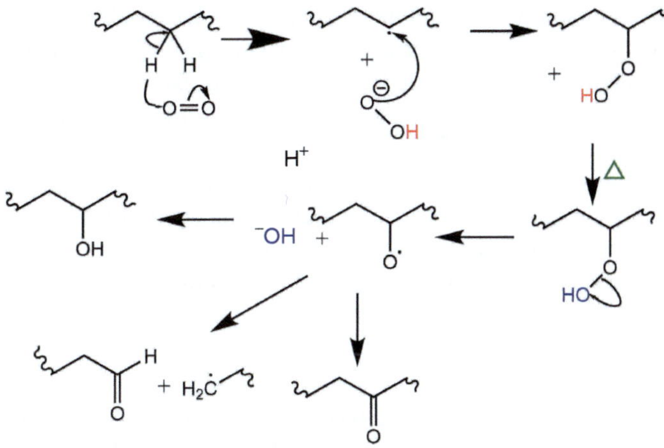

Figure 2.6 Thermal oxidation of PE. Source: The mechanism is adapted from Refs. [85, 135].

oxidation also works similarly (Figure 2.6). The following oxidation products are then primed for further bioassimilation. Research behind LDPE shows a 7% higher weight loss for thermally pretreated (80 °C for 10 days) polymers as compared to control via aerobic microbes (e.g. *Bacillus* sp.) [6].

Along with thermal and UV exposure, oxidative chemicals and surfactants may be used to pretreat recalcitrant polymers. Oxidative chemicals include nitric or sulfuric acid, which promote hydrophilicity in the polymer. Surfactants reduce the surface tension of a substrate. Some of these surfactants have very low critical micelle concentrations (CMCs) and therefore have a higher ability to solubilize and biodegrade the hydrocarbon polymer. Surfactants can be utilized at higher concentrations than their CMCs to get even more solubilizations, which may lead to higher levels of degradation on the PE surface. Other surfactants such as sodium dodecyl sulfate (SDS) oxidizes PE and sodium dodecyl benzene sulfonate (SDBS) hydrolyzes it [132]. Additionally, biosurfactants produced by *Lysinibacillus fusiformis*, for example, have the ability to solubilize the hydrocarbons because they are amphiphilic. In essence, these biosurfactants ensure smoother and easier microbial adhesion. Although it should be noted that the surfactants themselves must be biodegradable, otherwise, they would defeat the purpose [6, 132].

Additives can be used to increase the oxidation rate. Normal PE is recalcitrant. However, PE fitted with oxidation catalysts (i.e. iron complexes) can undergo abiotic transformation and subsequently support microbial growth [136]. Oxo-degradable additives may help pretreatment oxidization (e.g. photo oxidation) via UV radiation [137]. The additive rapidly reduces the molecular weight of the polymer, making it more susceptible to the second step of the process – the consumption of oligomers and monomers by microbes [137].

Although the pretreatment strategies discussed above were shown to promote biodegradation of the PE variants, microbial hydrolysis of UHMWPE is still quite challenging. In theory, UHMWPE should still be susceptible to biodegradation only if the molecular chains can be fragmented enough for bacterial or fungal consumption. However, because these chains are extremely long (high DP) and the scission mechanism operates on a stochastic basis, it will take far longer for any pretreatment strategy to oxidize and cleave the polymer into an appropriate size for bioassimilation. In addition, the high crystallinity of this PE variant prevents it from extracellular enzyme activity as these enzymes typically target the amorphous regions. Coupled with its chemical inertness, it is highly unlikely that UHMWPE can sustain biodegradation. Because UHMWPE is used in such highly specialized applications, it is not part of the overarching plastic pollution problem.

2.6.2 Polypropylene

PP is a vinyl polymer with methyl groups on alternating carbons. Polymerization yields three variants based on the orientation of the methyl branch in the polymer chain. These isomers can either be isotactic (same orientation), syndiotactic (alternating pendent groups), or atactic (random orientation of pendent groups) [138]. These methyl branches add a layer of steric hindrance which further disallows biodegradation. Typically, PP waste exposed to natural conditions degrade over very long time scales [139].

Pretreatments are also required for PP degradation because of its high molecular weight, chemical inertness, and lack of hydrophilicity on its chains, PP must be subjected through similar pretreatment strategies to be oxidized and fragmented. In the PP chain, there are three positional carbons (e.g. primary, secondary, and tertiary), where the radical can form in the presence of oxygen to promote oxidation [131]. This oxidation introduces ester, carboxyl, and carbonyl functional groups [139]. These functional groups promote further microbial attachment to polymer surface in the form of biofilm. For example, thermal pretreatment promotes formation of radicals that propagate to form hydrophilic groups, resulting in chain scission [131]. This fragmentation allows for greater weight loss through microbial bioassimilation. One study showed that the thermally pretreated PP (80 °C for 10 days in hot air oven) exhibited 7.1% weight loss in six months through aerobic microbes, while a similar untreated sample showed only a dismal 0.42% degradation by soil consortia [6]. When comparing LDPE and PP with the same pretreatment conditions (i.e. 100 KGy of gamma radiation), PP had a 1.2% mass loss after fungal inoculation, whereas LDPE had no mass loss, whatsoever [140].

Irradiation is a promising pretreatment. UV exposure of PP works similar to thermal exposure as in the case for PE. For example, UV-treated PP is more biodegradable than LDPE [133]. Other techniques such as gamma irradiation can induce photo-oxidation, which is a form of pretreatment before the microorganisms begin their role by embrittling the polymer and introducing polar groups to promote hydrophilicity. However, dosages are important as lower doses of gamma rays can sterilize plastics, while higher doses serve to allow more microbial adhesion onto the plastic films [140].

Bacteria are generally known to be less efficient in degrading PP. This is likely because the continuously repeating methylene units in PP increase hydrophobicity on the polymer's surface, which hampers biofilm formation by bacteria [131]. Despite this, a number of bacterial species such as *Pseudomonas azotoformans* and *B. subtilis* were shown to produce biofilms on pretreated PP films and enhance the polymer degradation, which further bolsters the idea behind synergistic effects between pretreatment and microbial degradation. In addition, *Bacillus flexus* was reported to produce the highest CFU/ml (colony forming units), regardless of any pretreatments in a 12 month period in minimal media, thereby indicating that is growth was directly attributable to PP biodegradation [141]. In addition to degrading pretreated PP, it also oxidized untreated PP, which shows the microorganism's promise as a candidate for further exploration on practical large-scale approaches.

On the other hand, fungi are promising biodegradation candidates because of their surplus of degrading enzymes in addition to their ability to survive in harsher ecosystems under low nutrient and moisture conditions. They also release plastic-degrading enzymes such as laccase, which is a phenol oxidase and part of a family of multi-copper enzymes [142]. For example, *Lasiodiplodia theobromae* and *Psychotria flavida* could degrade the PP film that was irradiated with different doses of gamma radiation. Moreover, fungi can adhere onto the surface of PP. *P. flavida* and *Humboldtia brunonis* are two known endophytic fungi that grow very well on plastic surfaces [140]. Endophytic fungi can extend their hyphae into holes and crevices.

2.6.3 Polystyrene

Polystyrene is a widely used polymer because of its low cost in materials and processing, its variety of forms tailored for versatile applications. In addition to its stability, moisture, and thermal resistance, it is also very hard to biodegrade primarily because of its aromaticity. Efforts to engage the polymer in this process have yielded extremely long time scales – a single PS sheet buried in soil lasted up to 32 years with no sign of significant decay. The bulky nature of the polymer does not allow for flexible polymer chains that can engage with extracellular enzymes released by bacteria in the biofilm. In fact, microorganisms have a difficult time attaching to the polymer surface because of the hydrophobicity of PS. Any enzyme release acts only superficially as the enzymes themselves are too large to penetrate the polymer network. However, it has been recently shown that microbes can catabolize styrene without further breakdown. It is well understood that the oxidation of styrene is performed through a series of steps with styrene monooxygenase and styrene oxide isomerase, to name a few. The vinyl group in styrene is first converted to an oxide, which is then oxidized further into an aldehyde. After more oxidation, the resulting carboxylate functional group then bonds with the phenylacetyl coenzyme A ligase and is then brought into the Krebs cycle (Figure 2.7) [143].

PS must also be oxidized to improve its hydrophilicity. Similar to the pretreatment strategies of other two polyolefins, PS oxidation is accomplished after thermal or UV exposure. The addition of oxidation catalysts such as metal ions such as cobalt, nickel, and manganese can improve the oxidation rates as a pretreatment step that ultimately improves the hydrophilicity of the polymer surface [143]. Chemical catalysis is the most common method for abiotic degradation yielding 70–90%

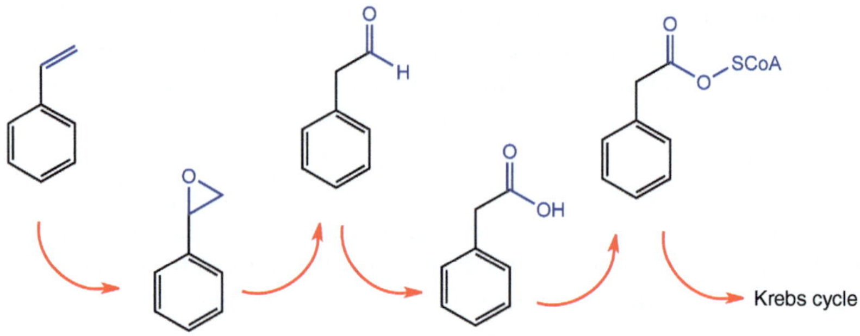

Figure 2.7 The metabolism pathway of styrene. Source: The mechanism is adapted from Ref. [136].

conversion, although liquefying PS at 240 °C does degrade the polymer into its monomers and, often, oligomers [144].

Combining pretreatment with bioassimilation is the best strategy. As with most non-biodegradable polymers, a synergistic effect must be achieved between the breakdown step and the following microbial assimilation of polymer fragments. In fact, this was exploited by Savoldelli et al. when implementing a two-step strategy involving liquefying and, consequently, degrading PS into its constituents and cooled down to allow for microbial inoculation that would further break the monomer down. *P. putida* and *Salmonella* have been effective in this process especially because of the increased surface area of the well-mixed liquid system. The temperature used in the first step was 240 °C, which is just around the melting point of PS but not quite high enough for the thermal degradation of PS. If it were, then the biodegradation step would be redundant and unnecessary all the while being more energy intensive. This method showed 100% degradation in a matter of days [144].

Typically, most evaluation of plastic degradation occurs on the lab scale. Regardless, there are some records of the extent of littering begotten by the population [145]. This may benefit abiotic processes as plastics are more exposed to oxygen and UV radiation, both of which enhance oxidation. However, if not under controlled laboratory conditions, the harmful chemical hazards emanating from the degradation process may harm the environment [85]. Plastics are also found in marine environments and are typically hindered in oxidation. Those exposed to seawater faced lower degradation rates than those of land [146]. The complete degradation may also be affected because of the varying conditions of the range of marine environments, which can lead to decades of persistence in deep water regions [147].

2.7 Biodegradation of Halogenated Polymers

Two of the most common halogenated polymers are polyvinyl chloride (PVC) and polytetrafluoroethylene (PTFE). PVC is a vinyl polymer with chlorine groups attached to every other carbon. PTFE, often known by its trade name Teflon, is also a vinyl polymer but is composed entirely of a (C–C) chain and attached fluorine atoms. Both polymers are extremely difficult to degrade enzymatically.

2.7.1 Polyvinyl Chloride

PVC can harbor bacterial and fungal surface colonization, which typically require a few weeks to become established [148–150] and can persist after the addition of germicides [151]. These observations may motivate the future studies of PVC disposal. A number of studies described weight loss of plasticized PVC films because of the action of microorganisms. However, it is not clear if this weight loss is attributed to PVC or the plasticizers [152, 153]. For example, in one study, researchers exposed plasticized polyvinyl chloride (pPVC) to the atmosphere for two years, where various bacteria and fungi were discovered colonizing its surface. Three fungi, namely, *Aureobasidium pullulans*, *Rhodotorula aurantiaca*, and *Kluyveromyces* spp., could grow independently on pieces of pPVC containing plasticizers dioctyl phthalate and dioctyl adipate (DOA). They showed that these fungi can degrade DOA and survive on it as its sole carbon source. However, no bacterial colonization was observed [149]. Although it is unclear if microorganisms are capable of degrading PVC, there are certainly microorganisms that consume its monomer vinyl chloride (VC). For example, aerobic cultures of bacteria isolated from contaminated groundwater and soil have been observed to dechlorinate VC to ethylene or assimilating it directly into their biomass [154, 155], while another bacterium, *Dehalococcoides ethenogenes*, can use VC as a terminal electron acceptor instead of oxygen in anaerobic cultures [156]. Thus far, there are no published studies exposing these bacterial species to the polymerized form of the substrate, but their current abilities may be reassuring for the future of PVC biodegradation.

2.7.2 Polytetrafluoroethylene

PTFE facilitates little to no biofilm formation and is frequently used to protect surfaces from microbial colonization, such as on stainless steel used in food processing [157] or biomedical implants [158]. This resistance to microbial surface colonization directly contrasts with PVC, possibly because of PTFE's hydrophobicity and slippery surface properties. One study targeted modified PTFE as a scaffold for bacterial immobilization. In that study, by creating a thin PTFE film via sputtered deposition onto polyethylene, researchers induced the formation of chemical functional groups in the PTFE surface such as hydroxyl, carboxyl, or amine groups, which facilitated bacterial biofilm formation [159]. Among identified bacteria forming biofilm on the surface of PTFE, none of them could degrade the material.

2.8 Biodegradation of Polyethers

2.8.1 Polyethylene Glycol

Polyethylene glycol (PEG) perhaps is the most commonly known polyether. PEG is susceptible to oxidation via free radicals and its degradation is hindered through antioxidant additions. Activation energy for hydrogen abstraction is reduced and formation and propagation of radicals are induced. The resulting radicals then form carboxylates with reactive oxygen species such as hydrogen peroxide or hydroxyl groups. Heating PEG at 80 °C in aerobic environments leads to similar random chain scission events and the lowering of the molecular weight distribution, the melting point, and the heat of fusion. In fact, biodegradation is stifled when antioxidants have been incorporated into the polymer [160]. In addition, PEG fragments more in excess air because of higher degrees of oxidation from radical additions with peroxides. This fragmentation results in formic esters from terminal cleavage.

Fragmentation can also occur through enzymatic action. This process begins with enzymes oxidizing the terminal alcohols from PEG fragments via PEG alcohol dehydrogenase [161]. The resulting aldehyde is further oxidized to a carboxylate by PEG aldehyde dehydrogenase [162]. Alternatively, these enzymes act to oxidize and cleave the polymer into smaller fragments in the same manner as the aforementioned abiotic degradation. Certain strains can grow on PEG depending on the degree of polymerization, mostly in the form of activated sludges. Some of the most prominent genera of PEG-degrading bacteria are *Pseudomonas*, *Sphingomonas*, and *Gluconobacter*, and some methanogens, from which several enzymes capable of degrading the PEG chains have been purified [161]. For example, a PEG dehydrogenase was found in PEG-utilizing sphingomonads, which show high growth on PEG 6000 and 20,000. Moreover, PEG-degrading microbes can metabolize PEG fragments in periplasmic space or in the cytoplasm, which indicates transport of these oligomers past cell walls. Despite this, PEG metabolites have no close relatives in nature [163].

2.8.2 Polyurethane

With the advent of polyurethane (PU) synthesis beginning in the 1950s, its versatile uses in the furniture and automotive industries have prompted a manufacturing boon. Thus, various strategies have been implemented to mitigate the burgeoning problem of PU waste accumulation. On the industrial scale, up to 30% of PU foams is ground and incorporated into new foams [164]. Consequently, the microionization of these foams allows for better chemical hydrolysis similar to PET. However, it cannot treat waste generated from the resulting PU disposal [164]. Although this class of polymers are defined by the urethane bond, it only accounts for a small proportion of the bonds in the polymer to, sometimes, none at all [165]. These bonds act as linkages connecting polyisocyanates and polyols whose ratio of "hard" and "soft" segments can be modulated in order to generate myriad characteristics such as elasticity and tensile strength [166]. In fact, PU can be classified as either polyester or polyether depending on the polyol used. Polyethers are more represented industrially and are typically used for thermal insulation [164]. Hard segments lend PU its insolubility in water or commercial solvents and thermal resistance. Therefore, common pretreatment strategies cannot be employed to the same extent. A common polyol, or "soft" segment (as it is often referred), is poly(propylene glycol) and typically undergoes increased chain extension through the use of additives [167]. It itself is biodegradable by myriad microorganisms and subject to the same hindrances such as lack of nutrient availability [168]. In fact, biodegradation may actually endanger the environment through the release of toxic chemicals depending on the additives present in the polymer [164].

Initial studies conducted in the late 1960s demonstrated great resistance to abiotic and biotic degradation conditions [164]. However, sunlight may promote photolysis [167]. Further research has focused on accentuating abiotic conditions that would stimulate microbial growth *in natura* [164]. Although polyester-based PUs have been shown to degrade in aerobic conditions [167], polyether-based PU has shown no discernable breakdown in either aerobic or anaerobic conditions [164]. Similarly, in laboratory conditions, polyester-based PU has shown remarkable degradation ability from a variety of organisms such as *Delftia acidovorans*, which can degrade 2–3 g of PU every four to six days [164, 169]. Alternatively, many microbes struggle to grow on polyether-based PU with only *Dietzia maris* and *Exophiala jeanselmei* degrading the polymer to any meaningful extent. In fact, less than a third of PU foam was degraded during a month of incubation [170]. Although there is a mismatch in biodegradation results between polyester and polyether-based PU, the biochemical pathways to degradation are suspected to be similar despite the inherent recalcitrance of the ether bond [164].

Enzymes comprise the first wave of attack in biodegradation but are hard to narrow down [168, 171]. The membrane-bound polyurethanase adheres to the PU surface and hydrolyzes the urethane linkage, thereby releasing the polymer constituents [172]. The resulting monomers are still attached to the membrane-bound enzymes, which allow for polyurethanases released to grind the PU surface, which augments the contact surface, thereby improving microbial adhesion [171]. PU's increasing waste accumulation is promulgated by its inherent difficulties in biodegradation. This can be mitigated somewhat by further exploration into degrading microorganisms especially for polyether-based PU.

2.9 Application of Biodegradation

We earlier discussed how enzymatic degradation can be exploited and used along mechanical pretreatments to enable near-full depolymerization and repolymerization of PET waste within industrial setting. This approach can also be employed for other hydrolysable polymers such as PLA once more efficient enzymes are characterized through engineering or novel discoveries in nature. However, it is important to note that more efficient technologies should also be established for recovery and purification of monomers from the rest of the components in the bioreactor.

Biodegradation can also be used for energy recovery. Coupled with microfiltration, wastewater streams from many industries can be treated to separate value-rich products such as microbial lipids that can be used for energy consumption [173]. Hydrogen and methane can be produced via anaerobic digestion from organic materials. One- and two-stage systems both yielded around 14.20 kJ/kg of organic material [174]. Aerobic bioreactors also produce biogas at 85% and often out-compete its counterpart, which has a 10% less methane yield at the same residence time. In fact, the combination of the two could result in a more potent energy recovery system reaching up to 140 MJ/m^3 of wastewater [175]. Although both treatment types are found concurrently, their mechanism differs. Namely, aerobic treatments proceed through organic waste oxidation, whereas anaerobic treatments occur through hydrolysis, fermentation, and methanogenesis [176]. Methanogenesis can also occur through biodegradation in oil fields because of the preponderance of anaerobes in low-nutrient environments. Anaerobic cultures degraded *n*-alkanes completely at 30 °C in 208 days. On the other hand, aerobic microcosms could fully degrade shorter chain *n*-alkanes (up to C33) in a tenth of the time [177].

In the case of upcycling, studies have shown that various forms of PHA have been found in municipal waste sludge. In fact, the synthesis of PHAs was observed in the context of wastewater treatment [178]. Although fermented food waste was used to yield an extrapolated 63.5 tons of PHA per day, the economic cost would be too high to implement this strategy [179]. The argument can be made that PHA, whose monomers are natural metabolites of bacteria present in various consortia, can be used to promote true upcycling processes. This is only possible because of the inherent biodegradability of PHAs. As other polymers, such as PE and PP, accumulate in waste streams, difficulties arise in verifying whether these fragments are bioassimilated via PHA production. Despite the many efforts of filtering out plastic contaminants, it still persists in the environment [180].

2.10 Current Challenges and Future Prospects for Biodegradation of Plastics Wastes

Additional polymers with continuous C–C polymer chain, namely, PE, PP, and PS, are highly recalcitrant to enzymatic degradation. Their surface hydrophobicity hinders the growth of biofilms that is needed for biodegradation. Hydrolysis requires oxidation of the chain, for example, by air or alternative oxidant, eventually assisted by UV pretreatment or oxidation catalysts/enzymes. Although a number of enzymes have been identified to partly contribute to hydrolysis of these polymers, complete mechanisms are still unknown.

On the other hand, condensation polymers including natural polymers (e.g. cellulose and starch) and polyesters are susceptible to enzymatic degradation. Among polyesters, however, PET is a challenging target for enzymatic attack. Despite the numerous studies of PET biodegradation, most accomplishments have been limited to degradation of amorphous or low-crystalline (<15%) PET products and yields for hydrolysis of high-crystalline PET bottles and textiles have been very slow. Given these challenges, it is therefore necessary to identify and characterize new efficient enzymes from the environment to tackle these limitations. Additionally, rational or semirational enzyme engineering should be highly considered by researchers to create variants of the existing PET hydrolases and new identified enzymes, which can exhibit enhanced thermostability and depolymerization activity against high-crystalline PET products. Finally, the application of thermomechanical pretreatments, such as extrusion and microionization, and UV radiation on polymers appear to help enhance their overall biodegradation as we showcased one example in the PET depolymerization case study (see Section 2.5.4).

It is important to note that biodegradability is not always desirable. Yet, for certain applications, such as single-use bags and mulch films, it can be enhanced to offset some of the more apathetic behaviors from people. For instance, designing LDPE bags with pro-oxidants (e.g. TiO_2) improve biofilm formation, albeit, at a rather slow pace. It would be better to phase out polyolefins in general in favor of polymers with constituents that are natural metabolites for a wide array of microbes, thereby reducing the time it spends as garbage. This would have obvious limitations, however, for products designed for longevity such as plastic containers (e.g. Tupperware) and sewer piping. If scientists can solve the cost-effectiveness problem of producing polyesters on the same scale as hydrocarbons, it can overtake PE and PP and mitigate further plastic accumulation.

The growth rate of biodegradable polymers is expected to be around 15% per annum [28]. However, there could be a catch-22 in that promulgating biodegradable plastics may actually lead to increased littering as consumers often incur less responsibility because they overestimate deterioration rates [84]. One could also make the argument that promoting further biodegradation may upset the balance of nature by further augmenting CO_2 levels [181]. In other cases, such as food packaging, biodegradability might not be desired. Food packaging, in part, is designed to prevent degradative changes in food and separate microbiota present within and outside the package. Any contamination, and subsequent bioassimilation of the plastic, will defeat the purpose of the packaging [182].

Designing the ideal biodegradable polymer depends on the desired time of functionality. Shorter time scales can benefit from the incorporation of starch with PLA or PHA that can readily hydrolyze and bioassimilate the monomers via the plethora of microbes available. Longer functional times would require aliphatic polyesters in the form of polymer alloys, which can compensate for the subpar mechanical properties found in each polymer [183].

2.A Detailed Mechanism of PET Hydrolysis

Catalytic mechanism of PET hydrolysis. Polyesters are almost always hydrolyzed by endo-type hydrolases. Likewise, PET degradation is thought to occur through endolytic cleavage of its internal ester bonds. The product of PET hydrolysis reaction is mainly mono(2-hydoxyethyl)terephthalate (MHET) along with minor amounts of TPA and bis(2-hydroxyethyl)terephthalate (BHET). The ratio of these products, however, can vary based on incubation temperature, reaction length, and enzyme concentration [88]. Most PET-degrading enzymes are also able to degrade BHET to MHET to EG and TPA. As noted earlier, thermophilic cutinases from actinomycetes (e.g. TfH, Cut190, and LCC) and HiC from *H. insolens* fungus are the main known PET hydrolases. These enzymes can significantly degrade the main body of amorphous and low-crystallinity PET chains at temperatures above PET's T_g.

Despite their high hydrolysis activity, no detailed molecular mechanism has been proposed for these enzymes. Instead, most attention has been drawn toward IsPETase, partly because IsPETase shows very potent activity against PET near room temperature compared to cutinases and its host bacterium, *I. sakaiensis*, can assimilate the degraded PET chains.

Although PET hydrolysis ability of IsPETase and PET hydrolases differ extensively in their optimal conditions, the structures of the enzymes are very similar. In particular, the catalytic triad in these enzymes are almost identical, and the residues constituting the hydrophobic cleft, known as the PET binding site, are highly conserved (see Figure 2.A.4b), suggesting that similar PET hydrolysis mechanisms may be employed by these enzymes. One group recently proposed a detailed mechanism for degradation of PET by IsPETase [121]. Although this mechanism was primarily suggested for a minor PET-degrading enzyme, it can hint at some molecular aspects of PET hydrolysis by main PET hydrolases. Using *in silico* docking experiments with 2-hydroxyethyl-(mono-hydroxyethyl terephthalate)$_4$ (2-HE(MHET)$_4$), regarded as a model substrate for PET chain, and mutagenesis analysis, authors showed that the PET chain binds at two sites along the IsPETase flat hydrophobic surface – one MHET moiety is bound to subsite I and three adjacent MHET moieties are bound to subsite II. According to their model, once MHET moieties are positioned at these two subsites, degradation is initiated by cleavage of the scissile ester bond located between subsite I and subsite II near the catalytic Ser^{160} residue (see Figure 2.A.4a,c). This hydrolysis reaction releases one PET chain with TPA terminal (^{TPA}PET) from subsite I and another one with HE terminal (^{HE}PET) from subsite II.

In the next step, two released PET chains are successively broken down into smaller molecules via two partially different mechanisms. In the first mechanism, the scissile ester bonds in both chains are cleaved as described above, producing one MHET monomer from the ^{HE}PET-terminal chain and one TPA molecule from the ^{TPA}PET-terminal chain along with $^{HE}PET_{n-1}$ from both chains. Degradation of the resultant $^{HE}PET_{n-1}$ continues following the same process. In the second mechanism, which is thought to be less efficient than the previous one, ^{HE}PET and ^{TPA}PET chains and the enzyme interact in the reverse direction. Here, one or two MHET moieties are positioned at subsite II instead of subsite I, and the cleavage of the sessile ester bond located between subsites I and II is catalyzed with help of Ser^{160} residue. Continuation of this process then produces various sizes of PET monomers and dimers, such as 2-HE(MHET)$_2$, (MHET)$_2$, MHET, and BHET. Moreover, BHET can be broken down into MHET, TPA, and EG, and eventually, three molecules, MHET, TPA, and EG, are accumulated as the result of PET degradation.

References

1 Ragaert, K., Delva, L., and Van Geem, K. (2017). Mechanical and chemical recycling of solid plastic waste. *Waste Management* 69: 24–58.

2 Villain, F., Coudane, J., and Vert, M. (1995). Thermal degradation of polyethylene terephthalate: study of polymer stabilization. *Polymer Degradation and Stability* 49 (3): 393–397.

3 Sinha, V., Patel, M.R., and Patel, J.V. (2010). Pet waste management by chemical recycling: a review. *Journal of Polymers and the Environment* 18: 8–25.
4 Kawai, F., Kawabata, T., and Oda, M. (2020). Current state and perspectives related to the polyethylene terephthalate hydrolases available for biorecycling. *ACS Sustainable Chemistry & Engineering* 8 (24): 8894–8908.
5 Larrañaga, A. and Lizundia, E. (2019). A review on the thermomechanical properties and biodegradation behaviour of polyesters. *European Polymer Journal* 121.
6 Arkatkar, A., Arutchelvi, J., Sudhakar, M. et al. (2009). Approaches to enhance the biodegradation of polyolefins. *Open Environmental Engineering Journal* 2 (1): 68–80.
7 Chandra, R. and Rustgi, R. (1998). Biodegradable polymers. *Progress in Polymer Science* 23 (7): 1273–1335.
8 Dilkes-Hoffman, L.S., Lant, P.A., Laycock, B., and Pratt, S. (2019). The rate of biodegradation of PHA bioplastics in the marine environment: a meta-study. *Marine Pollution Bulletin* 142: 15–24.
9 Coughlan, M.P. (1990). Cellulose degradation by fungi. In: *Microbial Enzymes and Biotechnology* (ed. W.M. Fogarty et al.), 1–36. Dordrecht: Springer.
10 Wang, M., Liu, K., Dai, L. et al. (2013). The structural and biochemical basis for cellulose biodegradation. *Journal of Chemical Technology and Biotechnology* 88 (4): 491–500.
11 Pérez, S. and Samain, D. (2010). Structure and engineering of celluloses. In: *Advances in Carbohydrate Chemistry and Biochemistry* (ed. D. Horton), 25–116. Academic Press.
12 Coughlan, M.P. (1992). Enzymic hydrolysis of cellulose: an overview. *Bioresource Technology* 39 (2): 107–115.
13 Béguin, P. (1990). Molecular biology of cellulose degradation. *Annual Review of Microbiology* 44: 219–248.
14 Wang, M., Li, Z., Fang, X. et al. (2012). Cellulolytic enzyme production and enzymatic hydrolysis for second-generation bioethanol production. *Advances in Biochemical Engineering/Biotechnology* 128: 1–24.
15 Klyosov, A.A. (1990). Trends in biochemistry and enzymology of cellulose degradation. *Biochemistry* 29 (47): 10577–10585.
16 Béguin, P. and Aubert, J.P. (1994). The biological degradation of cellulose. *FEMS Microbiology Reviews* 13 (1): 25–58.
17 Pozzo, T., Pasten, J.L., Karlsson, E.N., and Logan, D.T. (2010). Structural and functional analyses of β-glucosidase 3B from *Thermotoga neapolitana*: a thermostable three-domain representative of glycoside hydrolase 3. *Journal of Molecular Biology* 397 (3): 724–739.
18 Prawitwong, P., Waeonukul, R., Tachaapaikoon, C. et al. (2013). Direct glucose production from lignocellulose using *Clostridium thermocellum* cultures supplemented with a thermostable β-glucosidase. *Biotechnology for Biofuels* 6 (1).
19 Ljungdahl, L.G. and Eriksson, K.-E. (1985). Ecology of microbial cellulose degradation. In: 237–299.

20 Holwerda, E.K., Thorne, P.G., Olson, D.G. et al. (2014). The exometabolome of *Clostridium thermocellum* reveals overflow metabolism at high cellulose loading. *Biotechnology for Biofuels* 7 (1).

21 You, C., Zhang, X.Z., Sathitsuksanoh, N. et al. (2012). Enhanced microbial utilization of recalcitrant cellulose by an ex vivo cellulosome-microbe complex. *Applied and Environmental Microbiology* 78 (5): 1437–1444.

22 Ou, Y., Huang, Q., and Chen, J. (2001). Degradation of new cellulose-based complex material. *Journal of Applied Polymer Science* 81 (4): 809–812.

23 Mohanty, A.K., Misra, M., and Hinrichsen, G. (2000). Biofibres, biodegradable polymers and biocomposites: an overview. *Macromolecular Materials and Engineering* 1–24.

24 Gallant, D.J., Bouchet, B., Buleon, A., and Perez, S. (1992). Physical characteristics of starch granules and susceptibility to enzymatic degradation. *European Journal of Clinical Nutrition* .

25 Oates, C.G. (1997). Towards an understanding of starch granule structure and hydrolysis. *Trends in Food Science and Technology* 8: 375–382.

26 Thoma, J.A. (1968). A possible mechanism for amylase catalysis. *Journal of Theoretical Biology* 19 (3): 297–310.

27 Moreno-Chulim, M.V., Barahona-Perez, F., and Canche-Escamilla, G. (2003). Biodegradation of starch and acrylic-grafted starch by *Aspergillus niger*. *Journal of Applied Polymer Science* 89 (10): 2764–2770.

28 Künkel, A., Becker, J., Börger, L. et al. (2016). Polymers, biodegradable. In: *Ullmann's Encyclopedia of Industrial Chemistry*, 1–29. Wiley.

29 Zhong, Y., Godwin, P., Jin, Y., and Xiao, H. (2020). Biodegradable polymers and green-based antimicrobial packaging materials: a mini-review. *Advanced Industrial and Engineering Polymer Research* 3 (1): 27–35.

30 Danjaji, I.D., Nawang, R., Ishiaku, U.S. et al. (2002). Degradation studies and moisture uptake of sago-starch-filled linear low-density polyethylene composites. *Polymer Testing* 21 (1): 75–81.

31 Vieyra, H., Aguilar-Méndez, M.A., and San Martín-Martínez, E. (2013). Study of biodegradation evolution during composting of polyethylene-starch blends using scanning electron microscopy. *Journal of Applied Polymer Science* 127 (2): 845–853.

32 Sivan, A. (2011). New perspectives in plastic biodegradation. *Current Opinion in Biotechnology* 22: 422–426.

33 Laycock, B., Nikolić, M., Colwell, J.M. et al. (2017). Lifetime prediction of biodegradable polymers. *Progress in Polymer Science* 71: 144–189.

34 Danso, D., Chow, J., and Streita, W.R. (2019). Plastics: environmental and biotechnological perspectives on microbial degradation. *Applied and Environmental Microbiology* 85.

35 Tosa, T. and Chibata, I. (1965). Utilization of cyclic amides and formation of ω-amino acids by microorganisms. *Journal of Bacteriology* 89 (3): 919–920.

36 Takehara, I., Fujii, T., Tanimoto, Y. et al. (2018). Metabolic pathway of 6-aminohexanoate in the nylon oligomer-degrading bacterium *Arthrobacter* sp. KI72: identification of the enzymes responsible for the conversion of

6-aminohexanoate to adipate. *Applied Microbiology and Biotechnology* 102 (2): 801–814.

37 Sudhakar, M., Priyadarshini, C., Doble, M. et al. (2007). Marine bacteria mediated degradation of nylon 66 and 6. *International Biodeterioration and Biodegradation* 60 (3): 144–151.

38 Kanagawa, K., Oishi, M., Negoro, S. et al. (1993). Characterization of the 6-aminohexanoate-dimer hydrolase from *Pseudomonas* sp. NK87. *Journal of General Microbiology* 139 (4): 787–795.

39 Deguchi, T., Kitaoka, Y., Kakezawa, M., and Nishida, T. (1998). Purification and characterization of a nylon-degrading enzyme. *Applied and Environmental Microbiology* 64 (4): 1366–1371.

40 Kawasaki, N., Nakayama, A., Yamano, N. et al. (2005). Synthesis, thermal and mechanical properties and biodegradation of branched polyamide 4. *Polymer (Guildf)* 46 (23): 9987–9993.

41 Yamano, N., Kawasaki, N., Ida, S., and Nakayama, A. (2019). Biodegradation of polyamide 4 in seawater. *Polymer Degradation and Stability* 166: 230–236.

42 Yamano, N., Kawasaki, N., Ida, S. et al. (2017). Biodegradation of polyamide 4 in vivo. *Polymer Degradation and Stability* 137: 281–288.

43 Turki, S., Ben, K.I., Weeckers, F. et al. (2009). Isolation of bioactive peptides from tryptone that modulate lipase production in *Yarrowia lipolytica*. *Bioresource Technology* 100 (10): 2724–2731.

44 John, R.P., Nampoothiri, K.M., and Pandey, A. (2007). Fermentative production of lactic acid from biomass: an overview on process developments and future perspectives. *Applied Microbiology and Biotechnology* 74: 524–534.

45 De Jong, S.J., Arias, E.R., Rijkers, D.T.S. et al. (2001). New insights into the hydrolytic degradation of poly(lactic acid): participation of the alcohol terminus. *Polymer (Guildf)* 42 (7): 2795–2802.

46 Song, J.H., Murphy, R.J., Narayan, R., and Davies, G.B.H. (2009). Biodegradable and compostable alternatives to conventional plastics. *Philosophical Transactions of the Royal Society of London, Series B: Biological Sciences* 364 (1526): 2127–2139.

47 Nova Institute (2020). Biodegradable polymers in various Environments, 1.

48 Krause, M.J. and Townsend, T.G. (2016). Life-cycle assumptions of landfilled polylactic acid underpredict methane generation. *Environmental Science & Technology Letters* 3 (4): 166–169.

49 Kolstad, J.J., Vink, E.T.H., De Wilde, B., and Debeer, L. (2012). Assessment of anaerobic degradation of Ingeo™ polylactides under accelerated landfill conditions. *Polymer Degradation and Stability* 97 (7): 1131–1141.

50 Vargas, L.F., Welt, B.A., Teixeira, A. et al. (2009). Biodegradation of treated polylactic acid (PLA) under anaerobic conditions. *Transactions of the ASABE* 52 (3): 1025–1030.

51 Saeidlou, S., Huneault, M.A., Li, H., and Park, C.B. (2012). Poly(lactic acid) crystallization. *Progress in Polymer Science* 37: 1657–1677.

52 Tsuji, H. (2002). Autocatalytic hydrolysis of amorphous-made polylactides: effects of L-lactide content, tacticity, and enantiomeric polymer blending. *Polymer (Guildf)* 43 (6): 1789–1796.

53 Tsuji, H. and Miyauchi, S. (2001). Poly(L-lactide): VI. Effects of crystallinity on enzymatic hydrolysis of poly(L-lactide) without free amorphous region. *Polymer Degradation and Stability* 71 (3): 415–424.

54 Iwata, T. and Doi, Y. (1998). Morphology and enzymatic degradation of poly(L-lactic acid) single crystals. *Macromolecules* 31 (8): 2461–2467.

55 Cai, H., Dave, V., Gross, R.A., and McCarthy, S.P. (1996). Effects of physical aging, crystallinity, and orientation on the enzymatic degradation of poly(lactic acid). *Journal of Polymer Science Part B: Polymer Physics* 34 (16): 2701–2708.

56 Jarerat, A. and Tokiwa, Y. (2001). Degradation of poly(L-lactide) by a fungus. *Macromolecular Bioscience* 1 (4): 136–140.

57 Tokiwa, Y. and Calabia, B.P. (2006). Biodegradability and biodegradation of poly(lactide). *Applied Microbiology and Biotechnology* 72: 244–251.

58 Pranamuda, H., Tsuchii, A., and Tokiwa, Y. (2001). Poly(L-lactide)-degrading enzyme produced by *Amycolatopsis* sp. *Macromolecular Bioscience* 1 (1): 25–29.

59 Lomthong, T., Hanphakphoom, S., Yoksan, R., and Kitpreechavanich, V. (2015). Co-production of poly(L-lactide)-degrading enzyme and raw starch-degrading enzyme by *Laceyella sacchari* LP175 using agricultural products as substrate, and their efficiency on biodegradation of poly(L-lactide)/thermoplastic starch blend film. *International Biodeterioration and Biodegradation* 104: 401–410.

60 Lipsa, R., Tudorachi, N., Darie-Nita, R.N. et al. (2016). Biodegradation of poly(lactic acid) and some of its based systems with *Trichoderma viride*. *International Journal of Biological Macromolecules* 88: 515–526.

61 Maharana, T., Mohanty, B., and Negi, Y.S. (2009). Melt-solid polycondensation of lactic acid and its biodegradability. *Progress in Polymer Science (Oxford)* 34: 99–124.

62 Tokiwa, Y. and Jarerat, A. (2004). Biodegradation of poly(L-lactide). *Biotechnology Letters* 26: 771–777.

63 Kawai, F., Nakadai, K., Nishioka, E. et al. (2011). Different enantioselectivity of two types of poly(lactic acid) depolymerases toward poly(L-lactic acid) and poly(D-lactic acid). *Polymer Degradation and Stability* 96 (7): 1342–1348.

64 Kazuyo, D.S., Aki Sasashige, T., Kanamoto, T., and Hyon, S.H. (2003). Preparation of oriented β-form poly(L-lactic acid) by solid-state coextrusion: effect of extrusion variables. *Macromolecules* 36 (10): 3601–3605.

65 David, L., Cheah, E., Cygler, M. et al. (1992). The alpha/beta hydrolase fold. *Protein Engineering* 5 (3): 197–211.

66 Perona, J.J. and Craik, C.S. (1997). Evolutionary divergence of substrate specificity within the chymotrypsin-like serine protease fold. *Journal of Biological Chemistry* 272: 29987–29990.

67 Qi, X., Ren, Y., and Wang, X. (2017). New advances in the biodegradation of poly(lactic) acid. *International Biodeterioration and Biodegradation* 117: 215–223.

68 Copinet, A., Bertrand, C., Govindin, S. et al. (2004). Effects of ultraviolet light (315 nm), temperature and relative humidity on the degradation of polylactic acid plastic films. *Chemosphere* 55 (5): 763–773.

69 Ray, S.S. and Bousmina, M. (2005). Biodegradable polymers and their layered silicate nanocomposites: In greening the 21st century materials world. *Progress in Materials Science* 50: 962–1079.

70 Jambeck, J.R., Geyer, R., Wilcox, C. et al. (2015). Plastic waste inputs from land into the ocean. *Science (80-)* 347 (6223): 768–771.

71 Iwamoto, A. and Tokiwa, Y. (1994). Enzymatic degradation of plastics containing polycaprolactone. *Polymer Degradation and Stability* 45 (2): 205–213.

72 Murphy, C.A., Cameron, J.A., Huang, S.J., and Vinopal, R.T. (1996). Fusarium polycaprolactone depolymerase is cutinase. *Applied and Environmental Microbiology* 62 (2): 456–460.

73 Abdel-Motaal, F.F., El-Sayed, M.A., El-Zayat, S.A., and Ito, S.-i. (2014). Biodegradation of poly(ε-caprolactone) (PCL) film and foam plastic by *Pseudozyma japonica* sp. nov., a novel cutinolytic ustilaginomycetous yeast species. *3 Biotech* 4 (5): 507–512.

74 Krasowska, K., Heimowska, A., and Morawska, M. (2016). Environmental degradability of polycaprolactone under natural conditions. *E3S Web of Conferences*, Kraków, Poland, 17–19 May 2016 (SEED 2016), https://doi.org/10.1051/e3sconf/20161000048.

75 Sun, H., Mei, L., Song, C. et al. (2006). The in vivo degradation, absorption and excretion of PCL-based implant. *Biomaterials* 27 (9): 1735–1740.

76 Al Hosni, A.S., Pittman, J.K., and Robson, G.D. (2019). Microbial degradation of four biodegradable polymers in soil and compost demonstrating polycaprolactone as an ideal compostable plastic. *Waste Management* 97: 105–114.

77 Wendlandt, K.D., Jechorek, M., Helm, J., and Stottmeister, U. (1998). Production of PHB with a high molecular mass from methane. *Polymer Degradation and Stability* 59 (1–3): 191–194.

78 Verlinden, R.A.J., Hill, D.J., Kenward, M.A. et al. (2007). Bacterial synthesis of biodegradable polyhydroxyalkanoates. *Journal of Applied Microbiology* 102: 1437–1449.

79 Vroman, I. and Tighzert, L. (2009). Biodegradable polymers. *Materials* 2: 307–344.

80 Sin, M.C., Gan, S.N., Annuar, M.S.M., and Tan, I.K.P. (2010). Thermodegradation of medium-chain-length poly(3-hydroxyalkanoates) produced by *Pseudomonas putida* from oleic acid. *Polymer Degradation and Stability* 95 (12): 2334–2342.

81 Sashiwa, H., Fukuda, R., Okura, T. et al. (2018). Microbial degradation behavior in seawater of polyester blends containing poly(3-hydroxybutyrate-co-3-hydroxyhexanoate) (PHBHHx). *Marine Drugs* 16 (1).

82 Weng, Y.X., Wang, X.L., and Wang, Y.Z. (2011). Biodegradation behavior of PHAs with different chemical structures under controlled composting conditions. *Polymer Testing* 30 (4): 372–380.

83 Ho, Y.H., Gan, S.N., and Tan, I.K.P. (2002). Biodegradation of a medium-chain-length polyhydroxyalkanoate in tropical river water. *Applied Biochemistry and Biotechnology - Part A Enzyme Engineering and Biotechnology* 337–347.

84 Haider, T.P., Völker, C., Kramm, J. et al. (2019). Plastics of the future? The impact of biodegradable polymers on the environment and on society. *Angewandte Chemie International Edition* 58: 50–62.

85 Gewert, B., Plassmann, M.M., and Macleod, M. (2015). Pathways for degradation of plastic polymers floating in the marine environment. *Environmental Science: Processes and Impacts* 17: 1513–1521.

86 Kikkawa, Y., Fujita, M., Abe, H., and Doi, Y. (2004). Effect of water on the surface molecular mobility of poly(lactide) thin film: an atomic force microscopy study. *Biomacromolecules* 5 (4): 1187–1193.

87 Jog, J.P. (1995). Crystallization of polyethyleneterephthalate. *Journal of Macromolecular Science, Part C: Polymer Reviews* 35 (3): 531–553.

88 Yoshida, S., Hiraga, K., Takehana, T. et al. (2016). A bacterium that degrades and assimilates poly(ethylene terephthalate). *Science (80-)* 351 (6278): 1196–1199.

89 Tokiwa, Y., Calabia, B.P., Ugwu, C.U., and Aiba, S. (2009). Biodegradability of plastics. *International Journal of Molecular Sciences* 10: 3722–3742.

90 Gamerith, C., Zartl, B., Pellis, A. et al. (2017). Enzymatic recovery of polyester building blocks from polymer blends. *Process Biochemistry* 59: 58–64.

91 Müller, R.J., Schrader, H., Profe, J. et al. (2005). Enzymatic degradation of poly(ethylene terephthalate): rapid hydrolyse using a hydrolase from *T. fusca*. *Macromolecular Rapid Communications* 26 (17): 1400–1405.

92 Dresler, K., Van Den Heuvel, J., Müller, R.J., and Deckwer, W.D. (2006). Production of a recombinant polyester-cleaving hydrolase from *Thermobifida fusca* in *Escherichia coli*. *Bioprocess and Biosystems Engineering* 29 (3): 169–183.

93 Lykidis, A., Mavromatis, K., Ivanova, N. et al. (2007). Genome sequence and analysis of the soil cellulolytic actinomycete *Thermobifida fusca* YX. *Journal of Bacteriology* 189 (6): 2477–2486.

94 Chen, S., Tong, X., Woodard, R.W. et al. (2008). Identification and characterization of bacterial cutinase. *The Journal of Biological Chemistry* 283 (38): 25854–25862.

95 Herrero Acero, E., Ribitsch, D., Steinkellner, G. et al. (2011). Enzymatic surface hydrolysis of PET: effect of structural diversity on kinetic properties of cutinases from *Thermobifida*. *Macromolecules* 44 (12): 4632–4640.

96 Thumarat, U., Kawabata, T., Nakajima, M. et al. (2015). Comparison of genetic structures and biochemical properties of tandem cutinase-type polyesterases from *Thermobifida alba* AHK119. *Journal of Bioscience and Bioengineering* 120 (5): 491–497.

97 Hu, X., Thumarat, U., Zhang, X. et al. (2010). Diversity of polyester-degrading bacteria in compost and molecular analysis of a thermoactive esterase from *Thermobifida alba* AHK119. *Applied Microbiology and Biotechnology* 87 (2): 771–779.

98 Thumarat, U., Nakamura, R., Kawabata, T. et al. (2012). Biochemical and genetic analysis of a cutinase-type polyesterase from a thermophilic *Thermobifida alba* AHK119. *Applied Microbiology and Biotechnology* 95 (2): 419–430.

99 Brueckner, T., Eberl, A., Heumann, S. et al. (2008). Enzymatic and chemical hydrolysis of polyethylene terephthalate fabrics. *Journal of Polymer Science, Part A: Polymer Chemistry* 46 (19): 6435–6443.

100 Ribitsch, D., Acero, E.H., Greimel, K. et al. (2012). A new esterase from *Thermobifida halotolerans* hydrolyses polyethylene terephthalate (PET) and polylactic acid (PLA). *Polymers (Basel)* 4 (1): 617–629.

101 Huang, Y.C., Chen, G.H., Chen, Y.F. et al. (2010). Heterologous expression of thermostable acetylxylan esterase gene from *Thermobifida fusca* and its synergistic action with xylanase for the production of xylooligosaccharides. *Biochemical and Biophysical Research Communications* 400 (4): 718–723.

102 Sulaiman, S., Yamato, S., Kanaya, E. et al. (2012). Isolation of a novel cutinase homolog with polyethylene terephthalate-degrading activity from leaf-branch compost by using a metagenomic approach. *Applied and Environmental Microbiology* 78 (5): 1556–1562.

103 Hegde, K. and Veeranki, V.D. (2013). Production optimization and characterization of recombinant cutinases from *Thermobifida fusca* sp. NRRL B-8184. *Applied Biochemistry and Biotechnology* 170 (3): 654–675.

104 Chertkov, O., Sikorski, J., Nolan, M. et al. (2011). Complete genome sequence of *Thermomonospora curvata* type strain (B9 T). *Standards in Genomic Sciences* 4 (1): 13–22.

105 Ladner, C., Ehninger, G., Gey, K.F., and Clemens, M.R. (1989). Effect of etoposide (VP16-213) on lipid peroxidation and antioxidant status in a high-dose radiochemotherapy regimen. *Cancer Chemotherapy and Pharmacology* 25 (3): 210–212.

106 Oda, M., Yamagami, Y., Inaba, S. et al. (2018). Enzymatic hydrolysis of PET: functional roles of three Ca2+ ions bound to a cutinase-like enzyme, Cut190*, and its engineering for improved activity. *Applied Microbiology and Biotechnology* 102 (23): 10067–10077.

107 Ronkvist, Å.M., Xie, W., Lu, W., and Gross, R.A. (2009). Cutinase-catalyzed hydrolysis of poly(ethylene terephthalate). *Macromolecules* 42 (14): 5128–5138.

108 Edgar, R.C. (2004). MUSCLE: a multiple sequence alignment method with reduced time and space complexity. *BMC Bioinformatics* 5.

109 Chen, S., Su, L., Chen, J., and Wu, J. (2013). Cutinase: characteristics, preparation, and application. *Biotechnology Advances* 31: 1754–1767.

110 Chen, S., Su, L., Billig, S. et al. (2010). Biochemical characterization of the cutinases from *Thermobifida fusca*. *Journal of Molecular Catalysis B: Enzymatic* 63 (3–4): 121–127.

111 Ribitsch, D., Heumann, S., Trotscha, E. et al. (2011). Hydrolysis of polyethyleneterephthalate by p-nitrobenzylesterase from *Bacillus subtilis*. *Biotechnology Progress* 27 (4): 951–960.

112 Perz, V., Baumschlager, A., Bleymaier, K. et al. (2016). Hydrolysis of synthetic polyesters by *Clostridium botulinum* esterases. *Biotechnology and Bioengineering* 113 (5): 1024–1034.

113 Sebastian, J. and Kolattukudy, P.E. (1988). Purification and characterization of cutinase from a fluorescent *Pseudomonas putida* bacterial strain isolated from phyllosphere. *Archives of Biochemistry and Biophysics* 263 (1): 77–85.

114 Wei, R., Oeser, T., Schmidt, J. et al. (2016). Engineered bacterial polyester hydrolases efficiently degrade polyethylene terephthalate due to relieved product inhibition. *Biotechnology and Bioengineering* 113 (8): 1658–1665.

115 Sulaiman, S., You, D.J., Kanaya, E. et al. (2014). Crystal structure and thermodynamic and kinetic stability of metagenome-derived LC-cutinase. *Biochemistry* 53 (11): 1858–1869.

116 Carr, P.D. and Ollis, D.L. (2009). Alpha/beta hydrolase fold: an update. *Protein and Peptide Letters* 16 (10): 1137–1148.

117 Kold, D., Dauter, Z., Laustsen, A.K. et al. (2014). Thermodynamic and structural investigation of the specific SDS binding of *Humicola insolens* cutinase. *Protein Science* 23 (8): 1023–1035.

118 Miyakawa, T., Mizushima, H., Ohtsuka, J. et al. (2015). Structural basis for the Ca2+-enhanced thermostability and activity of PET-degrading cutinase-like enzyme from *Saccharomonospora viridis* AHK190. *Applied Microbiology and Biotechnology* 99 (10): 4297–4307.

119 Numoto, N., Kamiya, N., Bekker, G.J. et al. (2018). Structural dynamics of the PET-degrading cutinase-like enzyme from *Saccharomonospora viridis* AHK190 in substrate-bound states elucidates the Ca2+-driven catalytic cycle. *Biochemistry* 57 (36): 5289–5300.

120 Tournier, V., Topham, C.M., Gilles, A. et al. (2020). An engineered PET depolymerase to break down and recycle plastic bottles. *Nature* 580 (7802): 216–219.

121 Joo, S., Cho, I.J., Seo, H. et al. (2018). Structural insight into molecular mechanism of poly(ethylene terephthalate) degradation. *Nature Communications* 9 (1).

122 Cui, Y., Chen, Y., Liu, X. et al. (2019). Computational redesign of a PETase for plastic biodegradation by the GRAPE strategy. *bioRxiv*.

123 Gusakov, A.V., Salanovich, T.N., Antonov, A.I. et al. (2007). Design of highly efficient cellulase mixtures for enzymatic hydrolysis of cellulose. *Biotechnology and Bioengineering* 97 (5): 1028–1038.

124 Spychaj, T. (2002). Chemical recycling of PET: methods and products. In: *Handbook of Thermoplastic Polyesters* (ed. S. Fakirov), (chapter 27), 1252–1290. Wiley.

125 Al-Sabagh, A.M., Yehia, F.Z., Eshaq, G. et al. (2016). Greener routes for recycling of polyethylene terephthalate. *Egyptian Journal of Petroleum* 25: 53–64.

126 Geyer, B., Lorenz, G., and Kandelbauer, A. (2016). Recycling of poly(ethylene terephthalate) – a review focusing on chemical methods. *Express Polymer Letters* 10: 559–586.

127 Ügdüler, S., Van Geem, K.M., Denolf, R. et al. (2020). Towards closed-loop recycling of multilayer and coloured PET plastic waste by alkaline hydrolysis. *Green Chemistry* 22 (16): 5376–5394.

128 Sobieraj, M.C. and Rimnac, C.M. (2009). Ultra high molecular weight polyethylene: mechanics, morphology, and clinical behavior. *Journal of the Mechanical Behavior of Biomedical Materials* 2: 433–443.

129 Carraher, C.E. (2017). *Introduction to Polymer Chemistry*, 4e, 1–560. CRC Press/Taylor & Francis group.

130 Raddadi, N. and Fava, F. (2019). Biodegradation of oil-based plastics in the environment: existing knowledge and needs of research and innovation. *Science of the Total Environment* 679: 148–158.

131 Arkatkar, A., Arutchelvi, J., Bhaduri, S. et al. (2009). Degradation of unpretreated and thermally pretreated polypropylene by soil consortia. *International Biodeterioration and Biodegradation* 63 (1): 106–111.

132 Mukherjee, S., RoyChaudhuri, U., and Kundu, P.P. (2017). Anionic surfactant induced oxidation of low density polyethylene followed by its microbial bio-degradation. *International Biodeterioration and Biodegradation* 117: 255–268.

133 Arutchelvi, J., Sudhakar, M., Arkatkar, A. et al. (2008). Biodegradation of polyethylene and polypropylene. *Indian Journal of Biotechnology* 7: 9–22.

134 Costa, L. and Bracco, P. (2016). Mechanisms of cross-linking, oxidative degradation, and stabilization of UHMWPE. In: *UHMWPE Biomaterials Handbook: Ultra High Molecular Weight Polyethylene in Total Joint Replacement and Medical Devices*, 3e (ed. S.M. Kurtz), 467–487. William Andrew Publishing.

135 Gardette, M., Perthue, A., Gardette, J.L. et al. (2013). Photo- and thermal-oxidation of polyethylene: comparison of mechanisms and influence of unsaturation content. *Polymer Degradation and Stability* 98 (11): 2383–2390.

136 Bonhomme, S., Cuer, A., Delort, A.M. et al. (2003). Environmental biodegradation of polyethylene. *Polymer Degradation and Stability* 81 (3): 441–452.

137 Portillo, F., Yashchuk, O., and Hermida, É. (2016). Evaluation of the rate of abiotic and biotic degradation of oxo-degradable polyethylene. *Polymer Testing* 53: 58–69.

138 Ojijo, V. and Sadiku, E.R. (2017). Improving wear resistance of polyolefins. In: *Polyolefin Fibres: Structure, Properties and Industrial Applications*, 2e (ed. S.C.O. Ugbolue), 391–420. Woodhead Publishing.

139 Arkatkar, A., Juwarkar, A.A., Bhaduri, S. et al. (2010). Growth of *Pseudomonas* and *Bacillus* biofilms on pretreated polypropylene surface. *International Biodeterioration and Biodegradation* 64 (6): 530–536.

140 Sheik, S., Chandrashekar, K.R., Swaroop, K., and Somashekarappa, H.M. (2015). Biodegradation of gamma irradiated low density polyethylene and polypropylene by endophytic fungi. *International Biodeterioration and Biodegradation* 105: 21–29.

141 Aravinthan, A., Arkatkar, A., Juwarkar, A.A., and Doble, M. (2016). Synergistic growth of *Bacillus* and *Pseudomonas* and its degradation potential on pretreated polypropylene. *Preparative Biochemistry & Biotechnology* 46 (2): 109–115.

142 Santo, M., Weitsman, R., and Sivan, A. (2013). The role of the copper-binding enzyme - laccase - in the biodegradation of polyethylene by the actinomycete *Rhodococcus ruber*. *International Biodeterioration and Biodegradation* 84: 204–210.

143 Ho, B.T., Roberts, T.K., and Lucas, S. (2018). An overview on biodegradation of polystyrene and modified polystyrene: the microbial approach. *Critical Reviews in Biotechnology* 38: 308–320.

144 Savoldelli, J., Tomback, D., and Savoldelli, H. (2017). Breaking down polystyrene through the application of a two-step thermal degradation and bacterial method to produce usable byproducts. *Waste Management* 60: 123–126.

145 Jayasiri, H.B., Purushothaman, C.S., and Vennila, A. (2013). Quantitative analysis of plastic debris on recreational beaches in Mumbai, India. *Marine Pollution Bulletin* 77 (1–2): 107–112.

146 Volova, T.G., Boyandin, A.N., Vasiliev, A.D. et al. (2010). Biodegradation of polyhydroxyalkanoates (PHAs) in tropical coastal waters and identification of PHA-degrading bacteria. *Polymer Degradation and Stability* 95 (12): 2350–2359.

147 Harrison, J.P., Boardman, C., O'Callaghan, K. et al. (2018). Biodegradability standards for carrier bags and plastic films in aquatic environments: a critical review. *Royal Society Open Science* 5.

148 Janjaroen, D., Ling, F., Monroy, G. et al. (2013). Roles of ionic strength and biofilm roughness on adhesion kinetics of *Escherichia coli* onto groundwater biofilm grown on PVC surfaces. *Water Research* 47 (7): 2531–2542.

149 Webb, J.S., Nixon, M., Eastwood, I.M. et al. (2000). Fungal colonization and biodeterioration of plasticized polyvinyl chloride. *Applied and Environmental Microbiology* 66 (8): 3194–3200.

150 Lin, W., Yu, Z., Chen, X. et al. (2013). Molecular characterization of natural biofilms from household taps with different materials: PVC, stainless steel, and cast iron in drinking water distribution system. *Applied Microbiology and Biotechnology* 97 (18): 8393–8401.

151 Vess, R.W., Anderson, R.L., Carr, J.H. et al. (1993). The colonization of solid PVC surfaces and the acquisition of resistance to germicides by water micro-organisms. *Journal of Applied Bacteriology* 74: 215–221.

152 Sharpe, A.N. and Woodrow, M.N. (1971). A rapid test for biodegradability of PVC film by pseudomonads. *The Journal of Applied Bacteriology* 34 (2): 485–489.

153 Booth, G.H., Cooper, A.W., and Robb, J.A. (1968). Bacterial degradation of plasticized PVC. *The Journal of Applied Bacteriology* 31 (3): 305–310.

154 Hartmans, S. and De Bont, J.A.M. (1992). Aerobic vinyl chloride metabolism in *Mycobacterium aurum* L1. *Applied and Environmental Microbiology* 58 (4): 1220–1226.

155 Verce, M.F., Ulrich, R.L., and Freedman, D.L. (2000). Characterization of an isolate that uses vinyl chloride as a growth substrate under aerobic conditions. *Applied and Environmental Microbiology* 66 (8): 3535–3542.

156 Cupples, A.M., Spormann, A.M., and PL, M.C. (2003). Growth of a Dehalococcoides-like microorganism on vinyl chloride and cis-dichloroethene as electron acceptors as determined by competitive PCR. *Applied and Environmental Microbiology* 69 (2): 953–959.

157 Huang, K., McLandsborough, L.A., and Goddard, J.M. (2016). Adhesion and removal kinetics of *Bacillus cereus* biofilms on Ni-PTFE modified stainless steel. *Biofouling* 32 (5): 523–533.

158 Bergamini, T.M., Bandyk, D.F., Govostis, D. et al. (1988). Infection of vascular prostheses caused by bacterial biofilms. *Journal of Vascular Surgery* 7 (1): 21–30.

159 Branco, R., Sousa, T., Piedade, A.P., and Morais, P.V. (2016). Immobilization of *Ochrobactrum tritici* As5 on PTFE thin films for arsenite biofiltration. *Chemosphere* 146: 330–337.

160 Han, S., Kim, C., and Kwon, D. (1997). Thermal/oxidative degradation and stabilization of polyethylene glycol. *Polymer (Guildf)* 38 (2): 317–323.

161 Sugimoto, M., Tanabe, M., Hataya, M. et al. (2001). The first step in polyethylene glycol degradation by sphingomonads proceeds via a flavoprotein alcohol dehydrogenase containing flavin adenine dinucleotide. *Journal of Bacteriology* 183 (22): 6694–6698.

162 Herold, D.A., Keil, K., and Bruns, D.E. (1989). Oxidation of polyethylene glycols by alcohol dehydrogenase. *Biochemical Pharmacology* 38 (1): 73–76.

163 Kawai, F. (2002). Microbial degradation of polyethers. *Applied Microbiology and Biotechnology* 58: 30–38.

164 Cregut, M., Bedas, M., Durand, M.J., and Thouand, G. (2013). New insights into polyurethane biodegradation and realistic prospects for the development of a sustainable waste recycling process. *Biotechnology Advances* 31: 1634–1647.

165 Howard, G.T. (2012). Polyurethane biodegradation. In: *Microbial Degradation of Xenobiotics. Environmental Science and Engineering*. Chapter 14 (ed. S. Singh), 371–394.

166 Zhang, X., Xu, R., Wu, Z., and Zhou, C. (2003). The synthesis and characterization of polyurethane/clay nanocomposites. *Polymer International* 52 (5): 790–794.

167 Urgun-Demirtas, M., Singh, D., and Pagilla, K. (2007). Laboratory investigation of biodegradability of a polyurethane foam under anaerobic conditions. *Polymer Degradation and Stability* 92 (8): 1599–1610.

168 Toscano, G., Cavalca, L., Letizia Colarieti, M. et al. (2013). Aerobic biodegradation of propylene glycol by soil bacteria. *Biodegradation* 24 (5): 603–613.

169 Nakajima-Kambe, T., Onuma, F., Kimpara, N., and Nakahara, T. (1995). Isolation and characterization of a bacterium which utilizes polyester polyurethane as a sole carbon and nitrogen source. *FEMS Microbiology Letters* 129 (1): 39–42.

170 Cregut, M., Bedas, M., Assaf, A. et al. (2014). Applying Raman spectroscopy to the assessment of the biodegradation of industrial polyurethanes wastes. *Environmental Science and Pollution Research* 21 (16): 9538–9544.

171 Nakajima-Kambe, T., Shigeno-Akutsu, Y., Nomura, N. et al. (1999). Microbial degradation of polyurethane, polyester polyurethanes and polyether polyurethanes. *Applied Microbiology and Biotechnology* 51: 134–140.

172 Jansen, B., Schumacher-Perdreau, F., Peters, G., and Pulverer, G. (1991). Evidence for degradation of synthetic polyurethanes by *Staphylococcus epidermidis*. *Zentralblatt fur Bakteriologie* 276 (1): 36–45.

173 Goswami, L., Kumar, R.V., Pakshirajan, K., and Pugazhenthi, G. (2019). A novel integrated biodegradation—microfiltration system for sustainable wastewater treatment and energy recovery. *Journal of Hazardous Materials* 365: 707–715.

174 Schievano, A., Tenca, A., Scaglia, B. et al. (2012). Two-stage vs single-stage thermophilic anaerobic digestion: comparison of energy production and biodegradation efficiencies. *Environmental Science & Technology* 46 (15): 8502–8510.

175 Gadow, S.I. and Li, Y.Y. (2020). Development of an integrated anaerobic/aerobic bioreactor for biodegradation of recalcitrant azo dye and bioenergy recovery: HRT effects and functional resilience. *Bioresource Technology Reports* 9.

176 Cakir, F.Y. and Stenstrom, M.K. (2005). Greenhouse gas production: a comparison between aerobic and anaerobic wastewater treatment technology. *Water Research* 39 (17): 4197–4203.

177 Cai, M., Jiménez, N., Krüger, M. et al. (2015). Potential for aerobic and methanogenic oil biodegradation in a water flooded oil field (Dagang oil field). *Fuel* 141: 143–153.

178 Serafim, L.S., Lemos, P.C., MGE, A., and MAM, R. (2008). Strategies for PHA production by mixed cultures and renewable waste materials. *Applied Microbiology and Biotechnology* 81: 615–628.

179 Rhu, D.H., Lee, W.H., Kim, J.Y., and Choi, E. (2003). Polyhydroxyalkanoate (PHA) production from waste. *Water Science and Technology* 221–228.

180 Okoffo, E.D., O'Brien, S., O'Brien, J.W. et al. (2019). Wastewater treatment plants as a source of plastics in the environment: a review of occurrence, methods for identification, quantification and fate. *Environmental Science: Water Research & Technology* 5: 1908–1931.

181 Albertsson, A.C. (1993). Degradable polymers. *Journal of Macromolecular Science, Part A Pure and Applied Chemistry* 30 (9–10): 757–765.

182 Petersen, K., Væggemose Nielsen, P., Bertelsen, G. et al. (1999). Potential of biobased materials for food packaging. *Trends in Food Science and Technology* 10: 52–68.

183 Luckachan, G.E. and Pillai, C.K.S. (2011). Biodegradable polymers- a review on recent trends and emerging perspectives. *Journal of Polymers and the Environment* 19 (3): 637–676.

3

Plastic Pollution. The Role of (Bio)Degradable Plastics and Other Solutions

Lei Tian[1], *Robert-Jan van Putten*[1], *and Gert-Jan M. Gruter*[1,2]

[1] *University of Amsterdam, Industrial Sustainable Chemistry, van 't Hoff Institute for Molecular Sciences (HIMS), Faculty of Science, Science Park 904, Amsterdam 1098 XH, The Netherlands*
[2] *Avantium Corporate Technologies, Zekeringstraat 29, Amsterdam 1014 BV, The Netherlands*

3.1 Introduction and Problem Definition

The issue of plastic waste in the environment has been obvious and problematic for some time, and the extent of health effects as a result of prolonged human exposure to micro- (0.1 μm to 5 mm) and nano-plastics (<0.1 μm) via the air we breathe, the food and drinks we consume, and via contact with our skin is still uncertain today.

When mass production of plastics began after the 1950s, it quickly found application into all facets of modern daily life. However, there was never a serious consideration about the fate of plastics in the environment, the resulting almost endless accumulation (due to very slow degradation), and of the short- and long-term effects that plastics will exert on living organisms when produced at today's enormous and still-expanding scale.

Despite all societal benefits, "plastics" are rightfully challenged by society because of their carbon footprint and because of the plastic litter that is accumulating on land, in rivers, and in oceans. It is however important to note that plastic littering is not an inherent plastic property. It is the society itself that did not design proper use and after-use protocols for safely handling plastic materials. This includes both careless consumers littering the environment and the release of plastic particles upon normal use of products such as textiles and tires. Plastics in the environment are of increasing concern because of their persistence and universal occurrence.

Plastics have been found in large amounts everywhere on land, on the highest mountain tops, and the most remote locations such as the poles, in the air, in rivers and on coast lines, in arctic sea ice, and at the sea surface and on the sea floor. Seven hundred marine species are known to have ingested or become entangled in macroplastics [1, 2], and evidence also shows that plastics are ingested by many terrestrial organisms [3]. Weathering and biodegradation of plastic debris cause

Biodegradable Polymers in the Circular Plastics Economy, First Edition.
Edited by Michiel Dusselier and Jean-Paul Lange.

fragmentation into particles that even small marine invertebrates may ingest. Its small size also renders this debris untraceable to its source and impossible to remove from marine environments. This suggests that the most effective mitigation strategies must involve redesigning plastics and reducing outputs to the environment. Even with the implementation of the best strategies to mitigate plastic waste leaking into the environment it will be impossible to prevent contamination altogether. It is therefore important to evaluate if we can also do something about the plastic waste lifetime in the environment (see Figure 3.1).

Many of the current large-volume plastics degrade extremely slowly in the environment or virtually not at all. The environmental conditions also play a role here. Polyolefins (polyethylene [PE], polypropylene [PP], polystyrene [PS]) have a low density (0.8–0.9 g/cm^3) and, as a consequence, they will float on seawater (density 1.03 g/cm^3). Floating plastics are exposed to oxygen, heat, and UV radiation from the sun, three factors that have been shown to accelerate non-biological plastic degradation. Other plastics such as polyvinyl chloride (PVC), polyesters such as polyethylene terephthalate (PET), and polyamides such as nylon-6 and -66 have densities higher than 1 g/cm^3, and as a consequence, they will migrate to the sea floor where it is dark, and typically in a lower temperature, but as our own research has shown, there can be typically more microbial activity in the sediment than in the seawater itself.

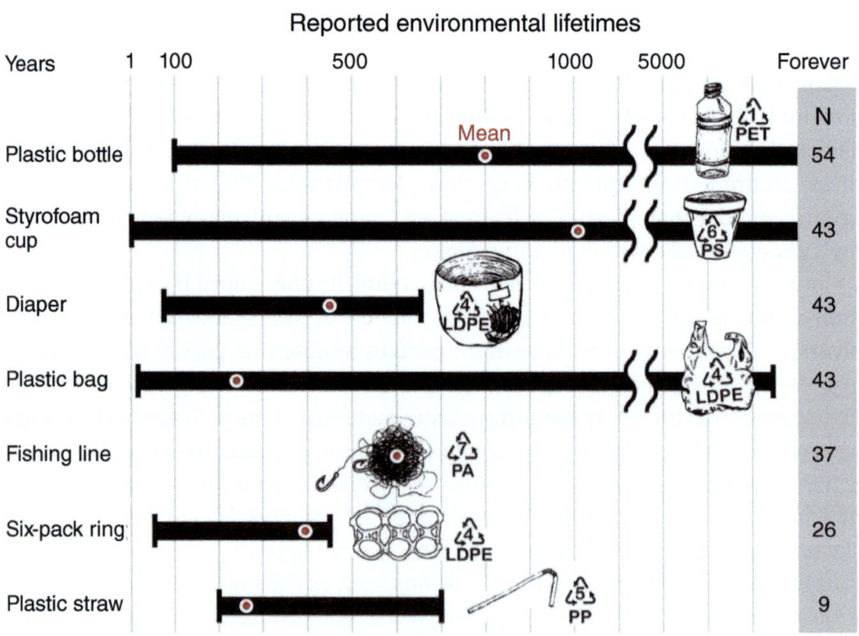

Figure 3.1 Review of 57 information graphics and documents that report environmental lifetimes of common plastic consumer goods. The bars represent the range of estimates, the red circles represent the mean of estimates, and the number of estimates for each plastic good (*N*) is provided on the right (*N* = 255 in total). PET = polyethylene terephthalate, PS = polystyrene, LDPE = low-density polyethylene, PA = polyamide, and PP = polypropylene. Source: From ref. [4]. Image credit: Natalie Reiner.

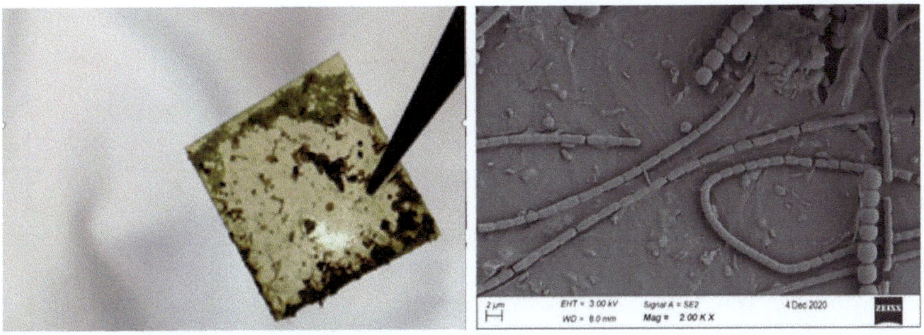

(a) (b)

Figure 3.2 Example of a biofilm formed on a 2 × 2 cm plastic PEF film (a) and an electron microscopic image showing fungi and bacteria (b). Source: Pictures Yue Wang; Industrial Sustainable Chemistry, HIMS, University of Amsterdam.

Plastic densities are not always stable. Plastic species in a marine environment will be covered by a community of microorganisms, invertebrates, and algae, together called a biofilm (see Figure 3.2). As a consequence, the particle density can increase and polyolefin debris may at some point in time sink.

Growing alongside the volume of global production and consumption of plastics are the diverse concerns about their impacts on the ecosystem and on human health, particularly as micro- and nanoplastics (MNPs) [5]. The concerns about MNPs are diverse [6, 7]. The first concern is about the MNPs irritation of body cells and tissues potentially leading to infections. Secondly, MNPs may contain residual monomers and additives (e.g. plasticizers, anti-oxidants, UV stabilizers) that can slowly diffuse into the body and eventually intervene in the metabolism. Finally, MNPs tend to accumulate toxic hydrophobic contaminants present in the air or in the water, such as polychlorinated biphenyls (PCBs). These contaminants can be released in the organism and affect its metabolism.

3.2 Sources of Macroplastics and MNPs

3.2.1 Mismanagement of Waste

Since mass production of plastics started in the 1950s, production has grown exponentially to over 370 million tons in 2019 (Figure 3.3), and this does not even include the production of fibers, thermosets, and rubber [8]. Contrary to the trend, this remained at a comparable number in 2020, which is explained by the crisis caused by the COVID-19 outbreak (https://phys.org/news/2021-06-global-plastics-production-falls.html). Only a very small group of polymers makes up the vast majority of these plastics: PE (low-density polyethylene [LDPE], linear low-density polyethylene [LLDPE], high-density polyethylene [HDPE]), PP, PET, PS, and PVC. In practical terms these materials should be considered non-biodegradable. They are, however, susceptible to wear, leading to the formation of small particulates of these plastics. When we relate this to the amounts of mismanaged plastic waste

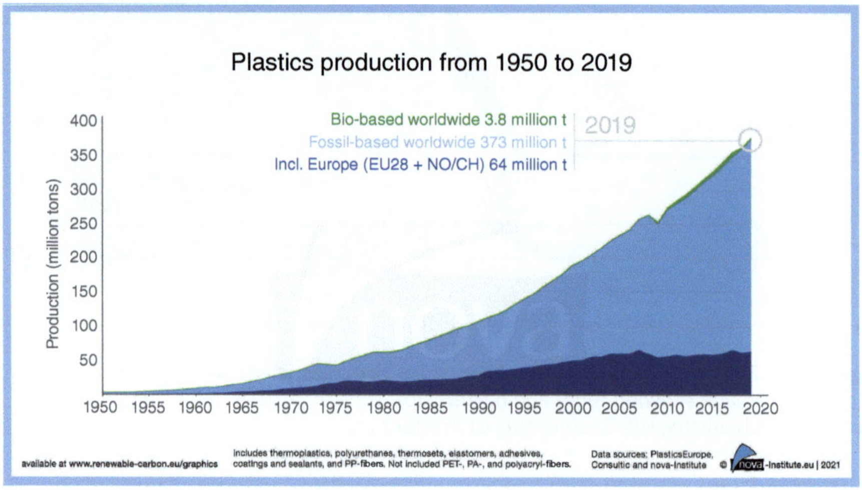

Figure 3.3 Worldwide plastics production since 1950 [8]. Source: nova-Institute GmbH, press release, 2021-01-28.

generated annually (>30 Mt), it becomes clear that millions of tons of macroplastics and MNPs enter the environment from simply mismanaging waste. Given the steep growth of plastics consumption this amount is only increasing.

In 2010, an estimated 2.5 billion tons of municipal solid waste was generated by 6.4 billion people living in 192 coastal countries (93% of the global population). Approximately 11% (275 million tons) of this waste was plastic, which roughly followed 2010 plastic resin production (270 million tons), with differences resulting from the time lag in disposal of durable goods (lifetime of years to decades). It was estimated that about 100 Mt of plastic waste was generated in coastal regions (population living within 50 km of the coast). Of this, about 32 million tons was classified as mismanaged and an estimated 4.8–12.7 million tons entered the ocean in 2010, i.e. 1.7–4.6% of the total plastic waste generated in those countries [9]. Next to this macroplastics pollution, 1.5 Mt of primary microplastics [10] enter the ocean annually after normal wear of consumer products such as fibers in clothing and tires, or because these microplastic were intentionally added to consumer product, e.g. as abrasives in cosmetics. The macroplastics in the environment can also break down into microplastics; these are referred to as secondary microplastics. If plastic production and waste generation continue to grow as we have seen in the last decades, the annual amount of plastic pollution may more than double by 2050 [11, 12] resulting in a 10-fold increase in the cumulative amount of ocean plastic in the period 2010–2025 relative to the cumulative plastic pollution in the 60 years of plastic production and use prior to 2010 (Figure 3.4) [9]. To visualize these amounts: 9.5 million tons of plastics entering our oceans every year is 300 kg/s. As a garbage truck fits 20 m^3 waste plastic with an average bulk density of 50 kg/m^3, this amount is the same as 18 full garbage trucks spilling their contents into the oceans every minute (Figure 3.5).

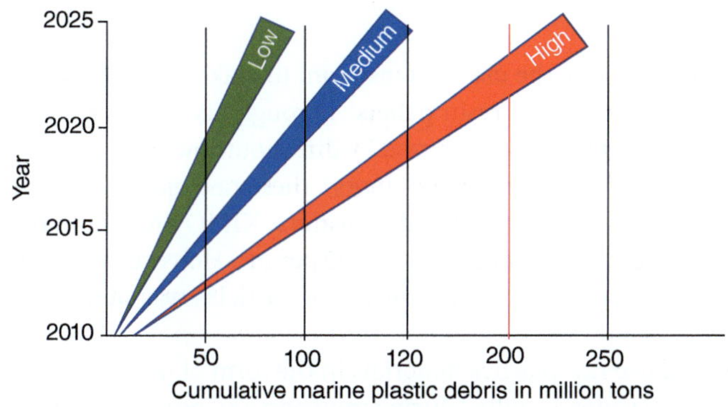

Figure 3.4 Estimated mass of mismanaged plastic waste input to the ocean (in millions of metric tons) by populations living within 50 km of a coast in 192 countries, plotted as a cumulative sum from 2010 to 2025. Estimates reflect assumed conversion rates of mismanaged plastic waste to marine debris (high, 40%; mid, 25%; low, 15%) [9].

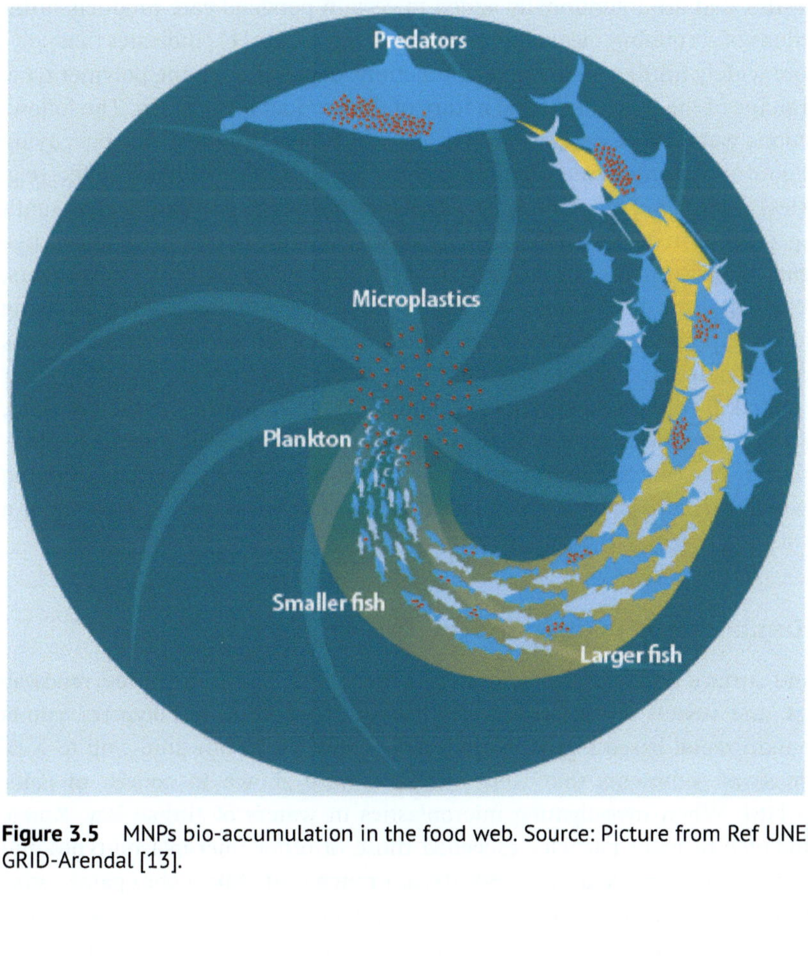

Figure 3.5 MNPs bio-accumulation in the food web. Source: Picture from Ref UNEP and GRID-Arendal [13].

3.2.2 Accidental Release

The primary production of plastics to be used downstream to make products is typically in the form of granules and small resin pellets. Through accidental spillage during land or sea transport, inappropriate use, and by direct outflow from processing plants (spillage and clean-up by hosing down floors), these raw material pellets enter the environment. In an assessment of Swedish waters, KIMO Sweden found plastic particle amounts in the typical range of 150–2400 per m^3. However, in a harbor adjacent to a plastic production plant, the amount of particles was more than 100 000 per m^3 [14, 15].

Many industrial sites that use raw plastics materials in the form of pellets are frequently located near bodies of water. If spilled during transport and processing, these materials may enter the surrounding environment, polluting waterways [16].

3.2.3 MNPs in Products

Microplastics that are intentionally added to certain personal care products fulfill the functions of scrubbing, cleansing, and exfoliating. Leslie [17] indicates that polymers more widely fulfill a wide range of functions depending on the polymer type, size and shape of the ingredient, and nature of the product it is used in. The following functions were listed: viscosity regulators, emulsifiers, film formers, opacifying agents, liquid absorbents binders, bulking agents, for an "optical blurring" effect (e.g. of wrinkles), glitters, skin conditioning, exfoliants, abrasives, oral care such as tooth polishing, gellants in denture adhesives, for controlled time release of various active ingredients, sorptive phase for delivery of fragrances, vitamins, oils, moisturizers, insect repellents, sun filters, and a variety of other active ingredients, prolonging shelf life by trapping degradable active ingredients in the porous particle matrix (effectively shielding the active ingredient from bacteria, which are too big to enter particle pores). By design these types of products result in littering of microplastics, as these polymer particles are essentially intended to be flushed out with the wastewater. This can be seen as significantly different from most of the plastic-littering issues, which are caused by unintended littering, sloppiness, or the absence of a proper waste-management infrastructure.

3.2.4 Degradation of Outdoor Objects

Paints and surface coatings also contain polymers, and painted surfaces, roadway markings, and vessels are subject to weathering. In addition to polymers, paints often contain metal-based pigments such as Cu and Zn. In one study, up to 0.2% of the mass of sediments that were analyzed were shown to consist of paint particles [18]. When investigating microplastics in waters of Jinhae Bay, Korea, the abundance of paint particles exceeded those of other microplastic types and that size frequencies peaked in the 50–100 μm range [19]. Alkyd ship paint resins and poly(acrylate/styrene) from fiberglass resins were dominant polymers. Similar findings were reported for the Incheon/Kyeonggi coastal region (Korea) [20].

In 2019, surface slicks (meandering lines of smooth water on the ocean surface that accumulates floating objects and organisms) were shown to contain a plastics-to-larval fish ratio that was 60-fold higher than that of adjacent waters [21]. Microplastics and paint particle loads were reported in beach sediments from an Italian subalpine lake in 2016 [22]. The paint particles typically were smaller than other types of microplastics, mostly 1–50 μm, likely due to their brittleness.

While terrestrial sources are the most important, large amounts of plastics enter the oceans from wear of sea-based sources every year. The fishery sector uses large amounts of plastics. Losses in the fisheries sector include loss of fishing gear (nets, ropes, floats, lines, gloves, fish boxes, strapping bands), and release of fibers and other fragments due to normal wear and tear (e.g. use of ground ropes). Fishing gear may be lost at sea by accident, abandonment, or deliberate disposal (together termed "abandoned, lost, or otherwise discarded fishing gear" or ALDFG), and likely is the largest category in terms of volume and potential impact of all the ocean sources. ALDFG can have a significant impact on depleting fish and shellfish and can cause unnecessary impacts on non-target species and habitats.

Expanded polystyrene (EPS) buoys are used extensively in Asia for the hanging culture of mussels and oysters. It is estimated that approximately 1.8 million are discarded into the marine environment annually [23]. Each EPS buoy (~60 l) can generate 7.6 million micro-size EPS fragments of <2.5 mm diameter, or 7.6×10^{21} nanoparticles of <250 nm diameter [24]. Consequently, EPS buoys and fragments accounted for >10% of marine debris on 94 Korean beaches in 2008 [23]. In 2012, the Korean Ministry of Ocean and Fisheries provided financial support to fishermen to replace old buoys with high-density, less easily degraded buoys.

3.2.5 Wear (Tires, Clothing)

Apart from littering and other unintended spillage of plastics into the environment, the wear of many products during normal use also causes the release of synthetic material into the environment. These are materials that are used in daily life, often for durable applications, and just wear down over time: carpets, clothing, tires, shoes, artificial turf. In most cases these contain polymers of some kind (rubbers, thermosets, and thermoplastics) that will therefore be released into the environment in the form of small particles. The total amount of these primary microplastics entering the environment is estimated to be well over three million tons per year [10]. For such applications, alternatives should be sought further upstream, in the design phase.

The most common and closest-to-home example of microplastic release into the environment comes from textile. Currently the majority of textiles used are synthetic, most commonly nylon (polyamide) and (PET) polyester. During each wash a single garment could lose close to 2000 fibers [25]. Other studies indicate losses as high as 700 000 fibers for every 6 kg of laundry [26]. These fibers end up in the wash water and are thus flushed into the sewage systems. Filtering is often absent at any point of the water-management chain or it is not very efficient. This means that polyamide and polyester microplastics are constantly being released into the environment, and this type of waste is building up over time. Laundry of synthetic fabric is the largest source of all primary microplastic pollution, amounting to 30–35% of microplastic pollution. The severity of this problem, e.g. the impacts on aqueous water ecosystems, is still largely unknown and may thus be underestimated.

It is estimated that annually approximately 1.3 Mt of tire wear is generated on European roads [27, 28]. That is equal to over 2 kg per person per year. Tire wear is estimated to contribute to >50% of microplastic emissions in Denmark and Norway and approximately 30% in Germany [28]. There is no reason to assume that these values would differ much in other areas of the world, and tire wear is therefore estimated to be responsible for almost 30% of all primary MNP pollution. To make matters worse, tire wear plays an important role in the particulate matter burden (as a large contributor to air pollution) in urban areas.

Tire tread typically consists of natural or synthetic rubber (40–50 wt%), filler (soot, carbon black, silica, chalk; 30–35 wt%), softener (oil and resin; 15 wt%), vulcanization agents (sulfur and ZnO; 2–5 wt%) and additives (plasticizers, antioxidants, preservatives, etc.; 5–10 wt%) [28, 29].

Rubbers have been the focus of biodegradability research for over a century. This is especially the case for natural rubbers, which are based on *cis*-1,4-isoprene. Initially this was an undesired property regarding durability of the material. A century later this is now desirable for the opposite reasons of trying to understand the fate of these materials in nature. A lot of the research focuses on the isolation of certain organisms, both bacteria and fungi, and researching their effect on the breakdown of these rubbers [28, 30, 31]. Often significant mass losses and decrease in molecular weight are observed in a matter of weeks or months. It appears that also vulcanized natural rubbers, as well as synthetic *cis*-1,4-isoprene-based rubbers are susceptible to biodegradation [31]. A lot is known about the organisms and enzymes responsible for the biodegradation, as well as the mechanisms, yet given that a lot of research focuses on isolating and applying certain organisms, rather than random degradation in soil or aquatic environments, it is difficult to make statements regarding the timeframe of these materials breaking down in various natural environments.

Today plastic pollution is ubiquitous: macroplastic can be seen everywhere, but especially in less prosperous parts of the globe and there is an exponential growth. There is, however, also the invisible pollution of microplastics, which are found in every corner of the world. Microplastics, especially from fibers and tire and road particles, typically have densities higher than 1 g/cm^3, which means they sink to the bottom of rivers, lakes, and oceans. They are ingested by living organisms and thus find their way up the food chain.

3.2.6 Waste and Wastewater Management (Water/Wind)

Rivers represent a key entry point of macro- and MNPs to the ocean. From the limited data available, it would appear that river catchments, especially those draining areas with high population densities and industrial development, can carry a significant plastic load to the ocean. However, there is a great lack of information on the quantities entering the ocean globally via this entry point, which sources are most important, what measures may be effective at controlling these sources, and how all these aspects differ regionally. The effectiveness of wastewater and solid waste management will be an important factor in modifying the input to waterways, whatever the nature of the land-based sources concerned. For these reasons, significant regional differences may be expected. Extreme flooding events have the potential to mobilize plastic that would not be transported to the ocean otherwise. The effects of heavy rainfall are exacerbated by unsustainable land-use practices (e.g. deforestation, compacted soils). There is evidence that extreme events are becoming more common as a consequence of global warming.

3.3 Impacts of Macroplastics and MNPs

This section is focused on the impact of MNPs on ecosystems and economics. Emphasis has been given to MNPs as the ultimate ecological impact of MNPs is not well understood thus far. Furthermore, most of the published studies have focused on the marine surface and the actual amount of MNPs in different environmental compartments (terrestrial, marine, freshwater, and atmospheric), and their impacts have not been widely discussed. This is due to the difficulties in assessing the quantity of MNPs (because of the small size of MNPs and the lack of adequate analytical approaches) and the fact that little is known about the (bio)chemical mechanisms in organisms regarding MNPs (due to the immensity and diversity of ecological species).

3.3.1 Ecological Impact of Macroplastics (Entanglement and Ingestion)

Entanglement has been recognized as the most visible ecological impact of macroplastics, especially on marine organisms (e.g. mammals, fish, and birds). Entanglements are often related to fishing gear, and they can lead to acute and chronic injury or death. It is estimated that between 57 000 and 135 000 pinnipeds and baleen whales globally are entangled each year, and up to 50% of humpback whales in US waters show scars from entanglement [26].

Ingestion of macroplastics is another important ecological problem and has been reported for a wide range of marine organisms. Large quantities of plastic sheeting and plastic bags are frequently found in gut compartments of turtles and toothed whales. Ingestion of plastic debris has been reported in seabirds such as albatrosses and in 46 cetacean (whale) species: for some species in as much as 31% of the cases. The different feeding habits of closely related species can influence their susceptibility. Chemicals leached from the plastic debris can cause additional adverse health effects on marine organisms.

3.3.2 Economic Impact of Macroplastics

It is estimated that the impact of plastic pollution in the oceans alone is at least US$8 billion per year [32]. The economic impact of macroplastics can be direct and indirect. For example, macroplastics can lead to economic costs in the commercial shipping sector due to damage caused by entanglement or collision. Indirect impacts include loss of target species due to ghost fishing and the loss of attractiveness of beaches and shorelines for recreational purposes. Decrease of tourism leads to a loss of revenue and jobs in the local and regional economy. Also, clean-up costs can be significant and pose an undue burden on local authorities. Given the difficulty of cleaning and the low production cost of plastics, removing plastics is likely significantly more expensive than producing them.

3.3.3 Ecological Impacts of MNPs

The biological effects of MNPs involve physical interactions (e.g. ingestion of MNPs), chemical exposure to plastic-related compounds (e.g. additives), to sorbed toxins (synthetic or natural), and to surface-associated organisms (e.g. pathogens). The impacts of MNPs on biota in the aquatic environment, especially the marine environment, have been widely investigated since the 1970s. However, investigations on the ecological impacts of MNPs on soil and atmosphere only started in the last 20 years.

3.3.3.1 Aquatic Environment

The MNPs, from different sources (direct release, washed by surface run-off, blown by wind, etc.), eventually end up in the aquatic environment. MNPs were detected in rivers, estuaries, and oceans. Numerous studies have proven the ingestion of MNPs by various aquatic organisms [33] and is considered the most common cause for MNP uptake by marine species. It often starts with small aquatic biota, like plankton, and subsequently moves up the food chain to filter organisms, small fish and whale sharks, affecting the entire food chain [13]. It has been reported that 36% of pelagic (bottom feeding) and demersal (open water) fish collected from the English Channel had MNPs in their digestive systems [34]. MNPs have been found in the digestive tract of all the specimens sampled along the British coast [35]. Aquatic biota may take MNPs in as their food due to misidentification or indiscriminate consumption, and/or ingest indirectly as a result of trophic transfer along the food web [33]. In addition, the adherence of MNPs to soft tissues of mussels has been reported as a novel way for MNPs uptake beyond ingestion, which contributes about 50% of the MNPs uptake in mussels [36]. The shore crab (*Carcinus maenas*) can even take up MNPs through inspiration across the gills [37].

MNPs can accumulate in the tissues of blue mussel (*Mytilus edulis*) and translocate from the gut to the circulatory system within three days [38]. Ingestion or tissue translocation of MNPs can induce several adverse effects in aquatic biota, including digestive system blockage, tissue damage, behavioral changes, growth delay, decrease in reproductivity, immune response, and even death [39]. For example, the feeding capacity of the copepod *Calanus helgolandicus* was significantly altered after the prolonged exposure to PS [40]. For limnic zooplankton (*Daphnia magna*) ingestion of MNPs led to immobilization under laboratory conditions [41]. Exposure to MNPs led to histopathological changes and changes in blood biochemical parameters in juvenile African catfish (*Clarias gariepinus*) [42]. To make matters worse there is increasing evidence that MNPs can sorb pathogens and pollutants, and these pathogens and pollutants can also lead to growth delay, decrease in reproductivity, and increase in mortality of aquatic organisms [43]. In addition, plastic-related chemicals (PRCs) are present. These are substances related to plastics, including residual monomers, antioxidants, additives, or the degradation products of plastics. PRCs such as bisphenol analogues (e.g. bisphenol A [BPA], bisphenol F [BPF], bisphenol S [BPS], and bisphenol AF [BPAF]), 4-nonylphenol, and some phthalates have been detected in bottled water, water, seafood, vegetables, and many different types of packaged food. Many of these chemicals can cause adverse health effects in humans, including endocrine disruption, changes in neurobehavioral development, and metabolic diseases such as diabetes and obesity, raising concerns of consumer safety.

PRCs can enter the environment and food through pathways such as the discharge of industrial wastes to the environment, irrigation with reclaimed water, or through the application of polymers in agriculture and food or as food contact material [43].

3.3.3.2 Terrestrial Environment

Soil appears to be a sink of MNPs [33]. These MNPs may impact the geochemistry of soil and interact with soil biota [44]. The adverse effects of MNPs on terrestrial environment have, however, not been investigated as much as the impact on the aquatic environment. Most of the research was done through laboratory experiments, and earthworms and soil collembolans (springtails: wingless primitive insects) are generally used to investigate the ingestion of MNPs.

It is understood that earthworms and soil collembolans can ingest MNPs as well as transport them in the soil [33]. Earthworms are ingested by other animals, chickens among others, which in turn also ingest these MNPs. This has been shown in research on MNPs in soil, earthworm casts, and the feces, crops, and gizzards of chickens [45]. Polymer additives can accumulate in house crickets (*Acheta domesticus*) exposed to polyurethane MNPs [46]. This shows the risk of MNPs carrying these adsorbed chemicals into the food chain. In addition to the impacts on terrestrial organisms, MNPs may also facilitate the accumulation of high-molecular-weight humin-like material and stimulate enzymatic activity [47].

3.3.3.3 Atmosphere

MNPs in the form of fibers (mostly resulting from washing of textiles) can be airborne, as has been measured in many places and will be inhaled by humans and animals. So far, the ecological impact of airborne MNPs on atmosphere (climate and global warming, atmospheric chemistry, etc.) and wildlife has not been well understood. The inhaled or ingested MNPs may accumulate and exert localized particle toxicity by inducing or enhancing an immune response [6]. It is important to realize that airborne MNPs can be transported everywhere, and it is therefore understandable that microfibers are found in the most unexpected places, far from the sea and far from humanity, but can also be transported to aquatic and terrestrial environments, and thus, induce adverse impacts on aquatic organisms as well.

3.3.4 Threat to Human Health

3.3.4.1 MNPs in the Human Food Chain

As mentioned in Section 3.1, MNPs have been detected in hundreds of species of marine organisms, and it is safe to assume that they are present in the seafood sold for human consumption. It is estimated that humans ingest up to several thousand MNP particles per year through seafood consumption, with numbers varying depending on eating habits and living locations [13]. The most likely route for MNPs ingestion by human is the consumption of filter-feeding invertebrates like mussels and oysters because people eat the whole organism of shellfish (and sometimes the small fish), and many studies have reported the presence of MNPs in their tissues and gut (Section 3.3.3.1).

Furthermore, MNPs have been found in beer, sugar, honey, and table salt [48]. Food should therefore be considered an important source of MNPs exposure for humans. As MNPs may contain or absorb contaminants, ingestion of MNPs could be a health threat to humans. Furthermore, MNPs in nano-size (less than 100 nm) have the potential to be absorbed by the body through the digestive system as evidenced by medical studies where nanospheres are used for drug delivery [13].

3.3.4.2 Plastic-Related Contaminants

In addition to the physical effects of MNP ingestion, the chemical additives and monomers that may be released from MNPs can have adverse health effects on human. Some of the plastic additives, like flame retardants, antioxidants, UV-stabilizers, and plasticizers, are toxic to humans. For example, phthalates and BPA show endocrine disruption potential in humans, and animal studies have also suggested their carcinogenicity and the negative impact on reproductive functions even at very low doses [13]. More importantly, some of the additives, such as phthalates and parabens, do not degrade quickly and may thus be biomagnified in the food web. These can, in the end, be ingested by humans.

3.3.4.3 Other Contaminants

MNPs may function as a vector for carrying contaminants and pathogens. For example, MNPs can absorb environmental pollutants, including PCBs, dichlorodiphenyldichloroethylene, nonylphenols, polycyclic aromatic hydrocarbons (PAHs), and metals. Besides, MNPs can also absorb pathogens and transport them inside organisms via MNPs ingestion [49]. In addition, MNPs can concentrate these contaminants to orders of magnitude higher than in the surrounding environment due to their high surface-area to volume ratio [13]. These contaminants have the potential to enter the food web and accumulate through the ingestion of MNPs carrying them. Consequently, they can end up in the human diet. Although human exposure to PCBs and PAHs via MNPs ingestion is much lower than from other pathways (e.g. ingestion of crops treated with herbicide, burning of waste, and industrial exposure) [13], concerning the increasing occurrence of MNPs in food as well as the toxicity of the mentioned contaminants, ingestion of MNPs still poses a risk to human health.

3.3.5 Socio-Economic Impacts of MNPs

MNPs can impact social and economic activities. To date, there is no evaluation of the exact bill for MNPs removal or the income loss due to MNPs in the form of marine and land litter. This is mainly because the economic and social costs are difficult to fully assess as the full extent of the impact of MNPs on ecosystems and human health is still unknown. It is typically more costly to clean up pollution than to prevent it, and in this case that should also be expected due to the small size and high level of dilution. It makes much more sense to deal with this at the source. In addition, the cost for MNPs analysis and monitoring should also be considered.

Apart from the economic costs, there are social costs which include reduced opportunities for recreational activities, health risks to visitors, and psychological benefits of access to nature (e.g. concerns and stress about the presence of [large] amounts of visible plastic or about the ingestion/inhalation of MNPs). However, the social cost cannot be estimated due to the lack of credible research evidence.

3.4 Plastic Biodegradability

The mechanism of biodegradation and the impact of polymer type and environment on biodegradability is discussed in detail in Chapter 4 of this book. Biodegradable plastics are polymers that (simply said) degrade to CO_2, water, and biomass (in aerobic conditions) by means of microorganisms. There are huge differences in biodegradation rates for different biodegradable polymers and of the biodegradation rate of a single polymer in various environments. It is therefore important to report the environmental conditions in which degradability occurs (e.g. home or industrially compostable). Home-compostable plastics are polymers that biodegrade at a similar rate or faster than cellulose (wood) at ambient temperature in 12 months in soil (e.g. polyhydroxyalkanoates [PHA's]). At the time of writing, there is no European standard for home compostability, and the French standard AFNOR NF T 51-800 should be considered for home-compostability requirements. Industrially compostable plastics are polymers that biodegrade at >50 °C in six months in soil (e.g. polylactic acid [PLA]). EN 13432 or equivalent standards (e.g. ISO 18606) should be considered for industrial compostability requirements.

It is important to realize that there are also plastics that are not biodegradable according to home- or industrial compostability standards, but that degrade (much) faster than conventional polymers in nature. For example, our own (yet unpublished) research shows that PLA and polyethylene furanoate (PEF) degrade under ambient conditions within years, while PET shows no degradation in these timeframes in centuries. With novel bio-based plastics we now have an option to avoid these mistakes by designing/selecting our future plastics for better closed-loop recycling and a better fate-in-nature [50].

Polymers which will biodegrade in the terrestrial environment, under favorable conditions (e.g. cellulose acetate, polybutylene succinate [PBS], polycaprolactone

[PCL], polyethersulfone [PES], polyvinyl alcohol [PVA]), also biodegrade in the marine environment, but much more slowly and their widespread use is likely to lead to continuing littering problems and undesirable impacts.

Some of the claims and counterclaims about particular types of polymer, and their propensity to biodegrade in the environment, appear to be influenced by commercial interests. Some evidence, albeit limited, suggests that public perceptions about whether an item is biodegradable can influence littering behavior; i.e. if a bag is marked as biodegradable it is more likely to be discarded inappropriately. On the balance of the available evidence, biodegradable plastics will not play a significant role in reducing marine litter [51].

Oxo-degradable materials contain a pro-oxidant that induces chemical degradation (degeneration) under favorable conditions. In this process CO_2 is not formed, so this polymer property should not be considered to be similar to biodegradation. Complete breakdown of the polymers and biodegradation still have to be proven.

3.5 Solutions

It is evident that solutions must be found for the large-scale pollution of the earth with plastic waste. Given the exponential growth of plastic consumption, with in its wake the littering, the problem is increasing and already getting out of hand. As is the case for many problems: if it could be solved easily, the problem would not be there, or it would at least be relatively under control. Banning synthetic polymers and moving away from plastics is an undesirable scenario, for many alternatives for these materials would lead to other major environmental issues. Replacing synthetic garments with cotton, for example, appears to be a relatively straightforward solution; the cultivation of cotton, however, is often environmentally and socially taxing.

3.5.1 Cleaning Up

The Ocean Cleanup project (www.oceancleanup.com), initiated by Dutchman Bojan Slat, uses collection systems to filter floating plastic waste out of the Oceans. As of May 2021, the total amount of plastic waste collected by Ocean Cleanup is 464 920 kg. This project has garnered a lot of attention and really put a spotlight on marine-littering issues. The amount of waste they are able to process, however, is very small compared to the immense size of the problem, as millions of tons of plastic enter the oceans every year. Such clean-up projects could play an important role in dealing with already-existing litter and preventing that from persisting in nature for centuries, but it will be futile unless we stop adding to the problem at this alarming rate and also when one looks at the size of the operation run by Ocean Cleanup and the costs that come with it. When plastic has to be collected from the ocean by these systems and then transported, potentially sorted and recycled, or dealt with otherwise, it is highly likely that this will be much more expensive than producing and using that same amount of plastic sustainably. This in turn means that the unsustainable production and use of plastics are much more expensive than we are led to believe.

3.5.2 Waste Mitigation

To tackle the plastic waste problem, a combination of measures is required. The best solutions to mitigate plastic pollution often start with managing plastic waste, which varies significantly depending on geographical and social circumstances [52]. Many different solutions have been suggested, often at regional or national level [53, 54]. Suggested options include post-consumption waste management [55, 56], reducing plastic, reuse, and the implementation of new delivery models [57], bans on single-use non-recyclable plastics (e.g. EU directive 2019/904) and microbeads in cosmetic products [58, 59]. The Basel convention was amended to regulate international plastic waste trading [60].

Littering is generally caused by the fact that plastics are considered cheap and often used only for single-use application. Treating plastic as a valuable material would put an incentive on collection for reuse or recycle. In general, it would make sense to reuse materials more and to design materials to be fit for reuse. To facilitate the reuse of plastics in the economy all products should be designed with an after-use pathway in mind and without planned obsolescence.

3.5.3 Material Design

Next to the earlier-mentioned design of products for reuse (thermal and caustic resistance) and design for recycling (from multilayer to monolayer, avoiding pigments, fillers, and additives that hamper recycling), from polyolefins to closed-loop recyclable polyesters, special attention needs to be given to designing plastics that do not contain toxic chemical additives such as plasticizers, flame retardants, and pigments, as this undermines their potential for secondary use and creates health and ecological risks. Chemicals regulations need implementation and reform to phase out toxic chemicals through substitution and circular economy solutions. For some applications, non-plastic materials may provide innovative, cost-effective, and competitive alternatives with beneficial outcomes (e.g. paper straws). Such substitutions and alternatives should be explored, researched, and developed, alongside the re-design of old-generation plastic products to improve their reparability and recyclability.

3.5.4 Bringing It All Together

In the last decade, many studies and reports have advanced the understanding of the (ocean) plastic pollution challenges. At the same time, an evidence-based roadmap that describes available mitigation pathways as well as an actionable plan was still missing. In 2020, the PEW Charitable Trusts with SYSTEMIQ published a report called "Breaking the Plastic Wave," one of the first comprehensive assessments of pathways toward stopping plastic pollution [14]. The findings of their analysis were also published in *Science* [61]. The report evaluated very different responses to the plastic pollution crisis, from laminating plastics to turning it into fuels, from developing biodegradable substitutes to closed- and open-loop recycling. From this, **10 critical findings** were proposed to reduce the future plastic pollution by 80% without compromising social and economic benefits:

1. Without action, the annual flow of plastic into the ocean will triple by 2040 to 29 Mt/yr.
2. New policies and voluntary commitments can reduce annual plastic leakage with 7% (narrow focus and typically implemented in low-leakage countries).
3. There is no single solution. Upstream and downstream solutions must be deployed together.
4. Solutions to reduce ocean pollution by 80% by 2040 exist today: reduce growth; substitute; design for recycling; expand waste collection in middle/low-income countries; develop plastic to plastic; build facilities to convert non-recyclable plastic; reduce plastic waste exports; implement known solutions for microplastics (MPs) from tires, textiles, and cosmetics.
5. Additional innovation across value chain. Upstream: replacing multilayer and other multi-material plastics requiring new materials and new delivery models. Biodegradable materials can play a role here. Expected investments: $100 billion per year by 2040.
6. Major reduction (from $2500 billion in 2021 to $1200 billion in 2040) and redirection of capital investments in the plastic industry.
7. 80% Reduction of plastic leakage will be an opportunity for a new circular plastics economy.
8. Differentiate implementation priorities regionally. High-income countries focus on microplastic leakage, while middle/low-income countries target reduce, collection, sorting, and recycling.
9. Plastic leakage reduction also benefits climate, economy (jobs), health, and environment.
10. The time is now! An implementation delay of five years results in an additional 80 million tons plastic leaking into the ocean by 2040.

Point 5 of the 10 critical findings mentions the development of new materials, which we addressed above (Section 3.5.3). At first glance an obvious solution for plastic in the environment, and one that is often embraced by consumers, is to make all packaging material biodegradable. If we were to continue the current way of consuming plastic, this would appear to make sense, since a lot of applications are single use and if these plastics would be biodegradable, littered material would degrade faster. This, however, ignores many key issues regarding the pollution: the amount of plastics used is increasing, so even if these plastics are biodegradable, if littering is not stopped, the waste will still be there. Then there is the issue of the rate of biodegradation: how fast should something degrade? Your material cannot degrade faster than the time needed for the application. Furthermore, the integrity of the material will already be negatively affected long before it has been completely degraded. Realistically one must think of degradation times of two to five years. This timeframe is still not enough when considering the life cycle of many animals and the issue of animals ingesting these plastics. Furthermore, if littering increases, even though materials degrade, the pollution still increases. Then there is the question of how well these materials would degrade under less favorable conditions: cold temperatures, anaerobic conditions, marine environment.

The next issue is related to CO_2 emissions: if all 350–400 Mt of plastics produced every year, using 8% of fossil resources production, are biodegradable and degrade in a timeframe of five years, it would essentially increase CO_2 output by 8%. This would mean that these materials should be made from either biomass or CO_2-sourced building blocks. This biomass needs to regrow to recapture this CO_2. Bio-based and CO_2-based monomers typically have a higher fraction of heteroatoms in their molecular structure, unless one seeks to make drop-in chemicals from biomass (bioPE, bioPET). Given how inexpensive oil-based monomers are, this does not make sense from an economic point of view. The most common source of fixed carbon is biomass, and carbohydrates form the majority of this biomass. About half the mass of carbohydrates consists of oxygen atoms, and it is difficult and therefore expensive to remove this to make for instance BioPE. PE is mostly used for single-use applications and does not have the mechanical properties to make it suitable for reuse. It therefore makes very little sense to produce bioPE, except for specialized applications.

As has been mentioned the current polymer market is dominated by only five different materials. Since these are very inexpensive, typically between 0.8 and 2.5$/kg, they are often used because they are good enough for the application and cheap. Assuming an average cost of $1.50/kg for the almost 400 million tons produced per year comes down to a total market size of around $800 billion. With a world population of around seven billion people, this means the average person spends around $100 per year on plastic. This makes the "it is too expensive to change to sustainable" argument a weak one, especially considering the price societies are paying for the consequences of this pollution, both in economic and social costs.

Looking at the future and the pollution issues regarding plastics, both from greenhouse emissions and littering, it would make more sense to look at designing materials with sustainability and durability in mind. Enormous strides can be made by simply reusing materials more often: reusing a bottle 10 times would by definition reduce the waste to 10% compared to single-use. This means this bottle needs to be designed to be fit for reuse, and materials such as PE, PP, and PET are not necessarily suitable for this, given their intrinsic chemical and mechanical properties. After an application is not anymore suitable for reuse, it should be recycled as much as possible (see Chapter 11). Applying deposit systems to add economic incentive would stimulate collection and reduce littering.

The intrinsic chemical structure of biomass and the monomers that can be produced from this offer opportunities that fossil-based monomers and thus polymers do not provide (Figure 3.6). Since novel materials are required to be designed to fit future needs in terms of reuse, recycle, and designed degradability of plastics, it

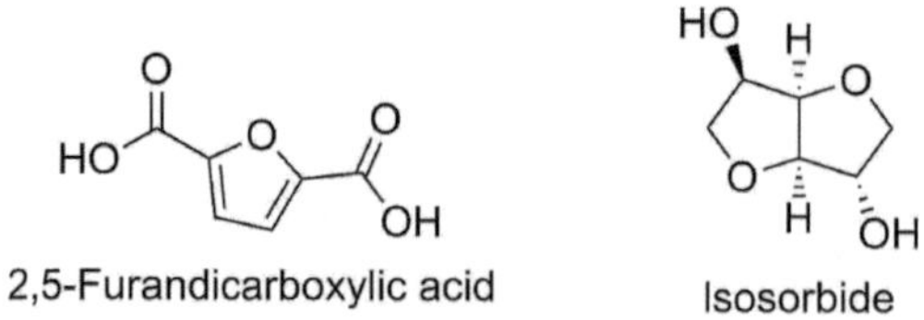

Figure 3.6 2,5-Furandicarboxylic acid and isosorbide: rigid monomers that in principle can be made only from biomass.

would make sense to produce them from sustainable sources: biomass, CO_2, and other carbon already "above ground" (i.e. waste PET).

Currently recycling is a complicated process for most plastic applications, due to multilayers, colorants, and blends. In Japan, for instance, it is prohibited to add colorants to plastics to facilitate recycling. The suggestion to add even more (novel) plastic materials into the mix may seem to steer away from facilitating our current recycling. However, we should realize that the transition to sustainable circular materials has hardly started! Only two million tons of our current plastics (0.5%) is (partly) bio-based, and we must evaluate novel materials that can better fulfill the requirements for reuse, closed-loop recycling, and fate-in-nature than the materials that we use today. Unfortunately it is not possible today to predict the winning (sustainability, performance, and cost) plastic materials of the future.

The microplastic issue is a different issue to deal with. Primary microplastic pollution is mainly caused by wear of durable materials (tires, clothing), which means these materials should last, something that conflicts with a high level of biodegradability. With regard to microfibers from clothing, it would make most sense to fit washing machines with appropriate filters.

Data must be generated and used to inform citizens. Smartphone applications have facilitated local authorities and non-government organizations (NGOs) in collecting local data on marine litter (e.g. the European economic area [EEA] Marine Litter Watch app, Trashhunters app). Citizens and public procurement officers can also be empowered with better data on the products they buy (e.g. the Beat the Microbead app, the Good Scrub Guide), finding opportunities to reduce their contribution to marine litter. The United Nations Environment Program massive open online course (MOOC) in 2015–2016 on marine litter provided free public access to the latest research. A second edition of the MOOC in 2017 will continue this (https://www.marinelittermooc.org).

3.5.5 Policies and Legislation

Structural waste should be avoided in all economic sectors. Producers should consider shifting from selling goods to providing services or access to (rather than ownership of) goods where this can increase product durability and reduce material demand and waste from the manufacturing stage through to product end of life. Business models based on reusable packaging exist in the context of B2B (e.g. Svenska Retursystem, operating a pool of reusable packaging for the whole retail sector) and B2C (e.g. Splosh and Replenish, shipping active-cleaning ingredients to be used in refillable bottles; and Repack, developing reusable transport packaging for e-commerce). Other innovative delivery models such as The Disappearing Package avoid packaging altogether, rethinking the entire packaging concept.

3.6 Conclusions

Plastic debris in the form of macroplastics or MNPs are everywhere, especially in the marine environment. These plastics originate from a multitude of sources and are composed of a great variety of polymers and copolymers as well as additives and, optionally, adsorbed pollutants. We can conclude that more information is required about the sources, distribution, fate, and potential impact of plastics on both human and animal health, on land, in the air, and in the marine environment. This is particularly true in the case of MNPs, as we lack adequate knowledge of their potential physical and chemical effects. Information is needed at all scales, as sources, circumstances, and mitigation strategies at each scale will vary.

If plastic is treated as a valuable resource, rather than just as a cheap waste product, opportunities to create a secondary value for the material after its first intended use will provide economic incentives for collection and reprocessing. All efforts to design future products for reuse or better recycling will add to the waste value of plastics. Non-recyclable plastics will have lower value and will end up in the environment more frequently. In addition, solutions need to be part of comprehensive programs to improve waste management generally: that is, waste collection and disposal infrastructure, waste-management practices, and enforcement. Such programs could include improved design and application of single-use plastics, increased consumer awareness and behavioral changes, improved recycling and re-use, and the introduction of economic instruments to reduce littering and promote secondary uses of plastic debris [61].

Innovative technologies in recycling can lead to recycling of a greater proportion of waste. The concept of extended producer responsibility, according to which a producer's responsibility for a product is extended to the postconsumer stage of the product's life cycle [62], could lead to more responsible packaging design.

Successful global management of the marine litter problem will require the development and implementation of effective policies and measures, supported by international and regional treaties and conventions. Decision makers must give marine litter a higher profile in national environmental protection regulations and development plans. It will be especially important to use education and outreach programs to encourage key user groups, industry sectors, and the general public to modify behavior and assume greater personal responsibility for their actions. Tackling the plastic waste issue will move us toward a cleaner ocean, will reduce the many pressures and impacts on biodiversity and, at the same time, will greatly reduce related social and economic costs [63].

Finally, the economic and social costs of plastic litter are difficult to fully assess as the full extent of the impact of MNPs on ecosystems and human health is still unknown. It is typically more costly to clean up pollution than to prevent it. It thus makes much more sense to deal with this at the source.

References

1 Gall, S.C. and Thompson, R.C. (2015). *Marine Pollution Bulletin* 92: 170–179.
2 Rochman, C.M., Browne, M.A., Underwood, A.J. et al. (2016). *Ecology* 97: 302–312.
3 Lwanga, E.H., Vega, J.M., Quej, V.K. et al. (2017). *Scientific Reports* 7: 14071.
4 Ward, C.P. and Reddy, C.M. (2020). *Proceedings of the National Academy of Sciences of the United States of America* 117: 14618–14621.
5 Law, K.L. (2017). *Annual Review of Marine Science* 9: 205.
6 Wright, S.L. and Kelly, F.J. (2017). *Environmental Science and Technology* 51: 6634–6647.
7 Yong, C.Q.Y., Valiyaveetill, S., and Tang, B.L. (2020). *International Journal of Environmental Research and Public Health* 17: 1509.
8 Skoczinski, P., M. Carus, D. de Guzman et al. (2021) Bio-based building blocks and polymers – global capacities, production and trends 2020–2025 – Graphics: Figure 1 (PNG). https://renewable-carbon.eu/publications/product/bio-based-building-blocks-and-polymers-global-capacities-production-and-trends-2020-2025-graphics-figure-1-png/. https://phys.org/news/2021-06-global-plastics-production-falls.html (accessed 28 December 2021).
9 Jambeck, J.R., Geyer, R., Wilcox, C. et al. (2015). *Science* 347: 768.
10 Boucher, J. and Friot, D. (2017). *Primary Microplastics in the Oceans: A Global Evaluation of Sources*, vol. 10. Gland, Switzerland: IUCN.
11 Geyer, R., Jambeck, J.R., and Law, K.L. (2017). *Science Advances* 3: e1700782.
12 Lebreton, L. and Andrady, A. (2019). *Palgrave Communications* 5: 6.
13 UNEP and GRID-Arendal (2016). *Marine Litter Vital Graphics*. United Nations Environment Programme and GRID-Arendal https://www.unep.org/resources/report/marine-litter-vital-graphics (accessed 28 December 2021).
14 Cole, M., Lindeque, P., Halsband, C. et al. (2011). *Marine Pollution Bulletin* 62 (12): 2588–2597.
15 Lozano, R.L. and Mouat, J. (2009). *Marine Litter in the North-East Atlantic Region: Assessment and Priorities for Response*. KIMO International.
16 Sundt, P. and Schulze, P.-E. (2015). *Sources of Microplastic-Pollution to the Marine Environment*. Mepex for the Norwegian Environment Agency.
17 Leslie, H.A. (2014). *IVM Institute for Environmental Studies* 476: 1–33.
18 Takahashi, C.K., Turner, A., Millward, G.E. et al. (2012). *Marine Pollution Bulletin* 64: 133–137.
19 Song, Y.K., Hong, S.H., Jang, M. et al. (2015). *Archives of Environmental Contamination and Toxicology* 69 (3): 279–287.
20 Chae, D.H. (2015). *Archives of Environmental Contamination and Toxicology* 69 (3): 269–278.
21 Gove, J.M. (2019). *Proceedings of the National Academy of Sciences of the United States of America* 116 (48): 24143–24149.
22 Imhof, H.K. (2016). *Water Research* 98: 64–74.
23 Lee, J., Lee, J.S., Jang, Y.C. et al. (2015). *Archives of Environmental Contamination and Toxicology* 69 (3): 288–298.

24 UNEP (2016). *Marine Plastic Debris and Microplastics – Global Lessons and Research to Inspire Action and Guide Policy Change*. Nairobi: United Nations Environment Programme https://ec.europa.eu/environment/marine/good-environmental-status/descriptor-10/pdf/Marine_plastic_debris_and_microplastic_technical_report_advance_copy.pdf.

25 Browne, M.A., Crump, P., Niven, S.J. et al. (2011). *Environmental Science and Technology* 45: 9175–9179.

26 Napper, I.E. and Thompson, R.C. (2016). *Marine Pollution Bulletin* 112 (1): 39–45.

27 Wagner, S., Hüffer, T., Klöckner, P. et al. (2018). *Water Research* 139: 83–100.

28 Baensch-Baltruschat, B., Kocher, B., Stock, F. et al. (2020). *Science of the Total Environment* 733: 137823.

29 Sommer, F., Dietze, V., Baum, A. et al. (2018). *Aerosol and Air Qualitative Research* 18: 2014–2028.

30 Shah, A.A. (2013). *International Biodeterioration and Biodegradation* 83: 145–157.

31 Linos, A., Berekaa, M.M., Reichelt, R. et al. (2000). *Environmental Microbiology* 66 (4): 1639–1645.

32 UNEP (2014). *Valuing Plastics: Business Case for Measuring, Managing and Disclosing Plastic Use in the Consumer Goods Industry*. Nairobi: United Nations Environment Programme (UNEP).

33 Akdogan, Z. and Guven, B. (2019). *Environmental Pollution* 254: 113011.

34 Lusher, A.L., McHugh, M., and Thompson, R.C. (2013). *Marine Pollution Bulletin* 67 (1, 2): 94–99.

35 Nelms, S.E., Barnett, J., Brownlow, A. et al. (2019). *Scientific Reports* 9 (1): 1–8.

36 Kolandhasamy, P., Su, L., Li, J. et al. (2018). *Science of the Total Environment* 610, 611: 635–640.

37 Watts, A.J.R., Lewis, C., Goodhead, R.M. et al. (2014). *Environmental Science and Technology* 48: 8823–8830.

38 Browne, M.A., Dissanayake, A., Galloway, T.S. et al. (2008). *Environmental Science and Technology* 42: 5026–5031.

39 Hale, R.C., Seeley, M.E., La Guardia, M.J. et al. (2020). *Journal of Geophysical Research: Oceans* 125 (1): e2018JC014719.

40 Cole, M., Lindeque, P., Fileman, E. et al. (2015). *Environmental Science and Technology* 49: 1130–1137.

41 Rehse, S., Kloas, W., and Zarfl, C. (2016). *Chemosphere* 153: 91–99.

42 Karami, A., Romano, N., Galloway, T. et al. (2016). *Environmental Research* 151: 58–70.

43 Tian, L., Zheng, J., Goodyer, C.G., and Bayen, S. (2020). *Food Chemistry* 326: 126942.

44 de Souza Machado, A.A., Kloas, W., Zarfl, C. et al. (2018). *Global Change Biology* 24: 1405–1416.

45 Lwanga, E.H., Vega, J.M., Quej, V.K. et al. (2017). *Scientific Reports* 7 (1): 1–7.

46 Gaylor, M.O., Harvey, E., and Hale, R.C. (2012). *Chemosphere* 86: 500–505.

47 Liu, H., Yang, X., Liu, G. et al. (2017). *Chemosphere* 185: 907–917.

48 Lusher, A., Hollman, P., and Mendoza-Hill, J. (2017). *Microplastics in Fisheries and Aquaculture: Status of Knowledge on Their Occurrence and Implications for Aquatic Organisms and Food Safety*. FAO.
49 Viršek, M.K., Lovšin, M.N., Koren, Š. et al. (2017). *Marine Pollution Bulletin* 125 (1, 2): 301–309.
50 Ahmada, S., Wong, K.Y., Tseng, M.L. et al. (2018). *Resources, Conservation and Recycling* 132: 49–61.
51 UNEP (2015). *Biodegradable Plastics and Marine Litter. Misconceptions, Concerns and Impacts on Marine Environments*. Nairobi: United Nations Environment Programme (UNEP).
52 Velis, C.A., Lerpiniere, D., and Tsakona, M. (2017). *Prevent Marine Plastic Litter – Now!* International Solid Waste Association https://marinelitter.iswa.org/reports (accessed 28 December 2021).
53 Xanthos, D. and Walker, T.R. (2017). *Marine Pollution Bulletin* 118 (1, 2): 17–26.
54 Abbott, J.K. and Sumaila, U.R. (2019). *Review of Environmental Economics and Policy* 13 (2): 327–336.
55 Cordier, M. and Uehara, T. (2019). *Science of the Total Environment* 670: 789–799.
56 Napper, I.E. and Thompson, R.C. (2019). *Environmental Science and Technology* 53: 4775–4783.
57 World Economic Forum (2016). *The New Plastics Economy – Rethinking the Future of Plastics*. Ellen MacArthur Foundation/McKinsey & Company http://www.ellenmacarthurfoundation.org/publications (accessed 28 December 2021).
58 Schnurr, R.E., Alboiu, V., Chaudhary, M. et al. (2018). *Marine Pollution Bulletin* 137: 157–171.
59 Dauvergne, P. (2018). *Environmental Politics* 27: 579–597.
60 Directive (EU) 2019/904 of the European Parliament and of the Council of 5 June 2019 on the reduction of the impact of certain plastic products on the environment. https://eur-lex.europa.eu/legal-content/EN/TXT/PDF/?uri=CELEX:32019L0904 (accessed 28 December 2021).
61 Lau, W.W.Y., Shiran, Y., Bailey, R.M. et al. (2020). *Science* 369: 1455–1461.
62 Walls, Margaret (2006), Extended Producer Responsibility and Product Design: Economic Theory and Selected Case Studies. http://dx.doi.org/10.2139/ssrn.901661.
63 UNEP. (2011). Emerging issues in our global environment. https://wedocs.unep.org/handle/20.500.11822/8276 (accessed 28 December 2021).

3.A Definitions

Degradation

The partial or complete breakdown of a polymer as a result of, e.g. UV radiation, oxygen attack, biological attack. This implies alteration of the properties, such as discoloration, surface cracking, and fragmentation.

Biodegradation

Biological process of organic matter, which is completely or partially converted to water, CO_2/methane, energy and new biomass by microorganisms (bacteria and fungi).

Mineralization

Definition in the context of polymer degradation, as the complete breakdown of a polymer as a result of the combined abiotic and microbial activity, into CO_2, water, methane, hydrogen, ammonia, and other simple inorganic compounds.

Biodegradable

Capable of being biodegraded.

Compostable

Capable of being biodegraded at elevated temperatures in soil under specified conditions and time scales, usually only encountered in an industrial composter (standards apply).

Oxo-degradable

Containing a pro-oxidant that induces degradation under favorable conditions. Complete breakdown of the polymers and biodegradation still have to be proven.

4

Tutorial on Polymers – Manufacture, Properties, and Applications

Gert-Jan M. Gruter[1,2] *and Jean-Paul Lange*[3,4]

[1] *Avantium Corporate Technologies, Zekeringstraat 29, Amsterdam 1014 BV, The Netherlands*
[2] *University of Amsterdam, Industrial Sustainable Chemistry, van 't Hoff Institute for Molecular Sciences (HIMS), Science Park 904, PO Box 94157, Amsterdam 1090 GD, The Netherlands*
[3] *Shell Global Solutions International B.V., Grasweg 31, Amsterdam 1031 HW, Netherlands*
[4] *University of Twente, Sustainable Process Technology, Faculty of Science and Technology, PO Box 217, Enschede 7500 AE, Netherlands*

4.1 Introduction

4.1.1 Today's Petrochemical Industry

Today's petrochemical industry is reported to consume about 10% of fossil carbon sources as feedstock and another 7% as energy to drive its chemical processes [1]. The feedstock is primarily used to produce olefins and aromatics, also called base chemicals. The base chemicals are for a major part feedstock for our plastics which are currently (2021) produced at a scale of about 400 million tons per year (Mt/yr). This number excludes fibers (textiles and carpets), thermosets, rubber (tires), and recycled plastic (Figure 4.1a). The largest fraction of these base chemicals is used to produce hydrocarbon polymers such as polyolefins (110 Mt/yr polyethylene [PE]; 66 Mt/yr polypropylene [PP]) and polystyrene (PS) (16 Mt/yr), while a smaller fraction is converted to intermediates by insertion of oxygen, nitrogen, or chlorine. Most of these intermediates are also used for making polymers, namely high-performance and high-value (engineering) polymers such as polyethylene terephthalate (PET) (80 Mt/yr of which 52 Mt/yr for fibers), polyvinylchloride (PVC) (70 Mt/yr), nylon (10 Mt/yr), polyurethane (PUR) (26 Mt/yr), and polycarbonates (PCs) (6 Mt/yr). Only a minor fraction is converted to solvents, detergents, plasticizers, and a variety of products for the fine chemical industry.

Since their discoveries in the 1930s, polymers have been developed with an impressive diversity of technical properties (strengths, light-weight, click-flexibility/elasticity, transparency, thermal/electrical insulation, etc.) that they combine with low cost (Table 4.1) and fashionable design possibilities (shape, color, gloss, etc.). Hence, they rapidly penetrated all facets of today's life, from housing to health, clothing, sport, transport, food, water, and many more. The impressive market

Biodegradable Polymers in the Circular Plastics Economy, First Edition.
Edited by Michiel Dusselier and Jean-Paul Lange.

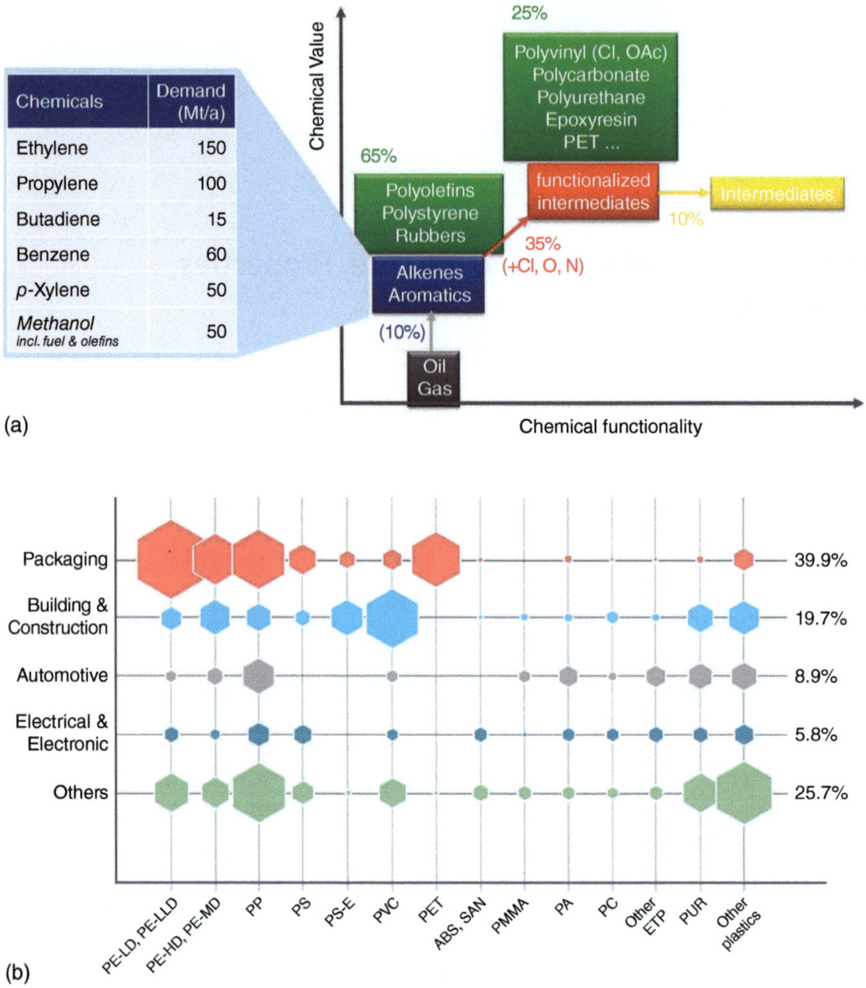

Figure 4.1 Structure of the global petrochemical industry (a) and markets for polymer products in Europe (b) (the size of each hexagon represent the size of the application for the given polymer). Source: With permission from Plastic Europe [2].

growth witnessed in the past is expected to continue in the future; likely with a factor 3–4 between 2000 and 2050; faster than global wealth (gross domestic product [GDP]) and global energy.

Today, the majority (40% in Europe) of the polymers is used in packaging and consists of PE, PP, and PET (Figure 4.1b). Other markets include construction (20%; based on PVC, PE, PS, and PUR), automotive (9%; based on PP and PUR), electrical/electronic (6%), and many others [2, 3]. More on properties and applications below.

Table 4.1 Market price of major polymers for the period 2010–2020.

Polymers	Market price (€/kg)
PE, PP, PET, PVC	1.00–1.50
Acrylonitrile–butadiene–styrene (ABS), polyoxomethylene (POM), bio-PET (20% bio-based)[a]	1.50–2.00
Polycarbonate (PC), polyamide-6 (PA6), polycaprolactone (PCL), bio-PE	2.00–3.00
Polyamide-66 (PA66), poly(methylmetacrylate) (PMMA)	3.00–4.00
PLA, PHA, PBS	>4.00

a) Industry uses 30% bio-based content as they consider all oxygen connected to bio-based carbon to be bio-based. Scientifically this is not correct as this partly bio-based PET is produced from fossil *para*-xylene and bio-based ethylene and none of the oxygen is bio-based.

4.1.2 Today's Bio-based Plastic Industry

Currently, bio-based plastics represent less than 1% of the 400 million tons of plastic produced annually. As demand is rising, and with the development of more sophisticated bio-based polymers, with novel applications and products emerging, the market for bio-based plastics is continuously growing and diversifying. According to market data compiled by European Bioplastics in cooperation with the Nova-Institute, global bio-based plastics production capacity is expected to increase with 15% from around 2.1 million tons in 2019 to 2.4 million tons in 2024 [4]. This growth is mainly found in niche applications, for bio-based polymers are typically more expensive because their production scale is still today much smaller than those for the well-known fossil plastics and because the processes for the bio-based materials have not been fully optimized yet.

4.1.3 Environmental and Climate Challenges

The issue of plastic waste in the environment has been obvious and problematic for some time and only recently resulted in serious action in the form of certain single-use bans (straws, cutlery, plastic plates, etc.), even though the extent of health effects as a result of the pervasive and ubiquitous human exposure to plastics constituents is today still uncertain. When mass production of plastics began after the middle of the twentieth century, it quickly found application into all facets of modern daily life. However, there was never a serious consideration of the (long-term) effects plastics will exert on the environment (and on living organisms) when produced at today's enormous and still-expanding scale (see Chapter 3 for an in-depth discussion). With novel bio-based and biodegradable plastics we now have an option to avoid these mistakes by designing/selecting our future plastics for better closed-loop recycling and a better fate-in-nature [5].

Despite all societal benefits, “plastics” are presently vilified by a significant part of society because of the plastic litter that is accumulating on lands, in rivers, and in oceans. Globally, 58% of plastic waste was discarded or landfilled, and only 18% was recycled in 2015. It is estimated that 4.8–12.7 Mt of plastic waste entered the ocean in 2010 [6].

Growing alongside the volume of global production and consumption of plastics are the diverse concerns on their impacts on the ecosystem and human health, particularly as microplastics [7]. Although not fully understood yet, the threat of microplastics seems to crystallize into a few forms [8, 9]. Firstly, there is the mechanical irritation of microplastic on body cells and tissues that leads to infections. Secondly, some microplastics may contain residual monomers and additives (e.g. plasticizers) that can slowly diffuse into the body and, eventually, intervene in the metabolism. Thirdly, microplastics are often hydrophobic and, thereby, tend to accumulate hydrophobic contaminants present in the air or in the water. These contaminants can be released in the organism and affect its metabolism.

Moreover, the production of plastics is also accompanied by the generation of process wastes, including CO_2 [10–12]. To be circular and sustainable, the industry and society will need to better manage these process waste streams, ensure responsible end-of-life of the product, and, eventually, switch to renewable material and energy sources.

According to the Paris agreement of December 2015, Europe needs to reduce its current annual 4 Gt/yr of CO_2 emissions to 1 billion tons by 2050 (a 75% reduction). The energy sector has many renewable alternatives to fossil resources: solar, wind, geothermal, nuclear, etc. For chemicals and materials, however, the only source of renewable carbon is biomass and CO_2. We will remain brief on the latter and focus on the former. The contribution of the plastic-related CO_2 emission to climate change is particularly gaining attention today. Conventional (fossil fuel–based) plastics produced in 2015 emitted 1.8 Gt_{CO2e} over their life cycle (excluding any carbon credits from recycling), which corresponds to 3.8% of the 47 Gt_{CO2e} emitted globally that year [13]. As the plastic production is expected to triple by 2050, the life-cycle green house gas (GHG) emissions could grow to 6.5 Gt_{CO2e}, by maintaining the current energy mix, and even to 8.0 Gt_{CO2e} if all plastic waste is incinerated (with energy generation). It is obvious that a tripling of these emissions by 2050 may then exceed the total carbon budget as the world agreed in 2016 in Paris [14] to reduce carbon emissions with 90% by 2050 to keep global temperature rise below 2 °C. The routes from biomass are in advanced development stage, while those based on CO_2 are at very early stage of development.

4.2 Production of Polymers

Most monomers are presently derived from crude oil. We will therefore review the diversity of fossil routes to polymers. However, we’ll still devote a short section on bio-based polymers and refer to the literature for more extensive discussion.

Most monomers are produced by *steam cracking* of an aliphatic cut of crude oil into olefins, mainly ethylene and propylene but also some butenes (a monomer for certain linear-low-density polyethylenes [LLDPEs]) and butadiene (a monomer for SBRs [styrene–butadiene rubbers] and styrene butadiene styrene [SBS] fibers) as well as liquid product rich in aromatics. Olefins can also be produced by steam cracking or catalytic dehydrogenation of ethane and propane. Some more chemicals are extracted from oil-refining processes: fluid catalytic cracking (FCC) delivers some propylene and aromatics, while catalytic reforming produces aromatics.

4.2.1 Addition Polymers

Ethylene and propylene can be polymerized as to various grades of PE and PP via radical polymerization low-density polyethylene (LDPE) or coordination polymerization (so called Ziegler–Natta polymerization to, e.g. high-density polyethylene [HDPE], LLDPE, ethylene–propylene–diene monomer [EPDM], and *i*-PP). The polymerization is highly exothermic and irreversible, making thereby recycling by selective depolymerization back to the monomer impossible (see Chapter 12 on recycling). The main chains of these polyolefins consist of C—C bonds only and cannot be biologically degraded.

Ethylene and propylene can also be upgraded to various "vinyl monomers" with the general structure of $H_2C{=}CH{-}X$ (Figure 4.2). For instance, ethylene can be converted to styrene (X = aryl; by alkylation of benzene to ethylbenzene, followed by dehydrogenation), to vinyl chloride (X = Cl; by addition of Cl_2 followed by HCl

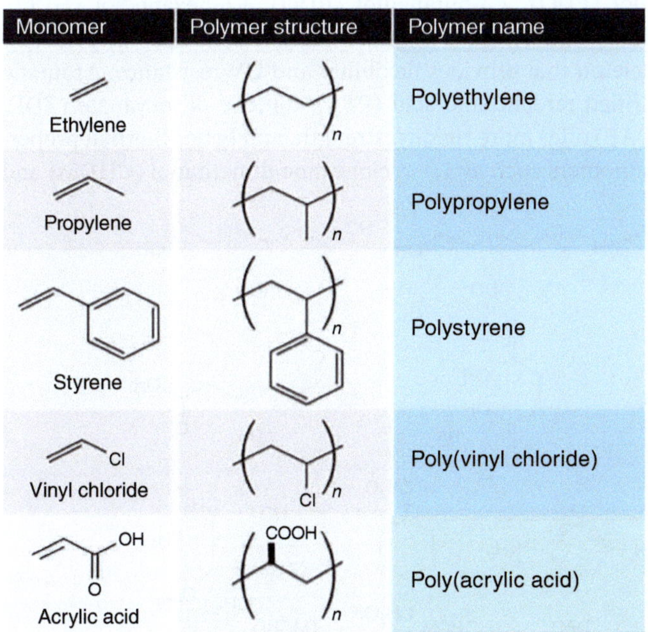

Figure 4.2 Major vinyl monomers and polymers.

elimination), or to vinyl acetate (X = acetate; by oxidative addition of acetic acid). On the other hand, propylene can be converted to acrylic acid (X = COOH; by oxidation) or to acrylonitrile (X = CN; by ammoxidation). Much like the olefins, these "vinyl monomers" can in their turn be polymerized by radical or coordination polymerization. The polymerization reaction is here again highly exothermic and irreversible. The polarity introduced by the X-side groups allows the chain to better interact with one another and with other surfaces. It thereby leads to polymers with higher strength and improved adhesion properties (e.g. for coatings and for multi-layer films for packaging).

4.2.2 Condensation Polymers

Polarity, which is advantageous for many applications for vinyl polymers, can also be built into the main polymer chain rather than the side chains. This is achieved by linking monomers through ester bonds (—C(O)O—), amide bonds (—C(O)—NH—), urethane or carbamate bonds (—C(NH)—O—), or carbonate bonds (—O—C(O)—O—). These polar bridges are made through reversible reactions and, therefore, open the path to recycling by depolymerization (see Chapter 12 on recycling).

The monomers generally consist of diols, diacids, diamines, hydroxy acids, or amino acids (Figure 4.3) that are connected in AABB or ABAB chains (Figure 4.4). Ideally, the two functional groups are placed in α–ω positions, i.e. at "each end" of the monomer (at so-called primary carbons), although deviations from this rule become more frequent. Aliphatic monomers such as mono ethylene glycol (MEG), 1,3-propane diol (PDO), 1,4-butanediol (BDO), 1,6-hexanediol (HDO), hexamethylene diamine (HMDA), succinic acid (SA), and adipic acid (ADA) contain an aliphatic skeleton that provides flexibility and UV resistance. Aromatic monomers such as purified terephthalic acid (PTA), toluene diisocyanate (TDI), 1,4-diaminobenzene (DAB) offer more rigidity, strength, and better thermal properties. Cyclic aliphatic monomers such as 1,4-cyclohexane dimethanol (CHDM) and

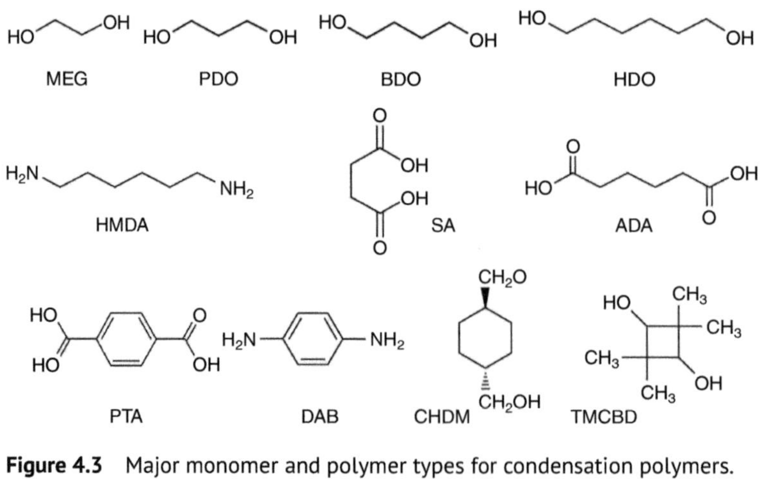

Figure 4.3 Major monomer and polymer types for condensation polymers.

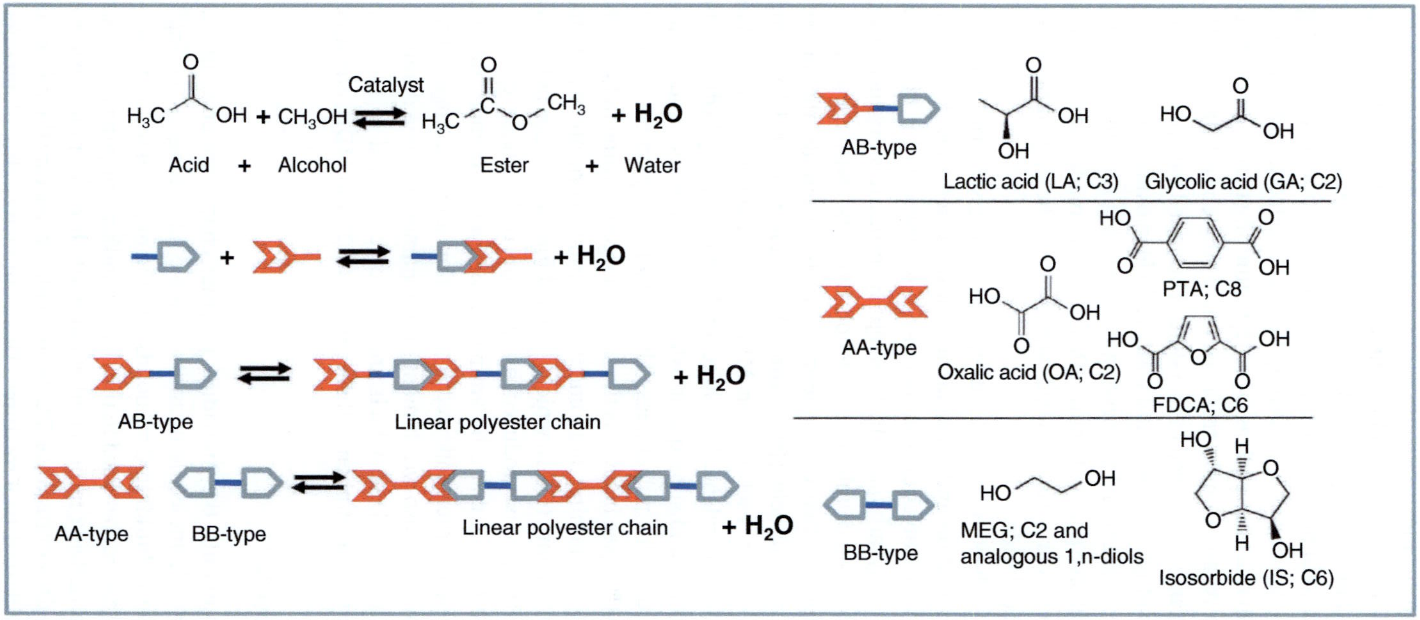

Figure 4.4 Basic polyester construction options and representative AB and AA + BB type polyester monomers.

2,2,4,4-tetramethyl 1,3-cyclobutanediol (TMCBD) have also emerged to combine rigidity and UV resistance.

The chemistry required to produce the monomers for condensation polymers is more diverse and more complex than for the olefins described above. When starting from fossil hydrocarbons, extensive and selective functionalization is required. The economic advantage of this is that weight is often added to the molecules by incorporation of cheap reactants such as O_2 or H_2O, limiting thereby to overall cost per kg of the monomer. As an example, from one ton of ethylene (at 100% yield), 2.14 ton of ethylene glycol can be produced in two steps. The added weight originates from oxygen (ethylene to ethylene oxide) and water (ethylene oxide to ethylene glycol). This largely offsets the additional processing cost, making thereby ethylene glycol not much more expensive per kg than the ethylene it was made from. It should be noted that similar benefits are also achieved with some of the vinyl monomers discussed above, e.g. vinyl chloride.

As for the diols, the simplest one, "mono" ethylene glycol, (MEG), is derived from ethylene via epoxidation and subsequent hydration (see above). MEG is one of the constituents of PET polyester. Another very important diol is bisphenol-A (BPA), which is used in polycarbonate (PC) and epoxy resins. It is produced by converting benzene to phenol and coupling two phenols tail-to-tail with acetone.

As for the diacids, two components seem to dominate the market, namely "purified" terephthalic acid (PTA) and adipic acid (ADA). PTA is mainly used in PET polyester after reaction with MEG. It is produced by air oxidation of the methyl-groups of *para*-xylene. ADA is generally produced by hydrogenation of benzene to cyclohexane and subsequent oxidation cyclohexanol and eventually ADA. ADA is mainly used for polyamides. The polyamide nylon-66, which is obtained from ADA + HMDA (see below), is the second largest volume nylon in the market.

Among the N-containing monomers, two monomers are worth mentioning, HMDA and caprolactam (CPL), which are used for polyamides nylon 66 and nylon 6, respectively. HMDA is produced by addition of hydrogen cyanide (HCN) to butadiene while CPL is made by amination of cyclohexanone, the intermediate of ADA mentioned earlier, followed by oxidation and rearrangement. Two other aromatic diamines are also worth mentioning, namely toluene diamine (TDA) and methylene–dianiline (MDA), which are further converted to the corresponding toluene- and methylene diphenyl diisocyanates (TDI and MDI), and used in PURs.

4.2.3 Thermosets

In thermoset materials, the individual polymer chains are cross-linked with one another into a single molecule in an irreversible hardening or "curing" step. One approach consists of building C=C bonds into the main chains and cross-link them via –S– bridges (e.g. rubber used in tires) or via radical coupling (e.g. in unsaturated polyesters). Alternatively, minor components with three (or more) reactive heads are built into the chain. For instance, PURs are made by reacting the diisocyanates mentioned earlier with diol/triol mixtures. This approach is also adopted for the epoxy resins.

In contrast to the thermoplastics discussed in Sections 4.2.1 and 4.2.2 above, thermosets cannot be melt-processed, and therefore the products are often molded into the desired shape before curing. They are also difficult to recycle, as discussed in Chapters 11 (on vitrimers) and 12 (on managing plastic waste).

4.2.4 Renewable Monomers

The discussion on bio-based polymers needs to start with a discussion on the suitable biomass feedstocks, their composition, their availability, and sustainability. Such discussion is clearly outside the scope of this chapter and will be reduced to a minimum, referring the reader to appropriate literature for further information. With the risk of over-simplification, biomass feedstock mainly consists of vegetable oil (e.g. soy or palm), "accessible" sugars (e.g. cane, beet, or starch), and "non-accessible" structural lignocellulose (e.g. wood, straw), which is a composite of cellulose fibers (a crystalline glucose sugar polymer) bound in a matrix of hemicellulose (an amorphous mixed sugar polymer) and lignin (an amorphous phenolic polymer).

These feedstocks are generally rich in oxygen and might, therefore, not be ideal to produce olefins, that are today's backbone of the petrochemical industry. They may be more suited to deliver oxygenated intermediates. As an example, producing one ton of ethylene from glucose requires 3.21 ton of glucose at 100% yield: fermentation to two ethanol and two CO_2 followed by dehydration of ethanol to ethylene and water: $C_6H_{12}O_6 \rightarrow 2\ C_2H_4 + 2\ CO_2 + 2\ H_2O$. In contrast, one ton of ethylene glycol requires just 0.97 ton of glucose when made via direct hydrogenolysis: $C_6H_{12}O_6 + 3\ H_2 \rightarrow 3\ C_2H_4(OH)_2$. A poor mass efficiency is weighing heavily on the feedstock cost and the overall economics of the process [15]. For this reason, it is expected that with an increasing renewable polymer market share we will see a shift from polyolefins to condensation polymers.

These specific monomers may require complex chemistry when starting for bio-based feedstocks such as sugar. These require now extensive and selective defunctionalization. However, bio-based feedstocks are well suited for producing novel monomers with the same reactive functionalities, hydroxyls, and acids, but different backbones. This opens the door to novel condensation polymers with novel properties. This complex chemistry will be illustrated by the following paragraphs, starting with the fossil routes and closing with the emerging sugar-based routes. More detailed discussion can be found in the literature.

4.2.4.1 Oils-Based Monomers

Global oilseed production was around 600 Mt/yr globally in 2019 [16], with large land requirement. Two hundred Mt per year consists of oil that covers fatty acids and triglycerides that are obtained either from various terrestrial oil plants such as palm, coconut, rape seed, sunflower, and soybean. Although marginal today, oils are also getting harvested from aquatic biomasses, typically microalgae. Beyond their use in food (80%), vegetable oils are mainly used for biodiesel and detergent. Their industrial use in polymers is limited to niche applications, such as commercialized bifunctional building blocks for very-small-volume bio-based plastics, namely sebacic acid and 11-aminoundecanoic acid, both from castor oil and azelaic acid derived from oleic acid. In addition, dimerized fatty acids are primarily used for polyamide resins and polyamide hot-melt adhesives [17, 18].

4.2.4.2 Sugar-Based Monomers

Accessible sugars are presently produced at the total scale of ~4 Gt/yr, i.e. ~3 Gt/yr of starch from cereals and tubers and ~1 Gt/yr of sugars from cane and beet [19], i.e. about 10× the total volume of chemicals. Hence, this so-called first-generation biomass could replace a significant part of crude oil in making chemicals and polymers, after having met the needs for food and feed. Sugars indeed represent an emerging and promising feedstock for manufacturing chemicals intermediates and, eventually, bio-based plastics. From a chemical point of view beet- and cane sugar (200 MT/a) is a disaccharide, consisting of two monosaccharides: glucose and fructose. Sucrose can be fermented to ethanol and to certain monomers such lactic acid, but for most chemical conversions either one of glucose or fructose needs to be used because both molecules hugely differ in structure and reactivity. This requires sucrose hydrolysis and chromatographic separation (using simulated moving bed [SMB] or simulated moving bed technology). In contrast, starch consists of glucose only and, as such is a more desired feedstock for many chemical conversions (such as sorbitol and MEG) and is also very suitable for fermentations.

In contrast with the conventional fossil hydrocarbons, sugars already contain oxygen functionalities and are, therefore, well suited to produce oxygenated monomers [11, 15, 20, 21]. In fact, sugars are particularly suitable for producing diols, diacids, and hydroxy acids needed for condensation polymers, as illustrated in Figure 4.5. Sugars can occasionally be converted to well-known monomers such as MEG, heavier diols, and succinic acid. However, sugars can also deliver new chemical intermediates that are difficult to produce from fossil hydrocarbons. The most promising ones likely include lactic acid, furan dicarboxylic acid (FDCA), and isosorbide.

Of course, sugars can also be fully deoxygenated to produce ethylene and butadiene (e.g. via ethanol) or even aromatics. But such deep deoxygenation is destroying quite some value, as discussed above and in the literature [11, 15, 20, 21].

For illustration, the companies associated with the authors are active in a few of these routes. Avantium is piloting the acid-catalyzed dehydration of fructose to FDCA [23], the hydrogenolysis of glucose to ethylene glycol (MEG) [24], and the polyester polyethylene furanoate (PEF) that is made from these two monomers [25]. Shell is exploring the solid-state fermentation of sugar crops to ethanol and lactic acid [26] and the production of bio-based pentenoate and valerate esters [27, 28], acrylic acid [29], MEG [30], as well as BDO/tetrahydrofuran (THF) [31].

Some of these monomers can then be used as drop-in substitute of present fossil monomers. This is, e.g. the case for MEG and BDO. Drop-in monomers benefit from the presence of a well-developed market. However, the performance being the same as the fossil equivalent, drop-in monomers need to compete on price to drive the market introduction. In contrast, new monomers present new structures that allow the development of new polymers with new properties. In this case a better performance can justify higher prices, which will benefit market introduction when

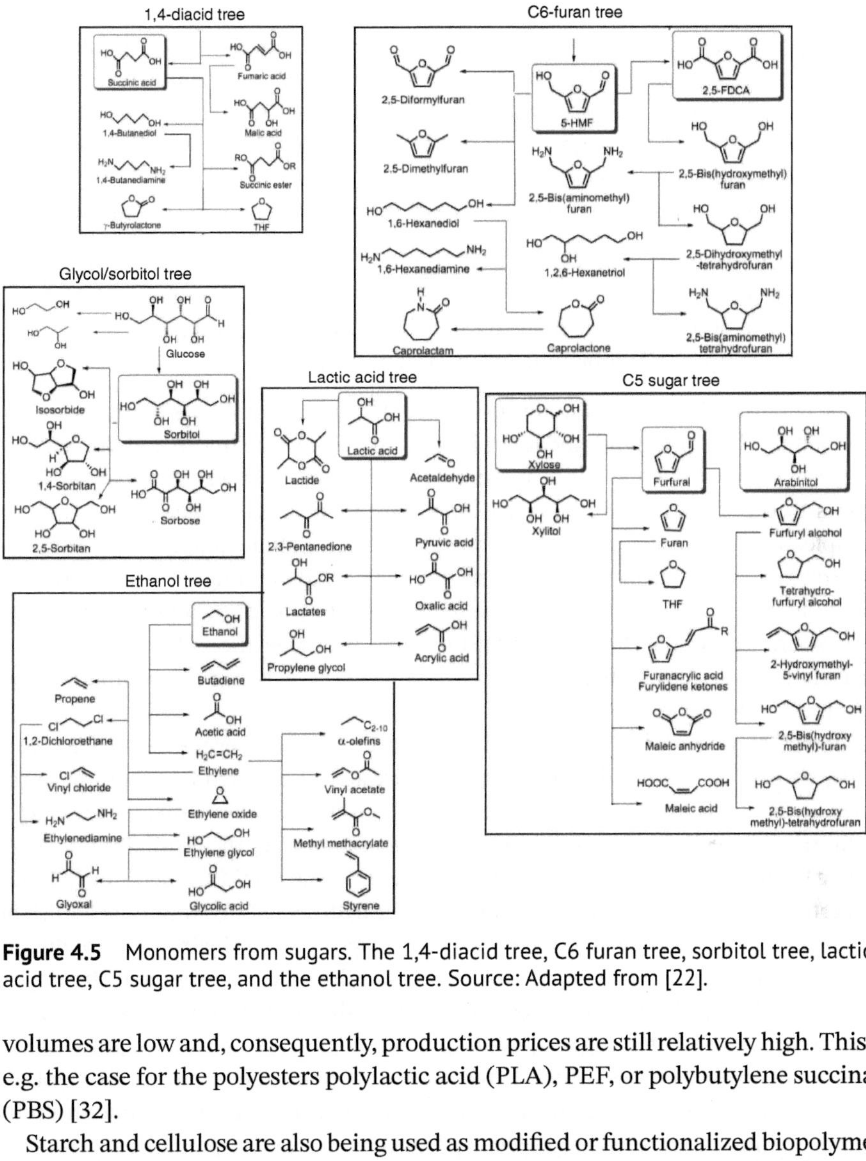

Figure 4.5 Monomers from sugars. The 1,4-diacid tree, C6 furan tree, sorbitol tree, lactic acid tree, C5 sugar tree, and the ethanol tree. Source: Adapted from [22].

volumes are low and, consequently, production prices are still relatively high. This is, e.g. the case for the polyesters polylactic acid (PLA), PEF, or polybutylene succinate (PBS) [32].

Starch and cellulose are also being used as modified or functionalized biopolymers to produce plastic materials, i.e. without hydrolysis to sugars (see Chapter 8). Starch is being used as additive or filler to synthetic polymers, and it is converted into a thermoplastic material (thermoplastic starch or TPS) by blending with plasticizers such as glycerol [33]. Cellulose acetate (CA) has been used for more than 150 years and is presently used at 1 Mt/yr scale [34]. Current applications are as a film base in photography, cigarette filters, and synthetic fibers.

4.2.4.3 Lignocellulose-Based Monomers

The first-generation and accessible sugars discussed above truly represent a steppingstone for lignocellulosic (second-generation) biomass and the "inaccessible sugars" it is made of. Lignocellulose is even available in much larger volumes: 200 Gt/yr of which 20 Gt/yr is potentially accessible [35] and can be produced with much lower, even marginal environmental impact [36]. Lignocellulose can indeed be harvested as "free" by-product of agriculture (e.g. straw) and forestry (thinning residue). But it can also be grown as dedicated "perennial" plants without mining the soil and using fertilizers.

Lignocellulose is composed of three main fractions, each about 1/3 of the dry mass:

- hemicellulose (a polysaccharide of different C5- and C6-sugars whose composition is species dependent),
- cellulose (a polysaccharide made up of cellulose for all species), and
- lignin (a complex alkylphenolic polymer network).

The cellulose forms fibers that provide the strength and stiffness of the plant and are held in place through imbedding in a matrix of hemicellulose and lignin. This purposely makes lignocellulose resistant to weather and microorganism to allow plants to stand up the whole year, sometimes even for centuries (trees). Hence, complex and expensive deconstruction technologies, often called fractionation or pre-treatment, are needed to extract its valuable sugars and lignin prior to further processing. As we need relatively pure glucose as a starting material for most monomers, the main challenge for a biorefinery to feed the polymer industry is to isolate glucose from lignocellulose, ideally without being contaminated with hemicellulosic sugars.

Several fractionation technologies have been developed over the years. They generally consist of cooking the lignocellulose in water at elevated temperature in the presence of an acid or a base to swell and hydrolyze the most digestible component, the hemicellulose, and free the cellulosic fibers from the lignin binders [37]. Other technologies are using organic solvents or ionic liquid as medium [38, 39]. In certain cases, the fractionation is done under hydrogenation conditions [40]. The isolated cellulose is then intendedly used for paper pulp or hydrolyzed to glucose by means of enzymes, allowing thereby to exploit the sugar conversion technologies discussed above. The hemicellulose hydrolysate that is rich in various sugars can also be used as feed for sugar-upgrading technologies. The residual lignin stream is generally used as process fuel. However, it is increasingly explored for valorization to aromatic fuel and chemical components. For instance, fractionation under hydrogenation conditions is reported to produce mono- and di-phenolic species with reasonable yield. Such stream could be considered for further upgrading to phenolic or aromatics monomers [41]. The other pre-treatment approaches deliver lignin with much higher molecular. This lignin appears to be recalcitrant to further depolymerization into well-defined monomeric components. However, it shows promises for applications as filler or as component for thermoset resins [42, 43]. Lignin plastics are discussed in depth in Chapter 10.

These fractionation technologies have a lot of potential, but so far, the industry failed to deliver on this promise. Since the first commercial-scale lignocellulosic ethanol plant in 2013, at least eight more have opened around the world, of which some, not surprisingly, are failing, while others are progressing with much difficulties. Most are in North America, two are in Brazil, and one is in Europe, all target cellulosic ethanol [44]. Although not required for ethanol, none of these technologies can produce glucose that is pure enough for use in downstream chemical conversions to high-purity monomers.

The companies associated with the authors are also very active in this area. Avantium is piloting an improvement of the 100-year-old Bergius technology that hydrolyzes the hemicellulose with 35% HCl at ambient temperature and pressure and, subsequently, hydrolyzes the cellulose with 42% HCl. In the end, all that remains in the reactor is pure lignin that can be valorized via cascading. A commercial plant using an improved version of this technology in Delfzijl, the Netherlands, has currently entered the detailed engineering phase by Avantium and partners [45]. The plant will produce the first commercial product in 2023. Shell is developing an acid-hydrolysis process based on concentrated "reversible" acid for easy recovery and recycle of the acid. The reversible acid is made by reversible association of an aldehyde with SO_2 and water to form a hydroxy sulfonic acid R–CH(OH)–SO_3H [46].

4.2.4.4 CO_2-Based Monomers

Some industrial processes already use CO_2 for the synthesis of organic products. In the production of urea yearly 100 Mt of CO_2 per year is consumed in the reaction with ammonia, which accounts for 0.25% of the global annual anthropogenic emission of CO_2. The use of CO_2 as chemical feedstock has recently seen an increased interest. The synthesis of polymers such as PURs and polycarbonates from CO_2 as a raw material has now reached industrial reality at companies such as Covestro and Novomer.

Electrochemical reduction of CO_2 represents a possible means of producing chemicals or fuels, including monomers such as oxalic acid, glycolic acid, ethylene, and ethylene glycol. Princeton spin-out Liquid Light, now part of Avantium, has during the last decade focused on several of these targets. Predictions say that in the next decades, up to 20% of CO_2 emissions can be utilized when the energy efficiency (selectivity) and the productivity (CO_2 converted per hour per m^2 electrode, expressed as current density) are optimized. This is undoubtedly an area with immense potential and opportunity.

4.3 Main Polymers Applications

The popularity of polymers is the fruit of a combination of key properties, which are illustrated in Figure 4.6:

- *Light weight*: Polymers typically show modest densities between 0.7 kg/l for polyolefins and 1.4 kg/l for PET, PVC, and typical engineering plastics (Figure 4.6b). They are heavier than wood (0.6–0.9 kg/l) but lighter than glass (2.5 kg/l) or steel (8 kg/l). They can also be made much lighter, e.g. <0.1 kg/l, when shaped as foam. But this will be discussed later.

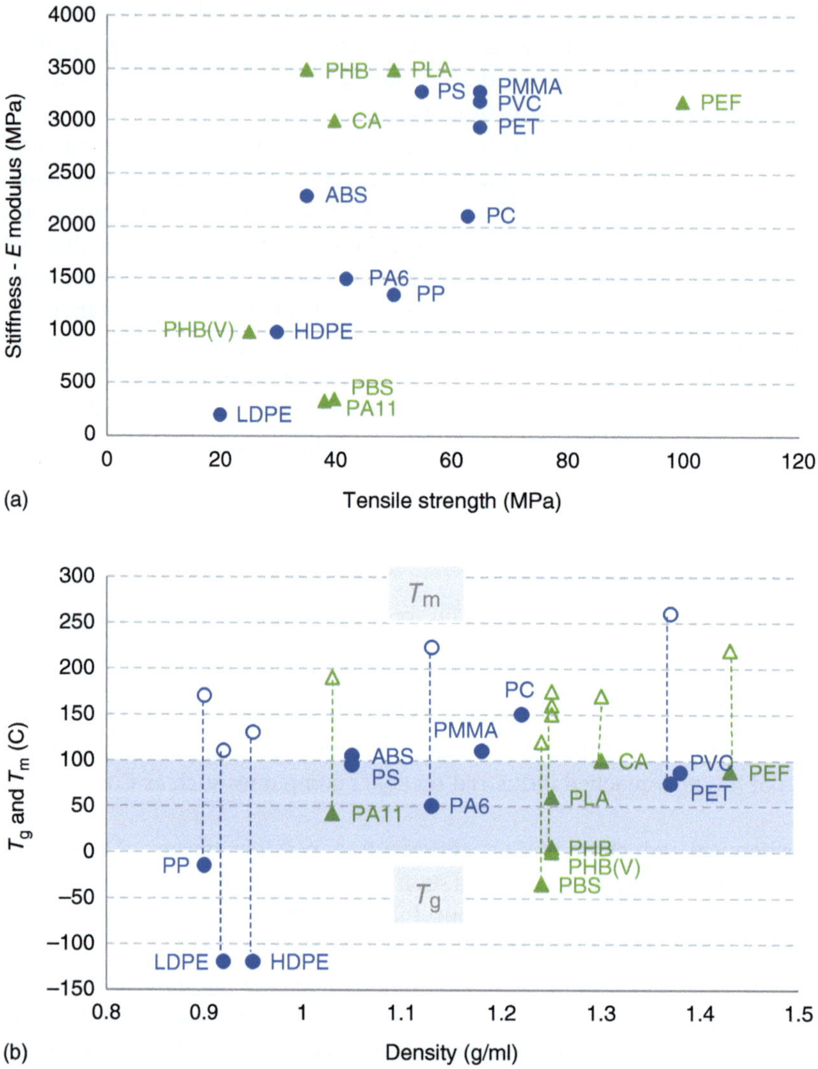

Figure 4.6 Key properties of major fossil-based (blue) and bio-based (green) polymers. In (a) the E-modulus is plotted versus the tensile strength. In (b) the T_g and T_m are plotted versus the density. The blue shaded bar represents the typical application temperature window that should lie below the T_m of crystalline polymers and below the T_g of amorphous ones. Source: Based on data from [47, 48].

- *Strength*: Polymers offer good (tensile or stretch) strength up to 100 MPa (Figure 4.6a). Demanding applications will request high-performance materials such as polyester (PET) and polycarbonate (PC), or composite materials reinforced with glass- or carbon fibers. In contrast, simpler applications will use the cheaper polyolefins. Polymers are thereby stronger than wood (2 MPa) and glass (7 MPa) but weaker than steel (>250 MPa). This gap disappears when correcting for their 6–8× lower density and expressing strength per weight of the material (so-called specific modulus).

- *Elasticity*: Polymers show elasticity that allows screw-less clicking of solid bodies, flexible films, rubber sealants, and elastics. This convenience is offered by the tunable stiffness of plastics (expressed as Young or E modulus), which ranges between 0.1 and 4 GPa for typical unfilled plastics (Figure 4.6a). For comparison, alternative materials are much stiffer with a modulus of 10 GPa for wood, 50–90 GPa for glass, and ~200 GPa for steel.

However, an important limitation of polymer materials is the temperature window within which they can be used. They generally soften above 100–150 °C, i.e. above their glass transition temperature T_g or their melting temperature T_m, depending on whether the polymer is amorphous or crystalline, respectively (Figure 4.6b).

With the risk of oversimplification, the tensile strength is favored by high-molecular-weight and strong molecular interactions between the chain, e.g. via dipolar and/or hydrogen bonding. High polymer density provides a first indication of strong molecular interactions, as can be deduced from the data displayed in Figure 4.6. However, other factors need to be considered as well. Orientation, resulting from stretching the material in one or two directions (fibers, mono/bi-axially stretched films, and injection blow-molded bottles), aligns polymer chains and consequently increases molecular interactions. The flexibility increases with decreasing molecular interactions and decreasing rigidity of the monomers. There is therefore a certain degree of trade-off between strength and flexibility (Figure 4.6a).

Bio-based polymers are generally rich in oxygen and, thereby, present relatively high density, high strength, and high stiffness, as illustrated in (Figure 4.6, green points). Interestingly, the combination of key properties may differ from those present in today's fossil-based polymer and present, thereby, new possibilities.

These key properties are exploited is the various applications, i.e. rigid materials, films, fibers, foams, coatings, adhesives, sealants, elastomers (CASE) and composites, which will be discussed in the following sections.

4.3.1 Rigids

Most polymers are applied as rigid materials, e.g. as containers (bottles and boxes), pipes, furniture, or body parts (cars, appliances). Applications that do not demand high strength but value good elasticity, e.g. for conveniently clicking them in place, e.g. as lid on a container or to other part of a body, may select cheap polyolefins. In contrast, application that require higher strength may prefer engineering polymers such as polymethylmetacrylate (PMMA), PC, or PET.

Beyond strength and elasticity, some containers (e.g. for food) and piping may require good barrier properties, e.g. to protect goods from O_2 ingress (e.g. beer and fruit juices) or from CO_2 release (e.g. carbonated soft drinks, sparkling water, and beer). These will require polymers with higher barrier such as PET (for bottles) or PVC (for piping). The 100% bio-based polyester PEF can be mentioned here as it has

10× better barrier than PET for O_2 and six times better barrier for CO_2. High polymer density can be used as first indicator for good barrier properties.

Other desirable properties may include operating temperature (e.g. for sterilization) or transparency/opaqueness, which is achieved with glassy or crystalline polymers, respectively.

Overall, bio-based monomers are advantageously prepared with high oxygen content and are, thereby, prone to deliver polymers with good strength and barrier properties. Flexibility may be more challenging, however. It can be favored by incorporating foreign monomers that disturb the order and stacking of the chains. This is, e.g. achieved by incorporating isophthalic acid in PET, hydroxyvalerate (a C_5 hydroxyacid) into polyhydroxyburyrate (a C_4 hydroxyacid), or by incorporating D-lactic acid or glycolic acid into poly-L-lactide (see Chapter 7 for ring-opening routes to such polyesters).

4.3.2 Films

Films are thin plastic membranes used for separation, for holding items, barrier layers, or printable surfaces. Main applications are packaging, plastic bags, building and construction, electrical wrapping, photographic film, etc. Many of the requirements discussed for bottles under rigids above also hold for film applications (lightweight, strength, barrier, and transparency), with the trade-off between strength and stiffness being further affected by the thickness of the film. Almost all plastics can be formed into thin film, but LDPE, HDPE, LLDPE, PP, PET, and nylon are most common. For special applications requiring extra strength, biaxially oriented films are used (e.g. BoPET or BoPP).

Most films are produced by extrusion (allowing stretching, thinning, and orientation) or blown processes (e.g. for plastic bags). Other production processes include casting, solution deposition, and calendaring.

For certain properties, one film may not meet all criteria. In that case, films can be co-extruded (extruding two or more layers of dissimilar polymers into a single film). Multilayer films can often not be mechanically recycled. This opens new opportunities for novel materials from renewable feedstock that can be made of a single layer thanks to better properties than the current materials.

Bio-based cellophane and cellulose acetate are plastics made from cellulose that were developed even before the modern plastic films. Here, a fraction of the hydroxyl groups of the cellulose are capped to reduce H-bonding (see Chapter 8) and the stiffness that is characteristic of cellulosic fibers. Emerging bio-based polymers include PLA for single-use food packaging, where shelf life and barrier are less important. PEF films are also being developed for long shelf-life applications (barrier for carbonated soft drinks and beer) and for durable optical films (optical properties).

4.3.3 Fibers

A fiber is a long thin drawn thread of a material. Polymer fibers are a subset of man-made fibers and are mostly made from PET polyester, polyamide (nylon), or polyolefins. Nylon fibers were developed in the 1930s (by DuPont) as a replacement for silk (e.g. for women stockings), parachutes, and (military) ropes. Today, the total polymer fiber production is around 60 MT/a. Textiles for apparel and carpets are the largest volume applications.

Spinning is a manufacturing process for creating polymer fibers. It is a specialized form of extrusion that uses a spinneret to form multiple continuous filaments. There are many types of spinning: wet, dry, dry jet-wet, melt, gel, and electrospinning. As an example, melt spinning is used for polymers that can be melted. The polymer solidifies by cooling after being extruded from the spinneret. Nylon and polyester fibers for textiles and carpets are produced via this process. Finally, the fibers are drawn to increase strength and orientation. This may be done while the polymer is still solidifying or after it has completely cooled. For many applications, such as textiles, texturizing is a process step during which the fibers are modified to change their physical appearance. Texturizing techniques can include bulking (where thermoplastic fibers are twisted, heat set, and untwisted), crimping and coiling, among others. Texturizing takes advantage of the thermoplastic nature of synthetic fibers and uses it to set texturized features in place.

Synthetic fibers are more durable than most natural fibers, and they are readily dyed in many different colors. Many fibers offer consumer-friendly functions such as stretching, waterproofing, and stain resistance. Sunlight, moisture, and oils from human skin cause all fibers to break down and wear away. Natural fibers tend to be much more sensitive than synthetic blends, mainly because natural products are biodegradable. Often a lot of water is needed for their growth and processing.

4.3.4 Foams

Foams are objects that were formed while blowing or trapping gas in pockets (bubbles) to deliver materials with either exquisite isolation properties (both thermal and acoustic), e.g. for construction and automotive or with comfortable rebound properties, e.g. for matrasses and cushions. The former ones consist of rigid foams and favor closed cells to entrap the gas. The latter ones consist of flexible foams and favor open cells to allow to gas to escape upon compression and reincorporate upon decompression. PUR foams are made by condensing polyols with aromatic diisocyanate (R–OH + O=C=N–R* → R–O–C(O)–NH–R*). In most cases, the void volume is very large, accounting, e.g. for 90% of the overall volume of the material.

Most common polymers used for foam applications are expanded polystyrene (EPS) and PUR. EPS applications are used for packaging of fragile goods such as electronics as solid blocks or "loose-fill peanuts," or as rigid panels for buildings insulation and food containers. PUR foam applications include rigid and flexible foams; the rigid for insulation panels, microcellular foam seals, and gaskets, spray foam; and the flexible for mattresses, seats, and carpet underlay.

The differentiating element in making foam may not be so much in the properties of monomer than the way to blow gas to produce the foam. In some cases, e.g. EPS and latex mattresses, the foam structure is provided by blowing a foreign gas through the polymer melt and quenching the material before it collapses back.

In other cases, the gas is produced inside the polymer melt by decomposing part of the monomers and, thereby, achieving better gas dispersion and very high gas volume fraction. PUR foams are typical example for the latter case, where a large fraction of the diisocyanates is reacted with water to release CO_2 and deliver an amine ($R{-}N{=}C{=}O + H_2O \rightarrow R{-}NH_2 + CO_2$) that can react with a neighboring isocyanate end of chain to form a very strong urea bond (R–NH–C(O)–NH–R).

The structure and properties of the PUR product depend on the type of the diisocyanate, the polyol, and the synthesis process. Long-chain diols lead to flexible foams, while short-chain triols lead to rigid ones. Aromatic diisocyanates tend to segregate into hard domains amid an aliphatic polyol matrix, offering thereby strength and rigidity. The diverse possibilities for chemical compositions of PUR provide an opportunity for the integration of novel, bio-based feedstocks. The total or partial replacement of fossil-derived polymer precursors (i.e. isocyanates, polyols, and other additives) with renewable, bio-based feedstocks is one route to improve the sustainability of these products. "Green polyols" that include polyols based on vegetable oils and polycarbonate polyols that incorporate CO_2 into the propylene oxide chain represented approximately 13% of the global $19 billion polyol market in 2015 [49]. The global polyol market is expected to exhibit the same strong growth as the PUR market, while the share of green polyols is expected to increase, indicating that bio-based polyols will represent an enormous opportunity for new growth.

A diverse range of polyols are currently used in PUR applications, some of which can be bio-based (e.g. propylene glycol, ethylene glycol, BDO). Commonly bio-based polyols are a rapidly growing source of polyols and currently include compounds derived from triglycerides and sugars such as glucose, sucrose, sorbitol, and glycerol that can be used as initiators for synthesis of polyether polyols. Lignins derived from chemical pulping processes and biorefining processes are promising candidates for use in PUR synthesis (generally as rigid aromatic polyol) if well-established challenges can be overcome. In addition to increasing the bio-based content of the PUR products, lignin incorporation into various PUR products has been shown in some cases to provide performance advantages that include enhanced crosslinking density, improved biodegradability, increased ultraviolet (UV) stability, antioxidant properties, and improved mechanical strength and thermal stability of the final product.

Bio-based diisocyanates, particularly aromatic ones, seem to be more challenging to produce. However, bio-based aliphatic diisocyanates such as penta-methylene diisocyanate have been recently commercialized to enable 100% bio-based PURs [50].

4.3.5 CASE (Coatings, Adhesives, Sealants, Elastomers)

CASE is a category of materials that includes coatings, adhesives, sealants, and elastomers. Across a range of industries, CASE materials improve the performance, operation, resistance, and appearance of products. When used in place of conventional fastening methods, CASE products provide exceptional durability for long-term use.

1) Coatings are materials that are applied onto another material or substrate. Depending on their intended purpose coatings can be functional and/or decorative. Functional coatings protect the substrate and improve the performance of the final product. In some cases, the surface properties of the product can be changed. Coatings can be applied on a wide variety of substrates such as paper, film and foil, wood and metal. Typical coating polymers are based on acrylate and vinyl ester monomers that can be polymerized on the substrate by radical initiators. The polarity of the polymer needs to match with that of the substrate for good adhesion.
2) Adhesives are somewhat similar to coatings. They contain a binding agent to hold multiple substrates together. Adhesives are also known as glue, cement, or paste. They can bind substrates of different materials effectively and can distribute stress evenly across joints. Examples of adhesives are polyacrylate and polyvinyl acetate, polyethylene-vinyl acetate, polychloroprene, and isoprene rubber (IR).
3) Sealants are applied to prevent fluids from passing through surfaces, joints, or openings in materials. In construction industries, sealants can block the flow of dust, heat, or sound. Sealants are applied to almost any type of material. Some also have adhesive properties. Sealants have lower strength and greater elongation (i.e. a more rubber-like) than adhesives, which can be important for certain applications. Typical sealants are dense (non-foamed) PUR and silicone.
4) An elastomer or a rubber is a polymer with viscoelasticity (i.e. having both viscosity and elasticity). They generally have a low Young's modulus and high failure strain when compared with other materials. A rubber is a thermoset which is obtained after vulcanization (cross-linking). This cross-linking allows them to be used above their glass transition temperature to allow significant molecular reconformation, without breaking any covalent bonds. Therefore, at ambient temperature, rubbers are relatively deformable. The primary use is for seals, adhesives, and molded flexible parts. The application areas for various types of rubber are diverse. They cover segments such as tires, soles for shoes, grips for tools, and in damping and insulating elements. The polymers that are most frequently used for elastomers and rubbers are IR (Isoprene Rubber), BR (butadiene rubber), SBR , NBR (nitrile–butadiene rubber), EPM rubber (ethylene–propylene monomer), and EPDM rubber. When polymerized in 1,2-position the diene monomers result after polymerization in a dangling unsaturated side-chain that can be reacted with sulfur for cross-linking (vulcanization). The global elastomer production volume is around 40 MT/a.

Coatings and adhesives offer opportunities for the application of bio-based polymers. Acrylate and vinyl-esters can be produced from sugars. Lignin has also been used as constituent of phenolic resins, e.g. in wood glue. There is also scope for bio-based rubbers. Of course, of the more than 28 Mt/yr of rubber produced in 2017, almost 50% was natural rubber. Natural rubber is an isoprene rubber. Finally, bio-based ethylene and propylene can be used to produce (partly) bio-based EP(D)M. Plastomers are ethylene with α-olefin (1-hexene or 1-octene) copolymers, which also can be (partially) bio-based.

4.3.6 Composites

A composite material is made from two or more constituent materials that show significantly different physical and/or chemical properties. When combined, the resulting composite has characteristics different from the individual components. The individual components remain separate and distinct within the finished structure. This differentiates composites from mixtures (blends) and from solid solutions. In this paragraph we consider mainly the fiber-reinforced plastics. The polymer matrix is in this case typically reinforced with fibers such as glass (fiberglass), carbon (carbon fibers), aramid, or basalt. Other fibers such as paper, wood, or asbestos have also been used. The polymer that is used is typically a thermoset such as epoxy, vinyl ester, or unsaturated polyester. In addition, phenol formaldehyde resins are also still in use. Fiber-reinforced polymers are commonly used in the aerospace, automotive, marine, and construction industries. They can also typically be found in ballistic armor. Fiber-reinforced thermoplastic composite based on PET or PVC are also growing applications because they are maintenance-lean alternatives to wood, such as wall panels or terraces.

The strength of composites strongly depends on the strength of the interface between the matrix and fibers. Much effort has been invested in modification of the matrix and/or the fibers to allow the formation of a stable dispersion of the fibers and, eventually, a secure chemical linkage between the two phases.

Bio-based composites are fiber-reinforced materials that are produced partly or completely from renewable raw materials. Both the matrix and the fiber reinforcement can be bio-based. Such composites are typically used in the construction industry, in the automotive sector and for electronics casings. The first-generation bio-based composites had various disadvantages in processing and also with regards to costs. This delayed their market introduction. However, with the increasing availability of bio-based resins and fibrous materials for composite applications, the quality and applicability of the bio-based composites has significantly increased over recent years.

Traditional fiber-reinforcing materials such as glass can also be replaced by bio-based fiber reinforcements such as natural fibers (flax, hemp, wood), viscose, and PLA. These bio-based materials can be used as long or short fibers, as non-woven fibers, and as fabrics. An important benefit of the use of bio-based fibers in composites is they can be used in thermal recycling. Bio-based plastics resins can be used for their specific properties because of their environmental benefits, including a reduction in CO_2 emissions.

4.4 End-of-Life and Biodegradation

While polymers have been mainly designed for the cost and performance required for the application, as discussed above, we increasingly recognize the need to include responsible end-of-life in the design as well. According to the waste hierarchy, we need to design the object for reuse, if not for mechanical recycling or for chemical recycling, and otherwise for energy recovery or biodegradation. These options will be discussed at length in a subsequent chapter (Chapter 12 recycling), and biodegradation will be subject of a dedicated chapter (Chapter 2 on biodegradation). We nevertheless wish to briefly address the end-of-life of the product here and relate it to the applications discussed above and refer to the dedicated chapter of this book for more detailed discussion and proper references.

4.4.1 Reuse and Recycling

Reuse and (mechanical) recycling require collection and sorting/cleaning of well-defined and homogeneous polymer products. These conditions are most easily met for rigid materials of a significant size, e.g. as encountered solid packaging such as bottles. Other applications may be more challenging, however.

Films are often contaminated and may come as multilayer materials. Technologies are being developed to disentangle the various films. However, these processes are expensive. Pyrolysis may be more cost effective after all.

Foams are prohibitive to transport due to their extremely low density and their bulkiness. Mechanical recycling may be limited to use as cheap filler. Depolymerization may be more promising.

Fibers in textiles (and carpets) are often used in combination of various interwoven fibers types. Recycling is therefore very challenging.

Finally, CASE materials may represent the highest recycling challenge of all. They are generally used in small amounts and are intimately bound to other materials. Collection and sorting seem an impossible task.

When all else fails, the residual materials that cannot be recovered in sufficient quality and quantity need to be incinerated. Here, the heterogeneity of the materials does not really matter anymore. Or should it be subjected to biodegradation?

4.4.2 Biodegradation

Biodegradation is the breakdown of organic matter by microorganisms, such as bacteria and fungi. When carried out aerobically, i.e. in the presence of oxygen as is the case for composting and litter, it produces mostly carbon dioxide, water, and new biomass, e.g. as growing fungi [51]. When carried out anaerobically, i.e. in absence of oxygen as in a sealed off landfill or in a biodigester, it produces biogas (an equimolar mixture of CH_4 and CO_2) together with growing biomass. The anaerobic digestion can be considered as a form of energy recovery since it delivers a fuel after all. In contrast, the aerobic digestion is just waste destruction as it destroys the chemical and energetic content of the waste without other benefit that producing a low-value "natural fertilizer" (compost). It therefore earns its last position in the waste hierarchy. But more on this residual value later.

In practice, almost all chemical compounds and materials are subject to biodegradation processes. The significance, however, is in the relative rates of such processes, such as days, weeks, years, or centuries. Several factors determine the rate at which this degradation of organic compounds occurs. Some factors are related to the polymeric structure such as composition, monomer distribution, crystallinity, molecular weight. But others are equally important and depend on environmental conditions such as light, water, oxygen, and temperature, which play a role.

Plastics that are biodegradable can be classified by being home compostable or industrially compostable. Industrial composting conditions require elevated temperature (55–60 °C) combined with a high relative humidity and the presence of oxygen. These conditions are more optimal compared to other everyday biodegradation conditions in soil, surface water, or marine water. At home-composting conditions, the temperature is lower and less constant compared to industrial composting. As a result, the degradation rate of a material is (much) slower compared to industrial composting or differs in duration of the lag phase due to the slower microbial enrichment of degrading microorganisms.

Even if we would design all future plastic materials for re-use and recycle (which is the EU plan for 2050), human behavior results in unplanned and undesired leakage of plastic into our environment, typically into our oceans. It is estimated that yearly we lose about eight million tons of plastics (or eight large football stadiums filled to the roof; assuming 1 ton/m^3) to the environment. Because of this, biodegradation can be very important for a plastic material's fate-in-nature, particularly for applications that are prone to losses such as small-size packaging or textiles microfibers. In this respect, the advantage of biodegradable/compostable plastics is that the materials that end up in nature do not endlessly accumulate as do conventional materials that often take centuries to break down.

Beyond environment preservation, biodegradation can also be valuable for specific application, e.g. by omitting the need to remove the product after use. This, e.g. applies to chirurgical materials or agricultural mulch films.

Since all materials degrade when given enough time, we need to come up with certain rules on how to classify and label a specific plastic material for "claiming" a certain biodegradability. Certain organizations have highjacked the term "biodegradable plastic" for materials that degrade as fast or faster than cellulose. Consequently, degradable materials are not a "biodegradable plastic" when it degrades slower than cellulose. Such definition is attractive as it removes the variation of "environmental factors." Nevertheless, in her recently published plastics strategy, the European Union is close to proposing standardized rules for defining and labeling biodegradable and compostable plastics to allow for accurate sorting and avoid false environmental claims (A European Strategy for Plastics in a Circular Economy, EU, 2018).

4.5 Conclusions

Within a few decades after their discovery, polymer materials have penetrated nearly all facets of our daily life. They can thank their success to a combination of valuable properties such as light weight, strength, and elasticity, as well as transparency or barrier properties. Most importantly, however, is the ease to produce polymeric materials in all possible forms, varying from tubes to bottles, to bags, fibers, foams, rings, coatings, complex molded shapes up to 3D-printed devices. This unique set of properties make them irreplaceable in the future.

However, the success of polymers is also leading to challenges, i.e. environmental challenges posed by wastes produced during their manufacture or by irresponsible disposal of spent product. We are not talking only about plastic litter here but also refer to climate change. The CO_2 emissions related to polymer manufacture and eventual incineration at end-of-life could indeed exceed those of the energy sector at some point in the second half of this century. We therefore need to revisit the way we produce and use plastics, as argued by various organizations, e.g. by the Ellen MacArthur Foundation [52]. This implies recycling spent plastics and switching from fossil to renewable carbons as feedstock for the industry in addition to using renewable energy to power all processes.

This chapter has tried to briefly review the manufacture and properties of today's main polymers to provide some guidance for research toward bio-based and biodegradable polymers. We hope it will help designing novel polymers, e.g. polyesters, which can easily and efficiently be made from bio-based feedstock and can eventually replace today's fossil polymers by providing a comparable or improved combination of properties.

4.A Definitions: Biopolymer vs. Bio-based Polymer and Relation to Biodegradation

Consumers are confused by the wealth of terminology around plastics and their inconsistent use. They assume that bio-based polymers are biodegradable and cherish it without understanding it, because biodegradability is considered a positive attribute [53]. To take away confusion, we first need to get our definitions right.

Let us look at the definition for ***biopolymer***. According to the Merriam-Webster dictionary, a biopolymer is "*a polymeric substance (such as a protein or polysaccharide) formed in a biological system*" and according to Wikipedia biopolymers are "*polymers produced by living organisms.*" Examples of biopolymers are DNA, insulin, starch, and cellulose. In specific cases, organisms can be "tuned" to produce specific biopolymers such as polyhydroxyalkanoates (PHAs). The biopolymers are under most general conditions also biodegradable (see definition below), with lignin being a notorious exception.

But we also have synthetic (man-made) polymers that can be made from bio-based feedstock. Most of these building blocks (monomers) are derived from sugars. Some examples are lactic acid, succinic acid, ethylene glycol, bio-ethylene, FDCA, and isosorbide. We call the resulting polymers **bio-based** polymers. Examples for this category are polylactic acid (PLA), PBS, (PET; only MEG bio-based), PEF, bio-polyethylene (bio-PE) and many others. Accordingly, biopolymers are a sub-set of the larger family of bio-based polymers.

In analogy a **bio-plastic** is a plastic derived from a biopolymer. PHAs and (modified) starches are examples. Similarly, bio-based plastics are plastics derived from man-made polymers obtained (partly) from bio-based monomers. Bio-PE, PLA, PEF, and PBS are examples for this class. However, unfortunately, these plastics are often also called bioplastics as organizations such as European bioplastics made up a new definition: bio-plastics are bio-based, biodegradable, or both. As a result, adding to the confusion, they even call fossil-based biodegradable polymers such as polycaprolactone (PCL) bioplastics (Figure 4.7).

This being, hopefully, clarified, we now need to include biodegradation into this discussion (Figure 4.8). We are now moving away from the source of the polymer to considering its fate in the environment, a property that is related to the chemical structure of the polymer and not to its source. As discussed earlier in this chapter, many bio-based polymers (bio-made and man-made) undergo biodegradation, some rapidly (e.g. PHA), other more slowly (e.g. PLA). However, fossil-based polymers of similar structure also undergo biodegradation, as observed for the polyester PCL, a C6 cousin of the C4 PHA and C3 PLA. For the very same reason, bio-based polymers may not undergo biodegradation at any significant rate. This is the case for bio-PE that shows the same structure as its fossil cousin and degrade at the same rate.

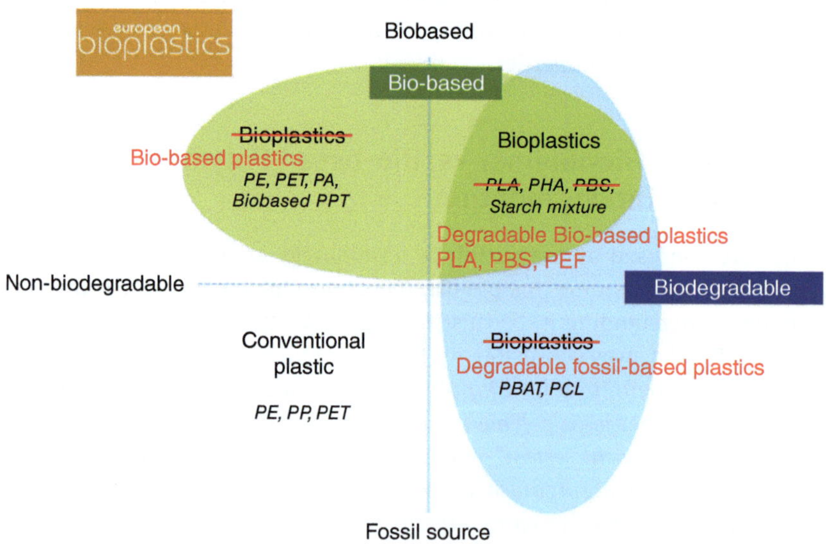

Figure 4.7 Scheme developed by European Bioplastics (www.european-bioplastics.org) and improved with suggestions in red to avoid more confusion going forward.

Figure 4.8 Plethora of environmental claims and labels.

List of Polymers

ABS	acrylonitrile–butadiene–styrene
BR	butadiene rubber; N/S BR = nitrile/styrene BR
CA	cellulose acetate
EPDM	ethylene–propylene–diene monomer rubber
PA6	polyamide 6 (polycaprolactam)
PA66	polyamide 66 (polyadipicacid hexamethylenediamine)
PBS	polybutylene succinate
PC	polycarbonate
PE	polyethylene; HD/LD/LLD PE = high/low/linear-low-density PE
PEF	polyethylenefuranoate
PET	polyethyleneterephthalate
PHA	polyhydroxyalkanoate
PLA	polylactic acid
PMMA	polymethylmetacrylate
POM	polyoxomethylene
PP	polypropylene
PS	polystyrene; EPS = expanded PS
PUR	polyurethane
PVC	polyvinylchloride
SBR	styrene butadiene rubber
SBS	styrene butadiene styrene

References

1 Valencia, R.C. (2013). *The Future of the Chemical Industry by 2050*. Weinheim: Wiley-VCH.

2 PlasticEurope (2016). *Plastics – The Facts 2016, An Analysis of the European Plastics Production, Demand and Waste Data*. European Bioplastics.

3 International Energy Agency (IEA) (2018). *The Future of Petrochemicals, Towards More Sustainable Plastics and Fertilizers*. Plastics Europe.

4 European Bioplastics. Bioplastics Market data 2021. https://www.european-bioplastics.org/market/ (accessed 31 January 2022).

5 Ahmada, S., Wong, K.Y., Tseng, M.L., and Wong, W.P. (2018). Sustainable product design and development: a review of tools, applications and research prospects. *Resources, Conservation and Recycling* 132: 49–61.

6 Jambeck, J.R., Geyer, R., Wilcox, C. et al. (2015). Plastic waste inputs from land into the oceans. *Science* 347: 768.

7 Law, K.L. (2017). Plastics in the marine environment. *Annual Review of Marine Science* 9: 205.

8 Wright, S.L. and Kelly, F.J. (2017). Plastic and human health: a micro issue? *Environmental Science and Technology* 51: 6634–6647.

9 Yong, C.Q.Y., Valiyaveetill, S., and Tang, B.L. (2020). Toxicity of microplastics and nanoplastics in mammalian systems. *International Journal of Environmental Research and Public Health* 17: 1509.

10 Lange, J.-P. (2002). Sustainable development: efficiency and recycling in chemicals manufacturing. *Green Chemistry* 4 (6): 546–550.

11 Lange, J.-P. (2010). Resource efficiency of chemical manufacturing chains: present and future. In: *Sustainable Development in the Process Industry – Cases and Impacts* (ed. J. Harmsen and J.B. Powell), 23–38. Wiley.

12 ICCA/IEA/DECHEMA Roadmap Catalysis (2013). Petrochemical Industry Energy & GHG Savings via Catalysis – Still a Large Opportunity. https://dechema.de/en/industrialcatalysis.html (accessed 17 November 2020).

13 Olivier J. G. J., K.M. Schure, J.A.H.W. Peters, et al. (2017). Trends in Global CO_2 and Total Greenhouse Gas Emissions: Summary of the 2017 report. PBL Netherlands Environmental Assessment Agency.

14 United Nations Climate Change. What is the Paris Agreement? https://unfccc.int/process-and-meetings/the-paris-agreement/the-paris-agreement (accessed 31 January 2022).

15 Lange, J.-P. (2016). Catalysis for biorefineries – performance criteria for industrial operation. *Catalysis Science and Technology* 6: 4759–4767. https://doi.org/10.1039/C6CY00431H.

16 https://apps.fas.usda.gov/psdonline/circulars/oilseeds.pdf (accessed 30 December 2021).

17 Hill, K.-H. (2007). Industrial development and application of biobased oleochemicals. In: *Catalysis for Renewables, from Feedstock to Energy Production* (ed. G. Centi and R.A. van Santen), 75–99. Weinheim: Wiley-VCH.

18 IEA Bioenergy. Bio-Based Chemicals, A 2020 Update. https://ieabioenergy.com/wp-content/uploads/2020/02/Bio-based-chemicals-a-2020-update-final-200213.pdf (accessed 31 January 2022).

19 Wolf, V. and Haß, M. World markets for sugar and starch: status and prospects. In: *Encyclopedia of Sustainability Science and Technology* (ed. R.A. Meyers). New York, NY: Springer https://doi.org/10.1007/978-1-4939-2493-6_997-1.

20 Lange, J.-P. (2007). Lignocellulose conversion: an introduction to chemistry, process and economics. In: *Catalysis for Renewables: From Feedstock to Energy Production*, vol. 238(1) (ed. G. Centi and R.A. van Santen), 21–51. Weinheim: Wiley-VCH.

21 Dusselier, M., Mascal, M., and Sels, B.F. (2014). Top chemical opportunities from carbohydrate biomass: a chemist's view of the biorefinery. *Topics in Current Chemistry* 353: 1–40. https://doi.org/10.1007/128_2014_544.

22 Isikgor, F.H. (2015). Lignocellulosic biomass: a sustainable platform for the production of bio-based chemicals and polymers. *Polymer Chemistry* 6: 4497–4559. https://doi.org/10.1039/C5PY00263J.

23 Munoz de Diego, C., Schammel, W.P., Dam, M.A., and Gruter, G.J.M (2011). Method for the preparation of 2,5-furan dicarboxylic acid and esters thereof. European Patent EP2486027B1. Priority date: 7 October 2009.

24 van der Waal, J.C., Gruter, G.J.M, Claassens-Dekker, P. (2016). Process for preparing ethylene glycol from a carbohydrate source. European patent EP3245183B1. Priority date: 13 January 2015.

25 Kolstad, J.J., Gruter, G.J.M., Dam, M.A. et al. (2015). Polyester and method for preparing such a polyester. European patent EP3116934B1. Priority date: 11 March 2014.

26 Gallagher, D., Parker, D., Allen, D.J., and Tsesmetzis, N. (2018). Dynamic bacterial and fungal microbiomes during sweet sorghum ensiling impact bioethanol production. *Bioresource Technology* 163–173.

27 Lange, J.-P., Vestering, J.Z., and Haan, R.J. (2007). Towards 'bio-based' nylon: conversion of c-valerolactone to methyl pentenoate under catalytic distillation conditions. *Chemical Communications* 3488–3490.

28 Lange, J.-P., Price, R., Ayoub, P.M. et al. (2010). Valeric biofuels: a platform of cellulosic transportation fuels. *Angewandte Chemie International Edition* 49: 4479–4483.

29 Pramod, V.C., Fauziah, R., Seshan, K., and Lange, J.-P. (2018). Bio-based acrylic acid from sugar via propylene glycol and allyl alcohol. *Catalysis Science and Technology* 8: 289–296.

30 Chewter, L.A., Huizenga, P., Lange, J.P. et al. (2017) Process for the preparation of glycols. WO2017097847A1 (Shell Int. Res. Maatschapij B.V. Priority date: 09 December 2015).

31 Lange, J.-P. and Wadman, S.H. (2020). Furfural to 1,4-butanediol/tetrahydrofuran – a detailed catalyst and process design. *ChemSusChem* 13: 5329–5337.

32 Xu, J. and Guo, B. (2010). Poly(butylene succinate) and its copolymers: research, development and industrialization. *Biotechnology Journal* 5: 1149.

33 Bastioli, C., Magistrali, P., and Garcia, S.G. (2014). Starch. In: *Bio-based Plastics, Materials and Applications* (ed. S. Kabasci), 9–33. Chichester: Wiley.

34 Ganster, J. and Fink, H.-P. (2014). Cellulose and cellulose acetate. In: *Bio-based Plastics, Materials and Applications* (ed. S. Kabasci), 35–62. Chichester: Wiley.

35 Kuhad, R.C. and Singh, A. (1993). Lignocellulose biotechnology: current and future prospects. *Critical Reviews in Biotechnology* 13: 151–172.

36 Lange, J.-P., Lewandowski, I., and Ayoub, P. (2010). Cellulosic biofuels: a sustainable option for transportation. In: *Sustainable Development in the Process Industry – Cases and Impacts* (ed. J. Harmsen and J.B. Powell), 171–208. Wiley.

37 Galbe, M. and Wallberg, O. (2019). Pretreatment for biorefineries: a review of common methods for efficient utilisation of lignocellulosic materials. *Biotechnology for Biofuels* 12: 294.

38 De Wild P.J., Huijgen W.J.J., van der Linden R., den Uil H., Snelders J., Benjelloun-Mlayah B. Organosolv fractionation of lignocellulosic biomass for an integrated biorefinery. ECN Report Feb 2015. https://publicaties.ecn.nl/PdfFetch.aspx?nr=ECN-V--14-010.

39 Zhu, S., Wu, Y., Chen, Q. et al. (2006). Dissolution of cellulose with ionic liquids and its application: a mini-review. *Green Chemistry* 8: 325.

40 van den Bosch, S., Schutyser, W., Vanholme, R. et al. (2015). Reductive lignocellulose fractionation into soluble lignin-derived phenolic monomers and dimers and processable carbohydrate pulps. *Energy and Environmental Science* 8: 1748.

41 Liao, Y., Koelewijn, S.-F., Van den Bossche, G. et al. (2020). A sustainable wood biorefinery for low–carbon footprint chemicals production. *Science* https://doi.org/10.1126/science.aau1567.

42 Glasser, W.G. and Sarkanen, S. Lignin, properties and materials. *Joint meeting of the 195th National Meeting of the American Chemical Society and the 3rd Chemical Congress of North America*, Toronto, Canada (5–11 June 1988).

43 Hu, T.Q. (2002). *Chemical Modification, Properties, and Usage of Lignin*. Springer.

44 Ernsting, A. and Smolker, R. (2018) Dead End Road: the False Promise of Cellulosic Biofuels. Biofuelwatch, September 2018. www.biofuelwatch.org.uk/wp-content/uploads/Cellulosic-biofuels-report-2.pdf (accessed 30 December 2021).

45 Avantium press release. Avantium to open pilot biorefinery. https://www.avantium.com/press-releases/avantium-to-open-pilot-biorefinery-sights-already-set-on-commercial-scale-up/ (accessed 31 January 2022).

46 Botella, C., Zhang, K., Baugh, A. et al. (2019). Reversible acid pretreatment scale up studies for the production of cellulosic ethanol from ensiled sweet sorghum. *Biochemical Engineering Journal* 150: 107266.

47 van der Vegt, A.K. (1991). *Polymeren, van keten tot kunstof*. Delft University Press.

48 Dolmans, R. RWTH Aachen 2013. PhD thesis, (University: RWTH Aachen). https://publications.rwth-aachen.de/record/229253 (accessed 30 December 2021).

49 Adhesives and sealants Industry (2017). Green and Bio Polyols Market Worth $4.71 Billion by 202 https://www.adhesivesmag.com/articles/95408-green-and-bio-polyols-market-worth-471-billion-by-2021 (accessed 30 December 2021).

50 Golling, F.E., Pires, R., Hecking, A. et al. (2019). Polyurethanes for coatings and adhesives–chemistry and applications. *Polymer International* 68: 848–855.

51 ITEH STANDARDS CEN/TR 15351:2006 (2006). Plastics - Guide for vocabulary in the field of degradable and biodegradable polymers and plastic items.

52 World Economic Forum (2016). *The New Plastics Economy – Rethinking the Future of Plastics.* Ellen MacArthur Foundation/McKinsey & Company http://www.ellenmacarthurfoundation.org/publications (accessed 30 December 2021).

53 Zwicker, M.V., Brick, C., Gruter, G.-J.M., and van Harreveld, F. (2021). (Not) Doing the right things for the wrong reasons: an investigation of consumer attitudes, perceptions, and willingness to pay for bio-based plastics. *Sustainability* 13: 6819.

5

Condensation Polyesters

Jules Stouten and Katrien V. Bernaerts

Maastricht University, Aachen-Maastricht Institute for Biobased Materials (AMIBM), Faculty of Science and Engineering, Brightlands Chemelot Campus, Urmonderbaan 22, Geleen, 6167 RD, The Netherlands

5.1 Introduction

Polyesters can be referred to as macromolecules in which the main chain segments are repetitively linked by ester units. This excludes polymers that contain ester linkages within the side groups of the repeating unit such as poly(vinyl acetate) and poly(meth)acrylates [1]. As will be discussed later, the main chain ester linkages play a key role in the biodegradability of polyesters. Within the polyester chain, a great variety exists with respect to the repeating units that are used, which includes linear aliphatic polyesters with varying spacer length (e.g. poly(butylene succinate) [PBS]), semi-aromatic polyesters containing at least one aromatic and one aliphatic unit (e.g. poly(ethylene terephthalate) [PET]), or wholly aromatic polyesters (e.g. poly(4-hydroxybenzoic acid)).

The condensation polyesters are one of the oldest classes of synthetic polymers. The first group of synthetic polyesters was the alkyds, which were commercially developed between 1910 and 1915 by General Electric Company [2]. Notably, a resin was obtained from the condensation reaction between glycerol and phthalic anhydride. Later in the twentieth century, in 1928, W.H. Carothers started his research toward condensation polyesters at DuPont. For the first time, a linear polyester was obtained from octadecanedioic acid and 1,3-propanediol with a molecular weight of 12 000 g/mol, and at the time, it was called a "super polyester." [3] The improvement in molecular weight was significantly higher than the previously obtained molecular weights of between 400 and 5000 g/mol. The research group of Carothers continued the work on (mainly aliphatic) polyesters, but this did not lead to any commercial development at the time. Later, the incorporation of terephthalic acid for the production of semi-aromatic polyesters was further investigated, which led to the discovery of PET fibers [4]. At the same time, other polyesters containing terephthalic acid and glycols with various spacer lengths were developed. Since then, large developments were made in the field of polyesters, and they are a prevalent class of polymers in the current plastics market. Still, nowadays, polyesters are continuously

Biodegradable Polymers in the Circular Plastics Economy, First Edition.
Edited by Michiel Dusselier and Jean-Paul Lange.

studied by scientists because a significant fraction can be considered as biobased or biodegradable polymers. This is now relevant more than ever with the concerns regarding the net increase in CO_2 emissions and plastic waste management, which are accompanied by the consumption of plastics.

The history of biodegradable plastics is quite recent, but (condensation) polyesters have played a major role so far. The first commercial biodegradable condensation polyesters were launched in the beginning of the 1990s. These were PBS and copolymers thereof and were sold under the trade name Bionolle®. In the late 1990s, biodegradable aromatic condensation polyesters based on terephthalic acid copolymers with high melting temperature were developed and later sold under the trade names Ecoflex®, Eastar Bio®, Origo-Bi®, and Biomax®. Most of the developments in the discovery of biodegradable condensation polyesters have been made in the 1990s and since then has focused on introducing new biobased and/or functional monomers, polymer blends with other biodegradable polymers, composites, and the copolymer composition–property relationship to achieve the optimal balance between the physical performance and degradability.

Nowadays, commodity plastics such as poly(propylene) and poly(ethylene), which make up the majority of the market, persist for many years in nature after disposal because of their non-degradability. Biodegradable plastics only make up a small fraction, approximately 0.5% of the total plastic production [5]. They mainly find applications in products where biodegradability is one of the main desired features, such as biomedical, pharmaceutical, packaging, and agricultural industries. However, with the current surge in interest in the circular economy and waste management, the biodegradable plastic market is expected to grow steadily in the coming years. A recent report of the Nova Institute predicted that the total production of biodegradable plastics will grow from 1.17 mt in 2019 to 1.33 mt in 2024 [5].

Another reason for the growing interest in condensation polyesters is their renewable characteristic. Many monomers that make up polyesters can also be derived from renewable resources, especially the aliphatic ones. On the other hand, investigations toward obtaining bio-derived terephthalic acid from renewable resources are currently ongoing [6].

In this chapter, we will elaborate more on the most important types of biodegradable condensation polyesters that are currently available and their synthesis. Furthermore, the most important mechanism of their degradation is further explained. This is followed by a section that addresses the currently available applications for biodegradable condensation polyesters. A brief discussion regarding the recycling of polyesters together with the concluding remarks is provided at the end of the chapter.

5.2 Preparative Methods

Condensation polyesters are prepared by condensation polymerization, which is a type of step growth polymerization. This involves the reaction between two molecules to form dimers, which in term react with more molecules to form trimers, tetramers, and so forth, eventually forming large structures. This occurs under the release of small molecules as a by-product, also known as the condensate. Condensation polymerization requires a constant supply of heat and optionally a vacuum in order to evaporate and distill the low boiling condensate. Reaction completion is required because of the relationship between molecular weight and conversion in step growth polymers. This relationship was first described by Carothers (Eq. (5.1)) [7].

$$\overline{X}_n = \frac{1 + r}{1 + r - 2rp} \tag{5.1}$$

The number-average degree of polymerization X_n is defined as the average number of structural units per polymer chain, r is the stoichiometric ratio of functional groups, and p is the conversion. Furthermore, in order to reach high molecular weights, the monomers should be present in stoichiometric balance. If one is added in excess, the theoretical attainable molecular weight decreases drastically.

The most common methods to produce aliphatic and semi-aromatic condensation polyesters are polycondensation and transesterification polycondensation. In a typical polyester synthesis, two bifunctional monomers containing functionalities AA and BB are reacted together in an AABB-type polycondensation. This usually involves a diacid and a diol releasing water as the condensate (Figure 5.1). Alternatively, it is also possible to obtain polyesters from AB-type monomers, which contain both an alcohol and a carboxylic acid group. Upon self-condensation, they form poly(hydroxy acids) of ABAB type. To speed up the reaction, homogeneous esterification catalysts such as antimony, titanium, tin, zirconium, and zinc are typically used [8–13].

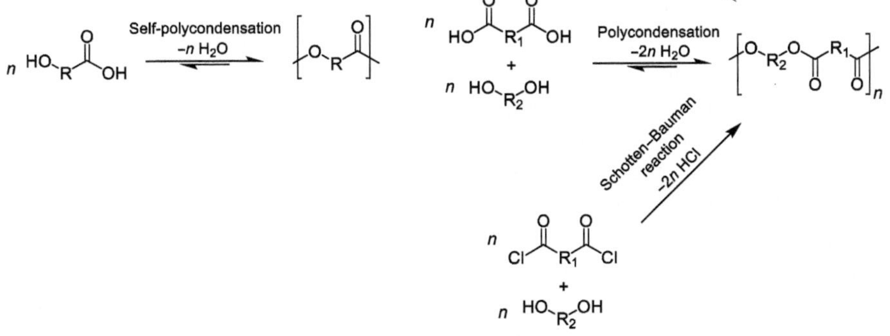

Figure 5.1 The main routes to obtain synthetic condensation polyesters.

The functionality of the monomers involved plays an important role in the outcome of the reaction. With the addition of a trifunctional or a higher functional monomer, a cross-linked network is obtained. When a monofunctional monomer is added, the molecular weight is reduced because the growing chain end is terminated. Therefore, the molecular weight and the cross-linking density depend on the functionality of the monomer and the ratio of each of the monomers in the polymerization.

Another polycondensation method is transesterification polycondensation. In this method, a diol is reacted with a diester. Usually, methyl or ethyl esters of the corresponding diacids are used (Figure 5.1). Transesterification polycondensation can have several advantages, for example, more facile removal of the condensate because of the higher volatility of methanol and ethanol compared to water, or instability of one of the monomers in the presence of carboxylic acid. The most prevalent example of transesterification polycondensation is the reaction between dimethyl terephthalate and ethylene glycol to produce PET.

Besides conventional metal-based catalysts, other catalytic methods can be employed, for example, enzyme-catalyzed polycondensation. In these methods, the reaction temperatures are low compared to melt polycondensation, but the reaction is generally performed in solvents. As catalysts, enzymes are used, which makes this procedure suitable for thermally or chemically unstable monomers. In the case of aromatic polyesters, usually, other types of polycondensation methods are employed. They are typically synthesized by reacting a diacylchloride with a diol releasing hydrogen chloride or by acetate exchange with aromatic diacids and phenyl esters. These reactions are performed in solutions and in bulk respectively and require modification of the monomer before polymerization [14, 15]. However, wholly aromatic polyesters are non-biodegradable and are therefore beyond the scope of this chapter [16].

5.3 Biodegradation of Polyesters

In general, biodegradation of polymers is achieved by scission of the main chain into progressively smaller segments and eventually undergoing decomposition into carbon dioxide, water, methane, and inorganic compounds. The latter could be as a result from the remaining catalysts or additives. In other words, long macromolecular chains are broken down to oligomers and then to monomers, which are then further metabolized by microbes. The whole process should be (at least partially) a result from the enzymatic action of microbes, such as bacteria or fungi. The most prevalent degradation mechanisms for polyesters are hydrolytic degradation and enzymatic degradation. This is because chain scission occurs at the ester bonds, which are susceptible to both hydrolysis and enzymatic attack. This results in the creation of a carboxylic acid and an alcohol and reduces the molecular weight, ultimately leading to the failure of the structural integrity of the plastic material.

The mobility of polyester chains is considered as one of the most important features that determines biodegradability [17]. This, in term, is correlated with a variety of polymer properties such as glass transition temperature (T_g), molecular weight, melting temperature, and degree of crystallinity. For example, most condensation polyesters are semicrystalline, which indicates that they are composed of crystals connected by amorphous segments. In degradation experiments, it was shown that first the amorphous segments are biodegraded because they possess higher mobility, especially above the T_g of the polymer [18]. For most aliphatic polyesters, the T_g value is low (typically well below the room temperature) and are thus susceptible for degradation under natural conditions. A prevalent example of an aliphatic polyester with a high T_g of around 59 °C is poly(lactic acid) (PLA). PLA is extremely slow to degrade under natural conditions but can degrade rapidly in composting facilities where elevated temperatures (up to 65 °C) and high humidity are used [19]. Another important factor is the hydrophilicity of the polyester, which is determined by monomer type and end group concentration. Hydrophilicity promotes biodegradability by increasing the interaction with the surrounding water molecules, which are essential in the most important modes of biodegradation [20]. The amount of end groups in the polymer is related to the molecular weight. Biodegradability can also be influenced by the synthesis procedure and processing method because these affect some polymer properties. For example, a relatively low molecular weight is obtained for condensation polyesters (typically 10^4–10^5 g/mol), whereas for polyesters obtained via ring opening polymerization (ROP), a higher molecular weight is typically obtained (10^5–10^6 g/mol). The molecular weight and the synthesis method influence properties such as crystallinity and end group concentration and monomer type.

As such, many commercial biodegradable polyesters are copolymers consisting of multiple types of monomers. The biodegradability of a copolyester can be tuned by incorporating monomers with various levels of rigidity and hydrophilicity. An additional concern that comes along is the degradability and toxicity of the liberated monomers. In general, aliphatic naturally occurring monomers, such as lactic acid, possess low toxicity for their environment [21]. On the other hand, aromatic monomers are typically more harmful than the aliphatic ones and persist for a longer time in the environment [22].

5.3.1 Hydrolytic Degradation

Hydrolytic degradation is the scission of macromolecular chains and accompanying lowering of the molecular weight because of the reaction with water molecules. In this process, H_2O is consumed, and as a result, chain scission occurs. The synthesis of a polyester is a reversible process. This indicates that in the presence of water, ester bonds can undergo reversible reaction into a carboxylic acid and an alcohol. This reversibility is a slow process at ambient temperatures. The susceptibility of ester bonds to undergo scission by hydrolysis is governed by the charge value of the reacting carbon [23]. Because strong aromatic compounds are able to delocalize the charge value on the involved carbon bonds, resistance to hydrolytic degradation is increased. This is one of the reasons that aromatic polyesters are less prone to undergo biodegradation than aliphatic polyesters.

Two main modes of hydrolytic degradation exist: bulk degradation and surface erosion. During bulk degradation, water is able to migrate into the polymer and cleave the ester bonds homogeneously throughout the material. However, in surface erosion, only macromolecular chains at the surface of the material are affected, indicating that the diffusion of water into the polymer is slower than the rate of bond hydrolysis [24]. This will have a heterogeneous breakdown of the macromolecular chains as a result. With the liberation and accumulation of carboxylic end groups upon bond cleavage, bulk autocatalytic hydrolysis occurs [25]. Several factors, such as the hydrophilicity, crystallinity, and geometry (thickness, surface area, porosity, and surface topology) of the polymer, are known to influence water diffusion [26].

5.3.2 Enzymatic Degradation

Enzymatic degradation of polyesters occurs in the presence of water, microorganisms, and their enzymes. Slightly stricter conditions are required for enzymatic degradation to take place in contrast to hydrolytic degradation, which is promoted in acidic or basic media and at high temperatures. Besides the presence of water, microbial degradation usually requires oxygen, nutrients, and mild temperatures. Furthermore, the process only occurs at the surface of the material because the large size of enzymes restricts them from diffusing in the polymer. For this, the enzyme needs to undergo an adsorptive binding to the substrate surface to start the degradation process [27]. Therefore, surface erosion is the only mechanism in the case of enzymatic degradation. Enzymatic hydrolysis of the polyester can proceed on the terminal ester bonds (exo-hydrolysis) resulting in monomers, or on the ester bonds within the macromolecular chain (endo-hydrolysis) resulting in oligomers and then monomers. The surface is continuously eroded, resulting in a linear mass loss [28]. Small macromolecular segments are expelled into the medium, which in term are further degraded and metabolized by other microbes present in the medium. With the slow decrease in mass because of degradation from the surface to the center of the material, other components present in the polyester are slowly expelled into the surrounding medium. That is why, these materials are particularly interesting for biomedical applications that require a controlled delivery process such as the release of drugs.

Similar to hydrolytic degradation, enzymatic degradation is affected by most of the features involving chain mobility and length. Furthermore, it is also influenced by the hydrophilic/hydrophobic balance in the main chain and molecular structure of the monomers [29]. A strong effect of the presence of aromatic groups exists. Aliphatic polyesters can be effectively degraded by lipases, but aromatic polyesters remain unaffected. Copolymers containing both aliphatic and aromatic monomers show biodegradability, which is mainly dictated by the amount of aliphatic units present. This is considered to be related to steric hindrance of the bulky aromatic units, which change the chain aspect ratio and limit the accessibility of the enzyme to the ester bond [30].

5.4 Aliphatic Polyesters

5.4.1 Poly(alkylene dicarboxylates)

Poly(alkylene dicarboxylates) are a family of aliphatic biodegradable polyesters. They are obtained from the polycondensation of diols such as ethylene glycol or butane diol with aliphatic dicarboxylic acids such as succinic acid and adipic acid. Typically, they exhibit low T_g because of the high conformational freedom of the aliphatic chains. However, they obtain their mechanical properties from their high crystallinity. The most prevalent examples are PBS and its copolyesters, and poly(ethylene adipate) (Figure 5.2).

PBS is obtained from the polycondensation of butane diol with succinic acid (or dimethyl succinate) and usually results in a molecular weight below 100 000 g/mol. Because PBS with a molecular weight below 100 000 g/mol is fairly brittle, with a low strain at break, higher molecular weights are usually desired to improve the ductility for processing. In order to achieve high molecular weights, chain extenders such as diisocyanates [31] and bisoxazoline [32] can be used during the synthesis. The chain extenders couple the chains by reacting with the end groups. This reaction is performed in a finalizing step to reach the desired high molecular weight.

Variations in the spacer length of the monomers result in different polymers such as poly(ethylene succinate) (PES), poly(butylene adipate) (PBA), and copolymers of PBS, namely, poly(butylene succinate-*co*-adipate) (PBSA) [33]. The nature of the comonomers that are used influences the properties and biodegradation rate [34]. For example, biodegradability improves when more adipic acid is incorporated because of the lower crystallinity and melting point [35].

Added value of PBS is its renewable character because the monomers can be synthesized from renewable resources. Both butane diol and succinic acid can be obtained from the fermentation of sugars [36]. In 2004, succinic acid was appointed by the US department of Energy (DoE) as one of the major biobased platform molecules [37]. However, currently, succinic acid is obtained from the hydrogenation of maleic anhydride, which is produced by oxidation of butane.

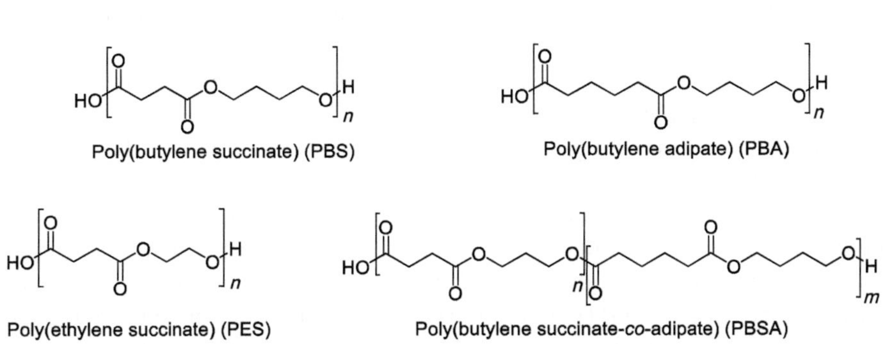

Figure 5.2 Chemical structures of some biodegradable poly(alkylene dicarboxylates).

PBS is a semicrystalline polymer with a melting temperature of around 90–120 °C and a T_g of between −45 and −10 °C [13, 34], fairly similar to those of poly(ethylene) and poly(propylene). Because of its high crystallinity, the biodegradation rate is relatively low, but PBS exhibits good mechanical properties. Furthermore, it has good processability, with film properties comparable to that of LDPE (low-density poly(ethylene)). The films of commercial PBS (Bionolle) have Young's modulus of about 0.5 GPa, tensile strength of 30 MPa, and strain at break of 710% [34]. The thermomechanical properties of PBS are superior to some other aliphatic polyesters, such as PLA. PLA has a higher Young's modulus and tensile strength than PBS but suffers in some applications because of its high brittleness, slow crystallization, and low heat distortion temperature [38].

An important application for aliphatic polyesters is in starch blends. This is done in order to reduce costs and improve biodegradability. For example, PBSA blends with starch content up to 30 wt% retain sufficient mechanical performance to be processed into films via film blowing extrusion [39]. Blending can also be performed in the presence of a compatibilizer in order to increase the amount of starch in the blend without significant loss in the mechanical properties. Compatibilized polyester/starch blends have typically higher water uptake and lower strain at break [40], but the tensile strength remains largely unaffected even at high starch content (up to 70 wt%) [41].

Another interesting class of poly(alkylene dicarboxylates) are obtained from long-chain diols and dicarboxylates. These monomers can be obtained from fatty acids, which naturally occur in plant oils and exist as triglyceride esters. Their corresponding polymers are structurally similar to polyethylene (PE) and have similar physical and mechanical properties [42]. The difference is that long aliphatic segments containing up to 24 carbon atoms are separated by ester bonds. Therefore, they are proposed as a biodegradable and biobased alternative for PE [43].

5.4.2 Poly(hydroxy acids)

Hydroxy acids can undergo self-condensation because they contain both a hydroxyl group and a carboxylic acid in their chemical structure. Upon polycondensation, a poly(hydroxy acid) is formed. They can be classified into α-, β-, ω-, and aromatic hydroxy acids. A separate class exists for hydroxy acids containing more than one hydroxyl or carboxyl group, which can yield cross-linked or branched polyesters. Recently, hydroxy acids have gained increased interest because they can be obtained from renewable resources.

The most studied and industrially relevant hydroxy acids for polymers are the α-hydroxy acids, particularly lactic acid and glycolic acid. Upon polymerization, they form PLA, poly(glycolic acid) (PGA) (Figure 5.3), and the copolymer poly(lactic-*co*-glycolic acid) (PLGA). However, in the industry, these polymers are obtained from the ROP of lactide and glycolide. Therefore, it is more appropriate to refer to these polymers as poly(lactide) and poly(glycolide). Direct condensation of lactic acid omits the synthesis and isolation of lactide, but polycondensation

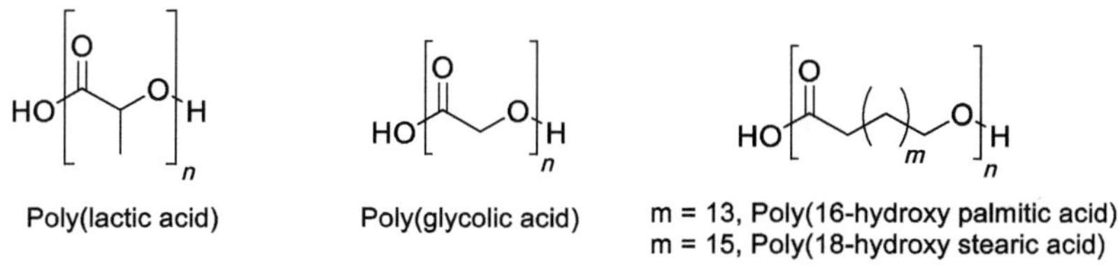

Figure 5.3 Chemical structures of some poly(hydroxy acids).

typically results in a much lower molecular weight. Polycondensation requires long reaction times and high temperature, in the presence of Brønsted acids, or tin or zinc based catalysts such as tin chloride and zinc oxide. In addition, this reaction generally suffers from racemization, transesterification, and discoloration compared to the polymers obtained in ROP [44].

The self-condensation of lactic acid results in an M_w of below 50 000 g/mol in most cases. To obtain high molecular weight PLA, binary catalysts comprising tin dichloride and *p*-toluenesulfonic acid have been proven to be effective [45]. Polycondensation is followed by solid-state polymerization to increase the molecular weight. Herein, the polymer is heated to above the T_g but below the melting point in a vacuum or nitrogen atmosphere. Through this way, PLA with an M_w of above 600 000 g/mol can be achieved [45]. Alternatively, water released during the condensation reaction can be azeotropically removed by means of a solvent. The reaction setup usually includes a Dean–Stark setup, which allows recovery of the dry solvent. Herein, the condensation polymerization occurs in solutions. In this process, high molecular weight PLA with an M_w of 230 000 g/mol can be reached [46, 47].

Other hydroxy acids that have attracted attention for the production of biodegradable polyesters are ω-hydroxy fatty acids (Figure 5.3), which have long aliphatic chains separating the carboxylic acid and the alcohol group. They in fact have similar properties as the long-chain poly(alkylene dicarboxylates), which were discussed in the previous section [48]. Similar to lactic acid, they can also undergo ROP from the corresponding lactone to yield the polyester. The most notable example is poly(pentadecalactone) [49, 50].

5.4.3 Cyclic Sugar-Based Monomers

Polyesters based on rigid cyclic monomers such as isosorbide and its isomers (diols) [51], as well as bicyclic diacetalized D-manno-, D-gluco-, galacto-alditols or aldirates (diols, diacids or diesters), are interesting because of their high T_g, non-toxicity, renewability, stereoselectivity, and biodegradability (Figure 5.4).

Incorporation of cyclic monomers in both aliphatic [8, 52–55] and aromatic [56, 57] polyesters has been reported. For isosorbide copolymers, generally, low molecular weight polymers are obtained because of the low reactivity of the sterically hindered secondary alcohols in isosorbide and its isomers. Nonetheless, they have received a considerable amount of attention from polymer scientists for the use in biobased and biodegradable polymers [51]. Acetalized alditols exhibit

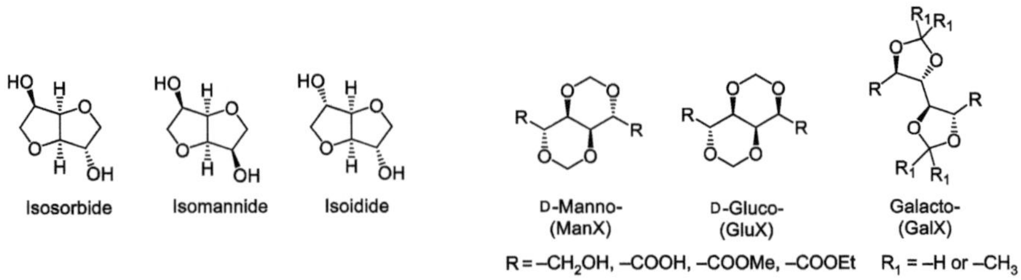

Figure 5.4 Chemical structure of some cyclic sugar-based monomers.

similar properties as isosorbide but with the benefit of being much more reactive in polycondensation, resulting in relatively higher molecular weights.

The group of Sebastian Muñoz-Guerra [53, 58] has shown that the hydrodegradability of polyesters is enhanced by the presence of bicyclic diacetalized D-manno, D-gluco, or galacto sugar derivatives. Moreover, not only the chemical hydrolysis was accelerated but also the sensitivity to enzymatic degradation with lipases was significantly increased when cycloaliphatic sugar-derived monomers were used instead of aromatic monomers.

With the incorporation of rigid sugar-based monomers in aliphatic biodegradable polyesters such as PBS, expansion of the performance window can be achieved without compromising the biodegradability, which is the case with aromatic monomers. Therefore, they are interesting for applications where good biodegradability and chain stiffness are required. For example, incorporation of isosorbide in PBS results in an increase in the T_g from −37 to 65 °C with 0 to 100 mol% isosorbide incorporation relative to succinic acid [59]. At the same time, the melting point and crystallinity decrease. Above 30 mol% incorporation relative to succinic acid, crystallization of the polyesters is troublesome. As a result, isosorbide PBS copolymers show a higher degree of weight loss than the pure PBS polymer in biodegradation studies [59]. Similar trends are observed for the incorporation of ManX-diol and GluX-diol in PBS with respect to biodegradability and thermal properties [60]. The tensile behavior showed that a higher Young's modulus and tensile strength were obtained for polymers containing more ManX-diol at the expense of the strain at break, which is typically very low for these type of polymers [54].

5.5 Semi-aromatic Polyesters

5.5.1 Poly(butylene adipate terephthalate) (PBAT)

Incorporation of aromatic monomers in the polymer chain improves the mechanical properties such as Young's modulus and tensile strength. However, polymers with high aromatic content such as PET are non-biodegradable. This is related to the previously mentioned charge delocalization and steric hindrance of the aromatic units, which hinder chemical hydrolysis and enzymatic degradation. Therefore, a range of semi-aromatic copolyesters were developed to balance the mechanical

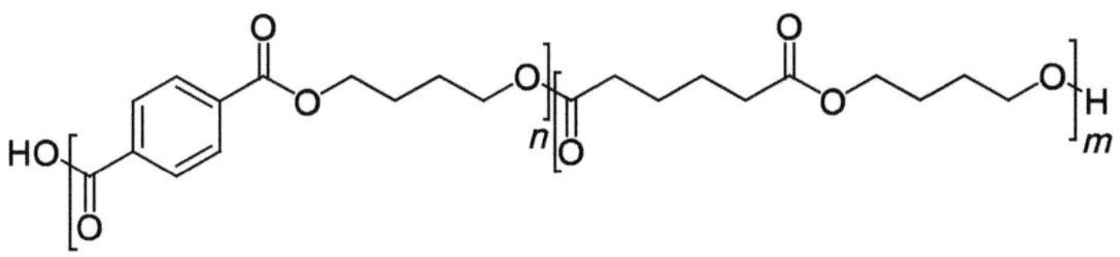

Poly(butylene adipate-*co*-terephthalate) (PBAT)

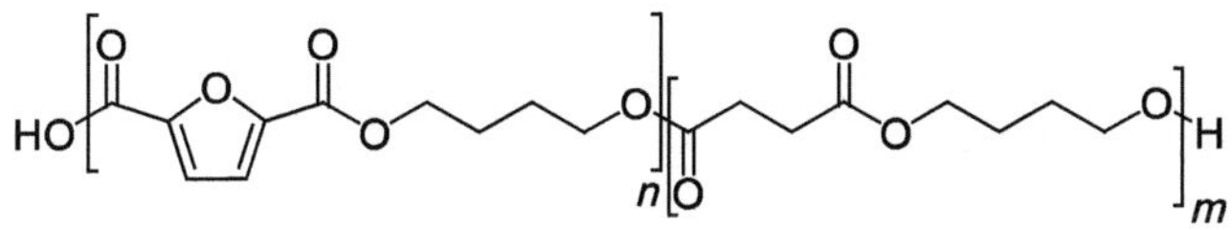

Poly(butylene succinate-*co*-furanoate) (PBSF)

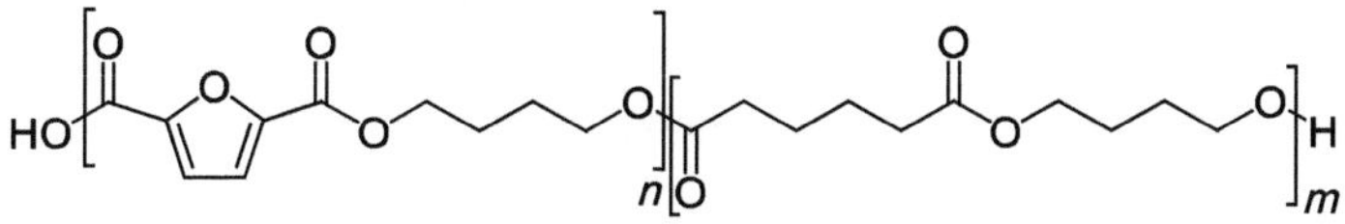

Poly(butylene adipate-*co*-furanoate) (PBAF)

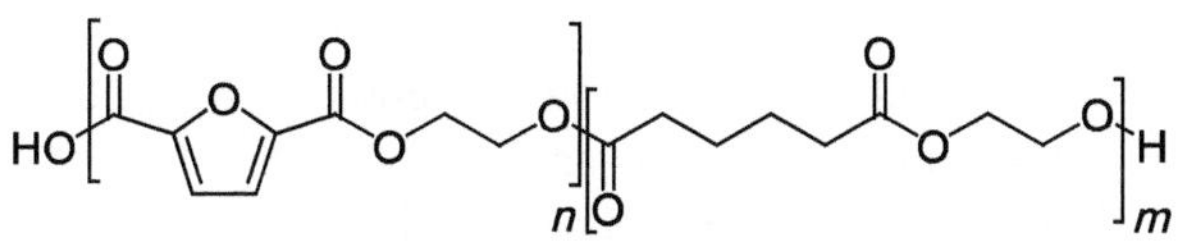

Poly(ethylene adipate-*co*-furanoate) (PEAF)

Figure 5.5 Chemical structures of some relevant biodegradable semi-aromatic polyesters.

and biodegradability properties. Typically, a mixture of aromatic and aliphatic diacids is copolymerized with an aliphatic diol. The resulting biodegradability is then a contribution from the incorporated aliphatic monomers, while the aromatic monomers improve the mechanical properties. The most prevalent biodegradable semi-aromatic polyester is poly(butylene adipate terephthalate) (PBAT) (Figure 5.5). It is prepared from the condensation copolymerization of terephthalic acid with adipic acid and butanediol.

The biodegradability of semi-aromatic copolyesters based on terephthalic acid, adipic acid, sebacic acid, and 1,3-propanediol was demonstrated for the first time in 1995 [62]. Since then, the development and biodegradation of semi-aromatic polyesters were well studied. In the case of PBAT, the biodegradation rate depends on the amount of terephthalic acid in the polymer. It was estimated that with a content ranging from 40 to 50 mol% terephthalic acid relative to adipic acid, the optimal balance between mechanical performance and biodegradation rate was reached [63]. Indeed, the typical terephthalic acid mole fractions in PBAT range from 35 to 60 mol% relative to adipic acid. In laboratory tests, it was shown that PBAT first degrades to oligomers and then to monomers. Furthermore, toxicological studies of the degradation products showed no adverse effects [64].

The mechanical properties of PBAT are similar to those of LDPE, but with a higher strain at break of around 800%. These properties make it a promising material for film applications. To reduce costs, the thickness of the film is reduced to improve cost competitiveness, which requires sufficient melt strength. In order to achieve the desired properties and stability during melt blowing, long-chain branching is introduced. This is achieved during the synthesis by introducing multifunctional branching agents such as polyhydric alcohols and multifunctional acids or epoxides [65].

Efforts in improving the mechanical performance have shown that increasing the terephthalic acid content improves Young's modulus but at the expense of the strain at break [66]. PBAT has a relatively low degree of crystallinity, which is caused by the copolymerization of terephthalate and adipate units. Because the monomers are randomly distributed, crystal formation is disrupted. Nucleating agents can be used during the synthesis to facilitate the crystallization. The melting points of PBAT range between 115 and 125 °C and the T_g between −9 and −33 °C for copolymers with a terephthalic acid content ranging between 40 and 60 mol% relative to adipic acid [66].

5.5.2 Furanoate Copolymers

With the continuing efforts in finding a route toward biobased terephthalic acid, the biobased alternative 2,5-furandicarboxylic acid (FDCA) has gained interest as well. Poly(ethylene furanoate) (PEF) has therefore been proposed as a PET replacement. Similar to PET, PEF is also considered non-biodegradable even though the enzymatic hydrolysis rate of its films is reported to be 1.7 times faster than PET [67]. Therefore, a range of (potentially) biobased semi-aromatic polyesters based on FDCA were developed recently (Figure 5.5) [61]. The monomers that are used for this polymerization are FDCA, 1,4-butanediol, ethylene glycol, succinic acid, and adipic acid. Polycondensation is performed by a two-step synthesis where first the monomers are heated together in an esterification step at 190 °C. Then, a polycondensation step is applied where the temperature is raised to around 230 °C [68]. Loos and coworkers prepared similar furan-based copolyesters via a two-step procedure using enzymatic transesterification polymerization in diphenyl ether at 95 °C [69].

The thermal properties of the copolymers largely depend on the amount of FDCA incorporation. Copolymers of poly(butylene succinate-*co*-furanoate) (PBSF) with different FDCA-to-succinic acid ratios yield a series of semicrystalline polyesters that crystallize slowly. Their melting points range between 93 and 170 °C and the T_g between −40 and 45 °C depending on the amount of FDCA incorporated. Similarly, the copolymers of poly(ethylene adipate-*co*-furanoate) (PEAF) show crystallinity only at low and high FDCA incorporation, which is the same for poly(butylene adipate-*co*-furanoate) (PBAF). The melting temperatures range between 50 and 190 °C and the T_g between −32 and 70 °C [70].

With respect to the mechanical properties, the incorporation of FDCA strongly influences the properties. Mechanical testing on semicrystalline PBAF with molar percentages of FDCA relative to adipic acid between 10 and 75% show a Young's modulus between 55 and 110 MPa. With 30, 40, and 50 mol% FDCA, the polymers are amorphous and therefore exhibit a lower Young's modulus but a high strain at break between 1040 and 1850% [71].

Semicrystalline polyesters based on FDCA and aliphatic diacids show a relatively good biodegradability. Depending on the FDCA content, PBAF shows a significant weight loss in the presence of a lipase at 37 °C in 28 days. Furthermore, the degradation rate is also influenced by the crystallinity and melting point [71]. The degradability of PEAF in enzymatic hydrolysis studies is complete within 30 days for copolymers with 10 mol% FDCA relative to adipic acid or less. The degradation rate is strongly decreased when the FDCA content is higher. The copolymers of PEAF containing more than 50 mol% FDCA relative to adipic acid are hardly susceptible to hydrolysis [70].

Recently, academic interest in biodegradable polyesters based of FDCA has surged. In these studies, the researchers aim to develop semi-aromatic polyesters with improved biodegradability or mechanical performance and study important parameters connected to these properties such as crystallinity (differential scanning calorimetry (DSC) and X-ray diffraction (XRD) studies) and morphology. In order to tune the degradability and mechanical properties, modification of poly(butylene furanoate) (PBF) and PEF with other biobased monomers is possible. In this way, modifications of semicrystalline polyesters with typical biodegradable and/or rigid monomers such as ε-caprolactone [72], lactic acid [73, 74], and isosorbide [75] were reported.

In summary, the combination of various aliphatic and aromatic monomers truly expands the possibilities for biodegradability and mechanical performance optimization. Related to the polymer structure, trends in biodegradability and mechanical performance based on monomer type and ratios can be observed. Besides chain rigidity, hydrophilicity is one of the important predictors for biodegradability [76]. This is related to the amount of polar (end) groups in the polymer structure such as esters, alcohols, and carboxylic acids. In aliphatic polyesters, usually, the presence of branches and the amount of methylene groups that separate the ester bonds negatively affect the hydrophilicity (Figure 5.6). Similarly, the presence and type of aromatic groups negatively affect the degradability as well. For example, the faster degradation of PBAF compared to PBAT originates from the disordered structure of FDCA in contrast to the symmetric terephthalic acid, which allows tight stacking of the polymer chains and thus retarding the degradation (Figure 5.6) [77].

As mentioned before, aromatic monomers can be incorporated to improve the rigidity and strength of the polymer. To obtain high-stiffness polyesters, incorporation of aromatic monomers must be high, but this is usually at the cost of ductility (Figure 5.7). Semi-aromatic polymers with alternating aromatic units such as PEF, PBT, and PET are considered non-biodegradable because of the high stiffness and aromatic content. This is represented in the high Young's modulus, unlike most of the biodegradable polyesters that typically exhibit a much lower Young's modulus but have high strain at break. This can be observed in Figure 5.7, which displays

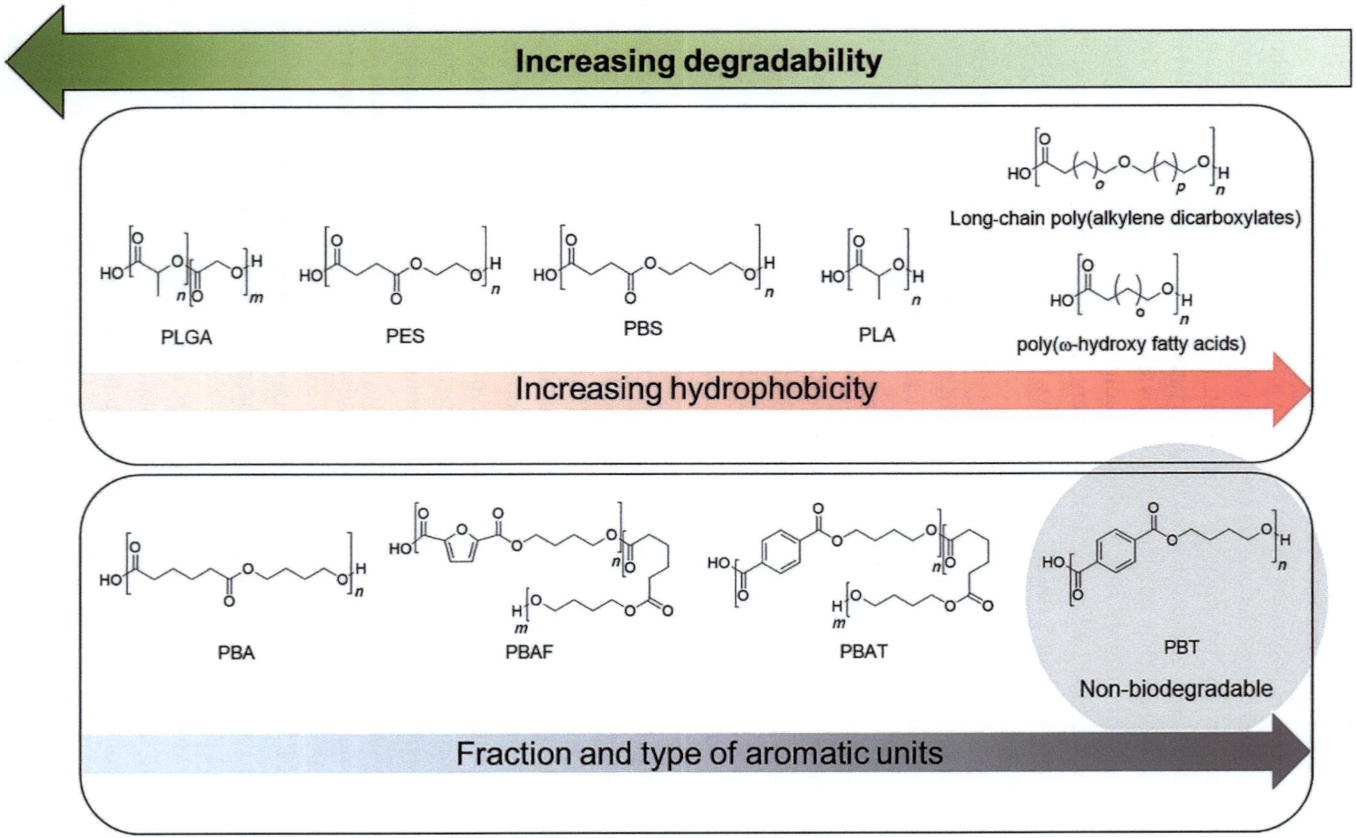

Figure 5.6 Schematic overview of the influence of hydrophobicity and aromatic units on the degradability of polyesters.

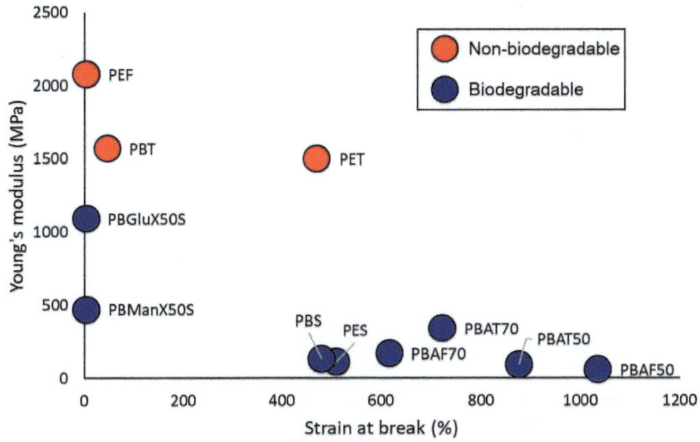

Figure 5.7 Young's modulus as a function of strain at break for several biodegradable (blue) and non-biodegradable (red) polyesters. Poly(ethylene furanoate) (PEF), poly(butylene terephthalate) (PBT), poly(ethylene terephthalate) (PET), poly(butylene GluX succinate) (PBGluXS), poly(butylene ManX succinate) (PBManXS), polybutylene succinate (PBS), poly(ethylene succinate) (PES), poly(butylene adipate furanoate) (PBAF), and poly(butylene adipate terephthalate) (PBAT) [78].

the Young's modulus as a function of strain at break for some biodegradable and non-biodegradable polymers. At the moment, it remains challenging to combine the properties of engineering plastics with good biodegradability.

5.6 Cross-linked Polyesters

Condensation polyesters are usually semicrystalline, which is required for the mechanical performance. When polymer chains are connected with each other via covalent bonds, the material is cross-linked and unable to form crystals and therefore remains amorphous. These cross-linked polyesters gain their mechanical performance from the 3D network that they form. Several strategies exist in order to obtain cross-linked condensation polyesters, which include the following:

- Cross-linked condensation polyesters can be prepared by adding multifunctional (number of functional groups $\geq$3) monomers.
- Functional groups can be added to the polymer structure by means of functional monomers, which are capable of undergoing (self-)cross-linking at a later stage.
- Native polyesters can be directly employed for cross-linking via radical reactions with the use of peroxide initiators.

These strategies will be discussed in the next section.

5.6.1 Multifunctional Alcohols or Carboxylic Acids

The addition of multifunctional monomers during polycondensation is an effective and inexpensive strategy to obtain cross-linked polyesters. Often, the monomers

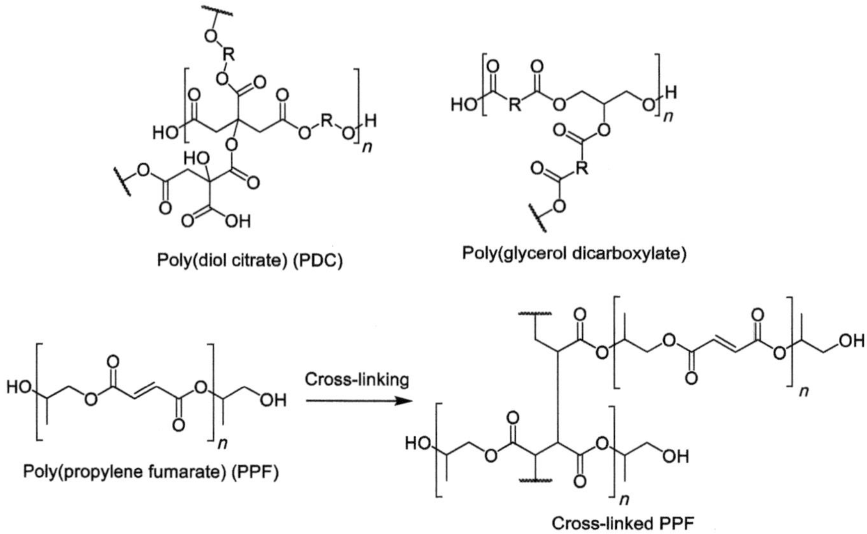

Figure 5.8 Chemical structures of some important biodegradable cross-linked condensation polyesters.

exhibit a simple structure, which can be multifunctional alcohols such as glycerol or pentaerythritol or acids such citric acid. This strategy is straightforward from a biodegradability perspective because the cross-links consist of the same degradable ester linkages as the main chain backbone, thus preventing heterogeneous degradation. Another advantage is that the polymers are usually biocompatible and produce non-toxic decomposition products. In this way, a range of poly(diol citrates) and poly(glycerol dicarboxylates) have been developed (Figure 5.8), which are especially interesting for soft tissue engineering applications. For the fabrication of materials, usually, first a pre-polymer is synthesized. The pre-polymer is then solubilized and solvent-cast in a mold. After evaporation of the solvent, a post-polymerization step is applied to form the cross-linked network.

An interesting example is fully biodegradable, plant-based, and non-toxic cross-linked resin foams obtained from the condensation between glycerol and citric acid at moderate temperatures (80 °C). The hardness, brittleness, and toughness of this resin can be controlled by changing the alcohol-to-acid ratio, the temperature, and the presence of additives. The resin is fully biodegradable with a degradation period ranging from a few days to a few months, depending on the degree of polymerization. Upon biodegradation in water, the original molecular components are formed back [79].

Cross-linked polyesters based on polyhydric alcohols and aliphatic dicarboxylic acids with various spacer lengths have also been investigated. The alcohols that were investigated were glycerol, pentaerythritol, 1,1,1-trimethylolpropane, 1,2,3,4-butanetetrol, and D-glucitol [80–82]. The resulting materials show a particular degradation behavior. Networks based on short spacer diacids show low sensitivity to enzymatic degradation, but upon increasing the spacer length, the weight loss as a result of biodegradation increased rapidly. When the spacer length consists of more than 14 methylene units, the degradability decreases rapidly. This is proposed to be related to the network structure and the concentration of degradable ester bonds. Also, an effect of the amount of hydroxyl groups of the polyhydric alcohol was observed. The degradation decreased when the amount of hydroxyl groups increased, which could be related to the cross-linking density. This shows a clear window of operation with respect to the possible monomer structures.

Besides the use of polyhydric alcohols, the use of multifunctional carboxylic acids to produce cross-linked condensation polyesters has also been investigated. One of the most interesting multifunctional carboxylic acids is citric acid. Citric acid is inexpensive, non-toxic, and produced from renewable resources at large scale, which contributes to the concept of circular economy. A range of cross-linked aliphatic polyesters based on citric acid and various diols have been investigated [83, 84]. It was found that the biodegradability could be tuned with the amount of methylene units present in the diol that was used.

5.6.2 Incorporation of Functional Monomers

Incorporation of functional monomers during the synthesis of condensation polyesters is a versatile method to obtain cross-linked materials. Incorporation of different types of functional groups may allow for different curing methods, such as UV curing, thermal curing, or with the aid of a cross-linker. One prerequisite is that the functional groups remain intact during the high temperatures typically involved during polycondensation; therefore, the monomers used must be thermally stable.

Monomers containing double bonds in the backbone have been employed to produce biodegradable cross-linked polyesters. Even though undesired cross-linking of double bond containing monomers during the polymerization can occur at elevated temperatures, a radical scavenger such as hydroquinone can be added to prevent cross-linking reactions. In this way, polyesters were prepared based on fumaric acid and 1,2-propylene diol to produce poly(propylene fumarate) (PPF) (Figure 5.8) [85–87]. Cross-linked PPF shows remarkable retention of the mechanical properties during degradation, which make them exceptionally useful in load-bearing applications [88]. Therefore, it is an important material for bone tissue engineering applications. In the presence of radical initiators, PPF can cross-link via radical polymerization with itself or together with monomers such as methyl methacrylate or *N*-vinyl pyrrolidinone. Also, biocompatible cross-linkers such as acrylated poly(ethylene glycol) can be used in order to perform an in situ cross-linking step after injection with PPF [87].

With respect to biodegradation of cross-linked PPF, chain scission occurs via the ester bonds and biodegradability depends on molecular weight, type of cross-linker, and cross-linking density. Furthermore, the T_g of non-cross-linked PPF is around 32 °C, which is below the body temperature. This promotes the biodegradability of non-load bearing scaffolds in biomedical applications [85]. For load bearing applications, cross-linked PPF is required for the improvement of the mechanical properties.

Similar to fumaric acid, long-chain unsaturated dicarboxylic acids based on oleic acid or castor oil can be used for cross-linked polyesters. The double bonds present in the polymer backbone can be transformed to epoxide groups and subsequently cross-linked using UV light and a cross-linker. However, these polymers showed only limited biodegradability because of the high hydrophobicity of the fatty acid monomer [89]. This can be relieved by decreasing the amount of functional fatty acid monomer but is at the expense of cross-linking density [90].

Also, modification of the existing biodegradable polyesters such as PBS is reported. Upon incorporation of functional monomers such as itaconic acid and maleic acid, cross-linking is enabled. As a result of the cross-linking, the biodegradability decreased, but some of it is retained [91].

5.6.3 Cross-linking of Native Polyesters

Finally, self-cross-linking of native polyesters is also investigated for biodegradable condensation polyesters. This is one of the most straightforward methods in cross-linking polyesters because the existing polymers can be directly employed in this method. One reason for cross-linking the existing polyesters such as PBS is to improve the mechanical strength.

The use of peroxide radical initiators in order to cross-link polymers is currently well established. Notably, it can be applied in saturated rubbers consisting of ethylene copolymers. Similarly, aliphatic polyesters containing ethylene moieties can also be cross-linked using this procedure. It involves heating the polymer together with the peroxide initiator to produce radicals, which can abstract hydrogen atoms from the polymer chains. The polymer radicals are then able to combine and form covalent bonds. During the process, chain scission reactions also occur as a result of the radical formation. This should be prevented to avoid a large decrease in molecular weight and loss of mechanical properties. Careful selection of the type of radical initiator is important because an adequate formation of radicals is required at the processing temperatures while avoiding premature cross-linking. Dicumyl peroxide is one of the most versatile radical initiators because it has a suitable decomposition range for polymers melting between 100 and 120 °C and shows a high hydrogen abstraction efficiency [92]. Using the method of radical formation, cross-linking of aliphatic polycondensates such as PBS [93, 94] and PBSA [95] can be achieved while retaining the biodegradability behavior. Furthermore, with the use of peroxide initiators, the grafting of monomers on polyester backbone and blend compatibilization with other biodegradable polyesters is enabled [96, 97].

5.7 Applications for Biodegradable Condensation Polyesters

In the past few decades, products developed from biodegradable condensation polyesters have been receiving a lot of attention. Recently, more applications for the investigated materials are being developed. In general, the good processability and mechanical properties of condensation polyesters allow for a variety of applications, especially where tunable properties are required. Processing of biodegradable condensate polyesters is usually possible in conventional polymer processing equipment and can occur via cast film extrusion, blown film extrusion, injection molding, hot pressing, and fiber spinning.

5.7.1 Biomedical Applications

One of the large advantages of polyesters, next to their biodegradability, is their biocompatibility, which expands the potential to biomedical applications. In addition, because degradation products are generally non-toxic, they are excellent candidates for biomedical uses such as (micro)capsules, implants, or sutures, which benefit from the breakdown of the temporary polymeric structure that provides mechanical support or carriage of bioactive molecules. In this context, biodegradability can enable the controlled release of drugs or prevent surgical intervention to remove implants because the material is able to undergo biodegradation and subsequently absorption or secretion of the degradation products by the body. Quantitatively, biomedical applications are only a minute fraction of the total biodegradable polymer market, but the effects on public health with respect to medical procedures are major. This therefore allows the relatively high price of biodegradable polyesters (e.g. >$10/kg).

Aliphatic condensation polyesters generally show a good biocompatibility *in vitro* and *in vivo* [98]. The use of PBS for tissue engineering applications has been extensively researched. PBS-based blends, composites, and copolymers display the potential for bone tissue engineering [99] and cardiac tissue engineering [100]. Furthermore, controlled drug release has been realized using PBS-based micro- and nanocapsules [101] and scaffolds [99]. On the other hand, semi-aromatic polyesters (i.e. PBAT and copolymers thereof) can possess a tunable balance between mechanical and biodegradable properties with the incorporation of terephthalic acid. Although being less investigated than aliphatic polyesters, they have been reported for the use in vascular applications [102] and bone tissue engineering applications [103].

Biomedical materials that are implanted in the mechanically dynamic environment of the body must be able to recover from deformations. Ideally, the material is able to undergo large deformations without imposing damage to the surrounding tissues. Specifically, elastomers exhibit this property. Therefore, some implants are aimed to mimic the extracellular matrix, which is a soft and elastomeric material providing support to tissues such as cartilage, blood vessels, tendons, and ligaments. Condensation polyesters are of particular interest especially because common monomers such as glycerol and sebacic acid have been approved by the US Food and Drug Administration for medical applications. In this way, tough biodegradable cross-linked polyesters based on glycerol and succinic acid [104] or sebacic acid [105] have been developed. The material shows a desirable mechanical performance such as high elasticity and toughness. As such, tubular scaffolds based on poly(glycerol sebacate) were developed and can be applied in blood vessel tissue engineering [106, 107]. The degradation rate and biointeraction can be tuned by grafting hydrophobic or bioactive molecules on the backbone.

Another issue with biomedical polymeric materials is the presence of catalysts, originating from the condensation polymerization. Metal-based catalysts can lead to unfavorable effects with respect to toxicity in the body. The low limit for metals present in polyesters such as tin excludes the use in biomedical applications. Currently, in polyester synthesis for biomedical applications, catalysts based on zinc, titanium, and zirconium are used. Nonetheless, there is a continuous trend toward the use of alternative (metal-free) catalysts and catalytic systems such as enzymatic polycondensation [108].

5.7.2 Agricultural Applications

One of the main applications of biodegradable condensation polyesters is in mulch films. They are used for controlling the soil temperature, limiting the weed growth, protection against harsh weather and pests, and improving water retention and crop yield. Mulch films are generally made from non-biodegradable plastics such as PE, which are difficult to recover because of the loss of the structural integrity after use. As a result, PE residues remain and accumulate in the soil, which negatively affects the crop yield [109, 110]. EU regulations require adequate waste disposal of mulch film by incineration or recycling. Therefore, employing a biodegradable film can be more cost-effective in some cases, even though the biodegradable mulch film is about twice as expensive as the PE mulch film. However, the benefit is that the waste does not need to be collected and can be tilled into the soil [111, 112]. It was estimated that approximately 161 000 tonnes of bioplastics were produced for the agricultural and horticultural applications, of which more than half consisted of PBAT and PBS [5].

5.7.3 Packaging Material

Because of the low cost and good properties, conventional plastics are widely used in packaging and plastic bag industry, which is a high-volume application. However, despite the higher costs, a growing awareness and interest of the consumers in environment-friendly products supports the implementation of biodegradable plastic carrier bags in retailers. Similar to compost bags, they can be processed in organic waste disposal. Both require good mechanical properties such as puncture and tear resistance, and they need to be able to support loads of 100 times their own weight [64]. Films made from PBAT are highly suitable for these applications. Another potential packaging application where biodegradability could bring added value is the small-size food packaging because of the high litter risk.

5.8 Polyester Recycling

With the drastic increase in the consumption of plastics in the past decades, the problems related to waste disposal and conservation of petrochemical resources have become more critical. In the transition from a linear to a circular economy, much more of the plastic waste we produce must be recycled. Recycling even has the preference over biodegradation, which is rather a litter management strategy. Thermoplastic polyesters represent an important class of recyclable plastics. The most significant example is PET. PET is the third most widely diffused polymer exploited in the packaging industry, monopolizing the bottles market for beverages, and covering almost 16% of the European plastic consumption in the packaging industry [113]. As mentioned earlier, PET is not biodegradable, but advanced progress is made in recycling PET, which allows us to reduce PET waste in an economical and environmentally acceptable way [114].

Currently, the dominant form of PET waste recycling worldwide is mechanical recycling. The process requires sorting, decontamination, and treating by washing, drying, and grinding before new plastic products can be obtained via melting and extrusion [115]. Thus, it is very difficult to recycle complex and contaminated PET wastes mechanically [116] because this process in fact requires homogeneous plastics and relatively clean material. Mechanical recycling is often called a "downcycling" route because the obtained recycled polymers have inferior properties (lower molecular weight, discoloration, risk of more impurities, etc.) compared to polymers made from virgin feedstocks [117].

Strategies to counterbalance the loss in the properties of mechanically recycled polymers upon reprocessing include intensive drying, reprocessing with vacuum degassing, and the use of chain extenders [118]. A good example in this respect is the increase of molecular weight of PET via additional solid-state polycondensation (SSP) [119]. SSP is usually done at a high temperature (below the melting temperature of the polymer) in the absence of oxygen and water, by means of either vacuum or purging with an inert gas to drive off the by-products of reactions [116].

In cases where plastic waste is not suitable for mechanical recycling, e.g. because of the complexity of the mixture and contaminations, incinerating the waste and using the heat to generate electricity can be applied, particularly if the CO_2 this generates can be captured sustainably rather than releasing to the atmosphere [117].

Chemical recycling plays an increasingly crucial role in extracting more value from waste plastic and is essential in creating a circular economy for plastics [120]. Chemical recycling is carried out either by solvolysis or by pyrolysis [121]. Pyrolysis occurs through degradation by heat in the absence of oxygen or air or under vacuum. However, pyrolysis of PET is seldom used as a method to depolymerize PET into its monomeric units on an industrial scale because pyrolysis generally leads to other liquid and gaseous side products, reducing process efficiency and necessitating costly separation steps [114, 116].

Solvolysis occurs through polymer degradation by different solvents such as water (hydrolysis), methanol (methanolysis), or glycol (glycolysis). In the case of polycondensation polyesters such as PET, this type of chemical recycling is based on the reversibility of polycondensation chemistry. This indicates that the polymer in the presence of a low molecular weight component (typical water, methanol, or glycol) and the required reaction conditions (temperature and catalyst) can be depolymerized back into the starting monomers or oligomers, namely, ethylene glycol, terephthalic acid (hydrolysis), dimethyl terephthalate (methanolysis), bis(hydroxylethylene) terephthalate (glycolysis), or polyols (glycolysis) [121]. The main advantage of this chemical route is the possibility of purifying (through distillation/crystallization steps or surface/vacuum treatments) monomers/oligomers by removing possible post-consumer contaminants (decontamination process) and consequently directly re-using these monomers for the production of recycled PET again for high-end applications with similar high molecular weights as the virgin polymers [116, 122]. The polyol oligomers can be reused in other polycondensates such as polyurethanes and epoxy resins. However, chemically recycled PET is more expensive than virgin PET because of its raw material cost, capital investment, and scale of operation [116].

The most recent inventions with respect to chemical recycling allow even the reuse, reprocessing, or reshaping of thermoset polymers. Conventional thermosets, widely used in composites, coatings, electronic packing materials, foams, etc., cannot be reprocessed, recycled, or reshaped under mild conditions. Traditional recycling of cross-linked polymers via mechanical reprocessing is impractical because their structures preclude flow, even at elevated temperatures. Their insolubility also precludes solution reprocessing. Although some of these materials are down-cycled into lower value products, most thermosets are incinerated, landfilled, or otherwise unaccounted for (i.e. leakage). This causes serious resource and environmental problems. Developing strategies to recycle these materials without compromising their performance represents a formidable challenge. Covalent Adaptable Networks (CANs) [123–125] are an emerging class of materials that contain exchangeable chemical bonds in the polymer network, leading to reversible cross-links. CANs combine the easy processing of thermoplastics that can flow upon heating and the high durability and resistance of thermosets. A subclass of CANs, called "vitrimers," has been developed in 2011 by Leibler and coworkers [126]. Vitrimers undergo stress relaxation upon associative bond exchanges, which are triggered by temperature, pH, or light. During this process, the cross-linking density stays intact because a new covalent bond needs to be formed either simultaneously or before breaking an existing covalent bond [127]. Vitrimers are thus permanent networks characterized by a combination of insolubility and gradual thermal viscosity behavior, in analogy with the thermal behavior of traditional glass. A typical class of vitrimers is based on transesterification chemistry, e.g. between hydroxyl and ester groups in the network at high temperatures in the presence of the right catalysts [126, 128–131].

5.9 Concluding Remarks

In this chapter, the development of condensation polyesters, from the first pioneering work by Carothers to today's advancements in biodegradability, is briefly reviewed. The preparative methods of condensation polyesters and fundamentals of polycondensation, developed over 80 years ago, are still relevant today and are discussed with focus on the synthesis of biodegradable polyesters.

Currently, much research is devoted to the incorporation of novel monomers with added functionality or improved properties. A main challenge remains to combine good mechanical performance with biodegradation properties. This leads to a surge in different polymer structures with tuned properties. Some potential disadvantages are connected to these developments. From a recyclability standpoint, the simplicity of plastic types in waste streams is desired to achieve efficient separation and purification of the plastics for recycling. The inherent reversibility of the condensation polymerization may also allow for facile recovery of monomers from plastic waste when recycling is not possible. Nonetheless, with the widening scope of applications (such as in the medical field), and because of increasing environmental awareness and legislation, the interest toward biodegradable condensation polyesters has expanded greatly in the past decades.

References

1 McIntyre, J. (2003). The historical development of polyesters. In: *Modern Polyesters: Chemistry and Technology of Polyesters and Copolyesters* (ed. J. Scheirs and T.E. Long), 3–28. Chichester: Wiley.

2 Kienle, R.H. and Hovey, A. (1929). The polyhydric alcohol-polybasic acid reaction. I. Glycerol-phthalic anhydride. *Journal of the American Chemical Society* 51 (2): 509–519.

3 Carothers, W.H. and Hill, J.W. (1932). Studies of polymerization and ring formation. XII. Linear Superpolyesters. *Journal of the American Chemical Society* 54 (4): 1559–1566.

4 Gilbert, M. (2017). Plastics materials: introduction and historical development. In: *Brydson's Plastics Materials* (ed. M. Gilbert), 1–18. Oxford: Butterworth-Heinemann.

5 European Bioplastics (2019). Report - Bioplastics market data 2019. Nova-Institute. https://docs.european-bioplastics.org/publications/market_data/Report_Bioplastics_Market_Data_2019.pdf.

6 Tachibana, Y. and Kimura, S. (2015). Kasuya K-i, Synthesis and verification of biobased terephthalic acid from furfural. *Scientific Reports* 5: 8249.

7 Carothers, W.H. (1936). Polymers and polyfunctionality. *Transactions of the Faraday Society* 32: 39–49.

8 Lavilla, C., Alla, A., Martínez de Ilarduya, A. et al. (2011). Carbohydrate-based polyesters made from bicyclic acetalized galactaric acid. *Biomacromolecules* 12 (7): 2642–2652.

9 Terzopoulou, Z., Karakatsianopoulou, E., Kasmi, N. et al. (2017). Effect of catalyst type on molecular weight increase and coloration of poly(ethylene furanoate) biobased polyester during melt polycondensation. *Polymer Chemistry* 8 (44): 6895–6908.

10 Price, J.A. and Stewart, M.J. (1972). Titanium compounds as polycondensation catalysts in preparation of linear polyesters. US Patent 3,644,291.

11 Thiele, U.K. (2001). The current status of catalysis and catalyst development for the industrial process of poly(ethylene terephthalate) polycondensation. *International Journal of Polymeric Materials* 50 (3–4): 387–394.

12 Duan, R.-T., He, Q.-X., Dong, X. et al. (2016). Renewable sugar-based diols with different rigid structure: comparable investigation on improving poly(butylene succinate) performance. *ACS Sustainable Chemistry & Engineering* 4 (1): 350–362.

13 Jacquel, N., Freyermouth, F., Fenouillot, F. et al. (2011). Synthesis and properties of poly(butylene succinate): efficiency of different transesterification catalysts. *Journal of Polymer Science Part A: Polymer Chemistry* 49 (24): 5301–5312.

14 Wilsens, C.H., Noordover, B.A., and Rastogi, S. (2014). Aromatic thermotropic polyesters based on 2,5-furandicarboxylic acid and vanillic acid. *Polymer* 55 (10): 2432–2439.

15 Conix, A. (1959). Thermoplastic polyesters from bisphenols. *Industrial and Engineering Chemistry* 51 (2): 147–150.

16 Müller, R.-J., Marten, E., and Deckwer, W.-D. (2001). Structure-biodegradability relationship of polyesters. In: *Biorelated Polymers* (ed. E. Chiellini, H. Gil, G. Braunegg, et al.), 303–311. Springer.

17 Marten, E., Müller, R.-J., and Deckwer, W.-D. (2005). Studies on the enzymatic hydrolysis of polyesters. II. Aliphatic–aromatic copolyesters. *Polymer Degradation and Stability* 88 (3): 371–381.

18 Gleadall, A., Pan, J., and Atkinson, H. (2012). A simplified theory of crystallisation induced by polymer chain scissions for biodegradable polyesters. *Polymer Degradation and Stability* 97 (9): 1616–1620.

19 Haider, T.P., Völker, C., Kramm, J. et al. (2019). Plastics of the future? The impact of biodegradable polymers on the environment and on society. *Angewandte Chemie International Edition* 58 (1): 50–62.

20 Bikiaris, D.N., Papageorgiou, G.Z., and Achilias, D.S. (2006). Synthesis and comparative biodegradability studies of three poly(alkylene succinate)s. *Polymer Degradation and Stability* 91 (1): 31–43.

21 Rychter, P., Biczak, R., Herman, B. et al. (2006). Environmental degradation of polyester blends containing atactic poly(3-hydroxybutyrate). Biodegradation in soil and ecotoxicological impact. *Biomacromolecules* 7 (11): 3125–3131.

22 Kim, M.-N., Lee, B.-Y., Lee, I.-M. et al. (2001). Toxicity and biodegradation of products from polyester hydrolysis. *Journal of Environmental Science and Health, Part A* 36 (4): 447–463.

23 Brannigan, R.P. and Dove, A.P. (2017). Synthesis, properties and biomedical applications of hydrolytically degradable materials based on aliphatic polyesters and polycarbonates. *Biomaterials Science* 5 (1): 9–21.

24 von Burkersroda, F., Schedl, L., and Göpferich, A. (2002). Why degradable polymers undergo surface erosion or bulk erosion. *Biomaterials* 23 (21): 4221–4231.

25 Lyu, S. and Untereker, D. (2009). Degradability of polymers for implantable biomedical devices. *International Journal of Molecular Sciences* 10 (9): 4033–4065.

26 Sisson, A.L., Schroeter, M., and Lendlein, A. (2011). Polyesters. In: *Handbook of Biodegradable Polymers: Synthesis, Characterization and Applications* (ed. A. Lendlein and A.L. Sisson), 1–21. Weinheim: Wiley-VCH.

27 Sarda, L., Verger, R., Carrière, F., and Ferrato, F. (1996). A critical reevaluation of the phenomenon of interfacial activation. *Methods in Enzymology* 286: 327–347.

28 Zhang, Z., Kuijer, R., Bulstra, S.K. et al. (2006). The in vivo and in vitro degradation behavior of poly(trimethylene carbonate). *Biomaterials* 27 (9): 1741–1748.

29 Montaudo, G. and Rizzarelli, P. (2000). Synthesis and enzymatic degradation of aliphatic copolyesters. *Polymer Degradation and Stability* 70 (2): 305–314.

30 Larrañaga, A. and Lizundia, E. (2019). A review on the thermomechanical properties and biodegradation behaviour of polyesters. *European Polymer Journal* 21: 109296.

31 Tserki, V., Matzinos, P., Pavlidou, E., and Panayiotou, C. (2006). Biodegradable aliphatic polyesters. Part II. Synthesis and characterization of chain extended poly(butylene succinate-*co*-butylene adipate). *Polymer Degradation and Stability* 91 (2): 377–384.

32 Huang, C.Q., Luo, S.Y., Xu, S.Y. et al. (2010). Catalyzed chain extension of poly(butylene adipate) and poly(butylene succinate) with 2, 2′-(1, 4-phenylene)-bis (2-oxazoline). *Journal of Applied Polymer Science* 115 (3): 1555–1565.

33 Okajima, S., Kondo, R., Toshima, K., and Matsumura, S. (2003). Lipase-catalyzed transformation of poly(butylene adipate) and poly(butylene succinate) into repolymerizable cyclic oligomers. *Biomacromolecules* 4 (6): 1514–1519.

34 Fujimaki, T. (1998). Processability and properties of aliphatic polyesters,'BIONOLLE', synthesized by polycondensation reaction. *Polymer Degradation and Stability* 59 (1–3): 209–214.

35 Ichikawa, Y. and Mizukoshi, T. (2011). Bionolle (polybutylenesuccinate). In: *Synthetic Biodegradable Polymers* (ed. B. Rieger, A. Künkel, G.W. Coates, et al.), 285–313. Berlin/Heidelberg: Springer-Verlag.

36 Song, H. and Lee, S.Y. (2006). Production of succinic acid by bacterial fermentation. *Enzyme and Microbial Technology* 39 (3): 352–361.

37 Werpy, T. and Petersen, G. (2004). Top Value Added Chemicals from Biomass: Volume I – Results of Screening for Potential Candidates from Sugars and Synthesis Gas. Golden, CO: National Renewable Energy Lab.

38 Xu, J. and Guo, B.H. (2010). Poly(butylene succinate) and its copolymers: research, development and industrialization. *Biotechnology Journal* 5 (11): 1149–1163.

39 Ratto, J.A., Stenhouse, P.J., Auerbach, M. et al. (1999). Processing, performance and biodegradability of a thermoplastic aliphatic polyester/starch system. *Polymer* 40 (24): 6777–6788.

40 Zeng, J.-B., Jiao, L., Li, Y.-D. et al. (2011). Bio-based blends of starch and poly(butylene succinate) with improved miscibility, mechanical properties, and reduced water absorption. *Carbohydrate Polymers* 83 (2): 762–768.

41 Mani, R. and Bhattacharya, M. (2001). Properties of injection moulded blends of starch and modified biodegradable polyesters. *European Polymer Journal* 37 (3): 515–526.

42 Cole-Hamilton, D.J. (2010). Nature's polyethylene. *Angewandte Chemie International Edition* 49 (46): 8564–8566.

43 Quinzler, D. and Mecking, S. (2010). Linear semicrystalline polyesters from fatty acids by complete feedstock molecule utilization. *Angewandte Chemie International Edition* 49 (25): 4306–4308.

44 Moon, S.I., Lee, C.W., Miyamoto, M., and Kimura, Y. (2000). Melt polycondensation of L-lactic acid with Sn(II) catalysts activated by various proton acids: a direct manufacturing route to high molecular weight poly(L-lactic acid). *Journal of Polymer Science Part A: Polymer Chemistry* 38 (9): 1673–1679.

45 Moon, S.-I., Taniguchi, I., Miyamoto, M. et al. (2001). Synthesis and properties of high-molecular-weight poly(L-lactic acid) by melt/solid polycondensation under different reaction conditions. *High Performance Polymers* 13 (2): S189–S196.

46 Ajioka, M., Enomoto, K., Suzuki, K., and Yamaguchi, A. (1995). Basic properties of polylactic acid produced by the direct condensation polymerization of lactic acid. *Bulletin of the Chemical Society of Japan* 68 (8): 2125–2131.

47 Kim, K.W. and Woo, S.I. (2002). Synthesis of high-molecular-weight poly(L-lactic acid) by direct polycondensation. *Macromolecular Chemistry and Physics* 203 (15): 2245–2250.

48 Jose, J., Pourfallah, G., Merkley, D. et al. (2014). Thermoplastic polyesters and co-polyesters derived from vegetable oil: synthesis and optimization of melt polycondensation for medium and long chain poly(ω-hydroxyfatty acid) s and their ester derivatives. *Polymer Chemistry* 5 (9): 3203–3213.

49 De Geus, M., van der Meulen, I., Goderis, B. et al. (2010). Performance polymers from renewable monomers: high molecular weight poly(pentadecalactone) for fiber applications. *Polymer Chemistry* 1 (4): 525–533.

50 Delgove, M.A., Luchies, J., Wauters, I. et al. (2017). Increasing the solubility range of polyesters by tuning their microstructure with comonomers. *Polymer Chemistry* 8 (32): 4696–4706.

51 Fenouillot, F., Rousseau, A., Colomines, G. et al. (2010). Polymers from renewable 1, 4: 3, 6-dianhydrohexitols (isosorbide, isomannide and isoidide): a review. *Progress in Polymer Science* 35 (5): 578–622.

52 Braun, D. and Bergmann, M. (1992). 1, 4: 3, 6-Dianhydrohexite als Bausteine für Polymere. *Journal fur Praktische Chemie - Chemiker Zeitung* 334 (4): 298–310.

53 Zakharova, E., Martínez de Ilarduya, A., León, S., and Muñoz-Guerra, S. (2017). Sugar-based bicyclic monomers for aliphatic polyesters: a comparative appraisal of acetalized alditols and isosorbide. *Designed Monomers and Polymers* 20 (1): 157–166.

54 Lavilla, C., Alla, A., Martínez de Ilarduya, A., and Muñoz-Guerra, S. (2013). High Tg bio-based aliphatic polyesters from bicyclic D-mannitol. *Biomacromolecules* 14 (3): 781–793.

55 Gavrila, I., Raffa, P., and Picchioni, F. (2018). Acetalised galactarate polyesters: interplay between chemical structure and polymerisation kinetics. *Polymers* 10 (3): 248.

56 Muñoz-Guerra, S., Lavilla, C., Japu, C., and de Ilarduya, A.M. (2014). Renewable terephthalate polyesters from carbohydrate-based bicyclic monomers. *Green Chemistry* 16 (4): 1716–1739.

57 Lavilla, C., Alla, A., Martínez de Ilarduya, A. et al. (2012). Biodegradable aromatic copolyesters made from bicyclic acetalized galactaric acid. *Journal of Polymer Science Part A: Polymer Chemistry* 50 (16): 3393–3406.

58 Japu, C., Martínez de Ilarduya, A., Alla, A. et al. (2015). Copolyesters made from 1,4-butanediol, sebacic acid, and D-glucose by melt and enzymatic polycondensation. *Biomacromolecules* 16 (3): 868–879.

59 Qi, J., Wu, J., Chen, J., and Wang, H. (2019). An investigation of the thermal and (bio) degradability of PBS copolyesters based on isosorbide. *Polymer Degradation and Stability* 160: 229–241.

60 Zakharova, E., Alla, A., de Ilarduya, A.M., and Muñoz-Guerra, S. (2015). Bio-based PBS copolyesters derived from a bicyclic D-glucitol. *RSC Advances* 5 (57): 46395–46404.

61 Wu, L., Mincheva, R., Xu, Y. et al. (2012). High molecular weight poly(butylene succinate-*co*-butylene furandicarboxylate) copolyesters: from catalyzed polycondensation reaction to thermomechanical properties. *Biomacromolecules* 13 (9): 2973–2981.

62 Witt, U., Müller, R., and Deckwer, W.-D. (1995). Biodegradation of polyester copolymers containing aromatic compounds. *Journal of Macromolecular Science, Part A: Pure and Applied Chemistry* 32 (4): 851–856.

63 Witt, U., Müller, R.-J., and Deckwer, W.-D. (1996). Evaluation of the biodegradability of copolyesters containing aromatic compounds by investigations of model oligomers. *Journal of Environmental Polymer Degradation* 4 (1): 9–20.

64 Siegenthaler, K., Künkel, A., Skupin, G., and Yamamoto, M. (2011). Ecoflex® and Ecovio®: biodegradable, performance-enabling plastics. In: *Synthetic Biodegradable Polymers* (ed. B. Rieger, A. Künkel, G.W. Coates, et al.), 91–136. Berlin/Heidelberg: Springer-Verlag.

65 Jian, J., Xiangbin, Z., and Xianbo, H. (2020). An overview on synthesis, properties and applications of poly(butylene-adipate-*co*-terephthalate)–PBAT. *Advanced Industrial and Engineering Polymer Research* 3 (1): 19–26.

66 Herrera, R., Franco, L., Rodríguez-Galán, A., and Puiggalí, J. (2002). Characterization and degradation behavior of poly(butylene adipate-*co*-terephthalate)s. *Journal of Polymer Science Part A: Polymer Chemistry* 40 (23): 4141–4157.

67 Weinberger, S., Haernvall, K., Scaini, D. et al. (2017). Enzymatic surface hydrolysis of poly(ethylene furanoate) thin films of various crystallinities. *Green Chemistry* 19 (22): 5381–5384.

68 Terzopoulou, Z., Tsanaktsis, V., Bikiaris, D.N. et al. (2016). Biobased poly(ethylene furanoate-*co*-ethylene succinate) copolyesters: solid state structure, melting point depression and biodegradability. *RSC Advances* 6 (87): 84003–84015.

69 Maniar, D., Jiang, Y., Woortman, A.J. et al. (2019). Furan-based copolyesters from renewable resources: enzymatic synthesis and properties. *ChemSusChem* 12 (5): 990–999.

70 Papadopoulos, L., Magaziotis, A., Nerantzaki, M. et al. (2018). Synthesis and characterization of novel poly(ethylene furanoate-*co*-adipate) random copolyesters with enhanced biodegradability. *Polymer Degradation and Stability* 156: 32–42.

71 Zhou, W., Wang, X., Yang, B. et al. (2013). Synthesis, physical properties and enzymatic degradation of bio-based poly(butylene adipate-*co*-butylene furandicarboxylate) copolyesters. *Polymer Degradation and Stability* 98 (11): 2177–2183.

72 Zheng, M., Zang, X., Wang, G. et al. (2017). Poly(butylene 2,5-furandicarboxylate-ε-caprolactone): a new bio-based elastomer with high strength and biodegradability. *Express Polymer Letters* 11 (8): 611–621.

73 Hu, H., Zhang, R., Shi, L. et al. (2018). Modification of poly(butylene 2,5-furandicarboxylate) with lactic acid for biodegradable copolyesters with good mechanical and barrier properties. *Industrial & Engineering Chemistry Research* 57 (32): 11020–11030.

74 Matos, M., Sousa, A.F., Fonseca, A.C. et al. (2014). A new generation of furanic copolyesters with enhanced degradability: poly(ethylene 2,5-furandicarboxylate)-*co*-poly(lactic acid) copolyesters. *Macromolecular Chemistry and Physics* 215 (22): 2175–2184.

75 Chebbi, Y., Kasmi, N., Majdoub, M. et al. (2019). Synthesis, characterization, and biodegradability of novel fully biobased poly(decamethylene-*co*-isosorbide 2,5-furandicarboxylate) copolyesters with enhanced mechanical properties. *ACS Sustainable Chemistry & Engineering* 7 (5): 5501–5514.

76 Min, K., Cuiffi, J.D., and Mathers, R.T. (2020). Ranking environmental degradation trends of plastic marine debris based on physical properties and molecular structure. *Nature Communications* 11 (1): 1–11.

77 Kim, H., Kim, T., Choi, S. et al. (2020). Remarkable elasticity and enzymatic degradation of bio-based poly(butylene adipate-*co*-furanoate): replacing terephthalate. *Green Chemistry* 22: 7778–7787.

78 Loos, K., Zhang, R., Pereira, I. et al. (2020). A perspective on PEF synthesis, properties, and end-life. *Frontiers in Chemistry* 8 (585).

79 Alberts, A.H. and Rothenberg, G. (2017). Plantics-GX: a biodegradable and cost-effective thermoset plastic that is 100% plant-based. *Faraday Discussions* 202: 111–120.

80 Nagata, M., Kiyotsukuri, T., Ibuki, H. et al. (1996). Synthesis and enzymatic degradation of regular network aliphatic polyesters. *Reactive and Functional Polymers* 30 (1–3): 165–171.

81 Nagata, M., Ibuki, H., Sakai, W., and Tsutsumi, N. (1997). Synthesis, characterization, and enzymatic degradation of novel regular network aliphatic polyesters based on pentaerythritol. *Macromolecules* 30 (21): 6525–6530.

82 Nagata, M., Machida, T., Sakai, W., and Tsutsumi, N. (1998). Synthesis, characterization, and enzymatic degradation studies on novel network aliphatic polyesters. *Macromolecules* 31 (19): 6450–6454.

83 Yang, J., Webb, A.R., Pickerill, S.J. et al. (2006). Synthesis and evaluation of poly(diol citrate) biodegradable elastomers. *Biomaterials* 27 (9): 1889–1898.

84 Yang, J., Webb, A.R., and Ameer, G.A. (2004). Novel citric acid-based biodegradable elastomers for tissue engineering. *Advanced Materials* 16 (6): 511–516.

85 Wang, S., Lu, L., and Yaszemski, M.J. (2006). Bone-tissue-engineering material poly(propylene fumarate): correlation between molecular weight, chain dimensions, and physical properties. *Biomacromolecules* 7 (6): 1976–1982.

86 He, S., Yaszemski, M.J., and Mikos, A.G. (2002). Poly(propylene fumarate) cross linked with Poly(ethylene glycol). US Patent 6,384,105.

87 He, S., Yaszemski, M.J., Yasko, A.W. et al. (2000). Injectable biodegradable polymer composites based on poly(propylene fumarate) crosslinked with poly(ethylene glycol)-dimethacrylate. *Biomaterials* 21 (23): 2389–2394.

88 Yaszemski, M.J., Payne, R.G., Hayes, W.C. et al. (1996). In vitro degradation of a poly(propylene fumarate)-based composite material. *Biomaterials* 17 (22): 2127–2130.

89 Roumanet, P.-J., Laflèche, F., Jarroux, N. et al. (2013). Novel aliphatic polyesters from an oleic acid based monomer. Synthesis, epoxidation, cross-linking and biodegradation. *European Polymer Journal* 49 (4): 813–822.

90 Sathiskumar, P. and Madras, G. (2011). Synthesis, characterization, degradation of biodegradable castor oil based polyesters. *Polymer Degradation and Stability* 96 (9): 1695–1704.

91 Teramoto, N., Ozeki, M., Fujiwara, I., and Shibata, M. (2005). Crosslinking and biodegradation of poly(butylene succinate) prepolymers containing itaconic or maleic acid units in the main chain. *Journal of Applied Polymer Science* 95 (6): 1473–1480.

92 Hu, G.-H., Flat, J.-J., and Lambla, M. (1997). Free-radical grafting of monomers onto polymers by reactive extrusion: principles and applications. In: *Reactive Modifiers for Polymers* (ed. S. Al-Malaika), 1–83. Dordrecht: Springer.

93 Kim, D., Kim, W., Lee, D. et al. (2001). Modification of poly(butylene succinate) with peroxide: crosslinking, physical and thermal properties, and biodegradation. *Journal of Applied Polymer Science* 81 (5): 1115–1124.

94 Huang, X., Li, C., Zhu, W. et al. (2011). Ultraviolet-induced crosslinking of poly(butylene succinate) and its thermal property, dynamic mechanical property, and biodegradability. *Polymers for Advanced Technologies* 22 (5): 648–656.

95 Kim, D., Kang, H., and Seo, K. (2001). Peroxide modification of poly(butylene adipate-co-succinate). *Journal of Applied Polymer Science* 81 (3): 637–645.

96 Signori, F., Boggioni, A., Righetti, M.C. et al. (2015). Evidences of transesterification, chain branching and cross-linking in a biopolyester commercial blend upon reaction with dicumyl peroxide in the melt. *Macromolecular Materials and Engineering* 300 (2): 153–160.

97 Persenaire, O., Quintana, R., Lemmouchi, Y. et al. (2014). Reactive compatibilization of poly(L-lactide)/poly(butylene succinate) blends through polyester maleation: from materials to properties. *Polymer International* 63 (9): 1724–1731.

98 Gigli, M., Fabbri, M., Lotti, N. et al. (2016). Poly(butylene succinate)-based polyesters for biomedical applications: a review. *European Polymer Journal* 75: 431–460.

99 Hariraksapitak, P., Suwantong, O., Pavasant, P., and Supaphol, P. (2008). Effectual drug-releasing porous scaffolds from 1,6-diisocyanatohexane-extended poly(1,4-butylene succinate) for bone tissue regeneration. *Polymer* 49 (11): 2678–2685.

100 Tallawi, M., Zebrowski, D.C., Rai, R. et al. (2015). Poly(glycerol sebacate)/poly(butylene succinate-butylene dilinoleate) fibrous scaffolds for cardiac tissue engineering. *Tissue Engineering Part C: Methods* 21 (6): 585–596.

101 Mohanraj, K., Sethuraman, S., and Krishnan, U.M. (2013). Development of poly(butylene succinate) microspheres for delivery of levodopa in the treatment of Parkinson's disease. *Journal of Biomedical Materials Research Part B: Applied Biomaterials* 101 (5): 840–847.

102 Wang, L., Wu, Y., Chen, L. et al. (2005). Fabrication and evaluation of tissue engineering vascular scaffolds based on biodegradable aliphatic–aromatic copolyesters. *Current Applied Physics* 5 (5): 557–560.

103 Santana-Melo, G.F., Rodrigues, B.V., da Silva, E. et al. (2017). Electrospun ultrathin PBAT/nHAp fibers influenced the in vitro and in vivo osteogenesis and improved the mechanical properties of neoformed bone. *Colloids and Surfaces B: Biointerfaces* 155: 544–552.

104 Wang, Y., Ameer, G.A., Sheppard, B.J., and Langer, R. (2002). A tough biodegradable elastomer. *Nature Biotechnology* 20 (6): 602–606.

105 Gao, J., Crapo, P.M., and Wang, Y. (2006). Macroporous elastomeric scaffolds with extensive micropores for soft tissue engineering. *Tissue Engineering* 12 (4): 917–925.

106 Gao, J., Crapo, P., Nerem, R., and Wang, Y. (2008). Co-expression of elastin and collagen leads to highly compliant engineered blood vessels. *Journal of Biomedical Materials Research Part A: An Official Journal of The Society for Biomaterials, The Japanese Society for Biomaterials, and The Australian Society for Biomaterials and the Korean Society for Biomaterials* 85 (4): 1120–1128.

107 Crapo, P.M., Gao, J., and Wang, Y. (2008). Seamless tubular poly(glycerol sebacate) scaffolds: high-yield fabrication and potential applications. *Journal of Biomedical Materials Research Part A: An Official Journal of The Society for Biomaterials, The Japanese Society for Biomaterials, and The Australian Society for Biomaterials and the Korean Society for Biomaterials* 86 (2): 354–363.

108 Yu, Y., Wu, D., Liu, C. et al. (2012). Lipase/esterase-catalyzed synthesis of aliphatic polyesters via polycondensation: a review. *Process Biochemistry* 47 (7): 1027–1036.

109 Sander, M. (2019). Biodegradation of polymeric mulch films in agricultural soils: concepts, knowledge gaps, and future research directions. *Environmental Science & Technology* 53 (5): 2304–2315.

110 Liu, E., He, W., and Yan, C. (2014). 'White revolution' to 'white pollution'—agricultural plastic film mulch in China. *Environmental Research Letters* 9 (9): 091001.

111 Velandia, M., Smith, A., Wszelaki, A. et al. (2018). The economics of adopting biodegradable plastic mulch films.

112 Sintim, H.Y. and Flury, M. (2017). Is biodegradable plastic mulch the solution to Agriculture's plastic problem? *Environmental Science & Technology* 51 (3): 1068–1069.

113 Nisticò, R. (2020). Polyethylene terephthalate (PET) in the packaging industry. *Polymer Testing* 90: 106707.

114 Thiounn, T. and Smith, R.C. (2020). Advances and approaches for chemical recycling of plastic waste. *Journal of Polymer Science* 58 (10): 1347–1364.

115 Raheem, A.B., Noor, Z.Z., Hassan, A. et al. (2019). Current developments in chemical recycling of post-consumer polyethylene terephthalate wastes for new materials production: a review. *Journal of Cleaner Production* 225: 1052–1064.

116 George, N. and Kurian, T. (2014). Recent developments in the chemical recycling of postconsumer poly(ethylene terephthalate) waste. *Industrial & Engineering Chemistry Research* 53 (37): 14185–14198.

117 Chemical Sciences and Society Summit (2020). Science to enable sustainable plastics – A white paper from the 8th Chemical Sciences and Society Summit (CS3). rsc.li/sustainable-plastics-report

118 Dimitris, S.A., Lefteris, A., Ioannis, A.K. et al. (2012). Recent advances in the chemical recycling of polymers (PP, PS, LDPE, HDPE, PVC, PC, nylon, PMMA). In: *Material Recycling - Trends and Perspectives* (ed. D.S. Achilias), 3–64. Rijeka: IntechOpen.

119 Vouyiouka, S.N., Karakatsani, E.K., and Papaspyrides, C.D. (2005). Solid state polymerization. *Progress in Polymer Science* 30 (1): 10–37.

120 Vollmer, I., Jenks, M.J.F., Roelands, M.C.P. et al. (2020). Beyond mechanical recycling: giving new life to plastic waste. *Angewandte Chemie International Edition* 59 (36): 15402–15423.

121 Fakirov, S. (2019). Condensation polymers: their chemical peculiarities offer great opportunities. *Progress in Polymer Science* 89: 1–18.

122 Wang, H., Li, Z., Liu, Y. et al. (2009). Degradation of poly(ethylene terephthalate) using ionic liquids. *Green Chemistry* 11 (10): 1568–1575.

123 Kloxin, C.J., Scott, T.F., Adzima, B.J., and Bowman, C.N. (2010). Covalent adaptable networks (CANs): a unique paradigm in cross-linked polymers. *Macromolecules* 43 (6): 2643–2653.

124 Bowman, C.N. and Kloxin, C.J. (2012). Covalent adaptable networks: reversible bond structures incorporated in polymer networks. *Angewandte Chemie International Edition* 51 (18): 4272–4274.

125 Kloxin, C.J. and Bowman, C.N. (2013). Covalent adaptable networks: smart, reconfigurable and responsive network systems. *Chemical Society Reviews* 42 (17): 7161–7173.

126 Montarnal, D., Capelot, M., Tournilhac, F., and Leibler, L. (2011). Silica-like malleable materials from permanent organic networks. *Science* 334 (6058): 965–968.

127 Denissen, W., Winne, J.M., and Du Prez, F.E. (2016). Vitrimers: permanent organic networks with glass-like fluidity. *Chemical Science* 7 (1): 30–38.

128 Brutman, J.P., Delgado, P.A., and Hillmyer, M.A. (2014). Polylactide vitrimers. *ACS Macro Letters* 3 (7): 607–610.

129 Frich, D., Goranov, K., Schneggenburger, L., and Economy, J. (1996). Novel high-temperature aromatic copolyester thermosets: synthesis, characterization, and physical properties. *Macromolecules* 29 (24): 7734–7739.

130 Frich, D., Economy, J., and Goranov, K. (1997). Aromatic copolyester thermosets: high temperature adhesive properties. *Polymer Engineering & Science* 37 (3): 541–548.

131 Meyer, J.L., Bakir, M., Lan, P. et al. (2019). Reversible bonding of aromatic thermosetting copolyesters for in-space assembly. *Macromolecular Materials and Engineering* 304 (4): 1800647.

6

Polyhydroxyalkanoates (PHAs) – Production, Properties, and Biodegradation

Martin Koller[1,2] *and Anindya Mukherjee*[3,4]

[1] *University of Graz, Research Management and Service, c/o Institute of Chemistry, NAWI Graz, Heinrichstrasse 28/IV, 8010 Graz, Austria*
[2] *ARENA – Association for Resource Efficient and Sustainable Technologies, Inffeldgasse 21b, 8010 Graz, Austria*
[3] *Global Organization for PHA (GO!PHA), Oudebrugsteeg 9, Amsterdam 1012JN, The Netherlands*
[4] *PHAXTEC, Inc., 12324 Hampton Way, Suite 201, Wake Forest, NC 27587, USA*

6.1 Introduction

Plastics are omnipresent in our daily lives; they are inexpensive, easy to process, and offer a multitude of functionalities. In the past 100 years, the industrial synthesis of fossil carbon-based chemosynthetic plastics and generally non-biodegradable plastics increased exponentially. Currently, 400 Megatons (Mt) of plastics is estimated to be produced and consumed annually. As a matter of fact, humankind uses and disposes fossil plastics with little afterthought on the implications; reduction, reuse, recycling, and litter elimination of spent plastics are urgently needed. The lack of biodegradability of the established plastics, which excludes their industrial composting, their lack of proper disposal methods, and therefore their leakage into diverse habitats, has resulted in rapid plastic accumulation in the environment. This has caused well-known detrimental effects such as the pollution of our oceans and ubiquitous generation of secondary microplastic particles in the environment and contaminating our food chain.

A seminal paper published in 2017 by Geyer et al. draws a comprehensive picture of the life cycle and destiny of all chemosynthetic plastics ever made. They calculated that until 2015, a total amount of about 10 Giga tons (Gt) of fossil plastics was synthesized globally; it is estimated that approximately 79% thereof has been deposited in landfills, or even simply released in the environment far more than the approximately 9% of plastics reported to have been recycled, or the roughly 12% of plastics which have been incinerated. The same study estimates that by 2050, 12 Gt of disposed plastic waste will be deposited in landfills or accumulated in the environment, which is double the quantity we have accumulated until 2015 [1]. In 2016, the Ellen MacArthur Foundation published a comprehensive report "The New Plastics Economy: Rethinking the future of plastics" highlighting the urgent need for applying the principles of circular economy to the production of plastics, especially those

Biodegradable Polymers in the Circular Plastics Economy, First Edition.
Edited by Michiel Dusselier and Jean-Paul Lange.

used for packaging. This report was based on contributions by experts from diverse industrial branches and various international organizations and presents a vision of an economy in which spent plastics should never become waste streams. The report concluded that in 2015, roughly 100 million tons of plastics was manufactured only for packaging purposes, making up the largest single source of plastic contamination. Moreover, they estimated that plastic manufacturing currently consumes about 6% of the total fossil mineral oil produced and causes approximately 1% of the total carbon emission that includes their incineration after their life span. At a first glance, these values do not seem to be significant. However, without serious efforts to change these trends, the study estimated that by 2050, plastic production would consume about 20% of the global fossil oil production and contribute to 15% of the atmospheric carbon emissions if the energy sector achieves its stated goal of reducing greenhouse gases by 80% by 2050 [2]. Although this report also mentions biodegradable/compostable materials as urgently needed, a comprehensive and authoritative study on how to incorporate such materials in our society is still lacking.

After the discovery of fossil fuels, such as coal, petroleum, and natural gas, which enabled the industrial revolution, carbon that was sequestered for hundreds of millions of years was suddenly released into the atmosphere. This resulted in the increase in atmospheric concentration of greenhouse gases (GHGs), especially CO_2, CH_4, and NO_x, which are the drivers of the climate change we are facing now. The increased temperatures due to increased GHGs have reduced CO_2 solubility in oceans, thus fueling further CO_2 evolution and further increasing the temperature. Plastics, typical products of crude oil and natural gas, have contributed to this development. At the very least, by accumulating plastics that do not decompose into CO_2, CH_4, or water in a reasonably short time frame, these fossil-based materials have other, even more destructive effects on our planet, including its oceans and all marine life forms, and eventually our primary sources of food [3]. Hence, according to the solutions by "Prevention is better than Cure" [4], plastic accumulation and contamination are the order of the day; however, this requires fundamentally new technological solutions applicable on a large scale and groundbreaking changes in the way the industry currently operates [5]. Although several approaches have been developed over decades and are currently being used, e.g. plastic collection followed by recycling or incineration, these procedures do not address the entire range of problems caused by plastic pollution as described above. This fact can be paraphrased by the famous quotation "less bad is not good enough" once coined by Schnitzer and Ulgiati when it comes to making industrial processes sustainable from a holistic point of view; therefore, we need to expend efforts to improve individual process steps as well as our fundamental approach to the types of industrial processes, feedstocks, and materials we should be using [6]. Although the current approaches for the treatment of spent plastics can be further enhanced and extended to capture significantly larger amounts of used plastics, there will always remain some sources of plastic pollution, which cannot be covered because of their de facto universal occurrence [1].

Recycling of plastic is essential, and the corresponding technologies need to be enhanced (cf. Chapter 6.5 in this book). As a downside, recycling, which includes chemical recycling in the future, will also come with significant expenses, including massive energy requirements (in turn causing further GHG emissions), problems associated with separation of different types of plastics, and quality of degradation products with every recycling step. Therefore, recycling alone is an insufficient tool to address the entire plastic waste predicament. Moreover, plastic recycling just postpones the problem of plastic pollution but does not prevent or cure it. Most of all, recycling can in no way lead to a solution to the microplastic problem; secondary microplastic particles with a size range from 0.1 μm (the size of a SARS-CoV-2 virus particle) to 5 mm (approximately the length of an ant) are created by disintegration of larger plastic materials visible to the naked eye; they are detached from textiles as microplastic fibers (as reported by Napper and Thompson, washing 6 kg cloths made of synthetic fibers in a washing machine releases, depending on the nature of the synthetic material [polyester/cotton blend, polyester, or acrylic], roughly 138 000, 496 000, and 728 000, 700 000 plastic fibers, respectively, per wash into the sewage! [7]), arise as abrasion from plastic shoe soles, and, as a major source of pollution, as abrasion from car tires. In addition, primary microplastics, which are by default added to diverse industrial products such as cosmetics, dental care products, abrasives, detergents, paints, polymeric citrus fruit coatings, adhesives, and others, are also introduced into the environment. Despite the current commitments of the industry to at least reduce the use of primary microplastics, it cannot be expected that they will completely be abolished in the foreseeable future. Moreover, the ratio of primary to secondary microplastics is negligible, the latter being the dominant source over time [1]. Animal studies have already confirmed the detrimental effects of microplastics on their intestinal systems [8]; microplastics can wander through various stages of metamorphosis in insects, which are then consumed by birds [9], and the residual microplastics then disrupt fine-tuned zoological systems [10]. Moreover, microplastic particles traverse the food chain starting from zooplankton or mollusks via crustacea and fish to ultimately enter the digestive systems of marine mammals or reach our dining tables. Recent studies report the occurrence of microplastics in most diverse types of food items such as drinking water [11], table salt [12], honey [13], beer [14], or seafood [15].

As a consequence of these issues, 127 nations have already prohibited plastics for single-use applications such as shopping bags made of fossil plastics [16]. Yet, outlawing plastic use or trusting the existing technologies solely to treat spent fossil plastics are no real solutions for eliminating the adverse ecological effects of plastic waste and GHG emissions, now and tomorrow. The dimension of the problem requires further, more radical, and innovative technologies and specifically solutions that are rooted in what nature teaches us through its natural cycles of up-cycling and biodegrading of all natural materials and best of all its well-established cycling of carbon [3]. This is exactly when such biological materials from renewable resources, which have properties in use matching those of fossil plastics, while, at the same time, remain embedded in circular and natural cycles, are brought into play.

6.1.1 General Aspects of Biodegradation of Polymers

Since the emergence of the very first life forms, biological materials have demonstrated their circular nature; this has kept GHG generation and consumption in a closed cycle. Following these successful principles of nature is the best approach when it comes to production of polymeric materials that act like plastics and prevent plastic pollution. This timeless strategy ensures from the very early stages of process development that accumulation of such xenobiotics, which have a fatal impact on the biosphere, can be obviated. This includes producing compounds from renewable sources such as GHGs like CO_2 and/or CH_4, such as biogas plus H_2O, and, after their life span, returning them to nature. Nature, in turn, transforms these compounds back into CO_2, CH_4, H_2O, and biomass during the biodegradation process, and the cycle gets closed. Biodegradability, and therefore industrial and home composting, needs and must be considered to be an important end-of-life option for single-use consumer plastics. Natural polymeric materials, including cellulose and polyhydroxyalkanoates (PHAs), typically biodegrade in the ecosphere under various conditions; they turn into CO_2, water, and biomass in the presence of O_2 and into CO_2, CH_4, water, and biomass during anaerobic digestion. Importantly, the process of biodegradation of natural materials is circular: photochemotroph-like plants use CO_2, water, and nutrients to grow, and, at their end of life, they biodegrade into the very products they consumed to thrive [3].

Technologically, biodegradation is already performed industrially by increasing the temperature and/or by adding specific microorganisms that consume the organic matter and turn them into CO_2, CH_4, water, and biomass or compost; these processes are called "Industrial Composting." In industrial composting facilities, compost temperatures exceed 50 °C because of the metabolic heat generated by the microbes thriving in it. This high temperature is a legislative requirement in many countries as it ensures sufficient pasteurization of the final product (compost). The same process of biodegrading natural materials can be carried out at lower temperatures; this process is known as "Home Composting." Here, temperatures as high as in industrial composters are typically not reached or reached only in the core, as the compost is agitated/turned less often and has a higher surface-to-volume ratio beneficial for cooling under ambient temperature. Because of a lower aeration in home composting, microbial growth is less intense, and less heat is generated. Timeline is also important in biodegradation and/or composting performance; higher temperature in Industrial Composting speeds up biodegradation, whereas the process of Home Composting takes longer; however, in both processes, biodegradation occurs [17].

6.1.2 General Aspects of Microbial Synthesis of PHAs

Indeed, there are myriads of materials being produced using CO_2 or CH_4 as renewable carbon sources, but only few of them have plastic-like properties and are also converted into CO_2, CH_4, H_2O, and biomass at their end of life by natural processes. PHAs, a group of microbial polyesters found in nature, are one such material family; remarkably, they also have many of the desired properties of the currently market-dominating types of fossil plastics, such as poly(ethylene terephthalate) (PET), polyethylene (PE), polypropylene (PP), polystyrene (PS), poly(vinyl chloride), or polyamides (e.g. nylon), and can be converted/processed using the existing equipment with only minimal modifications in processing conditions. This makes PHAs one of the most auspicious groups of compounds to substitute many of the commercial fossil plastics today, particularly those produced for single-use applications and packaging, which are difficult to collect or recycle [18]. However, the market presence of PHAs is still negligible; although the total global plastic production is estimated with almost 400 Mt/yr, PHA production does not even encompass 0.01% thereof [19]. The large variety of chemical, physical, and mechanical properties of PHAs arises from the numerous types of monomers (150 of them have been identified to date [20]) available to build them. Then, there is the question of polymer design using these 150 different monomers along with our vast knowledge of biology and polymer chemistry – random or block copolymers having various molecular weights and distributions – hence, literally an infinite number of different PHA molecules can be designed to mimic fossil plastics in their functionality and yet be produced from renewable carbon and be biodegradable as an end-of-life option. The discovery of over 150 different monomers that allows for innumerable types of PHA polymers resulted in the introduction of the terminus "PHAome" in the recent scientific literature by a leading research group in this field, analogous to the genome, proteome, transcriptome, lipidome, metallome, or metabolome of organisms. The term "PHAome" reflects the diversity of PHAs already found in nature and to those that still need to be discovered in biological samples [21]. Critically, different types of PHAs may also exhibit different biodegradability behaviors, adding to our ability to tune end-of-life options for PHA polymers [22].

Microbiologically, PHAs are a group of polyoxoesters that are naturally produced and degraded. They accumulate as inert intracellular inclusion bodies in the cytoplasm of around 40% of the world's microbial species, predominantly from prokaryotic genera belonging to the two domains Bacteria and Archaea [23]. Some extraordinary reports on eukaryotic wild-type yeasts as PHA producers occurred in the scientific literature only during the past few years. Their chemical structure resembles a linear polyoxoester of helical structure of the individual PHA chains, which are intracellularly organized in granules, also referred to as "carbonosomes," in which the polyester chains form a hydrophobic core part, surrounded by a hydrophilic layer mainly consisting of enzymes responsible for the cyclic processes for PHA biosynthesis and degradation plus structural proteins (phasins) responsible for the formation, size, shape, and location stabilization of PHA granules [24].

Based on the consumption of heterotrophic (mainly carbohydrates, alcohols, lipids, organic acids, and methane) or autotrophic (fixation of CO_2 and utilization of CO by conversion to CO_2) renewable substrates, microorganisms typically start to accumulate increased quantities of PHA as products of their secondary metabolism when they are exposed to certain nutritional insufficiencies or stressors, which impede the increase of cell density in the culture. Such strict separation of cell growth and PHA accumulation ("non-growth-associated PHA biosynthesis") is reported, e.g. for *Paraburkholderia fungorum* DSM 1749 or *Azotobacter vinelandii* when cultivated on carbohydrates; here, the switch from balanced growth to predominant PHA formation can conveniently be accomplished by stopping the supply of, e.g. the nitrogen source. In many cases, e.g. for the well-described PHA production strains *Cupriavidus necator* DSM 545 (an eubacterium) or *Haloferax mediterranei* DSM 1411 (a haloarchaeon), only low amounts of PHAs are formed during the phase of exponential growth, with PHA productivity suddenly increasing after the onset of the (nutritional) stress ("partially growth-associated PHA biosynthesis"). Some exceptions, however, are reported to synthesize significant fractions of PHAs in cell mass even under environmentally balanced conditions ("growth-associated PHA biosynthesis"), such as *Azohydromonas lata*, a species already used decades ago for (semi)industrial PHA production by Chemie Linz (Austria) or Imperial Chemical Industries (ICI, UK). When these microorganisms experience shortage of exogenous substrates/food, they remobilize and consume the accumulated PHAs as a source of energy and carbon, which gives them an advantage over polymer-free microbes when it comes to survival during periods of starvation (reviewed by [25]). In addition to their role as reserve materials, some intriguing novel functions of PHA were more recently uncovered, predominately substantiating that PHA biosynthesis and formation of their degradation products are the consequences of the "SOS-response" of microorganisms to diverse stressors [26–28] such as osmotic challenge caused by exposure to hypotonic or hypotonic media [29, 30], oxidative stress [31, 32], freezing [33, 34], thermal load [31], or UV irradiation [35].

6.1.3 Types and Properties of PHAs

The type and structure of PHAs that the microorganisms produce depend on their raw material source and the production organism. By 1995, 95 different types of PHA polymers and their corresponding unique building blocks (hydroxyalkanoic and some hydroxyalkenoic acids) were discovered, and to date, about 150 different types of PHA polymers and their corresponding building blocks have been described. Notably, many of these 150, often xenobiotic compounds were only polymerized in test tubes to PHAs by isolated PHA synthase enzymes and were never found in PHAs produced by living cells [36]. Nonetheless, all of them are produced using renewable carbon, and many of them are predicted to have the desired functionality for commercial use and most of them biodegrade at the end of life, the three essential criteria for circularity in use. Figure 6.1 shows the general chemical structure of PHAs and PHA granules ("carbonosomes") included in microbial cells.

To date, only around 15 different naturally occurring building blocks and their corresponding PHA polymers have been thoroughly studied. A lot of work has gone into elucidating their properties and applicability in industrial and consumer applications such as films, fibers, thermoformed and molded parts for use in packaging, food service, agriculture, medical devices, electronics, leisure industry, fabrics, paints and coatings, adhesives, etc. [37].

Like many types of natural polymers, PHAs can either consist of only one type of monomer, or two or more different building blocks. In the first case, the so-called homopolymers, poly(3-hydroxybutyrate) (PHB or P3HB) is the only important naturally occurring example discovered to date and the by far most frequently occurring and best-studied PHA from biological samples. However, its application is restricted

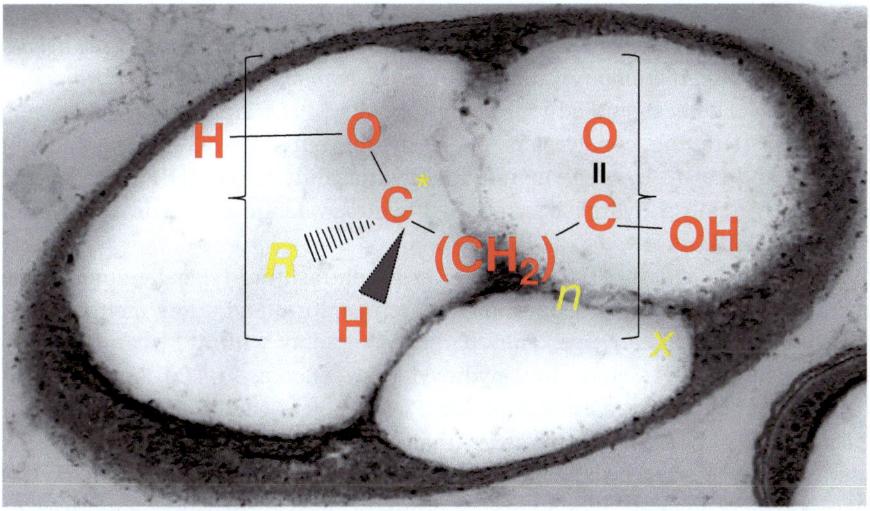

Figure 6.1 General formula for natural PHA biopolyesters; the asterisk (*) indicates the chiral center of most PHA constituents (important exception: achiral 4-hydroxybutyrate – 4HB). *n*: number of methylene groups in monomer's backbone; *x*: degree of polymerization (number of monomers building up the PHA polymer molecule); *R*: side chain of the monomer. Example: $R = CH_3$, $n = 1$ correspond to 3-hydroxybutyrate – 3HB. Gray background photograph: scanning transmission electron microscope (STEM) picture of a PHA-rich *Cupriavidus necator* cell cultivated on glucose (picture provided with courtesy by E. Inglolić, FELMI/ZFE, Graz, Austria); the white inclusion bodies, amounting to about 80% of the cell dry mass, are PHA granules. Source: Courtesy of E. Inglolić.

because of its high crystallinity, melting temperature close to its degradation temperature, and brittleness.

Better material properties and enhanced processability are achieved when interrupting this crystalline PHB matrix by inserting comonomers in the polymer backbone. The generated heteropolyesters are termed copolymers when two different building blocks are present, and terpolymers when three different monomers are present and infrequently quaterpolymers when four different monomers build up the PHAs. PHAs containing more than one comonomer generally have lower crystallinity, lower melting temperature, and enhanced processability in comparison to PHB (reviewed by [20]).

Depending on the chain length of monomers, PHAs can be divided into three major groups; PHAs consisting of monomers with three to five carbon atoms (3-hydroxypropionate [3HP; achiral], 3-hydroxybutyrate [3HB; chiral], 4-hydroxybutyrate [4HB; achiral], 3-hydroxyvalerate [3HV; chiral], 4-hydroxyvalerate [4HV; chiral], 5-hydroxyvalerate [5HV; achiral]) are termed short chain length PHAs (*scl*-PHA). Such *scl*-PHAs, with the exception of the highly flexible poly(4-hydroxybutyrate) (P4HB) and copolyesters of 3HB and 4HB with high 4HB fractions [38], are rather crystalline, have high melting points and glass transition temperature, and resemble typical thermoplastic materials from fossil carbon, such as polypropylene (PP). The most important *scl*-PHA production microorganisms are *C. necator*, *Burkholderia sacchari*, *Alcaligenes latus*, (today: *Azahydromonas lata* and *Azahydromonas australica*) or *Haloferax mediterranei*. The best-studied representatives of *scl*-PHA are the homopolyesters PHB and P4HB, the copolyesters poly(3-hydroxybutyrate-*co*-3-hydroxyvalerate) (PHBHV) and poly(3-hydroxybutyrate-*co*-4-hydroxybutyrate) (P3HB4HB), and the terpolyester poly(3-hydroxybutyrate-*co*-3-hydroxyvalerate-*co*-4-hydroxybutyrate) (P3HB3HV4HB) (reviewed by [20]).

Medium chain length PHAs (*mcl*-PHA), in contrast, are typically amorphous, rubber-like resins with low glass transition temperature and a low melting point or melting temperature range. They harbor monomers with 6–14 carbon atoms (3-hydroxyhexanoate, 3HHx, to 3-hydroxytetradecanoate, 3HTD) and are most frequently produced by members of the *Pseudomonas* genus. In addition, they typically have lower molecular mass than *scl*-PHA. Such *mcl*-PHAs are only experimentally produced in small quantities but are likely to enter the market as smart polymers such as "bio-latexes," thermo- or pressure-sensitive adhesives, or carriers for bioactive compounds [39, 40]. Using *mcl*-PHAs containing monomers with unsaturated side chains, readily biodegradable rubber-like polymers of superior product properties including those that are cross-linked can be produced by post-synthetic modification such as cross-linking [41].

The third group, long chain length PHAs (*lcl*-PHA) are scarcely described; these materials have monomers with more than 14 carbon atoms (3-hydroxypentadecanoate, 3-hydroxyhexadecanoate, 3-hydroxyheptadecanoate, 3-hydroxyoctadecanoate, unsaturated 3-hydroxyoctadecenoate, and others), with properties similar to those of *mcl*-PHAs with long side chain in monomers (reviewed by [20]).

6.2 Biosynthesis – Substrates and Strains

6.2.1 Principle Stoichiometry of PHA Biosynthesis

The following equations provide an overview about the principle stoichiometry of PHB biosynthesis under different conditions and theoretical substrate-to-PHB conversion yields:

(a) Chemoheterotrophic, aerobic conversion of glucose [42, 43]:

$$C_6H_{12}O_6\ (\text{glucose}) + (3p+1)\text{ADP} + (3p+1)\text{P}_\text{i} + (3/2)O_2$$
$$\rightarrow (1/n)(C_4H_6O_2)_n\ (\text{PHB}) + 2CO_2 + 3H_2O + (3p+1)\text{ATP}$$

p: P/O ratio; Pi: inorganic phosphate

Theoretical yield for aerobic PHB biosynthesis from glucose when converted via the Entner–Doudoroff (2-keto-3-desoxy-6-phosphogluconate [KDPG]) pathway: Based on the equation $Y = \frac{43\,x}{2(12x+y+16z)}$ (for glucose: $C_6H_{12}O_6$; $x = 6, y = 12, z = 6$) [41, 42]:

$$Y_{\text{PHB/glu (theor.)}} = 86/180\ \text{g/g} = 0.48\ \text{g/g}$$

Aerobic PHB formation from glucose is energetically favored; for the reaction of glucose to 3-hydroxybutyryl-CoA, the direct precursor of PHB, Pan et al. [44] report an energy gain of 193.72 kJ/mol, which is however lower than for the complete respiration of glucose to CO_2:

$$C_6H_{12}O_6\text{ (glucose)} + \text{free CoA} + 3NAD^+$$
$$\rightarrow 3\text{HB-CoA} + 2CO_2 + 3NADH + 3H^+;\ \Delta G^0 = -193.72\ \text{kJ/mol}$$

$$C_6H_{12}O_6\text{ (glucose)} + 6O_2 \rightarrow 6CO_2 + 6H_2O;\ \Delta G^0 = -2821.98\ \text{kJ/mol}$$

ΔG^0: Gibbs free energy change at standard conditions

(b) Chemoheterotrophic, aerobic conversion of fatty acids via β-oxidation (example: linoleic acid) [45]:

$$C_{18}H_{32}O_2 + 4.75O_2 \rightarrow 4.5\text{PHB} + 2.5H_2O$$

$$Y_{\text{PHB/acid (theor.)}} = 1.5\ \text{g/g}$$

(c) Chemoheterotrophic, aerobic conversion of glycerol [45]:

$$2C_3H_8O_3 + 2.5O_2 \rightarrow \text{PHB} + 2CO_2 + 5H_2O$$
$$Y_{\text{PHB/acid (theor.)}} = 0.46\ \text{g/g}$$

Energy generated by the formation of 1 mol 3HB-CoA from 2 mol glycerol: $\Delta G^0 = -996.15\ \text{kJ}$

(d) Chemoheterotrophic, aerobic conversion of methane by type-II methanotrophs via the serine pathway [43]:
$CH_4 + O_2 + NADH + H^+ \rightarrow CH_3OH + NAD^+$
$CH_3OH + \frac{1}{2}O_2 + ADP + P_i \rightarrow HCOH + ATP$
 (i) $HCOH + NAD^+ \rightarrow HCOOH + NADH;\ HCOOH + NAD^+ \rightarrow CO_2 + NADH$
 (ii) $HCOH + \text{glycine} \rightarrow \text{serine} \rightarrow\rightarrow\rightarrow \text{malyl-CoA} \rightarrow \text{acetyl-CoA}$
 (iii) $2\text{acetyl-CoA} + NADPH \rightarrow\rightarrow\rightarrow \text{PHB} + NADP^+$

$$\text{Sum}:\ 5CH_4 + 4O_2 + (5q-3)ADP + (5q-3)P_i \rightarrow \text{PHB} + CO_2 + 4H_2O + (5q-3)ATP$$

q: mol adenosine triphosphate (ATP) to be produced from the conversion of methanol to methanal; some ATP is utilized in the serine cycle to generate malolyl-CoA from malate

$$Y_{\text{PHB/CH4 (theor.)}} = 0.54\ \text{g/g}$$

Energy generated by the formation of 1 mol 3HB-CoA from 5 mol CH_4: $\Delta G^0 = -1148\ \text{kJ}$

(e) Photoautotrophic conversion of CO_2 by cyanobacteria via the Calvin cycle [46]:
 (i) Calvin cycle: $3CO_2 + 6NADPH + 9ATP + 5H_2O \rightarrow GAP + 6NADP^+ + 9ADP + 8P_i$
 (ii) $GAP + NAD^+ + ADP + P_i \rightarrow PEP + NADH + ATP$
 (iii) $PEP + ADP \rightarrow \text{pyruvate} + ATP \rightarrow \text{acetyl-CoA} + CO_2$

(iv) $2\text{Acetyl-CoA} + \text{NADPH} \rightarrow\rightarrow\rightarrow \text{PHB} + \text{NADP}^+$

$$\text{Sum}: 4\text{CO}_2 + 14\text{ATP} + 10\text{H}_2\text{O} + 13\text{NADPH} + 2\text{NAD}^+ \rightarrow \text{PHB}+ 14\text{ADP} + 14\text{P}_\text{i} + 13\text{NADP}^+ + 2\text{NADH}$$

GAP: glycerolaldehyde-3-phosphate; PEP: phosphoenolpyruvate

$$Y_{\text{PHB/CO2 (theor.)}} = 1.95\ \text{g/g}$$

(*Nota bene*: many side reactions take place for the formation of pigments, amino acids, lipids, glycogen, acetaldehyde, etc.!)
Energy generated by the formation of 1 mol 3HB-CoA from 4 mol CO_2: $\Delta G^0 = -4832.58$ kJ

6.2.2 Biosynthesis of *scl*- and *mcl*-PHAs

The mechanism of microbial PHA biosynthesis has been extensively studied. In principle, carbon sources are converted to oxoacyl-CoA thioesters, often to acetoacetyl-CoA generated by the condensation of the central metabolite acetyl-CoA; this reaction is catalyzed by the enzyme 3-ketothiolase (PhaA; a.k.a. acetyl-CoA – acetyltransferase, formerly known as β-ketothiolase; EC 2.3.1.9). After being catalyzed by the NADPH-dependent oxoacyl-CoA reductase (PhaB), the formed oxoacyl-CoA thioesters are reduced to the corresponding (*R*)-hydroxyacyl-CoA thioesters. Finally, polyester synthases (PhaC; former name: PHA polymerases) catalyze the linking of (*R*)-hydroxyacyl-CoA thioesters to form growing polyester chains; this polycondensation releases one molecule of free CoA per docked monomer (reviewed by [20]).

In microorganisms, four different classes (I–IV) of PHA synthases have been identified based on their subunits, amino acid sequences, and substrate specificity [47]. In general, high intracellular concentrations of ATP, NAD(P)H, and acetyl-CoA, indicators for a high energy charge of the cell, and low concentrations of free CoA favor the activity of enzymes involved in PHA biosynthesis; such conditions are found, e.g. when the tricarboxylic acid cycle (TCC) or other central metabolic pathways are interrupted due to the deprivation of a growth-essential nutrient, e.g. at the limitation of nitrogen or phosphate sources, in parallel to sufficient availability of exogenous carbon source. Mechanistically, the PHA biosynthesis provides the cells with a tool to regenerate $NADP^+$ by the above-described redox reaction between NADPH and acetoacetyl-CoA; this "pseudofermentation" generates (*R*)-hydroxybutyryl-CoA and $NADP^+$, the latter is used by the microbial cell for substrate oxidation, thus enabling nutrient consumption for energy generation even during conditions blocking the activity of TCC [20]. Figure 6.2 illustrates the cycle of PHA biosynthesis and biodegradation.

Class I PHA synthases, e.g. the enzyme found in the best know PHA producer *C. necator*, polymerize (*R*)-hydroxyacyl-CoA thioesters with the acyl group consisting of three to five carbon atoms. Moreover, class III PHA synthases, such as the one of

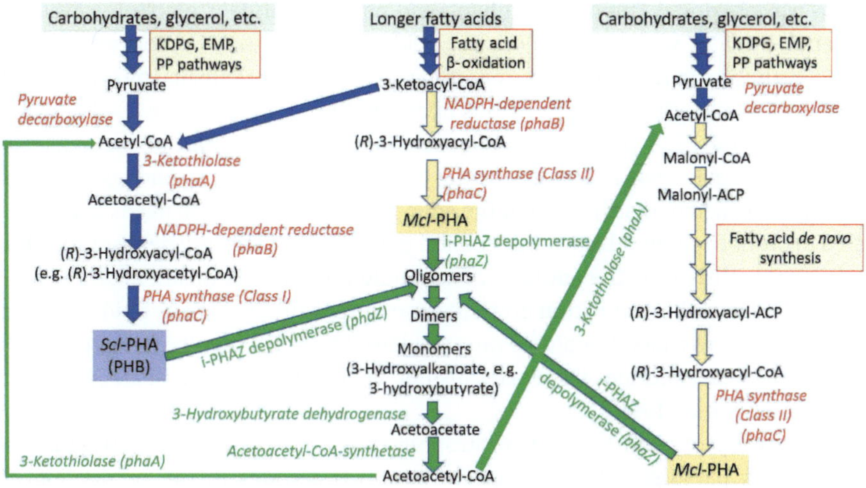

Figure 6.2 Cycle of intracellular PHA biosynthesis and degradation. Blue arrows lead to biosynthesis of short-chain-length PHAs (*scl*-PHA; e.g. PHB), beige arrows to biosynthesis of medium-chain-length PHA (*mcl*-PHA), and green arrows to remobilization of PHA reserves. Red font: PHA synthesis enzymes; green font: enzymes catalyzing PHA catabolism. EMP, Embden–Meyerhof–Parnas pathway.

Allochromatium vinosum, and class IV PHA synthases, e.g. in *Bacillus megaterium*, also generate *scl*-PHAs [20].

Class II PHA synthases preferentially polymerize longer (*R*)-hydroxyacyl-CoA thioesters with the acyl group consisting of 6–14 carbon atoms as substrates, thus forming *mcl*-PHAs [48]. Such compounds are often generated by β-oxidation of the respective fatty acids. This β-oxidation first generates (*S*)-3-hydroxyacyl-CoA, which cannot be directly incorporated into *mcl*-PHAs due to stereospecificity of the synthase's subunit. Therefore, specific isomerases that convert the (*S*)- into the (*R*)-isomers are active in such organisms. In addition, *mcl*-PHAs can also be produced from unrelated carbon sources such as sugars or glycerol via the fatty acid *de novo* synthesis. Different from β-oxidation, fatty acid *de novo* synthesis generates the (*R*)-isomers of acyl-CoAs, which are directly used as substrates for polymerization by class II synthases [39, 40]. Natural *mcl*-PHAs produced by wild-type organisms exclusively present heteropolyesters consisting of different building blocks. Homopolyesters, e.g. P(3HHx), poly(3-hydroxyoctanoate) (P3HO), or poly(3-hydroxydecanoate) (P3HD), can only be obtained by genetically engineered organism; in this case, the β-oxidation pathway needs to be weakened to avoid generation of diverse substrates for the PHA polymerase [47]. Similar to *mcl*-PHAs, the polymerization of *lcl*-PHA is also catalyzed by class II synthases (reviewed by [47]).

6.2.3 Heterotrophic Feedstocks

Typical substrates for PHA biosynthesis by heterotrophic microorganisms (dependent on organic substrates as carbon sources) are either hydrophilic compounds such as carbohydrates (mono-, di-, oligo-, and polysaccharides), alcohols (glycerol, methanol, etc.), or water-soluble organic acids. In such cases, homogeneous substrate distribution in the bioreactor system can easily be accomplished by established agitation systems (stirrer). Also, various hydrophobic compounds are used as starting materials for PHA production. For example, triacylglycerides or methylesters of fatty acids have been described. In the case of substrates that have poor water solubility, a sufficient distribution in the aqueous cultivation phase needs to be warranted in order to allow cell convenience and ready access to the nutrient source; this can be accomplished by the addition of adequate non-toxic emulsifiers and appropriate agitation of the cultivation broth to generate substrate (such as lipid) droplets with a maximum surface-to-volume ratio (reviewed by [25]). In some cases, the PHA production strain generates its own emulsifier, as shown for the *mcl*-PHA producer *Pseudomonas putida* [49].

The discovery of such novel heterotrophic substrates for bioplastic production has created a paradigm shift during the past two decades in the production of PHAs. Before, PHA production was predominately studied and carried out using "first-generation" substrates; hence, bioplastic manufacturing competes with food sources, such as glucose, sucrose, or edible oils. The use of "second-generation" feedstocks such as carbon-rich waste and surplus materials from diverse agricultural and industrial sources, has allowed biopolymer industry to overcome the "Food vs. Bioplastic" debate. The prime examples of such second-generation carbon sources for PHA production include hydrophilic, often complex, substrate cocktails such as ultra-filtrated surplus whey [50], molasses [51], crude glycerol from the biodiesel industry [52], hydrolyzed lignocellulose materials of diverse origins [53–55], sugar-rich waste from fruits and wine production [56, 57], yeast industrial wastewater [58], or hydrolyzed abundant chitin waste from crustacea aquaculture [59]. The lipophilic surplus feedstocks for PHA production are waste cooking oil [60], waste water from olive oil production [61], non-edible plant oils [62], or low-quality biodiesel fractions produced from lipophilic slaughtering waste [63]. Currently, numerous efforts are carried out globally to enhance PHA bioproduction from these diverse waste streams via genetically engineering of the production organisms [64].

An actual trend on the conversion of liquid waste streams with high biological oxygen demand (BOD) is its acidogenic conversion to a mixture of diverse organic acids, which is followed by conversion of these organic acids to PHA by mixed microbial cultures enriched in powerful PHA-producing microorganisms. Such enrichment is typically achieved by cyclic feast-and-famine feeding of mixed cultures; during the famine phases, PHA-rich cells have a survival advantage by their biopolymer storage and can refill this storage in the next feast phase; in contrast, the share of organisms not able to accumulate PHA will progressively decrease with time. Here, process conditions are highly decisive to achieve high productivity; the composition of the organic acid mix achieved during acidogenesis significantly impacts the composition of the PHA accumulated by the microbial community in the next step [65–67]. Among such liquid waste streams, the use of municipal solid waste [68], dairy waste [69], food waste [70], or crude glycerol [71] is reported in the literature.

6.2.4 Autotrophic Feedstocks

Beside heterotrophic PHA biosynthesis, several microbes are able to convert CO_2 as a carbon source in photoautotrophic (light as the energy source and inorganic compounds as the carbon source) or mixotrophic (supply of inorganic carbon source – CO_2 – plus organic compounds) cultivations. Especially, many cyanobacteria are highly versatile cell factories for solar-driven generation of biomass, biopolyesters, and other valuable bioproducts such as pigments, especially phycobiliproteins, in photoautotrophic setups. Such microorganisms fit well into the currently topical biorefinery concepts, which do not only utilize the valuable intracellular bioproducts but also upgrade the residual biomass for use as fertilizers or substrates for anaerobic biogas generation [72]. A range of diverse cyanobacteria from different genera have already been tested in lab scale for their potential as PHA producers in order to evaluate their productivity and dependence on light regime, cultivation vessels, macro- and micronutrient supply, feeding regime (autotrophic and mixotrophic), and type of CO_2 source. Among all these strains, representatives of the genera *Synechocystis*, *Synechococcus*, *Arthrospira* (*Spirulina*), *Oscillatoria*, *Calothrix*, and *Nostoc* currently appear most promising for large-scale processes, while among them, *Synechocystis* sp. was heavily subjected toward genetic engineering studies to improve PHA productivity and to close diverse metabolic bottlenecks [73]. High concentration of CO_2 can be found, e.g. in combustion flue gas from energy generation plants or "off gases" from different industrial-scale chemical processes such as cement production or fermentation industries (e.g. breweries). Mostly, such CO_2 sources need to be pretreated by different processes to remove chemical contaminants such as SO_x and NO_x before they can be supplied to the cyanobacterial culture [74]. However, this approach of using waste gas as the substrate combines mitigation of CO_2 from gaseous industrial effluents with bioproduction of marketable products such as PHA, as it was already successfully shown by Austrian researchers on a pilot plant-scale process operated for three years [75]. Unfortunately, such setups did not exceed yet pilot-scale processes; additional investigations need to be carried out to achieve adequate volumetric productivity by developing advanced photobioreactor setups, selection of cyanobacteria with excellent PHA accumulation capacity, and application of tools of genetic engineering to close metabolic bottlenecks [73]. If successfully optimized, such processes for production of PHAs, pigments, etc., based exclusively on the conversion of waste streams (flue gas for CO_2 supply plus agro-industrial effluents and wastewater and anaerobic digestates as nutrient sources), would constitute prime examples of "The Circular Bioeconomy" [76]. Research has successfully demonstrated that after extracting the PHA from *Spirulina* sp. LEB 18, the residual cyanobacterial biomass can be successfully recycled as the nutrient source to subsequent cultivations of the same organism [77]. To date, such cyanobacteria have been demonstrated to exclusively produce PHB homopolyesters from CO_2 as a sole carbon source, while copolyesters containing 3HV as comonomers were only obtained when cofeeding heterotrophic, structurally related cosubstrates such as valerate. This copolyester biosynthesis was demonstrated for the microorganism

Nostoc muscorum Argadh [78]. Recently, Tarawat et al. have shown that *Nostoc* sp. TISTR 9131 and TISTR 8164 and some *Anabaena* sp. were able to synthetize PHBHV from mixotrophic CO_2 and acetate supply, while *Anabaena spiroides* TISTR 8075 generated PHBHV even from photoautotrophic cultivation on CO_2 as a sole carbon source [79].

6.2.5 Syngas

Syngas, containing the carbon compounds CO and CO_2, can be used as a substrate for PHA bioproduction by *Rhodospirillum rubrum*, a Gram-negative, facultatively phototrophic purple non-sulfur bacterium. This organism is able to fix CO_2 via the well-known enzyme Ribulose-1,5-bisphosphate carboxylase/oxygenase (RuBisCo; EC 4.1.1.39) in the Calvin cycle and oxidize CO via a carbon monoxide dehydrogenase (EC 1.2.7.4) in the so-called "water-gas shift reaction" to CO_2 and H_2 (reviewed by [80]). Both artificial syngas (mixtures of CO, CO_2, H_2, and N_2) [81] and the pyrolysis products of organic waste materials were successfully applied for this purpose. Predominately, *scl*-PHAs were to this day obtained by syngas-based cultivations; only after genetic engineering of *R. rubrum*, it was possible to produce a rubber-like *mcl*-PHA bio-latex from syngas [82]. Organic waste converted to syngas for PHA production encompasses household waste [83], hydrocarbons [84], and food waste [85].

Recently, syngas was used as a substrate for anaerobic production of PHAs by obligate anaerobic acetogenic *Clostridium* strains, which are already routinely cultivated on the industrial scale for fodder production. Here, recombinant strains of *Clostridium coskatii* and *Clostridium ljungdahlii* expressing a new artificial PHA biosynthesis pathway were successfully tested for biopolymer production in flask-scale setups. The fermentation technology of gaseous substrates by wild-type Clostridia has considerably advanced during the past decade and has already reached multi-tons industrial scale cultivation setups in PR China. The authors of the study consider this process as an economically feasible process toward PHA production from waste and greenhouse gases [86].

6.2.6 Methane

Beside CO_2 and syngas, the greenhouse gas methane is widely available in natural gas at a volume share of greater than 90%, in emissions from landfills, and constitutes the flammable component in biogas originating from anaerobic digestion of organic matter (including spent PHAs or non-PHA residual biomass after biopolymer recovery!). Methanotrophic bacteria, a group of Gram-negative organisms of the methylotroph subgroup, are able to convert methane as a source of carbon and energy via the ribulose monophosphate pathway (RMP) or via the serine pathway; several methanotrophs are described to accumulate PHB. The *Methylocystis* genus encompasses various type II methanotrophic species, which resort to the serine pathway for PHB biosynthesis. Such organisms, such as *Methylocystis hirsuta*, can be found in various environments. For water-insoluble gases such as methane, mass transfer in bioreactors plays a crucial role for the cultivation process. In this context, Moradi et al. modeled the optimum conditions for PHB production from natural gas in a bubble column bioreactor using the COMSOL software, considering factors such as methane-to-air ratio, inlet gas rate, and gas hold up [87]. Rahnama et al. described PHB production by *M. hirsuta* from mixtures of natural gas and air in a simple bubble column and a forced liquid vertical tubular loop bioreactor, achieving an average PHB content in biomass of about 52 wt% [88]. This process was improved by Ghoddosi et al. who achieved increased methane uptake by the strain and high density by adding organic cosubstrates such as acetate, glucose, and methanol. In a bubble column bioreactor operated at optimum conditions, 8 g/l of biomass with a PHB content of 73.4 wt% was obtained [89].

From the technoeconomic point of view, methane-based PHA production is feasible and has been demonstrated to have the potential to be economically competitive. A study by Levett et al. reports a production price of PHA in a similar range like PHA biosynthesis from purified sugars; these authors claim high cooling costs as the decisive cost factor in this process, which compensates the gain from lower raw material costs [90]. However, they failed to consider more modern and advanced gas fermentation reactor designs that are already commercially available and are claimed to produce single-cell proteins (SCPs) economically using methane fermentation (commercialized by the companies Calysta [91] and Unibio [92]). In addition, the companies Newlight Technologies [93] and Mango Materials [94] have also successfully demonstrated PHB production using methane fermentation. The breakthrough in this regard is imminent because methane represents one of the most suitable and pure forms of carbon for industrial production of materials, and polyolefins from natural gas in North America represent one of the most inexpensive fossil plastics available today, primarily because of the currently low price of natural gas. Methane emissions from landfills and anaerobic digesters constitute a viable alternative to methane from natural gas; when used for PHA synthesis, this could theoretically replace about 20–30% of the total plastics produced annually today, but having the biodegradability characteristics that are so desired. Therefore, methane-based PHA production not only opens new avenues for biopolyester production but also mitigates the issues related to GHG emissions. As recently reviewed by Liu et al., factors such as the typically low efficiency of methanotrophs and often insufficient gas–liquid mass transfer in the bioreactors used need to be overcome. Moreover, most reports on PHA production to date are restricted to PHB homopolyester, which results in modest material properties. These authors therefore suggest the application of feast-famine feeding regimes, genetic engineering of microorganisms to increase specific PHA productivity, development of novel gas-permeable membrane bioreactors and two-phase partitioning bioreactors to overcome low solubility of methane and oxygen, and development of protocols for the production of PHA copolyesters with tailored molecular mass (by optimizing the feeding regime of 3HV-related precursors and selection of the appropriate growth-limiting factor) [95]. New methanotrophic strains as potential PHA producers are steadily isolated and described, as shown by studies of Khosravi-Darani et al. for the new *Microbacterium* sp. PBCC6 [96] or for Orita et al.'s *Methylobacterium extorquens* AM1, a strain that interestingly accumulates the PHB homopolyester during methylotrophic cultivation, but PHBHV copolyester under cobalt-limited conditions on methanol as the sole carbon source [97].

6.2.7 Production Strains

PHA biosynthesis is described for diverse groups of prokaryotes, such as Gram-negative and Gram-positive eubacteria, chemoheterotrophs, photoautotrophs, mixotrophs, and a plethora of typically extremophilic archaea. Currently, the quest for new powerful PHA production strains is steadily ongoing. Table 6.1 provides an updated overview of genera encompassing PHA-accumulating wild-type species.

In addition to prokaryotes, a very limited number of reports are available for PHA accumulation by eukaryotic wild-type strains. Oligomers of 3HB with low molecular mass, however, are found in almost all life forms, also in the human metabolism. Moreover, PHA biosynthesis genes were successfully transferred to different eukaryotes, which then displayed PHA accumulation in their biomass. Here, the pioneering work of Yves Poirier's research group from the past 30 years on photoautotrophic PHA accumulation by green plants, especially by the heavily studied genetic and plant-metabolic model organism *Arabidopsis thaliana*, needs to be emphasized. Starting from CO_2 fixation, these recombinant plants provide real phyto-factories, producing in parallel PHAs, carbohydrates, lipids, pigments, and diverse phytochemicals with the potential market value. Although first generation of transgenic plants for PHA production produced the biopolyester in the plant cells' cytoplasm, nucleus, and vacuole at rather low productivity [101], the second generation showed increased PHA yields by storing the material in the plastids [102]. For relevant reviews on *in planta* PHA biosynthesis, please refer to Refs. [103, 104].

Mcl-PHA biosynthesis was achieved in the peroxisomes of the yeast *Saccharomyces cerevisiae* when equipping the organism with *P. putida* PHA synthesis genes, which express the class II PHA synthase able to metabolize the intermediate of β-oxidation for *mcl*-PHA synthesis [105]. Similar experiments were also carried out with the yeast *Pichia pastoris*, which was equipped with *Pseudomonas aeruginosa* PHA synthase genes [106]. More recently, the oleaginous yeast *Yarrowia lipolytica* was genetically modified. It was shown that by variation of the yeast's β-oxidation multifunctional enzyme, it was possible to trigger the monomeric composition of the produced PHA; both homo- and heteropolyesters were obtained by this strategy [107]. When expressing the *P. aeruginosa* PAO1 *PhaC1* gene together with a PTS1 peroxisomal signal in *Y. lipolytica*, cultivation on oleic acid resulted in the accumulation of *mcl*-PHA heteropolyesters with monomers from C6 (3HHx) to C14 (3HTD) [108]. However, PHA production in eukaryotic cells such as yeasts is more complex compared to prokaryotes because of the compartmentation of the metabolic pathways in such higher developed organisms; hence, it would be beneficial to have wild-type yeasts available for this purpose. Indeed, it was reported that the wild-type isolate *Rhodotorula minuta* Y4 accumulated 2 wt% of PHBHV copolyester when cultivated on glucose and oleic acid [109].

Table 6.1 Compilation of PHA-accumulating microbial genera. Update from [98]

Eubacteria		Archaea
Gram-negative genera	**Gram-positive genera**	**Three-letter code for haloarchaea according to Litchfield [99]**
(A) *Accumulibacter*, *Acidovorax*, *Acinetobacter*, *Actinobacillus*, *Aeromonas*, *Alcaligenes*, *Alcanivorax*, *Alkalilimuicola*, *Allochromatium*, *Amphritea*, *Anabaena*[a], *Anacystis*[a], *Aphanocapsa*[a], *Aphanothece*[a], *Aquaspirillum*, *Aquitalea*, *Aromatoleum*, *Arthrospira*[a], *Asticcaulus*, *Aulosira*[a], *Aureobasidium*, *Axobacter*, *Azohydromonas*, *Azomonas*, *Azospirillum*, and *Azotobacter* **(B)** *Beggiatoa*, *Beijerinckia*, *Beneckea*, *Brachymonas*, *Bradyrhizobium*, *Brevundimonas*, and *Burkholderia* **(C)** *Caenibacterium*, *Caldimonas*, *Calothrix*[a], *Caryophanon*, *Caulobacter*, *Chelatococcus*, *Chloroflexus*[a], *Chlorogleopsis*[a], *Chroococcus*[a], *Chromatium*, *Chromobacterium*, *Chromohalobacter*, *Cobetia*, *Comamonas Competibacter*, *Cupriavidus* (formerly: *Alcaligenes*, *Ralstonia*, and *Wautersia*), *Cyanobacterium*[a], and *Cyanothece*[a] **(D)** *Dechloromonas*, *Defluviicoccus*, *Delftia*, *Desulfococcus*, and *Desulfobacterium* **(E)** *Ectothiorhodospira*, *Ensifer* (formerly: *Sinorhizobium*), and *Erwinia* **(F)** *Ferrobacillus*, *Fischerella*[a], and *Fundibacter* **(G)** *Gamphosphaeria*, *Gloeocapsa*[a], and *Gloeothece*[a] **(H)** *Haemophilus*, *Haererehalobacter*, *Hahella*, *Halomonas*, *Halorhodospira*, *Hapalosiphon*[a], *Hydrogenophaga*, *Herbaspirillum*, and *Hyphomicrobium*	**(A)** *Actinomycetes* and *Aneurinibacillus* **(B)** *Bacillus* and *Brochothrix* **(C)** *Clostridium* and *Corynebacterium* **(D)** *Derxia* **(I)** *Isoptericola* **(L)** *Lysinibacillus* **(M)** *Microbacterium*, *Micrococcus*, and *Microlunatus* **(N)** *Nocardia* **(P)** *Paucispirillum* and *Piscicoccus* **(R)** *Rhodococcus* **(S)** *Staphylococcus*, *Streptomyces*, and *Syntrophomonas* **Eukaryotes** **(R)** *Rhodotorula*[e] **(S)** *Scenedesmus*[f]	**(C)** *Cenarchaenum* **(H)** *Halalkalicoccus* (*Hac.*) *Halobacterium* (*Hbt.*) *Haloarcula* (*Har.*) *Halobiforma* (*Hbf.*) *Halococcus* (*Hcc.*) *Haloferax* (*Hfx.*) *Halogeometricum* (*Hgm.*) *Halogranum* (*Hgn.*) *Halomicrobium* (*Hmc.*) *Halopiger* (*Hpg.*) *Haloquadratum* (*Hqr.*) *Haloterrigena* (*Htg.*) *Halorhabdus* (*Hrd.*) *Halorubrum* (*Hrr.*) *Halostagnicola* (*Hst.*) **(N)** *Natrialba* (*Nab.*) *Natrinema* (*Nnm.*) *Natronobacterium* (*Nbt.*) *Natronococcus* (*Ncc.*) *Natronomonas* (*Nmn.*) *Natronorubrum* (*Nrr.*) *Nitrosopumilus*

Table 6.1 (Continued)

Eubacteria		Archaea
Gram-negative genera	**Gram-positive genera**	**Three-letter code for haloarchaea according to Litchfield [99]**
(I–K) *Ideonella*, *Iodobacter*, *Janthinobacterium*, *Jeongeupia*, *Klebsiella*, and *Kushneria*		
(L) *Labrenzia*, *Lampropedia*, *Leptothrix*, and *Legionella*		
(M) *Magnetospirillum*, *Marinobacter*, *Marinospirillum*, *Massilia*, *Mesorhizobium*, *Methanomonas*, *Methylarcula*, *Methylibium*, *Methylobacterium*, *Methylocystis*, *Methylosinus*, *Methylomonas*, *Methylovibrio*, *Microcoleus*[a], *Microcystis*[a], *Moraxella*, and *Mycoplana*		
(N) *Naxibacter* (today: *Massilia*), *Neptunomonas*, *Nitrobacter*, *Nitrococcus*, *Nodularia*[a], *Novosphingobium*, and *Nostoc*[a]		
(O) *Ocenicella*, *Oceanospirillum*, *Oligotropha*, and *Oscillatoria*[a]		
(P) *Pandoraea*, *Pannonibacter*, *Paracoccus*, *Paraburkholderia*, *Parafunghorum* (formerly: *Pseudomonas*), *Pedomicrobium*, *Pelomonas*, *Phormidium*[a], *Photobacterium*, *Plasticicumulans*, *Pleurocapsa*[a], *Polaromonas*, *Pleomorphobacterium* (=*Oceanicella*), *Protomonas*, *Pseudoanabaena*[a], *Pseudogulbenkiania*, and *Pseudomonas*		
(R) *Rhizobium*, *Rhodobacter*[b], *Rhodoferax*[b], *Rhodopseudomonas*, *Rhodospirillum*, *Rivularia*[a], *Rubrivivax*[b], *Roseateles*[c], and *Ruegeria*		
(S) *Saccharophagus*, *Salinivibrio*, *Schlegelella*, *Scytonema*[a], *Sinorhizobium*, *Sphaerotilus*, *Sphingomonas*, *Sphingopyxis*, *Spirillum*, *Spirulina*[a], *Stappia*, *Stella*, *Stenotrophomonas*, *Synechococcus*[a], and *Synechocystis*[a]		

Table 6.1 (Continued)

Eubacteria		Archaea
Gram-negative genera	**Gram-positive genera**	**Three-letter code for haloarchaea according to Litchfield [99]**
(T) Thauera, Thermus, Thiobacillus, Thiocapsa[d], *Thiococcus*[d], *Thiocystis*[d], *Thiodictyon*[d], *Thiopedia*[d], *Thiosphaera*[d], and *Tolypothrix*[a]		
(V) Variovorax, Vibrio		
(W) Wautersia (today: *Cupriavidus*), *Westielopsis*[a]		
(X) Xanthobacter		
(Y) Yangia, Yokenella		
(Z) Zobellella, Zoogloe		

a) Cyanobacteria.
b) Purple non-sulfur photosynthetic bacteria.
c) Aerobic photosynthetic bacteria with bacteriochlorophyll a.
d) Anoxygenic purple sulfur bacteria.
e) Yeasts.
f) Eukaryotic microalgae.
Source: Adaptation and extension from [98, 100].

Even animals were modified to produce PHA; in this context, Williams et al. achieved simultaneous expression of the dehydrase-domain mutant rat fatty acid synthase cDNA (provides *de novo* synthesis of the PHB precursor *R*-(−)-3-hydroxybutyryl-coenzyme A) and the *phbC* gene encoding class I PHA synthase from *C. nector* genes into insect cells of *Spodoptera frugiperda* (the fall armyworm, a butterfly); here, gene transfer was carried out by baculovirus vectors [110].

6.3 Bioengineering: Bioreactor Design and Feeding Regime

6.3.1 Feeding Regime

PHA production on a larger scale is performed under controlled conditions carried out in bioreactors, which can constitute batch (substrate feeding occurs only at the start, emptying the bioreactor for harvesting PHA-rich biomass only after stopping the process), fed-batch (substrate addition by pulse-feeding when needed according to consumption by the cells, harvest after stopping the process), and continuous (both substrate feed and product harvest occur during the entire process) processes. The feeding strategy and the bioreactor operation mode (batch, fed-batch, or continuous) need to be well adapted to the applied microbial strain, the substrate used, and the biomass and product formation kinetics [111, 112]. Processes for PHA production carried out in batch [113], repeated batch ("drain-and-fill" and "fill-and-empty" processes) [114], fed-batch [115], or cyclic fed-batch mode [116] display the typical disadvantage of insufficient controllability of PHA composition on the level of monomers and monomer distribution and are often of low volumetric productivity [117].

Figure 6.3 provides an overview of different cultivation regimes used to produce PHAs.

6.3.2 Continuously Operated Bioreactors for Liquid Feed

Detailed knowledge about the growth and PHA biosynthesis kinetics is required to develop efficient continuous production processes. To optimize such continuous processes, the operation mode (dilution rate) needs to be in accordance with the kinetics of biomass growth and PHA biosynthesis. Such continuous processes often enable high volumetric productivity, better control of the product composition and properties, and use of substrates, which may otherwise inhibit the organisms at low concentrations. The latter is typical for compounds such as propionic acid, valeric acid, or γ-butyrolactone (GBL), acting as precursors of desired PHA monomers such as 3HV or 4HB, respectively.

Single-, two-, and multi-stage bioreactor setups are reported for continuous processes for PHA biosynthesis. As mentioned before, PHAs are typical products of the secondary metabolism, synthesized by multi-stage metabolic processes; therefore, it is reasonable to switch to two-stage continuous processes to produce such biopolyesters, where catalytically active, almost PHA-free biomass is generated at high cell density in a first continuously operated bioreactor; this biomass is continuously transferred into the second continuously operated bioreactor, where nutritional conditions stop biomass growth but favor PHA biosynthesis (e.g. limitation in nitrogen source); in this second stage, the cells get filled with PHA granules at typically high productivity. In such two-stage continuous processes, both the growth-limiting component (e.g. nitrogen source) and the carbon source can be supplied to the system at feeding rates according to the consumption rate, which results in an almost complete consumption of substrates by the cells and allow the spent fermentation broth to be free of substrates. This, in turn, saves the substrate cost, while resulting in lower fraction of PHA in biomass compared to a surplus of carbon source in the second stage, which, however, needs to be disposed of with the spent fermentation broth.

In multistage continuous processes, the so-called "bioreactor cascades," a range of individual stirred tank bioreactors are serially connected; with increasing number of bioreactors, such setups mimic the tubular plug flow reactor characteristics [118]. It was demonstrated that in bioreactor cascades, volumetric productivity can be further increased in comparison to two-stage setups, as shown by Atlić et al., who reported the highest volumetric PHA productivity (1.85 g/(l·h) for the entire cascade, 3 g/(l·h) for vessel 4) in continuous processes ever for a continuously operated bioreactor cascade consisting of five bioreactor vessels. In this setup, the first bioreactor served to form PHA-poor biomass under nutritionally balanced conditions, while the second

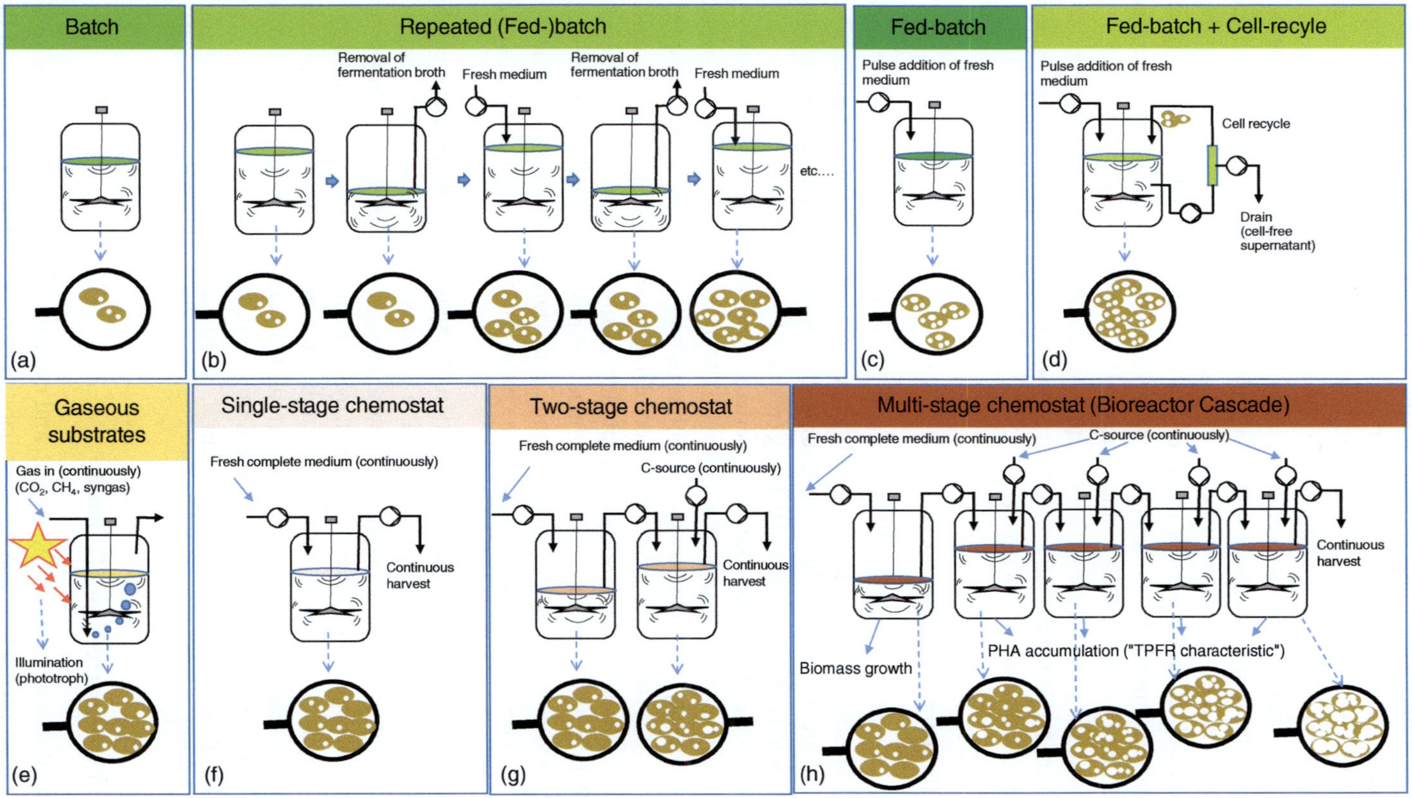

Figure 6.3 Different operation strategies to run PHA production processes in bioreactors. (a) batch, (b) repeated (fed-) batch, (c): fed-batch, (d) fed-batch with cell-recycle, (e) gaseous substrates, (f) single-stage chemostat, (g) two-stage chemostat, (h) multi-stage chemostat (bioreactor cascade). Source: [111] Reproduced with permission of MDPI publishers, Basel.

bioreactor (transient stage) served for depletion of residual nitrogen sources not utilized by the cells in bioreactor 1. In the vessels 3 to 5, increasing quantities of PHA were accumulated by the cells run under nitrogen-limited conditions with ample supply with the carbon source. In addition, such cascade setups enable the exact adaptation of the cultivation conditions in each individual bioreactor by triggering the substrate (main carbon source or precursor substrates) supply to each bioreactor; this allows designing blocky structured PHA with predefined compositions and properties [119]. Data from this cascade process were later used to develop mathematical models of the bioprocess; these models encompass formal kinetic [120] and high structured metabolic models [121], as well as footprint area analysis of imaged PHA-containing cells from individual reactors [122]; all these models provided new insights into the various screws that you have to turn to optimize cascade processes for PHA production, especially regarding nutrient supply to individual bioreactors and optimum number of cascade vessels [123].

6.3.3 Bioreactors for Gas Feed

Methane can be converted to PHA by methanotrophic production strains such as *Methylocystis* sp.; for such cases, high PHA yields of more than 0.5 g PHA per gram CH_4 are reported, which surmounts the theoretical yield of sugar-based heterotrophic PHA production of 0.48 g/g (reviewed by [124]).

For large-scale PHA production from methane, bioreactor systems with optimized capability to transfer the gas (air enriched with CH_4) to the microorganism is needed; various reactor types, including bubble columns and airlift bioreactors, come into play. A recent review reports the application of loop bioreactors for PHA production based on methane-rich natural gas; computational fluid dynamics and mathematical modeling were implemented to optimize the process. It was shown that gas holdup, shear stress, gas transfer coefficients, and liquid velocity vectors of the systems need to be investigated to conceive the interrelations between process parameters and PHA biosynthesis [96].

Syngas, in turn, is generated by gasification of organic waste materials; it can be metabolized into PHAs by, e.g. *R. rubrum*, a facultative anaerobic photosynthetic bacterium. However, using this substrate presents some safety risks regarding toxicity and risk of explosion. This was addressed by Karmann et al. who developed a platform of syngas fermentation, equipped with sophisticated technical safety precautions, tools to measure and control gas supply and consumption by cells, and modern process analytical technology (PAT), including an in-line flow cytometer to determine the cell density and PHA content in *R. rubrum* in fed-batch and continuous processes [125].

6.3.4 Photo-reactors for CO_2 Feed

Gaseous substrates such as CO_2, syngas, and methane are increasing being investigated for PHA production, given the availability of these carbon-rich gases and the importance in GHG reduction. The primary issue in gas-based substrate use is their lack of solubility in water, where the microorganisms and other nutrients reside during fermentation.

CO_2-based PHA biosynthesis occurs with photoautotrophic cyanobacteria, a group of prokaryotes formerly designated as "blue-green algae". Both discontinuously and continuously operated illuminated photobioreactors are reported in the literature, similar to photobioreactors used to cultivate eukaryotic microalgal species. The illumination regime (light intensity, wave length, dark–light cycles, etc.), geometry, mixing characteristics, and gassing/degassing behavior of such photobioreactors need to be optimized for efficient cultivation of cyanobacteria [126–129]. A recent review presents different photobioreactor types up to now tested to farm cyanobacteria, demonstrating that the most suitable photobioreactor setups predominantly depend on the specific light and nutrient requirements of the studied cyanobacterium [76]. Based on the results from pilot-scale runs with the cyanobacterium *Synechocystis* sp., this and other studies concluded that tubular photobioreactors might be the optimum solution for PHA production by such phototrophic production strains [75].

6.4 Downstream Processing for PHA Recovery

Downstream processing for the recovery of microbial PHA biopolyesters from biomass constitutes an integral part of the entire PHA production chain. This process is currently considered the major cost factor for PHA production besides feedstock cost. Moreover, the downstream processing is also contributing significantly to the ecological footprint of the product, e.g. by the use of solvent, chemicals, and energy. As comprehensively summarized in recent review articles, there are various current trends and ongoing efforts to optimize PHA recovery, aimed at reducing or even eliminating solvents and time expenditure, minimizing the use of chemicals in general and energy input; maximizing recovery yields and product purity, and maintaining the properties of native PHAs [130]. To an increasing extent, biological methods for PHA recovery replace chemical approaches, at least on a laboratory scale [131]. Moreover, the applied PHA recovery method is decisive for the molecular mass and purity (important for polymer processing and use of all kinds, including medical use) of the obtained product and the achievable recovery yield. Finally, the PHA recovery technique affects the PHA content recovered from biomass. For example, not all recovery techniques are applicable for all types of PHAs – crystalline *scl*-PHA requires different recovery types compared to amorphous *mcl*-PHA – or to all types of PHA-producing microorganisms – robust strains need other recovery techniques than organisms with fragile cell structures. In principle, solvent-based extraction methods, the use of ionic liquids (ILs) and supercritical solvents, chemical, mechanical, and biological methods for disintegration of residual biomass are described for PHA recovery. The applied recovery method not only determined product purity and recovery yield but also the fate of remaining non-PHA biomass [132].

Figure 6.4 provides a schematic illustration of diverse established and emerging methods for PHA recovery from microbial biomass.

6.4.1 Classical Solvents

The state-of-the-art in PHA recovery involves the use of halogenated solvents, especially chloroform, because it gives high extraction yields, excellent PHA polymer purity, and minimal loss of PHA polymer properties during melt processing. The downsides of this method are of course the use of high volumes of a toxic solvent that, despite the presence of state-of-the-art solvent recovery and recycling processes, can be hazardous and pose human and environmental risks. Besides, many of these outstanding PHA solvents such as chloroform have petrochemical origin, which clearly contradicts the paradigm of PHAs being renewable and benign green materials produced by sustainable process steps. Although resulting in high product recovery yields and purity, solvent-based PHA extraction typically destroys the natural (native) characteristics of PHA granules by reducing molecular mass via random and chain end scission, particularly at higher temperature and prolonged extraction time; this can obstruct their further processing [130].

After dissolving in halogenated solvents, the PHA requires a second precipitating solvent in order to obtain high-purity PHAs because simply evaporating the solvent would leave impurities in the PHA polymer. These "PHA anti-solvents" (ethanol, acetone, methanol, hexane, etc.) are typically added in excess and in their presence, together with lowering the temperature, dramatically reduces the PHA solubility in chloroform, thus precipitating the polymer. After precipitating the PHA polymer, the mixture of PHA solvent and "PHA anti-solvent" is removed and separated into their individual components via distillation, expending substantial energy. Typically, about 3 wt% PHA is dissolved after extraction in classical PHA solvents such as chloroform and need about 10-fold volume of anti-solvent for precipitation [132].

As a way out, a convenient, lower energy consuming separation process for the chloroform–ethanol mixture was developed by Rodolfo Bona in the research labs of Gerhart Braunegg, one of the pioneers of PHAs. This method is based on simply adding appropriate amounts of water. At the appropriate chloroform–water–ethanol ratio, a ternary three-component, two-phase system is created, which allowed recovery of the high-density chloroform at around 95% purity with the rest being ethanol. They showed that this chloroform containing some ethanol residue (ethanol is present as a stabilizer in commercial chloroform anyway) can be reused in subsequent PHA extraction cycles. Removal of minor water residues from this phase is possible by adding drying agents. The second phase, a low-density ethanol/water mixture, contains only minor quantities of chloroform (1%) and about 18% ethanol. The phase diagram of this three-component, two-phase system is presented in Figure 6.5. Despite the convenience and ease of using this separation technique, the authors of this chapter are not aware of its industrial implementation [132].

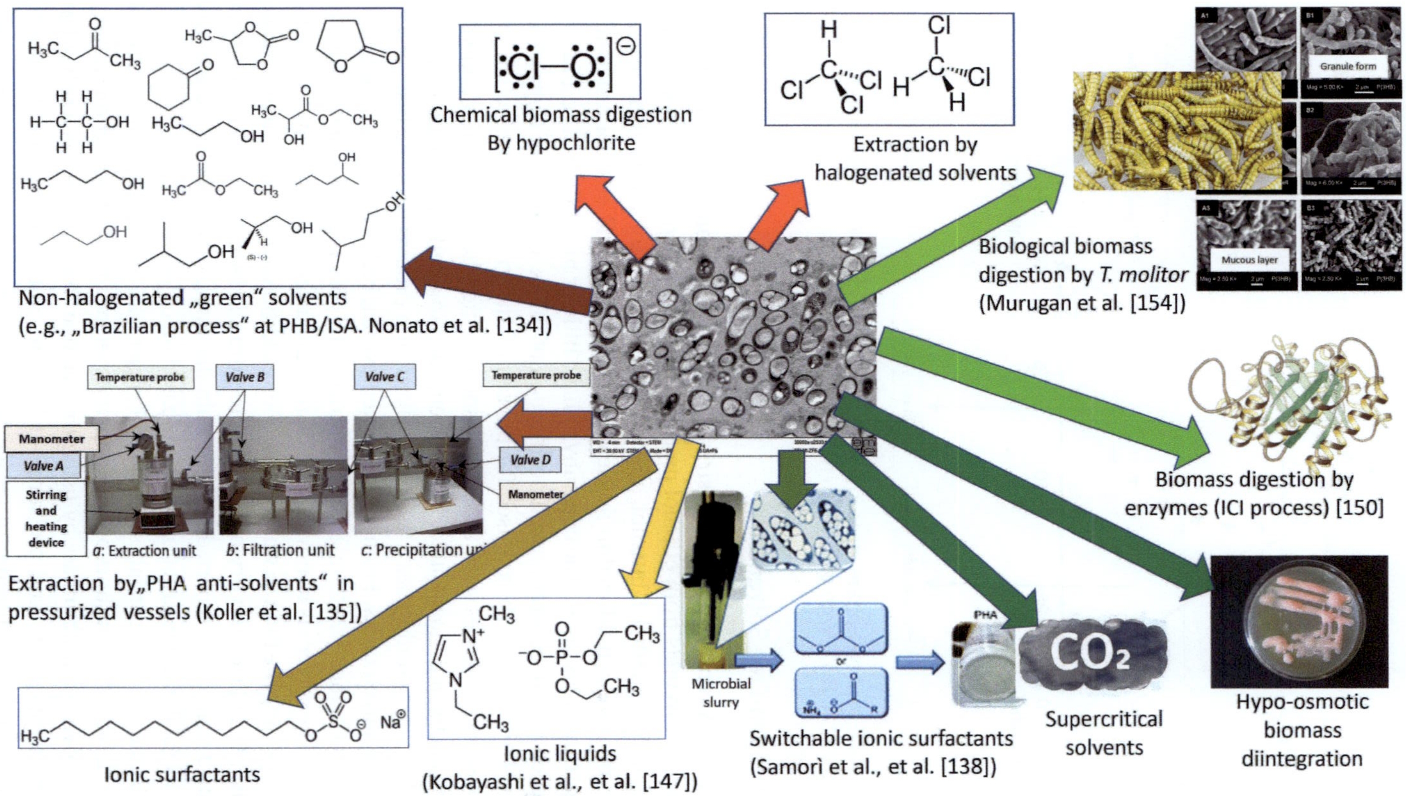

Figure 6.4 Compilation of established and emerging PHA recovery methods by chemical and biological methods; coloration of arrows provides a rough, maybe preliminary, estimation of the overall sustainability of individual methods, with red being not or hardly sustainable and green indicating high sustainability. Source: Koller [130]/with permission from Sciendo.

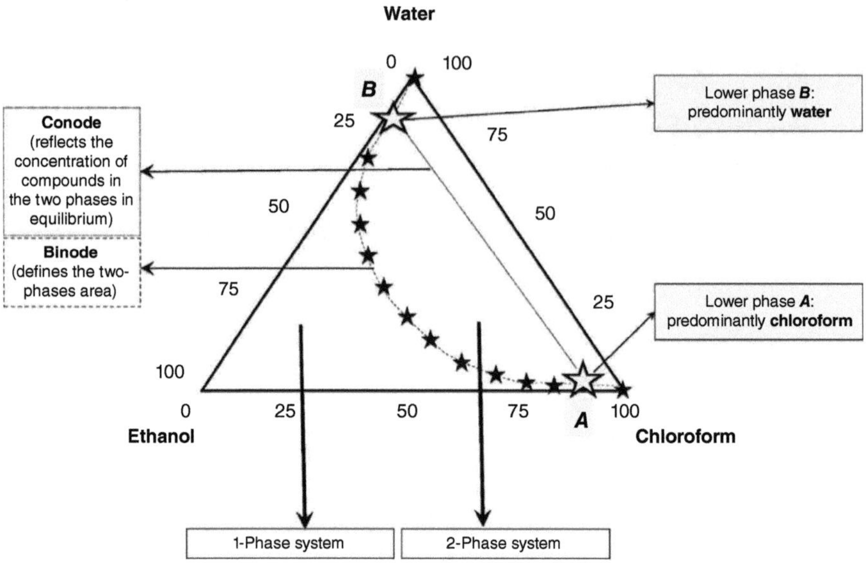

Figure 6.5 One- and two-phase areas in the ternary system chloroform–ethanol–water. Source: Koller et al. [132]/with permission from Sciendo.

6.4.2 Halogen-Free Solvents

A Brazilian PHA company, PHB Industrial S.A. (PHB/ISA), demonstrated the use of distillation by-products from ethanol production (fusel oils containing 1-propanol, iso-butanol, (*S*)-2-methyl-1-butanol, and isopentanol) as a solvent for PHA extraction. This mixture is produced in house during sugarcane and ethanol production from sugarcane. This allows the company to have little or no additional cost for solvents needed for the separation and purification of PHAs. Cultures of *Cupriavidus necator* or *Burkholderia* sp. are used for the semi-industrial PHA production process; sucrose and molasses serve as a carbon source. After the fermentation, cells are thermally inactivated, flocculated, and concentrated to a cell slurry of up to 300 g/l density. This slurry undergoes a multi-step extraction process with the fusel oils in continuously stirred tanks. From the generated extract, cell debris is removed, and, by cooling down, the PHA solution turns into a gel. By pressing and evaporation, the major part of the solvent is removed, and the residual fusel oil is withdrawn from the PHA by dispersing the PHA in water (subsequent additional distillation-based recovery of solvent) and vacuum drying the polymer into a PHA granulate. The purity of the obtained polymer is reported to be in a similar range (about 98%) to the purity obtained by established extraction methods based on chloroform [133].

In the frame of an industrial project with Partner BDI – BioEnergy International AG, Austria, one of the authors of this chapter developed a novel extraction strategy for PHA recovery from biomass by using the acetone that operates as a solvent at an elevated temperature and pressure but becomes anti-solvent upon cooling to room temperature. An aluminum device consisting of an extraction-, filtration-, and precipitation unit was designed and built to recover the PHA from freeze-dried

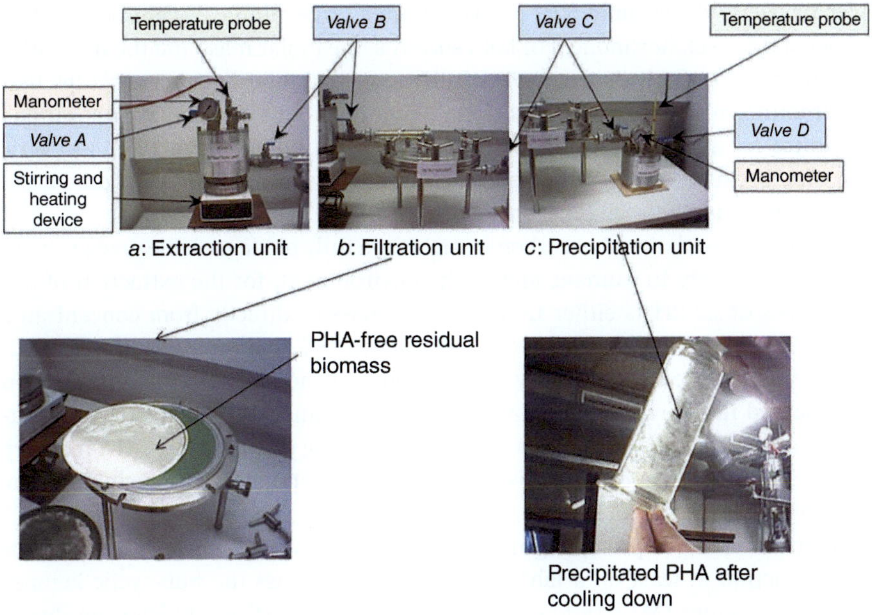

Figure 6.6 Setup for extraction of PHAs from microbial biomass with acetone under high p/T conditions. Modified from [72].

biomass of the haloarchaeon *Hfx. mediterranei* under high temperature (121 °C) and pressure (7 bar). It must be noted that acetone does not dissolve crystalline *scl*-PHAs as PHB or PHBHV under ambient temperature and pressure. The rest of the cell debris was then filtered out, and upon lowering the temperature and pressure, the PHA precipitated, leaving the lipids in acetone, which was then recycled after evaporation of the solvent. Prior to pressuring and raising the temperature of the acetone-PHA-cell debris mixture, the entire system was purged with nitrogen to remove oxygen and thus eliminating the risk of explosion. Recovery yields and product purity were competitive with results obtained for parallel control experimental setups that used chloroform [134]. Figure 6.6 illustrates this process.

In contrast to *scl*-PHAs, *mcl*-PHAs of low crystallinity can be extracted by acetone from biomass even at room temperature and ambient pressure; this was shown by Cerrone et al. for extraction of amorphous PHA copolyester consisting of 3-hydroxydodecanoate (3HDD), 3-hydroxydecanoate (3HD), 3-hydroxyoctanoate (3HO), and 3HHx (poly(3HDD-*co*-3HD-*co*-3HO-*co*-3HHx)) from lyophilized *Pseudomonas chlororaphis* biomass [135].

More recently, Jiang et al. studied the potential of cyclohexanone and GBL as extraction solvents for the *scl*-PHA PHB from *C. necator* biomass; extraction kinetics and its dependence on temperature were studied. Using cyclohexanone at 120 °C, 95% of present PHA was extracted from biomass within three minutes, while reduced temperature significantly decreased the extraction yield; purity was similar to control setups using chloroform. Using GBL at 120 °C, considerably lower recovery yields (only 50% recovered after three minutes) were obtained in comparison to cyclohexanone. For both solvents, the molar mass and the molecular mass distribution (polydispersity, *Đ*) of PHAs were similar to process setups that used chloroform as the extraction solvent. According to the authors, cyclohexanone constitutes an auspicious candidate for sustainable, halogen-free PHA recovery, *inter alia* because of the expedient recovery yields achieved. As a drawback, cyclohexanone is a petrochemical product [136].

Samorì et al. used dimethyl carbonate (DMC), a fully biodegradable green solvent with minor toxicity in humans and to the environment, for the extraction of different types of *scl*-PHAs either from dried biomass or directly from concentrated wet bacterial slurry. Polymer recovery yields above 85%, and product purity exceeding 95% was reported. Both for dry and wet biomass, no molecular mass reduction was observed in the recovered PHA. For these experiments, the PHB-rich *C. necator* biomass was treated for four hours at DMC's boiling point (90 °C); PHA recovery from solution was either accomplished via solvent evaporation or precipitation by addition of cooled ethanol [137].

Examples for other promising non-halogenated solvents, which were used for PHA extraction at least on a laboratory scale, encompass the non-cyclic ketones methyl ethyl ketone (MEK) and methyl isobutyl ketone (MIBK), which were shown to extract P(3HB-*co*-3HHx) copolyesters, both from wet and dry biomass, while ethyl acetate was suitable to extract PHAs from dry biomass. Extractions were carried out under temperatures of 100 °C in sealed screw top test tubes. Among them, especially MEK was suggested as a promising alternative to halogenated solvents because of its relatively low price and its low density, which facilitates separation of cell debris from the polymer solution. In addition, its higher boiling point makes it less dangerous than acetone [138].

6.4.3 Supercritical Solvents

Supercritical fluids used as agents for PHA extraction offer an excellent combination of solvation power similar to liquids plus expedient diffusion power (lower density and viscosity than liquids) similar to gases. Supercritical CO_2 (sCO_2), the best-studied supercritical fluid, exists above the critical point of CO_2 at 304 K and 74 bar. After the extraction, it is depressurized to evaporate as gaseous CO_2 and could be recompressed and recycled. No solvent residues remain in contrast to solvent-based PHA extraction processes [139]. For different intracellular bioproducts that could be extracted by sCO_2, Williams et al. offered various organic solvents such as alcohols that could act as modifiers for overcoming the polarity difference between CO_2 and the intracellular product to be extracted, thereby enhancing the solubility of sCO_2 in the highly polar intracellular space [140]. Hampson and Ashby reported that small amounts of chloroform indeed facilitated sCO_2-mediated recovery of *mcl*-PHAs from *Pseudomonas* sp. biomass [141]. This was later confirmed by Hejazi et al., who achieved excellent *scl*-PHA recovery yields of almost 90% when applying sCO_2 together with methanol as the modifier solvent. These authors reported that 0.2 ml of methanol was added to 4 ml of CO_2 along with an exposure time of 100 minutes at 40 °C and 200 atm as the optimum conditions for cell disintegration and PHB recovery [142]. Another study by the same team of researchers reported that an alkaline pretreatment of the lyophilized biomass of highly crystalline *scl*-PHAs can be recovered by sCO_2, at an even higher yield than what they showed possible by adding modifier solvents. It is noteworthy to also mention that sCO_2 can also be used to recover PHAs from wet PHA biomass, thus saving on PHA drying cost without sacrificing the quality of the recovered PHAs as shown by their retention of both molecular weight and distribution before and after the recovery process as compared to the recovery process that used lyophilized biomass. There was, however, slightly lower product purity noticed while using wet PHA biomass as the starting point as compared to dried lyophilized PHA biomass as the starting point [143].

6.4.4 Recovery by Chemical and Mechanical Disintegration of Biomass

Surfactants disrupt cell membranes, and intact PHA granules are set free. Samorì et al. used salts of longer fatty acids as surfactants; this process turned out to be applicable also to wet biomass, which in turn saves drying expenses. Polymer recovery yields above 99%, and a purity of more than 90% have been reported. In this investigation, ammonium laurate (dodecanoate) stood out as the so-called "switchable anionic surfactant"; it reversibly converts from a neutral (protonated) water-insoluble form at low pH-values to a deprotonated anionic water-soluble form at an elevated pH-value; the pH shift was done by adding and releasing CO_2. A suspension of *C. necator* biomass in an ammonium laurate solution was adjusted to pH-value 10, corresponding to the surfactant's anionic form. This dissolved the non-PHA bacterial matter and precipitated the PHA granules, which were then separated by centrifugation. The remaining supernatant liquid consisting of the ammonium laurate and disintegrated cells was neutralized by bubbling in CO_2, converting the laurate into protonated, water-insoluble lauric acid, which was then decanted from the resulting aqueous bicarbonate solution and recovered by centrifugation and reused in the next PHA recovery step. The remaining solution of dissolved cell constituents and NH_4HCO_3 can also be utilized as a nitrogen source in subsequent cultivations of the PHA production strain, thereby allowing for virtually no losses and complete circularity of the process [137].

Other well-established chemical methods based on the disintegration of the non-PHA cell material, leaving PHA granules intact, resort to the use of sodium hypochlorite (NaOCl) solution with sodium dodecyl sulfate (SDS). For example, Marudkla et al. reported a maximum PHA recovery of 78.7% when using 0.5% w/v SDS combined with 6% v/v NaOCl. However, SDS is an irritant, and NaOCl is a halogenated compound, contradicting the aim of abolishing chlorine from PHA manufacturing [144]. Researchers also observed that the presence of NaOCl reduced PHA molecular mass and generated different halogenated compounds (reviewed by [130]).

Several researchers have investigated using ILs; ILs are regarded as "molten salts" with melting points lower than 100 °C and often even below room temperature. Such

ILs have special properties, such as high thermal stability, insignificant vapor pressure, low flammability, and expedient ionic conductivity, and are considered superior to those reported for classical organic solvents. Fujita et al. have reported the direct dissolution of wet (95% water content!) biomass of the cyanobacterium *Synechocystis* sp. using the hydrophilic IL 1-ethyl-3-methylimidazolium methylphosphonate ([C2mim][MeO(H)PO2]). In this case, no heating or pretreatment was needed, and the authors successfully recovered "intracellular products such as PHA" [145]. This was verified by Kobayashi et al., who studied [C2mim][MeO(H)PO2] and other ILs for their potential to remove the non-PHA biomass of the cyanobacterium *Synechocystis* sp. PCC 6803, explicitly with the aim of releasing intact PHB granules. Again, [C2mim][MeO(H)PO2] stood out as the best IL to dissolve non-PHA matter while leaving PHB granules intact among the ILs tested. After cell disintegration, PHB was separated by filtration, with a recovery yield of more than 98%. ILs are generally recyclable, and this one in particular was readily recyclable by adding a small amount of solvent to precipitate the cyanobacterial debris, which were removed by filtration, and the mixture of IL and solvent was then evaporated at 40 °C for six hours for solvent removal. [146].

Haloarchaea, the extremely halophilic branch of the archaea domain, thrive at optimum salinities in the range between 2 and 5 M NaCl. This causes the generation of extraordinarily high inner-osmotic pressure to balance the high medium osmolarity outside the cells and opens the door for a convenient approach to release intracellular products from such organisms. Described for the first time in 1990 by Rodriguez-Valera and Lillo for PHA-rich *Hfx. mediterranei* biomass, the exposure of such extremely halophiles to hypotonic media (i.e. distilled water) rapidly and immediately disrupted cells and released the PHA granules without the addition of any organic extraction solvents. Due to the lower density of PHA granules in comparison to the non-PHA cell material, the thus treated broth formed a two-phase system, consisting of an upper layer of lower density PHA granule containing phase and a higher density cell debris phase at the bottom. The two phases are then separated by sedimentation or centrifugation. The top layer, "PHA cream," is then skimmed off, and the biopolymer further purified if required (reviewed by [147]).

Wild-type strains naturally accumulating high quantities of PHAs are equipped by nature to keep cell integrity at very high loads of intracellular storage materials. When expressing PHA synthesis genes in strains such as *Escherichia coli*, which are not natural PHA producers, the cell wall cannot stay intact at excessive loads of inclusion bodies such as PHAs. After cell harvest, high intracellular PHA fractions make cells burst even under mild biomass treatment. As an example, Choi and Lee transformed *E. coli* cells with *C. necator* PHA synthesis genes. These recombinant *E. coli* cells accumulated more than 70 wt% PHA in biomass. By simply stirring the biomass with 0.2 N NaOH at 30 °C for one hour, highly pure PHA (purity more than 95%) was recovered [148] primarily because of the poor cell integrity of the recombinant *E. coli* strains.

Advanced mechanical methods for cell disintegration for PHA recovery were also reported in the literature. Researchers have reported using high-pressure homogenization (HPH) where the disruption performance strongly depends on the biomass concentration and operates rather modestly at low biomass concentrations. Ghatnekar et al. described using HPH in the presence of SDS for recovery of PHB granules from *Methylobacterium* sp. V49. They obtained a maximum recovery yield of 98% and a product purity of 95% using 5% SDS solution and a pressure of 400 kg/cm^2 for HPH [149]. Other researchers studied the disruption performance of HPH for PHA-rich culture broth of *C. necator* with biomass concentrations in the range of 1–6 wt%. They investigated HPH both with and without pretreating the fermentation broth with NaOH and SDS solutions. Two cycles at 800 kg/cm^2 were carried out. The results indicated excellent performance of HPH in cell disintegration. Best results (>99.99%) were achieved by combined pretreatment with NaOH (strongly alkaline conditions of pH 12) and SDS (1%). The results obtained without any pretreatment were also very promising, and they achieved a degree of cell disruption of 91–99%. Separation of the released PHA granules was then carried out by dissolved air floatation (DAF) [132].

6.4.5 Biological PHA Recovery

Chemical methods described above to disintegrate the microorganisms are often expensive, toxic, and/or corrosive to machinery and humans. Frequently, these chemicals are not recyclable (such as sodium hypochlorite, strong bases or acids, or the irritating anionic surfactant SDS). The alternative, using green biocatalysts (enzymes), to disrupt the microorganisms is not necessarily cost-efficient because the enzymes are frequently lost after cell disrupture, the resulting PHA biomass has poor product purity, and the overall rates of reactions can be low, thus resulting high costs, as reported for industrial scale tests [133]. Initial large-scale enzymatic cell disintegration was developed, practiced, and reported decades ago by Imperial Chemical Industries (ICI), UK. They used enzyme mixtures containing proteases, phospholipases, lysozyme, and nucleases, which were typically applied, often in combination with surfactants and strong oxidants such as hydrogen peroxide for the subsequent chemical purification of the extracted PHA [150]. A novel enzymatic approach for digestion of non-PHA biomass has been presented by Kachrimanidiou et al. These authors prepared a crude enzyme cocktail via solid-state fermentation (SSF) of the fungus *Aspergillus oryzae*. This enzyme mixture was applied to lyse *C. necator* cells for PHA recovery. Temperature and pH-value were optimized for maximum *C. necator* lysis and reached 90% PHA purity and a recovery yield of 97–98% [151].

Typically, cell homogenization during PHA recovery generates high quantities of DNA, which increases the viscosity of the fermentation broth, thus complicating the recovery of the PHA granules. Another approach to improve PHA recovery involved expressing nuclease enzymes in PHA accumulating cells. Generally, viscosity reduction is achieved by heat treatment, addition of commercially available nucleases, or by adding sodium hypochlorite, all adding to the cost of PHA polymer recovery. Rodríguez Gamero et al. integrated *Staphylococcus aureus* nuclease genes into the PHA production strains *C. necator* and *Delftia acidovorans*. The nuclease enzyme has been readily expressed in both bacterial host strains without negatively affecting the PHA biosynthesis. They observed significant viscosity reduction in the rec. *C. necator* homogenate. This process of expressing the nuclease in the PHA producing microorganism could greatly reduce the overall production cost of PHA polymer recovery process [152].

A novel approach to PHA recovery involved the use of insects. *Tenebrio molitor* – meal worms in different stages of their life cycle were fed with freeze-dried PHA-rich *C. necator* biomass at the research group of Kumar Sudesh in Malaysia [153]. Residual biomass (proteins, etc.) was readily digested by the animals as SCP source, while the PHA granules were excreted by these worms in highly pure form. Subsequent purification steps using water, detergents, and heat treatment generated almost 100% pure PHA granules. Apart from PHA recovery, the cultivated meal worms might also serve as potential source of food or animal feed as an additional source of protein for human nutrition [154].

6.5 End-of-Life Options: Recycling and Biodegradation of PHAs

Just like other biopolymers such as thermoplastic starch (TPS), other polysaccharides such as chitin or cellulose, and proteins (e.g. gelatin, processed whey proteins), oxidative (aerobic) breakdown of PHA by bacteria or fungi, e.g. during composting, produces CO_2 and water, while anaerobic (anoxic) PHA degradation by living organisms, e.g. in biogas plants, generates CH_4, a potential substrate for new PHA production using methanotrophs, in addition to water and CO_2 [3, 155]. Biodegradability of PHA-based items is dependent on various factors, such as shape and thickness of polymer specimens (packaging films will biodegrade faster than bulky objects), material properties such as crystallinity, molecular mass, and composition of the PHA, and diverse environmental factors such as surrounding microflora, humidity, pH-value, temperature, or UV radiation; the details are elaborated below. This variability in the biodegradation of PHA matches all other biopolymers including starch, cellulose, proteins, or chitin, and elucidation of involved factors requires further investigation, especially to understand biodegradation rate for various types of PHAs, a subject with contradictory findings in the literature [156]. While most studies have reported that high crystallinity reduces the PHA degradation rate (which typically results in higher degradability of *mcl*-PHA than *scl*-PHA) [22], others report no impact of crystallinity on biodegradation [157]. Noteworthy, the scarcely studied homopolyester PHO, which has been biosynthesized in the laboratory and not found in nature, was even reported as not biodegradable at all in one study that has not yet been duplicated [158]. It can be authoritatively acknowledged that all types of PHAs currently produced in processable amounts are biodegradable [22]; as shown for commercially available PHAs like those produced by Kaneka, RWDC Industries, etc., they are not only biodegradable but also certified as "industrial compostable" and "home compostable" [3].

6.5.1 Recycling

Besides biodegradability, PHAs can also be mechanically recycled, chemically pyrolyzed into high value and widely used chemicals, and biologically recycled to generate methane (and compost); methane can then be upcycled into PHA again [159]. This post-use recycling of PHA is important to avoid the limitation of using PHA for low-value applications only. As is the case with all materials used and recycled today, PHAs would achieve that status of being recycled effectively once a significant quantity of spent PHA becomes available in the future, and it is selectively collected and not mixed with other polymers. Extrusion and injection molding were demonstrated as viable techniques for mechanical recycling of PHA, as shown by Zaverl et al. for the copolyester PHBHV, which did not show any significantly impaired material properties even after five recycling rounds [160], or by Zembouai et al. who recycled PHBHV/PLA blends six times without substantial quality loss [161]. A combination of melt extrusion and compression molding was used by Rivas et al. to process PHB homopolyester; however, significant quality loss was noticed already after two recycling cycles [162]. Nevertheless, we need to consider that mechanical recycling of the spent PHA is still in a very early stage of development, and further studies are needed to evaluate and understand the parameters determining PHA's recyclability, such as composition, crystallinity, and molecular mass. Vu et al. recently reviewed the current lack of knowledge about the interaction of type and composition of PHAs and accessibility of such differently composed types of PHA items to mechanical recycling strategies [159]. Understanding this correlation, however, is the key for the next steps in PHA post-use treatment; it will ultimately decide if spent bioplastic materials need to be separated at the source [163]. Moreover, we should consider that too many different types and grades of PHAs being on the market might complicate its mechanical recycling. For these reasons, PHA is the "Gold Standard" among classes of materials within the contemporary vision of a "Circular Economy," based on the principles of managing waste and pollution in a smart and sustainable way, keeping products and materials in use, and regenerating natural systems. This concept is currently pursued by numerous organizations worldwide [3].

As an alternative to mechanical recycling, spent PHAs can be upgraded chemically via pyrolysis. Known already since the early 1980s, this chemical recycling process generates, depending on the type of treated PHA, compounds such as crotonic acid and oligomers of 3HB (from PHB), and also 2-pentenoic acid (from pyrolyzing PHBHV) [164]. Pyrolysis of the *mcl*-PHA poly(3-hydroxydecanoate) in turn generates 2-decenoic acid [165]. Another advanced recycling approach is based on the microwave-supported reaction of PHA with solvents such as alkaline methanol; this way, the PHB homopolyester undergoes conversion to 3HB, 3-methoxybutyrate, and crotonic acid, as reported by Yang et al. [166]. In addition, poly(3-hydroxypropionate) (P3HP) yields 3HPA, the precursor to acrylic acid [167], while P4HB could yield GBL, 1,4-butane diol or even succinic acid: all important chemical intermediates [168]. This approach shows that several of the bulk chemicals produced today could eventually be produced from renewable sources via the production of PHA as the intermediate material, thereby injecting further circularity to our fossil fuel-dependent world. Moreover, spent PHA can also undergo hydrolysis to generate optically pure building blocks, which in turn can be used as chiral synthons for the synthesis of marketable compounds, such as pharmaceuticals, synthons for organic synthesis, or fragrances. Such chiral monomers, which have higher market values than the polymers themselves, could considerably contribute to making the entire PHA production process more valuable [169].

6.5.2 Incineration

When neither mechanical or chemical recycling nor composting of PHA is feasible, they can also be incinerated, which is the fate of more than 80% of collected petroleum plastics today in the European Union. Here, we need to consider that the incineration of PHA, based on its chemical composition, creates much less energy than incinerating fossil plastics such as polyolefins; hence, in the PHA case, incineration rather serves waste reduction than energy generation.

Figure 6.7 illustrates the cycle of PHA production and potential end-of-life options for the spent PHA.

6.5.3 Mechanistic Considerations of PHA Degradation

In nature, polymer biodegradation is catalyzed by depolymerases, a specific group of enzymes, mainly hydrolase enzymes. Depolymerases disintegrate the complex polymer molecules into constituent monomers to be used by microbes to generate energy for themselves and/or build up other cellar building blocks [170]. Such depolymerase enzymes are synthesized by numerous microorganisms; some secrete them extracellularly [171], while others have intracellular polymer

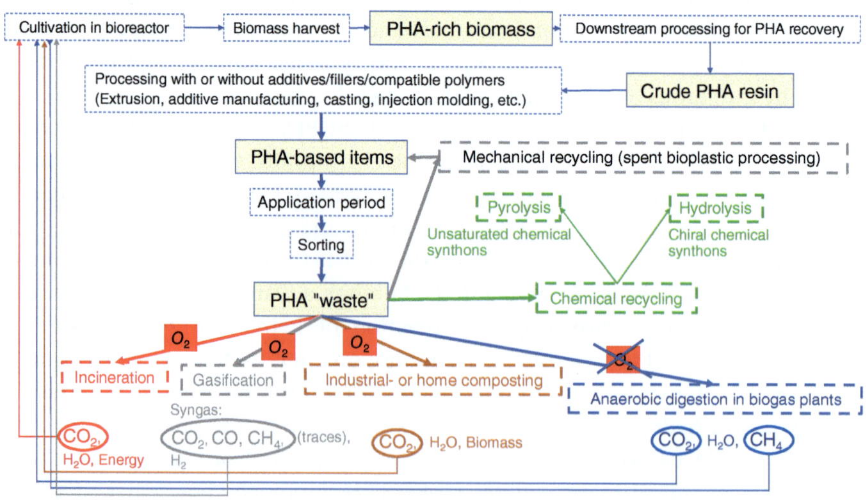

Figure 6.7 Cycle of PHA production, processing, sorting, and potential end-of-life options for spent PHA (PHA "waste"). CO_2 generated by incineration, composting, gasification, or anaerobic digestion can serve as a substrate for autotrophic PHA biosynthesis, e.g. by cyanobacteria. CO originating from gasification can be converted by *Rhodospirillum* sp. to CO_2 and converted to PHA, while CH_4 from biogas plants (anaerobic digestion) can be used for PHA biosynthesis by type II methanotrophs. Modified from [82].

degrading enzymes [172]. PHA depolymerases (PhaZs) specifically hydrolyze PHA polyoxoesters; these enzymes are encoded by *phaZ* genes. Two classes of PHA depolymerases are described, namely extracellular representatives (e-PhaZs), which are of low specificity and degrade almost all PHA polymers into microbially convertible oligomers and monomers, and their intracellular counterparts (i-PhaZs), which are specific to various types of PHA polymers. Typically, PHA are broken down *ex vivo* by extracellular depolymerases excreted by bacteria (e.g. streptomycetes) and fungi and by additional hydrolytic effects with a period of up to one year [173]. Further, e-PhaZs are typically characterized by low molecular mass with the pH optimum in the slightly alkaline range [170].

PHA biodegradation has been investigated over the past four decades and has been compared to various (semi)synthetic polymers. Many reviews summarize PHA biodegradation in soil, buffer solutions, freshwater, and seawater [169, 174]. The best studied PHA, PHB, showed a progressive non-biocatalytic molar mass reduction by almost 50% after 650 days of immersion at 37 °C even in pure water or phosphate-buffered saline [175].

Various factors impact the biodegradability of PHAs, such as [156, 176]

- composition (PHA homopolyesters such as PHB typically degrade slower than copolyesters)
- crystallinity (known to slow down PHA biodegradation compared to less-crystalline PHA polymers)
- molar mass (PHA polymers with lower molar mass typically biodegrade faster than those with higher molar mass)
- stereoregularity
- biodegradation conditions (temperature, humidity, pH-value, and the presence of nutrients for the degrading microorganisms).

In addition, extracellular depolymerases biodegrading *scl*-PHAs are different from those enzymes biodegrading *mcl*-PHA polymers. A range of *scl*-PHA depolymerases (EC 3.1.1.75) have been isolated and studied, while a limited number of *mcl*-PHA depolymerases (EC 3.1.1.76) have been characterized in detail. Specific *mcl*-PHA-degrading enzymes are rarer in microbes; they have been recently reviewed by Urbanek et al. and are known to exist mainly in some representatives of the genera *Pseudomonas* and *Streptomyces* and in some thermophiles [175]. Despite their diverse substrate specificity and amino acid sequence, all PHA depolymerases share a common α/β-hydrolase fold and a catalytic triad, which is also present in other α/β-hydrolases [177].

PHA polymers are also depolymerized by microorganisms intracellularly, catalyzed by i-PhaZs synthesized in the host cells; they mobilize the carbon and energy in the PHA reserves during periods when the microorganisms are lacking exogenous carbon source (see Figure 6.1). This *in vivo* degradation has its onset via i-PhaZs (EC 3.1.1.7x), generating (*R*)-3-hydroxyalkanoate monomers and the corresponding oligomers. The monomers are then reversibly oxidized to acetoacetate by the oxidoreductase (*R*)-3HB dehydrogenase (EC 1.1.1.30). Subsequently, the monomers are catalyzed by the transferase acetoacetyl-CoA synthetase (EC 2.3.1.194) into acetoacetate, which then is converted to acetoacetyl-CoA and finally hydrolyzed to acetyl-CoA, the central metabolite in all microorganisms [178]. This final reaction is catalyzed by the reversible enzyme 3-ketothiolase (EC 2.3.1.9) described previously in this chapter.

Remarkably, i-PhaZs are not able to hydrolyze isolated, extracellular PHAs, whereas e-PhaZs, when expressed *in vivo*, do not hydrolyze intracellular granules. Two reasons have been identified as the causes: the substantial differences in the physical structure and the crystallinity of native granules *in vivo* and denaturation of extracellular PHA polymers after extraction from biomass [179, 180]. Intracellular PHA depolymerization is known to occur in parallel to PHA accumulation even under conditions favoring PHA biosynthesis (excess carbon source in parallel to limitation of other growth essential nutrients). This causes continuous and simultaneous PHA biosynthesis and PHA biodegradation in living microbes; thus, the polyester chains undergo a steady modification and reorganization, which is frequently described in the literature as the "cyclic nature of the PHA metabolism" [181]. However, under conditions beneficial for PHA biosynthesis, i-PhaZs have considerably lower activity than the PHA synthases active in the same cell. *E. coli* cells generally do not have the i-PhaZs gene; therefore, recombinant *E. coli* containing PHA synthesis genes (*phaCAB*) are able to produce PHA polymers with ultra-high molar mass and low polydispersity [182]. This can be emulated in other natural PHA polymers producing microorganisms by deactivating (or knocking out) the PHA depolymerases from there in the genome, demonstrated recently by Adaya et al. in *A. vinelandii* [183].

The significant impact of PHA crystallinity on its extracellular biodegradation was demonstrated by Choi et al. They produced PHBHV copolyesters with different 3HV fractions produced by the bacterium *Alcaligenes* sp. MT-16 using glucose plus the 3HV-precursor levulinic acid. Increasing 3HV fractions in PHBHV decreased the PHA polymer's crystallinity and resulted in faster extracellular degradation when immersed in solutions of fungal (strain *Emericellopsis minima* W2) e-PhaZ, demonstrating the effect of crystallinity, which in turn depended on the PHA polymer's composition. Biodegradation experiments carried out with the same types of PHA for 20 weeks in non-biotic alkaline medium without the addition of enzymes did not result in any PHA degradation at all, demonstrating the need for these PHA depolymerases in the biodegradation process [184].

All described i- and e-PhaZs in the literature have been demonstrated to be specific to oxoester bonds; e.g. they do not cleave thioester bonds as present in polythioesters (PTEs), a group of sulfur-containing PHA analogs. PTE copolyesters of 3-hydroxyalkanotes and 3-mercaptoalknaotes, such as poly(3HB-*co*-3MP) or poly(3HB-*co*-3MB), are obtained by feeding PHA producers such as *C. necator* precursor compounds such as 3-mercaptobutyrate (3MB) or 3-mercaptovalerate (3MV), in addition to carbon sources such as gluconate. These sulfur-containing substrates are not natural products and are chemically produced, and only a limited number of bacteria such as *Schlegellella thermodepolymerans* biodegraded these PTE polyesters [185]. Interestingly, this strain microorganism was recently identified to produce PHA polymers from inexpensive, xylose-rich carbon sources [186].

6.6 Biodegradation – Added Value for Selected Applications

6.6.1 Packaging

The diversity of PHA monomers identified to date and the thermoplastic nature of these polymers make them a "Platform Polymer" capable of substituting numerous petroleum polymers for durable as well as single-use applications. Among the diverse areas of application, production of PHA-based biodegradable packaging materials is the most obvious one, especially for food packaging, where compostable, transparent packaging materials with high O_2, CO_2, and moisture barrier are desired. After consuming the food, the PHA-based packaging material, which is frequently contaminated with food residues, can simply be disposed of as bio-waste that can be home or industrially composted, reducing plastic waste considerably [187]. PHA polymers exhibit high O_2-barrier properties, which have attracted enormous interest in these materials for food packaging helping to reduce food waste due to oxidation [188]. PHA compares favorably to HDPE in the preservation of quality of packaged food [189]. As far back as 1996, PHBHV received European Union approval for food contact applications in trays, containers, and films for packaging food [190]. The entire food service packaging area could benefit. Recently, the beverage company Bacardi® announced to reduce about 3000 annual tons (corresponding to 80 million bottles) of PET currently used to bottle their alcoholic drinks; the company plans to use biodegradable poly(3-hydroxybutyrate-co-3-hydroxyhexanoate) (P(3HB-*co*-3HHx)), a highly auspicious and patented PHA copolyester having the trade name Nodax™ and produced by Danimer Scientific on a commercial scale. Production of this hybrid *scl-mcl*-PHA starts from bioconversion of palm, canola, or soy seed oil. According to Barcardi®, this "100% biodegradable spirit bottles" will biodegrade within 18 months in composting facilities, soil, freshwater, and seawater [191]; the rate of abiotic hydrolysis of PHA-based items in marine environment, however, depends on the shape, thickness, and monomeric composition.

One area that has captured attention recently includes coated paper for food service applications. Most of the food service paper is coated with a petroleum plastic such as PE. Such materials are difficult to recycle, thereby ending up in landfills or being incinerated. The renewable cellulose fiber could be recycled up to seven times if the coating could be separated economically from the fiber. With PE coating, this separation is difficult and often results in high cellulose fiber losses generating PE microplastics that clogs up recycling screens in paper mills or contaminates wastewater processing plants, thus making cellulose fiber recycling uneconomical [192]. The separated PHA polymer particles being biodegradable would aid in the denitrification of wastewater/sludge at wastewater treatment plants; PHA is already used as a denitrification agent in wastewater treatment plants [193]. Coating of paper with PHA can occur via solvent casting [194] or via compression molding of PHA nanocomposite films [195]. At the moment, Danimer Scientific is announcing its partnership with the company Kemira Pulp & Paper to develop repulpable and biodegradable paper and board coated with Nodax™ PHA instead of PE [196].

Moreover, PHA can also be used to produce mulching films for agricultural use; indeed, the major global PHA manufacturers such as Nafigate, Tianjin GreenBio Materials Co., Ltd. (GreenBio), or Shenzhen Ecomann Biotechnology Co., Ltd. currently produce PHA resins explicitly for manufacturing of mulch films and packaging purposes (reviewed by [197]). Another field of future application could be textile fibers obtained from electrospinning of PHA; this could also contribute to mitigating the microplastic problem when considering the high number of microplastic fibers typically released when washing, e.g. fleece pullovers [198].

6.6.2 Hygiene/Care/Cosmetics

Other applications for PHA polymers exist where recycling is not feasible, and used products generally end up in landfills or are incinerated, such as diapers, wet wipes, and other hygiene-related articles. Currently, many of these products use petroleum plastics that simply accumulate in landfills; however, when produced from PHA polymers, consumers or waste management companies would have multiple end-of-life options, especially in their ability to compost, recover, and recycle the carbon, making such products circular [3].

Moreover, the company NAFIGATE uses PHB produced from waste cooking oil to replace petrochemical primary microplastics for peeling by shower milks and substitutes polluting chemicals such as oxybenzone chemical UV filters in sunscreen creams; these products are already available on the market [199]. Similar products were also commercialized by the Italian company Bio-On under the trademark "Minerv-Bio Cosmetics" and contain microbeads for peeling emulsions such as body and face peelings, cleaning agents, or toothpastes. As micro-sized powder particles, only 5–20 μm in size, they give texture to creamy cosmetic products. Moreover, the lipophilic and hydrophobic nature of such PHA microparticles makes them especially interesting for skin care; they do not remove water from skin but provide greasy skin a natural matt appearance. In the sun protection sector, Bio-On produces sunscreens containing marine biodegradable PHA microparticles (brand name "MyKAI"); here, the high light-scattering effect of nano-sized PHA granules makes this biopolymer the material of choice for this application. In all these cases, PHA replaces petrochemical primary microplastic particles [200].

6.6.3 Medical – Drug Delivery

PHA's biodegradability under different environmental conditions especially *in vivo* makes this class of biopolymers useful candidates for drug carriers. As stated above, extracellular PHA depolymerases and other non-specific enzymes excreted by different microorganisms are thus able to disintegrate PHA polymers into soluble oligomers and monomers and eventually to CO_2 and water. The drug release rate of PHA-based drug delivery systems has been demonstrated to be tuned by the PHA polymer's composition and molecular mass. PHA polymers have shown to substantially influence the bioavailability of bioactive compounds, improved drug encapsulation with reduced toxicity compared to other biodegradable polymers [201, 202]. As an illustrative example for use of PHA in drug delivery, P(3HB-*co*-3HHx) nanoparticles of a mean size of about 150 nm were loaded with rhodamine-B. These nanoparticles were coated with subcytotoxic poly(ethylene imine) concentrations to support attachment to different cell types and the uptake by these cells. The reaction of cells to these loaded nanoparticles was investigated *in vivo* and *ex vivo*; the nanoparticles were transported along endolysosomal cell compartments, the endoplasmic reticulum, and the Golgi apparatus, without causing any negative effects on cell morphology or respiration [203]. In addition, P(3HB-*co*-3HHx) with 12 mol% 3HHx was processed by electrospinning together with graphene-coated silver nanoparticles. The generated fiber material was used to test its antibacterial activity against the test organisms *S. aureus* and *E. coli*. The results showed that the fibers with embedded nanoparticles strongly reduced the number of living cells; therefore, these new nanocomposites may be efficient in the dressing of chronic wounds and in sanitizing applications [204]. Other "smart wound dressing" based on non-woven films consisting of poly(3HB-*co*-4HB) were successfully tested in reducing inflammation, improving the angiogenic properties of skin and enhancing the wound healing process in rat model experiments. The presence of fibroblast cells on the biopolymer films accelerated wound healing, vascularization, and regeneration; it was shown that cells excreted proteins that formed a layer on the membrane surface and supported the movement of epidermal cells from wound-surrounding tissues [205]. Moreover, the adhesive properties of a *mcl*-PHA consisting of at least five different monomers, produced by the strain *Pseudomonas chlororaphis* from crude glycerol phase from the biodiesel production, were recently demonstrated by Pereira et al.; this semicrystalline material revealed very low melting temperature T_m (44 °C) and glass transition temperature T_g (−48 °C). Films made of this *mcl*-PHA were translucid, highly elastic, water repellent, and relatively permeable to O_2 and CO_2 and demonstrated excellent adhesion properties toward skin, which make them potential candidates for a new class of natural adhesives for wound dressings or closure [206].

Using biodegradable PHA polymers together with other biodegradable nanoparticles such as nanosized cellulose [19] or lignocellulose nanofibers [207] is already yielding promising solutions as bionanocomposites with specific features, such as enhanced gas barrier properties, which in turn can replace multilayered petroleum plastic-based packaging materials that would preserve food from oxidative degradation [208]. In addition, at end of life, such bionanocomposites could be composted to recover the carbon in the form of biogas for subsequent use as a model for a circular economy.

The mechanical properties of PHAs, such as the elastic modulus, tensile strain, and tensile strength of PHAs such as PHB and their composite materials are similar to those measured for human bones; therefore, they hold significant promise for application as implant materials, e.g. to cure femoral fractures. In comparison to polymers used in surgery, such as PLA, poly(glycolate) (PGA), or poly(lactid-*co*-glycolid) (PLGA), PHA-based implants have the added benefit of not reducing the local pH-value during *in vivo* degradation; this positive attribute makes PHAs compatible with human cells and the immune system. The low *in vivo* degradation rate of *scl*-PHA-based implants and its crystallinity need to be tuned to regulate the enzymatic degradation of the implants, as demonstrated for the remarkable recalcitrance of tiny bar-shaped PHB-based femoral implants against *in vivo* degradation in living rats [209]. Because of their versatility such as tunable mechanical properties, combined with excellent biocompatibility and *in vivo* degradability, PHA biopolyesters are among the most auspicious biomaterials for tissue engineering and are being used to replace and heal different types of hard or soft tissue; PHA-based tissue engineering is described for restoring cartilage, skin, cardiovascular tissues, bone marrow, and nerve conduits (recent reviews by [210, 211]). Moreover, PHAs were tested successfully for bone plates, bone plating systems, and bone marrow scaffolds (reviewed by [212]). Especially, we currently witness various established and new processing techniques in the biomedical field to design medical devices and products from PHAs, such as solvent casting, melt spinning, electrospinning, and additive manufacturing (reviewed by [213]).

6.6.4 Other Applications

For agricultural applications, blends consisting of PLA and PHA were successfully processed into mulches by non-woven textile technology. The authors of the study stated these new mulches to be useful as durable and compostable materials for long-term agricultural applications, such as row covers [214].

Currently, PHAs also start penetrating the toy sector. In 2018, the company Bio-On started commercializing their so-called "Minerv Supertoys," which are LEGO™-like toy bricks made of stained PHA [215].

The ban of plastic materials in various countries offered new chances for PHA to enter new markets; this was shown, e.g. by the US company Newlight, which produces PHAs from GHGs by marine microbes. The produced PHAs (commercialized as "AirCarbon" PHA) are processed to drinking straws and cutlery, which are sold as "Restore™" foodware products [216]. Moreover, the company uses this "AirCarbon" PHA as components of leather replacements in smartphone covers or fashion accessories and for production of eyewear frames [217].

A completely different application of PHAs is the conversion of PHB and *mcl*-PHA to methyl esters of the corresponding building blocks using acid methanolysis. The combustion heat of these products, a new class of biodiesel-like biofuels, was compared with established fuels and fuel additives such as gasoline, ethanol, or 1-butanol. The combustion heat of 3-hydroxyalkanoate methyl esters amounted to 20–30 kJ/g, similar to the value for ethanol (27 kJ/g) [218]. Later, it was shown that addition of 3HB-methyl ester to gasoline even outperforms the addition of ethanol in terms of oxygen content, flash point, boiling point, and dynamic viscosity; blending of gasoline with approximately 10% 3HB-methyl ester only insignificantly reduced the octane number [219]. Notably, PHAs used for the production of such novel biofuels do not need high molecular mass nor high purity; it can therefore be produced in simple, inexpensive, non-sterile setups by mixed microbial cultures using sewage water as a substrate [218].

6.7 Conclusions

The awareness of plastic pollution has thus far resulted in rigid regulations across the world to eliminate or strictly reduce the production and use of many articles from established petroleum plastics. These legislations have frequently banned the use of PHA polymers as well, without having conducted a system-based assessment of the circularity of renewable and biodegradable PHA polymers and the vast potential for benefits to society that these naturally occurring materials can have without the burden of end-of-life problems well known for petroleum plastics.

However, increasing awareness of the damage from fossil fuel use, emissions thereof, and the growing accumulation of plastics waste, especially in ecologically sensitive environment and their deleterious effects on our well-being are becoming increasingly clear to everybody, including consumers; brand owners are looking for solutions with limited success thus far. In this context, it should be emphasized that there is no generally valid recipe for an "ideal PHA" to be produced; as discussed in the text, different applications require different material properties, and PHAs offer a broad variety of such properties for most diverse final fields of application.

Several commercial efforts are underway worldwide to increase PHA polymer production and to translate them into applications to mitigate the negative effects of GHG emissions and plastics waste. Many of these PHAs have suitable attributes to replace fossil plastics. However, it must be recognized that the use of fossil plastics, their incineration and the resulting GHG emissions, and their accumulation in landfills and in our environment are issues that stand on their own. There needs to be a paradigm shift in terms of our use of fossil resource usage and disposal without which their consequences will take on catastrophic consequences. The use of PHA is one such solution. The primary drivers in the adaptation of any material are (i) their safety and suitability in use, (ii) their price, and (iii) their availability. Despite the pioneering work of many researchers since the late 1950s, industry has shown merely sporadic interest in their use. Therefore, PHA price and availability relative to fossil plastics have remained their greatest impediments. Current prices of PHA vary between US$ 4 and US$ 8 per kilogram, while fossil plastics they could replace are priced in the US$ 1 to US$ 2 per kilogram range. Their availability has also remained low; therefore, proliferation has been limited. Recently, Danimer Scientific has announced several joint development and commercialization agreements with Mars Wrigley, Pepsi, and Nestle. They have also announced significant volume ramp-up to several hundred thousand tons per year by 2025. The primary challenges in PHA production continue to be the cost and availability of carbon sources for the production of PHA polymers that do not compete with food production or do not destroy our fragile ecosystems any further. Other challenges in bringing down the overall cost of production are (i) innovating in highly productive microbial cell factories, including the use of mixed microbial cell cultures, (ii) new bioengineering concepts well adapted to cope with the requirements of these cell factories, (iii) challenging raw materials such as CO_2, CH_4, H_2, and other mixed carbon-rich waste streams, and (iv) the efficient and safe methods for PHA polymer recovery from microbial biomass.

The low fossil fuel prices due to the currently slowing global economy as a consequence of the COVID-19 pandemic, resulting in increased use of fossil plastics, in combination with their hygienic benefits, do not necessarily bode well for increased innovation and commercialization efforts in PHA polymers. However, rising public awareness of marine and terrestrial pollution by persistent petroleum plastics, the emerging ubiquitous occurrence of microplastic particles in the ecosphere, the commitment of many nations toward plastic pollution, and the fulfillment of climate targets nations have mutually agreed to, and the indisputable limitation of fossil raw materials will lead to a growing demand for the versatile, biodegradable PHA polymers. Finally, it should be emphasized that instable political situations in important oil-exporting countries can suddenly result in a rocketing of the crude oil price, as witnessed in late February 2022 after the Russian invasion in Ukraine. Such situations might significantly accelerate the switch towards renewables for polymer production.

References

1 Geyer, R., Jambeck, J.R., and Law, K.L. (2017). Production use and fate of all plastics ever made. *Science Advances* 3: e1700782. https://doi.org/10.1126/sciadv.1700782.

2 How to build a circular economy. https://www.ellenmacarthurfoundation.org/publications/the-new-plastics-economy-rethinking-the-future-ofplastics. Ellen MacArthur Foundation (accessed 11 November 2021).

3 Koller, M. and Mukherjee, A. (2020). Polyhydroxyalkanoates–linking properties, applications, and end-of-life options. *Chemical and Biochemical Engineering Quarterly* 34 (3): 115–129. https://doi.org/10.15255/CABEQ.2020.1819.

4 Anastas, P.T. and Warner, J.C. (1998). *Green Chemistry: Theory and Practice.* Oxford/New York: Oxford University Press ISBN 9780198502340.

5 Narodoslawsky, M. (2010). Structural prospects and challenges for bio commodity processes. *Food Technology and Biotechnology* 48 (3): 270–275. https://doaj.org/toc/1334-2606.

6 Schnitzer, H. and Ulgiati, S. (2007). Less bad is not good enough: approaching zero emissions techniques and systems. *Journal of Cleaner Production* 15 (13, 14): 1185–1189. https://doi.org/10.1016/j.jclepro.2006.08.001.

7 Napper, I.E. and Thompson, R.C. (2016). Release of synthetic microplastic plastic fibres from domestic washing machines: effects of fabric type and washing conditions. *Marine Pollution Bulletin* 112 (1, 2): 39–45. https://doi.org/10.1016/j.marpolbul.2016.09.025.

8 Lei, L., Wu, S., Liu, M. et al. (2018). Microplastic particles cause intestinal damage and other adverse effects in zebrafish *Danio rerio* and nematode *Caenorhabditis elegans. Science of the Total Environment* 619: 1–8. https://doi.org/10.1016/j.scitotenv.2017.11.103.

9 Al-Jaibachi, R., Cuthbert, R.N., and Callaghan, A. (2018). Up and away: ontogenic transference as a pathway for aerial dispersal of microplastics. *Biology Letters* 14 (9): 20180479. https://doi.org/10.1098/rsbl.2018.0479.

10 Trotter, B., Ramsperger, A.F.R.M., Raab, P., and Laforsch, C. (2019). Plastic waste interferes with chemical communication in aquatic ecosystems. *Scientific Reports* 9 (1): 5889. https://doi.org/10.1038/s41598-019-41677-1.

11 Eerkes-Medrano, D., Leslie, H.A., and Quinn, B. (2019). Microplastics in drinking water: a review and assessment. *Current Opinion in Environmental Science & Health* 7: 69–75. https://doi.org/10.1038/s41598-019-41677-1.

12 Peixoto, D., Pinheiro, C., Amorim, J. et al. (2019). Microplastic pollution in commercial salt for human consumption: a review. *Estuarine, Coastal and Shelf Science* 219: 161–168. https://doi.org/10.1016/j.ecss.2019.02.018.

13 Liebezeit, G., and Liebezeit, E. (2013). Non-pollen particulates in honey and sugar. *Food Additives & Contaminants: Part A* 30 (12), 2136–2140. https://doi .org/10.1080/19440049.2013.843025.

14 Liebezeit, G., and Liebezeit, E. (2014). Synthetic particles as contaminants in German beers. *Food Additives & Contaminants: Part A* 31 (9): 1574–1578. https://doi.org/10.1080/19440049.2014.945099.

15 Bitencourt, G.R., Mello, P.A., Flores, E.M. et al. (2020). Determination of microplastic content in seafood: an integrated approach combined with the determination of elemental contaminants. *Science of the Total Environment* 749: 142301. https://doi.org/10.1016/j.scifotenv.2020.142301 (accessed 11 November 2021).

16 https://www.wri.org/insights/127-countries-now-regulate-plastic-bags-why-arent-we-seeing-less-pollution. World Resources Institute (accessed 11 November 2021).

17 Barrena, R., Font, X., Gabarrell, X., and Sánchez, A. (2014). Home composting versus industrial composting: influence of composting system on compost quality with focus on compost stability. *Waste Management* 34 (7): 1109–1116. https://doi.org/10.1016/j.wasman.2014.02.008.

18 Kourmentza, C., Plácido, J., Venetsaneas, N. et al. (2017). Recent advances and challenges towards sustainable polyhydroxyalkanoate (PHA) production. *Bioengineering* 4: 55. https://doi.org/10.3390/bioengineering4020055.

19 Silva, F.A., Dourado, F., Gama, M., and Poças, F. (2020). Nanocellulose bio-based composites for food packaging. *Nanomaterials* 10 (10): 2041. https://doi.org/10.3390/nano10102041.

20 Koller, M. (2018). Chemical and biochemical engineering approaches in manufacturing polyhydroxyalkanoate (PHA) biopolyesters of tailored structure with focus on the diversity of building blocks. *Chemical and Biochemical Engineering Quarterly* 32 (4): 413–438. https://doi.org/10.15255/CABEQ.2018.1385.

21 Chen, G.Q. and Hajnal, I. (2015). The 'PHAome'. *Trends in Biotechnology* 33 (10): 559–564. https://doi.org/10.1016/j.tibtech.2015.07.006.

22 Doi, Y., Kanesawa, Y., Kunioka, M., and Saito, T. (1990). Biodegradation of microbial copolyesters: poly(3-hydroxybutyrate-*co*-3-hydroxyvalerate) and poly(3-hydroxybutyrate-*co*-4-hydroxybutyrate). *Macromolecules* 23: 26. https://doi.org/0024-9297/90/2223-0026802.50/0.

23 Koller, M. (2020). Advances in polyhydroxyalkanoate (PHA) production, volume 2. *Bioengineering* 7 (1): 24. https://doi.org/10.3390/bioengineering7010024.

24 Jendrossek, D. (2009). Polyhydroxyalkanoate granules are complex subcellular organelles (carbonosomes). *Journal of Bacteriology* 191 (10): 3195–3202. https://doi.org/10.1128/JB.01723-08.

25 Koller, M., Maršálek, L., Miranda de Sousa Dias, M., and Braunegg, G. (2017). Producing microbial polyhydroxyalkanoate (PHA) biopolyesters in a sustainable manner. *New Biotechnology* 37 (A): 24–38. https://doi.org/10.1016/j.nbt.2016.05.001.

26 Obruca, S., Sedlacek, P., Koller, M. et al. (2018). Involvement of polyhydroxyalkanoates in stress resistance of microbial cells: biotechnological consequences and applications. *Biotechnology Advances* 36 (3): 856–870. https://doi.org/10.1016/j.biotechadv.2017.12.006.

27 Obruca, S., Sedlacek, P., Slaninova, E. et al. (2020). Novel unexpected functions of PHA granules. *Applied Microbiology and Biotechnology* 104: 4795–4810. https://doi.org/10.1007/s00253-020-10568-1.

28 Müller-Santos, M., Koskimäki, J.J., Alves, L.P.S. et al. (2020). The protective role of PHB and its degradation products against stress situations in bacteria. *FEMS Microbiology Reviews* fuaa058. https://doi.org/10.1093/femsre/fuaa058.

29 Sedlacek, P., Slaninova, E., Koller, M. et al. (2019). PHA granules help bacterial cells to preserve cell integrity when exposed to sudden osmotic imbalances. *New Biotechnology* 49: 129–136. https://doi.org/10.1016/j.nbt.2018.10.005.

30 Pernicova, I., Novackova, I., Sedlacek, P. et al. (2020). Application of osmotic challenge for enrichment of microbial consortia in polyhydroxyalkanoates producing thermophilic and thermotolerant bacteria and their subsequent isolation. *International Journal of Biological Macromolecules* 144: 698–704. https://doi.org/10.1016/j.ijbiomac.2019.12.128.

31 Obruca, S., Sedlacek, P., Mravec, F. et al. (2016). Evaluation of 3-hydroxybutyrate as an enzyme-protective agent against heating and oxidative damage and its potential role in stress response of poly(3-hydroxybutyrate) accumulating cells. *Applied Microbiology and Biotechnology* 100 (3) https://doi.org/1365-1376: doi: 10.1007/s00253-015-7162-4.

32 Al Rowaihi, I.S., Paillier, A., Rasul, S. et al. (2018). Poly(3-hydroxybutyrate) production in an integrated electromicrobial setup: investigation under stress-inducing conditions. *PLoS One* 13 (4): e0196079. https://doi.org/10.1371/journal.pone.0196079.

33 Ayub, N.D., Tribelli, P.M., and López, N.I. (2009). Polyhydroxyalkanoates are essential for maintenance of redox state in the Antarctic bacterium *Pseudomonas* sp. 14-3 during low temperature adaptation. *Extremophiles* 13 (1): 59–66. https://doi.org/10.1007/s00792-008-0197-z.

34 Obruca, S., Sedlacek, P., Krzyzanek, V. et al. (2016). Accumulation of poly(3-hydroxybutyrate) helps bacterial cells to survive freezing. *PLoS One* 11 (6): e0157778. https://doi.org/10.1371/journal.pone.0157778.

35 Slaninova, E., Sedlacek, P., Mravec, F. et al. (2018). Light scattering on PHA granules protects bacterial cells against the harmful effects of UV radiation. *Applied Microbiology and Biotechnology* 102 (4): 1923–1931. https://doi.org/10.1007/s00253-018-8760-8.

36 Tan, G.-Y.A., Chen, C.L., Li, L. et al. (2014). Start a research on biopolymer polyhydroxyalkanoate (PHA): a review. *Polymers* 6: 706. https://doi.org/10.3390/polym6030706.

37 Chen, G.Q. (2009). A microbial polyhydroxyalkanoates (PHA) based bio-and materials industry. *Chemical Society Reviews* 38 (8): 2434–2446. https://doi.org/10.1039/B812677C.

38 Kucera, D., Novackova, I., Pernicova, I. et al. (2019). Biotechnological production of poly(3-hydroxybutyrate-*co*-4-hydroxybutyrate-*co*-3-hydroxyvalerate) terpolymer by *Cupriavidus* sp. DSM 19379. *Bioengineering* 6 (3): 74. https://doi.org/10.3390/bioengineering6030074.

39 Zinn, M. (2010). Biosynthesis of medium-chain-length poly[(*R*)-3-hydroxyalkanoates]. In: *Plastics from Bacteria* (ed. G.Q. Chen), 213–236. Berlin, Heidelberg: Springer https://doi.org/10.1007/978-3-642-03287-5_9.

40 Ramsay, B. (2016). Medium-chain-length Poly-3-hydroxyalkanoates. In: *Recent Advances in BiotechnologyMicrobiology, Feedstocks, and Metabolism*, vol. 1 (ed. M. Koller), 436–473. Sharjah: Bentham Science Publishers, ISSN 2468-5372(online), 2468-5364 (print). https://www.eurekaselect.net/ebook_volume/2006.

41 De Koning, G.J.M., Van Bilsen, H.M.M., Lemstra, P.J., Hazenberg, W., Witholt, B., Preusting, H., van der Galiën, J.G., Schirmer, A., Jendrossek, D. (1994). A biodegradable rubber by crosslinking poly(hydroxyalkanoate) from *Pseudomonas oleovorans*. *Polymer* 35: 2090-2097. https://doi.org/10.1016/0032-3861(94)90233-X.

42 Yamane, T. (1992). Cultivation engineering of microbial bioplastics production. *FEMS Microbiology Reviews* 9 (2–4): 257–264.

43 Yamane, T. (1993). Yield of poly-D(−)-3-hydroxybutyrate from various carbon sources: a theoretical study. *Biotechnology and Bioengineering* 41 (1): 165–170.

44 Pan, C., Ge, L., Lee, P.-H., and Tan, G.-J.A. (2021). An Introduction to the Thermodynamics Calculation of PHA Production in Microbes. In: *The Handbook of Polyhydroxyalkanoates*, Kinetics, Bioengineering, and Industrial Aspects, vol. II (ed. M. Koller), 3–19. CRC Press, Taylor & Francis Group ISBN 9780367275624. https://www.routledge.com/The-Handbook-of-Polyhydroxyalkanoates-Kinetics-Bioengineering-and-Industrial/Koller/p/book/9780367275624.

45 da Cruz Pradella, J.G., Ienczak, J.L., Delgado, C.R., and Taciro, M.K. (2012). Carbon source pulsed feeding to attain high yield and high productivity in poly(3-hydroxybutyrate) (PHB) production from soybean oil using *Cupriavidus necator*. *Biotechnology Letters* 34 (6): 1003–1007. https://doi.org/10.1007/s10529-012-0863-1.

46 Miao, R., Xie, H., Liu, X. et al. (2020). Current processes and future challenges of photoautotrophic production of acetyl-CoA-derived solar fuels and chemicals in cyanobacteria. *Current Opinion in Chemical Biology* 59: 69–76. https://doi.org/10.1016/j.cbpa.2020.04.013.

47 Rehm, B.H. (2003). Polyester synthases: natural catalysts for plastics. *Biochemical Journal* 376 (1): 15–33. https://doi.org/10.1042/bj20031254.

48 Wang, H.H., Zhou, X.R., Liu, Q., and Chen, G.Q. (2011). Biosynthesis of polyhydroxyalkanoate homopolymers by *Pseudomonas putida*. *Applied Microbiology and Biotechnology* 89 (5): 1497–1507. https://doi.org/10.1007/s00253-010-2964-x.

49 Bonilla, M., Olivaro, C., Corona, M. et al. (2005). Production and characterization of a new bioemulsifier from *Pseudomonas putida* ML2. *Journal of Applied Microbiology* 98 (2): 456–463. https://doi.org/10.1111/j.1365-2672.2004.02480.x.

50 Amaro, T.M., Rosa, D., Comi, G., and Iacumin, L. (2019). Prospects for the use of whey for polyhydroxyalkanoate (PHA) production. *Frontiers in Microbiology* 10: 992. https://doi.org/10.3389/fmicb.2019.00992.

51 Basnett, P., Marcello, E., Lukasiewicz, B. et al. (2020). Antimicrobial materials with lime oil and a poly(3-hydroxyalkanoate) produced via valorisation of sugar cane molasses. *Journal of Functional Biomaterials* 11 (2): 24. https://doi.org/10.3390/jfb11020024.

52 Takahashi, R.Y.U., Castilho, N.A.S., Silva, M.A.C.D. et al. (2017). Prospecting for marine bacteria for polyhydroxyalkanoate production on low-cost substrates. *Bioengineering* 4 (3): 60. https://doi.org/10.3390/bioengineering4030060.

53 Obruca, S., Benesova, P., Maršalek, L., and Marova, I. (2015). Use of lignocellulosic materials for PHA production. *Chemical and Biochemical Engineering Quarterly* 29 (2): 135–144. https://doi.org/10.15255/CABEQ.2014.2253.

54 González-García, Y., Grieve, J., Meza-Contreras, J.C. et al. (2019). Tequila agave bagasse hydrolysate for the production of polyhydroxybutyrate by *Burkholderia sacchari*. *Bioengineering* 6 (4): 115. https://doi.org/10.3390/bioengineering6040115.

55 Salgaonkar, B.B. and Bragança, J.M. (2017). Utilization of sugarcane bagasse by *Halogeometricum borinquense* strain E3 for biosynthesis of poly(3-hydroxybutyrate-*co*-3-hydroxyvalerate). *Bioengineering* 4 (2): 50. https://doi.org/10.3390/bioengineering4020050.

56 Rebocho, A.T., Pereira, J.R., Freitas, F. et al. (2019). Production of medium-chain length polyhydroxyalkanoates by *Pseudomonas citronellolis* grown in apple pulp waste. *Applied Food Biotechnology* 6 (1): 71–82. https://doi.org/10.22037/afb.v6i1.21793.

57 Kovalcik, A., Pernicova, I., Obruca, S. et al. (2020). Grape winery waste as a promising feedstock for the production of polyhydroxyalkanoates and other value-added products. *Food and Bioproducts Processing* 124: 1–10. https://doi.org/10.1016/j.fbp.2020.08.003.

58 Bhalerao, A., Banerjee, R., and Nogueira, R. (2020). Continuous cultivation strategy for yeast industrial wastewater-based polyhydroxyalkanoate production. *Journal of Bioscience and Bioengineering* 129 (5): 595–602. https://doi.org/10.1016/j.jbiosc.2019.11.006.

59 Gurav, R., Bhatia, S.K., Moon, Y.M. et al. (2019). One-pot exploitation of chitin biomass for simultaneous production of electricity, *N*-acetylglucosamine and polyhydroxyalkanoates in microbial fuel cell using novel marine bacterium *Arenibacter palladensis* YHY2. *Journal of Cleaner Production* 209: 324–332. https://doi.org/10.1016/j.jclepro.2018.10.252.

60 Kumar, P. and Kim, B.S. (2019). *Paracoccus* sp. strain LL1 as a single cell factory for the conversion of waste cooking oil to polyhydroxyalkanoates and carotenoids. *Applied Food Biotechnology* 6 (1): https://doi.org/10.22037/afb.v6i1.21628.

61 Talan, A., Kaur, R., Tyagi, R.D., and Drogui, P. (2020). Bioconversion of oily waste to polyhydroxyalkanoates: sustainable technology with circular bioeconomy approach and multidimensional impacts. *Bioresource Technology Reports* 11: 100496. https://doi.org/10.1016/j.biteb.2020.100496.

62 Bustamante, D., Tortajada, M., Ramon, D., and Rojas, A. (2019). Camelina oil as a promising substrate for *mcl*-PHA production in *Pseudomonas* sp. cultures. *Applied Food Biotechnology* 6 (1): 61–70. https://doi.org/10.22037/afb.v6i1.21635.

63 Koller, M., Shahzad, K., and Braunegg, G. (2018). Waste streams of the animal-processing industry as feedstocks to produce polyhydroxyalkanoate biopolyesters. *Applied Food Biotechnology* 5 (4): 193–203. https://doi.org/10.22037/afb.v5i4.18557.

64 Favaro, L., Basaglia, M., and Casella, S. (2019). Improving polyhydroxyalkanoate production from inexpensive carbon sources by genetic approaches: a review. *Biofuels, Bioproducts and Biorefining* 13 (1): 208–227. https://doi.org/10.1002/bbb.1944.

65 Reis, M.A.M., Serafim, L.S., Lemos, P.C. et al. (2003). Production of polyhydroxyalkanoates by mixed microbial cultures. *Bioprocess and Biosystems Engineering* 25 (6): 377–385. https://doi.org/10.1007/s00449-003-0322-4.

66 Oliveira, C.S., Silva, C.E., Carvalho, G., and Reis, M.A.M. (2017). Strategies for efficiently selecting PHA producing mixed microbial cultures using complex feedstocks: feast and famine regime and uncoupled carbon and nitrogen availabilities. *New Biotechnology* 37 (A): 69–79. https://doi.org/10.1016/j.nbt.2016.10.008.

67 Chan, C.M., Johansson, P., Magnusson, P. et al. (2017). Mixed culture polyhydroxyalkanoate-rich biomass assessment and quality control using thermogravimetric measurement methods. *Polymer Degradation and Stability* 144: 110–120. https://doi.org/10.1016/j.polymdegradstab.2017.07.029.

68 Valentino, F., Gottardo, M., Micolucci, F. et al. (2018). Organic fraction of municipal solid waste recovery by conversion into added-value polyhydroxyalkanoates and biogas. *ACS Sustainable Chemistry & Engineering* 6 (12): 16375–16385. https://doi.org/10.1021/acssuschemeng.8b03454.

69 Colombo, B., Sciarria, T.P., Reis, M. et al. (2016). Polyhydroxyalkanoates (PHAs) production from fermented cheese whey by using a mixed microbial culture. *Bioresource Technology* 218: 692–699. https://doi.org/10.1016/j.biortech.2016.07.024.

70 Moretto, G., Russo, I., Bolzonella, D. et al. (2020). An urban biorefinery for food waste and biological sludge conversion into polyhydroxyalkanoates and biogas. *Water Research* 170: 115371. https://doi.org/10.1016/j.watres.2019.115371.

71 Moita, R., Freches, A., and Lemos, P.C. (2014). Crude glycerol as feedstock for polyhydroxyalkanoates production by mixed microbial cultures. *Water Research* 58: 9–20. https://doi.org/10.1016/j.watres.2014.03.066.

72 Meixner, K., Kovalcik, A., Sykacek, E. et al. (2018). Cyanobacteria biorefinery – production of poly(3-hydroxybutyrate) with *Synechocystis salina* and utilisation of residual biomass. *Journal of Biotechnology* 265: 46–53. https://doi.org/10.1016/j.jbiotec.2017.10.020.

73 Kamravamanesh, D., Lackner, M., and Herwig, C. (2018). Bioprocess engineering aspects of sustainable polyhydroxyalkanoate production in cyanobacteria. *Bioengineering* 5 (4): 111. https://doi.org/10.3390/bioengineering5040111.

74 Price, S., Kuzhiumparambil, U., Pernice, M., and Ralph, P.J. (2020). Cyanobacterial polyhydroxybutyrate for sustainable bioplastic production: critical review and perspectives. *Journal of Environmental Chemical Engineering* 8 (4): 104007. https://doi.org/10.1016/j.jece.2020.104007.

75 Troschl, C., Meixner, K., and Drosg, B. (2017). Cyanobacterial PHA production – review of recent advances and a summary of three years' working experience running a pilot plant. *Bioengineering* 4 (2): 26. https://doi.org/10.3390/bioengineering4020026.

76 Gomes Gradíssimo, D., Pereira Xavier, L., and Valadares Santos, A. (2020). Cyanobacterial polyhydroxyalkanoates: a sustainable alternative in circular economy. *Molecules* 25 (18): 4331. https://doi.org/10.3390/molecules25184331.

77 da Silva, C.K., Costa, J.A.V., and de Morais, M.G. (2018). Polyhydroxybutyrate (PHB) synthesis by *Spirulina* sp. LEB 18 using biopolymer extraction waste. *Applied Biochemistry and Biotechnology* 185 (3): 822–833. https://doi.org/10.1007/s12010-017-2687-x.

78 Bhati, R. and Mallick, N. (2016). Carbon dioxide and poultry waste utilization for production of polyhydroxyalkanoate biopolymers by *Nostoc muscorum* Agardh: a sustainable approach. *Journal of Applied Phycology* 28: https://doi.org/161–168. doi: 10.1007/s10811-015-0573-x.

79 Tarawat, S., Incharoensakdi, A., and Monshupanee, T. (2020). Cyanobacterial production of poly(3-hydroxybutyrate-*co*-3-hydroxyvalerate) from carbon dioxide or a single organic substrate: improved polymer elongation with an extremely high 3-hydroxyvalerate mole proportion. *Journal of Applied Phycology* 32: 1095–1102. https://doi.org/10.1007/s10811-020-02040-4.

80 Amstutz, V. and Zinn, M. (2020). Syngas as a sustainable carbon source for PHA production. In: *The Handbook of Polyhydroxyalkanoates*, Microbial Biosynthesis and Feedstocks, vol. I (ed. M. Koller), 377–416. CRC Press, Taylor & Francis Group ISBN 9780367275594.

81 Karmann, S., Panke, S., and Zinn, M. (2019). Fed-batch cultivations of *Rhodospirillum rubrum* under multiple nutrient-limited growth conditions on syngas as a novel option to produce poly(3-hydroxybutyrate) (PHB). *Frontiers in Bioengineering and Biotechnology* 7: 59. https://doi.org/10.3389/fbioe.2019.00059.

82 Heinrich, D., Raberg, M., Fricke, P. et al. (2016). Synthesis gas (syngas)-derived medium-chain-length polyhydroxyalkanoate synthesis in engineered *Rhodospirillum rubrum*. *Applied and Environmental Microbiology* 82 (20): 6132–6140. https://doi.org/10.1128/AEM.01744-16.

83 Revelles, O., Beneroso, D., Menendez, J.A. et al. (2017). Syngas obtained by microwave pyrolysis of household wastes as feedstock for polyhydroxyalkanoate production in *Rhodospirillum rubrum*. *Microbial Biotechnology* 10 (6): 1412–1417. https://doi.org/10.1111/1751-7915.12411.

84 Sadykov, V., Bobrova, L., Pavlova, S. et al. (2012). *Syngas Generation from Hydrocarbons and Oxygenates with Structured Catalysts* (ed. A. Kurucz and I. Bencik), 53–140. New York: Nova Science Publishers, Inc. ISBN-13: 978-1608763238, ISBN-10: 1608763234.

85 Sharmila, G., Muthukumaran, C., Kumar, N.M. et al. (2020). Food waste valorization for biopolymer production. In: *Current Developments in Biotechnology and Bioengineering* (ed. S. Varjani, A. Pandey, E. Gnansounou, et al.), 233–249. Amsterdam: Elsevier B.V https://doi.org/10.1016/B978-0-444-64321-6.00012-4.

86 Flüchter, S., Follonier, S., Schiel-Bengelsdorf, B. et al. (2019). Anaerobic production of poly(3-hydroxybutyrate) and its precursor 3-hydroxybutyrate from synthesis gas by autotrophic clostridia. *Biomacromolecules* 20 (9): 3271–3282. https://doi.org/10.1021/acs.biomac.9b00342.

87 Moradi, M., Rashedi, H., Mofradnia, S.R. et al. (2019). Polyhydroxybutyrate production from natural gas in a bubble column bioreactor: simulation using COMSOL. *Bioengineering* 6 (3): 84. https://doi.org/10.3390/bioengineering6030084.

88 Rahnama, F., Vasheghani-Farahani, E., Yazdian, F., and Shojaosadati, S.A. (2012). PHB production by *Methylocystis hirsuta* from natural gas in a bubble column and a vertical loop bioreactor. *Biochemical Engineering Journal* 65: 51–56. https://doi.org/10.1016/j.bej.2012.03.014.

89 Ghoddosi, F., Golzar, H., Yazdian, F. et al. (2019). Effect of carbon sources for PHB production in bubble column bioreactor: emphasis on improvement of methane uptake. *Journal of Environmental Chemical Engineering* 7 (2): 102978. https://doi.org/10.1016/j.jece.2019.102978.

90 Levett, I., Birkett, G., Davies, N. et al. (2016). Techno-economic assessment of poly-3-hydroxybutyrate (PHB) production from methane – the case for thermophilic bioprocessing. *Journal of Environmental Chemical Engineering* 4 (4): 3724–3733. https://doi.org/10.1016/j.jece.2016.07.033.

91 https://www.calysta.com/our-technology. Calysta, Inc. (accessed 11 November 2021).

92 Jane Byrne. Single cell producer Unibio signs commercial deal with Chilean distribution giant. https://www.feednavigator.com/Article/2020/12/14/Single-cell-producer-Unibio-signs-commercial-deal-with-Chilean-distribution-giant (accessed 11 November 2021).

93 Mike Wither This company harvest CO_2 and methane to make carbon-negative plastics. https://nowthisnews.com/earth/this-company-harvest-co2-methane-to-make-carbon-negative-plastics (accessed 11 November 2021).

94 Mango materials moves on methane gas for bioplastics. https://www.mangomaterials.com/news-press/mango-materials-moves-on-methane-gas-for-bioplastics. Mango Materials (accessed 11 November 2021).

95 Liu, L.Y., Xie, G.J., Xing, D.F. et al. (2020). Biological conversion of methane to polyhydroxyalkaonate: current advances, challenges, and perspectives. *Environmental Science and Ecotechnology* 2: 100029. https://doi.org/10.1016/j.ese.2020.100029.

96 Khosravi-Darani, K., Yazdian, F., Babapour, F., and Amirsadeghi, A.R. (2019). Poly(3-hydroxybutyrate) production from natural gas by a methanotroph native bacterium in a bubble column bioreactor. *Chemical and Biochemical Engineering Quarterly* 33 (1): 69–77. https://doi.org/10.15255/CABEQ.2017.1263.

97 Orita, I., Nishikawa, K., Nakamura, S., and Fukui, T. (2014). Biosynthesis of polyhydroxyalkanoate copolymers from methanol by *Methylobacterium extorquens* AM1 and the engineered strains under cobalt-deficient conditions. *Applied Microbiology and Biotechnology* 98 (8): 3715–3725. https://doi.org/10.1007/s00253-013-5490-9.

98 Koller, M. (2020). Production, properties, and processing of microbial polyhydroxyalkanoate (PHA) biopolyesters. In: *Microbial and Natural Macromolecules – Synthesis and Applications* (ed. S. Das and H.R. Dash), 3–55. Amsterdam: Elsevier Inc ISBN: 978-0-12-820084-1. https://doi.org/10.1016/C2019-0-00320-2.

99 Litchfield, C.D. (2011). Potential for industrial products from the halophilic archaea. *Journal of Industrial Microbiology & Biotechnology* 38 (10): 1635. https://doi.org/10.1007/s10295-011-1021-9.

100 Koller, M., Salerno, A., Dias, M. et al. (2010). Modern biotechnological polymer synthesis: a review. *Food Technol Biotechnology* 48 (3): 255–269.

101 Poirier, Y., Dennis, D.E., Klomparens, K., and Somerville, C. (1992). Polyhydroxybutyrate, a biodegradable thermoplastic, produced in transgenic plants. *Science* 256 (5056): 520–523. https://doi.org/10.1126/science.256.5056.520.

102 Poirier, Y., Nawrath, C., and Somerville, C. (1995). Production of polyhydroxyalkanoates, a family of biodegradable plastics and elastomers, in bacteria and plants. *Bio/Technology* 13 (2): 142–150. https://doi.org/10.1038/nbt0295-142.

103 Poirier, Y. and Brumbley, S.M. (2010). Metabolic engineering of plants for the synthesis of polyhydroxyalkanaotes. In: *Plastics from Bacteria* (ed. G.-Q. Chen), 187–211. Berlin, Heidelberg: Springer https://doi.org/10.1007/978-3-642-03287-5_8. ISBN: 978-3-642-03287-5 (online), 978-3-642-03286-8 (print).

104 van Beilen, J.B. and Poirier, Y. (2012). Plants as factories for bioplastics and other novel biomaterials. In: *Plant Biotechnology and Agriculture* (ed. A. Altman and P.M. Hasegawa), 481–494. Amsterdam: Elsevier Inc. ISBN: 978-0-12-381466-1. https://doi.org/10.1016/C2009-0-61282-4.

105 Robert, J., Marchesini, S., Delessert, S., and Poirier, Y. (2005). Analysis of the β-oxidation of trans-unsaturated fatty acid in recombinant *Saccharomyces cerevisiae* expressing a peroxisomal PHA synthase reveals the involvement of a reductase-dependent pathway. *Biochimica et Biophysica Acta (BBA) – Molecular and Cell Biology of Lipids* 1734 (2): 169–177. https://doi.org/10.1016/j.bbalip.2005.02.010.

106 Poirier, Y., Erard, N., and MacDonald-Comber Petétot, J. (2002). Synthesis of polyhydroxyalkanoate in the peroxisome of *Pichia pastoris*. *FEMS Microbiology Letters* 207 (1): 97–102. https://doi.org/10.1111/j.1574-6968.2002.tb11035.x.

107 Haddouche, R., Poirier, Y., Delessert, S. et al. (2011). Engineering polyhydroxyalkanoate content and monomer composition in the oleaginous yeast *Yarrowia lipolytica* by modifying the ß-oxidation multifunctional protein. *Applied Microbiology and Biotechnology* 91 (5): 1327–1340. https://doi.org/10.1007/s00253-011-3331-2.

108 Gao, C., Qi, Q., Madzak, C., and Lin, C.S.K. (2015). Exploring medium-chain-length polyhydroxyalkanoates production in the engineered yeast *Yarrowia lipolytica*. *Journal of Industrial Microbiology & Biotechnology* 42 (9): 1255–1262. https://doi.org/10.1007/s10295-015-1649-y.

109 Abd-El-Haleem, D.A. (2009). Biosynthesis of polyhydroxyalkanotes in wild type yeasts. *Polish Journal of Microbiology* 58 (1): 37–41.

110 Williams, M.D., Rahn, J.A., and Sherman, D.H. (1996). Production of a polyhydroxyalkanoate biopolymer in insect cells with a modified eucaryotic fatty acid synthase. *Applied and Environmental Microbiology* 62 (7): 2540–2546. 0099-2240/96/$04.0010.

111 Koller, M. (2018). A review on established and emerging fermentation schemes for microbial production of polyhydroxyalkanoate (PHA) biopolyesters. *Fermentation* 4 (2): 30. https://doi.org/10.3390/fermentation4020030.

112 Blunt, W., Levin, D., and Cicek, N. (2018). Bioreactor operating strategies for improved polyhydroxyalkanoate (PHA) productivity. *Polymers* 10 (11): 1197. https://doi.org/10.3390/polym10111197.

113 Sindhu, R., Silviya, N., Binod, P., and Pandey, A. (2013). Pentose-rich hydrolysate from acid pretreated rice straw as a carbon source for the production of poly-3-hydroxybutyrate. *Biochemical Engineering Journal* 78: 67–72. https://doi.org/10.1016/j.bej.2012.12.015.

114 Gahlawat, G. and Srivastava, A.K. (2018). Enhancing the production of polyhydroxyalkanoate biopolymer by *Azohydromonas australica* using a simple empty and fill bioreactor cultivation strategy. *Chemical and Biochemical Engineering Quarterly* 31 (4): 479–485. https://doi.org/10.15255/CABEQ.2017.1148.

115 Penloglou, G., Vasileiadou, A., Chatzidoukas, C., and Kiparissides, C. (2017). Model-based intensification of a fed-batch microbial process for the maximization of polyhydroxybutyrate (PHB) production rate. *Bioprocess and Biosystems Engineering* 40: 1247–1260. https://doi.org/10.1007/s00449-017-1784-0.

116 Ibrahim, M.H. and Steinbüchel, A. (2010). High-cell-density cyclic fed-batch fermentation of a poly(3-hydroxybutyrate)-accumulating thermophile, *Chelatococcus* sp. strain MW10. *Applied and Environmental Microbiology* 76 (23): 7890–7895. https://doi.org/10.1128/AEM.01488-10.

117 Haas, C., El-Najjar, T., Virgolini, N. et al. High cell-density production of poly(3-hydroxybutyrate) in a membrane bioreactor. *New Biotechnology* 37 (A): 117–122. https://doi.org/10.1016/j.nbt.2016.06.1461.

118 Koller, M. and Muhr, A. (2014). Continuous production mode as a viable process-engineering tool for efficient poly(hydroxyalkanoate) (PHA) bio-production. *Chemical and Biochemical Engineering Quarterly* 28 (1): 65–77. https://hrcak.srce.hr/117989.

119 Atlić, A., Koller, M., Scherzer, D. et al. (2011). Continuous production of poly([*R*]-3-hydroxybutyrate) by *Cupriavidus necator* in a multistage bioreactor cascade. *Applied Microbiology and Biotechnology* 91 (2): 295–304. https://doi.org/10.1007/s00253-011-3260-0.

120 Horvat, P., Špoljarić, I.V., Lopar, M. et al. (2013). Mathematical modelling and process optimization of a continuous 5-stage bioreactor cascade for production of poly [–(*R*)-3-hydroxybutyrate] by *Cupriavidus necator*. *Bioprocess and Biosystems Engineering* 36 (9): 1235–1250. https://doi.org/10.1007/s00449-012-0852-8.

121 Lopar, M., Špoljarić, I.V., Atlić, A. et al. (2013). Five-step continuous production of PHB analyzed by elementary flux, modes, yield space analysis and high structured metabolic model. *Biochemical Engineering Journal* 79: 57–70. https://doi.org/10.1016/j.bej.2013.07.003.

122 Vadlja, D., Koller, M., Novak, M. et al. (2016). Footprint area analysis of binary imaged *Cupriavidus necator* cells to study PHB production at balanced, transient, and limited growth conditions in a cascade process. *Applied Microbiology and Biotechnology* 100 (23): 10065–10080. https://doi.org/10.1007/s00253-016-7844-6.

123 Koller, M., Vadlja, D., Braunegg, G. et al. (2017). Formal-and high-structured kinetic process modelling and footprint area analysis of binary imaged cells: tools to understand and optimize multistage-continuous PHA biosynthesis. *The EuroBiotech Journal* 1 (3): 203–211. https://doi.org/10.24190/ISSN2564-615X/2017/03.01.

124 Strong, P., Laycock, B., Mahamud, S. et al. (2016). The opportunity for high-performance biomaterials from methane. *Microorganisms* 4 (1): 11. https://doi.org/10.3390/microorganisms4010011.

125 Karmann, S., Follonier, S., Egger, D. et al. (2017). Tailor-made PAT platform for safe syngas fermentations in batch, fed-batch and chemostat mode with *Rhodospirillum rubrum*. *Microbial Biotechnology* 10 (6): 1365–1375. https://doi.org/10.1111/1751-7915.12727.

126 Singh, A.K. and Mallick, N. (2017). Advances in cyanobacterial polyhydroxyalkanoates production. *FEMS Microbiology Letters* 364 (20): fnx189. https://doi.org/10.1093/femsle/fnx189.

127 Koller, M., Khosravi-Darani, K., and Braunegg, G. (2017). Advanced photobioreactor systems for the efficient cultivation of cyanobacteria. In: *Photobioreactors Advancements, Applications and Research* (ed. Y.F. Tsang), 35–90. New York: Nova Science Publishers, Inc ISBN: 1536123544, 9781536123548.

128 Costa, J.A.V., Moreira, J.B., Lucas, B.F. et al. (2018). Recent advances and future perspectives of PHB production by cyanobacteria. *Industrial Biotechnology* 14 (5): 249–256. https://doi.org/10.1089/ind.2018.0017.

129 Drosg, B., Fritz, I., Gattermayr, F., and Silvestrini, L. (2015). Photo-autotrophic production of poly(hydroxyalkanoates) in cyanobacteria. *Chemical and Biochemical Engineering Quarterly* 29 (2): 145–156. https://doi.org/10.15255/CABEQ.2014.2254.

130 Koller, M. (2020). Established and advanced approaches for recovery of microbial polyhydroxyalkanoate (PHA) biopolyesters from surrounding microbial biomass. *The EuroBiotech Journal* 4 (3): 113–126. https://doi.org/10.2478/ebtj-2020-0013.

131 Gonzalez, K., Navia, R., Liu, S., and Cea, M. (2020). Biological approaches in polyhydroxyalkanoates recovery. *Current Microbiology* online ahead of print. https://doi.org/10.1007/s00284-020-02263-1.

132 Koller, M., Niebelschütz, H., and Braunegg, G. (2013). Strategies for recovery and purification of poly[(*R*)-3-hydroxyalkanoates] (PHA) biopolyesters from surrounding biomass. *Engineering in Life Sciences* 13 (6): 549–562. https://doi.org/10.1002/elsc.201300021.

133 Nonato, R., Mantelatto, P., and Rossell, C. (2001). Integrated production of biodegradable plastic, sugar and ethanol. *Applied Microbiology and Biotechnology* 57 (1, 2): 1–5. https://doi.org/10.1007/s002530100732.

134 Koller, M., Bona, R., Chiellini, E., and Braunegg, G. (2013). Extraction of short-chain-length poly-[(*R*)-hydroxyalkanoates] (*scl*-PHA) by the "anti-solvent" acetone under elevated temperature and pressure. *Biotechnology Letters* 35 (7): 1023–1028. https://doi.org/10.1007/s10529-013-1185-7.

135 Cerrone, F., Radivojevic, J., Nikodinovic-Runic, J. et al. (2019). Novel sodium alkyl-1, 3-disulfates, anionic biosurfactants produced from microbial polyesters. *Colloids and Surfaces B: Biointerfaces* 182: 110333. https://doi.org/10.1016/j.colsurfb.2019.06.062.

136 Jiang, G., Johnston, B., Townrow, D. et al. (2018). Biomass extraction using non-chlorinated solvents for biocompatibility improvement of polyhydroxyalkanoates. *Polymers* 10 (7): 731. https://doi.org/10.3390/polym10070731.

137 Samorì, C., Basaglia, M., Casella, S. et al. (2015). Dimethyl carbonate and switchable anionic surfactants: two effective tools for the extraction of polyhydroxyalkanoates from microbial biomass. *Green Chemistry* 17 (2): 1047–1056. https://doi.org/10.1039/C4GC01821D.

138 Riedel, S.L., Brigham, C.J., Budde, C.F. et al. (2013). Recovery of poly(3-hydroxybutyrate-*co*-3-hydroxyhexanoate) from *Ralstonia eutropha* cultures with non-halogenated solvents. *Biotechnology and Bioengineering* 110 (2): 461–470. https://doi.org/10.1002/bit.24713.

139 Reverchon, E. (1997). Supercritical fluid extraction and fractionation of essential oils and related products. *Journal of Supercritical Fluids* 10 (1): 1–37. https://doi.org/10.1016/S0896-8446(97)00014-4.

140 Williams, J.R., Clifford, A.A., and Al-Saidi, S.H. (2002). Supercritical fluids and their applications in biotechnology and related areas. *Molecular Biotechnology* 22 (3): 263. https://doi.org/10.1385/MB:22:3:263.

141 Hampson, J.W. and Ashby, R.D. (1999). Extraction of lipid-grown bacterial cells by supercritical fluid and organic solvent to obtain pure medium chain-length polyhydroxyalkanoates. *Journal of American Oil Chemistry Society* 76 (11): 1371–1374. https://doi.org/10.1007/s11746-999-0152-x.

142 Hejazi, P., Vasheghani-Farahani, E., and Yamini, Y. (2003). Supercritical fluid disruption of *Ralstonia eutropha* for poly(β-hydroxybutyrate) recovery. *Biotechnology Progress* 19 (5): 1519–1523. https://doi.org/10.1021/bp034010q.

143 Khosravi-Darani, K., Vasheghani-Farahani, E., Shojaosadati, S.A., and Yamini, Y. (2004). Effect of process variables on supercritical fluid disruption of *Ralstonia eutropha* cells for poly(*R*-hydroxybutyrate) recovery. *Biotechnology Progress* 20 (6): 1757–1765. https://doi.org/10.1021/bp0498037.

144 Marudkla, J., Patjawit, A., Chuensangjun, C., and Sirisansaneeyakul, S. (2018). Optimization of poly(3-hydroxybutyrate) extraction from *Cupriavidus necator* DSM 545 using sodium dodecyl sulfate and sodium hypochlorite. *Agriculture and Natural Resources* 52 (3): 266–273. https://doi.org/10.1016/j.anres.2018.09.009.

145 Fujita, K., Kobayashi, D., Nakamura, N., and Ohno, H. (2013). Direct dissolution of wet and saliferous marine microalgae by polar ionic liquids without heating. *Enzyme and Microbial Technology* 52 (3): 199–202. https://doi.org/10.1016/j.enzmictec.2012.12.004.

146 Kobayashi, D., Fujita, K., Nakamura, N., and Ohno, H. (2015). A simple recovery process for biodegradable plastics accumulated in cyanobacteria treated with ionic liquids. *Applied Microbiology and Biotechnology* 99 (4): 1647–1653. https://doi.org/10.1007/s00253-014-6234-1.

147 Koller, M. (2019). Polyhydroxyalkanoate biosynthesis at the edge of water activity-haloarchaea as biopolyester factories. *Bioengineering* 6 (2): 34. https://doi.org/10.3390/bioengineering6020034.

148 Choi, J.I. and Lee, S.Y. (2000). Efficient and economical recovery of poly(3-hydroxybutyrate) from recombinant *Escherichia coli* by simple digestion with chemicals. *Biotechnology and Bioengineering* 62: 546–553. https://doi.org/10.1002/(SICI)1097-0290(19990305)62:5<546::AID-BIT6>3.0.CO;2-0.

149 Ghatnekar, M.S., Pai, J.S., and Ganesh, M. (2003). Production and recovery of poly-3-hydroxybutyrate from *Methylobacterium* sp V49. *Journal of Chemical Technology and Biotechnology* 77: 444–448. https://doi.org/10.1002/jctb.570.

150 Holmes, P.A. and Lim, G.B. (1990). Separation process. US Patent 4910145-A, filed 30 June 1989 and issued 20 March 1990.

151 Kachrimanidou, V., Kopsahelis, N., Vlysidis, A. et al. (2016). Downstream separation of poly(hydroxyalkanoates) using crude enzyme consortia produced via solid state fermentation integrated in a biorefinery concept. *Foods and Bioproducts Processing* 100: 323–334. https://doi.org/10.1016/j.fbp.2016.08.002.

152 Rodríguez Gamero, J.E., Favaro, L., Pizzocchero, V. et al. (2018). Nuclease expression in efficient polyhydroxyalkanoates-producing bacteria could yield cost reduction during downstream processing. *Bioresource Technology* 261: 176–181. https://doi.org/10.1016/j.biortech.2018.04.021.

153 Murugan, P., Han, L., Gan, C.Y. et al. (2016). A new biological recovery approach for PHA using mealworm, *Tenebrio molitor*. *Journal of Biotechnology* 239: 98–105. https://doi.org/10.1016/j.jbiotec.2016.10.012.

154 Chee, J.Y., Lakshmanan, M., Jeepery, I.F. et al. (2019). The potential application of *Cupriavidus necator* as polyhydroxyalkanoates producer and single cell protein: a review on scientific, cultural and religious perspectives. *Applied Food Biotechnology* 6 (1): 19–34. https://doi.org/10.22037/afb.v6i1.22234.

155 Koller, M. (2019). Switching from fossil plastics to microbial polyhydroxyalkanoates (PHA): the biotechnological escape route of choice out of the plastic predicament? *The Euro-Biotech Journal* 3 (1): 32–44. https://doi.org/10.2478/ebtj-2019-0004.

156 Yutaka, T. and Buenaventurada, C. (2004). *Review* Degradation of microbial polyesters. *Biotechnology Letters* 26: 1181–1189. https://doi.org/10.1023/B:BILE.0000036599.15302.e5.

157 Li, Z., Lin, H., Ishii, N. et al. (2007). Study of enzymatic degradation of microbial copolyesters consisting of 3-hydroxybutyrate and medium-chain-length 3-hydroxyalkanoates. *Polymer Degradation and Stability* 92: 1708–1714. https://doi.org/10.1016/j.polymdegradstab.2007.06.001.

158 Renard, E., Walls, M., Guérin, P., and Langlois, V. (2004). Hydrolytic degradation of blends of polyhydroxyalkanoates and functionalized polyhydroxyalkanoates. *Polymer Degradation and Stability* 85: 779–787. https://doi.org/10.1016/j.polymdegradstab.2003.11.019.

159 Vu, D.H., Åkesson, D., Taherzadeh, M.J., and Ferreira, J.A. (2020). Recycling strategies for polyhydroxyalkanoate-based waste materials: an overview. *Bioresource Technology* 298: 122393. https://doi.org/10.1016/j.biortech.2019.122393.

160 Zaverl, M., Seydibeyoğlu, M.Ö., Misra, M., and Mohanty, A. (2012). Studies on recyclability of polyhydroxybutyrate-co-valerate bioplastic: multiple melt processing and performance evaluations. *Journal of Applied Polymer Science* 125: E324–E331. https://doi.org/10.1002/app.36840.

161 Zembouai, I., Bruzaud, S., Kaci, M. et al. (2014). Mechanical recycling of poly(3-hydroxybutyrate-*co*-3-hydroxyvalerate)/polylactide based blends. *Journal of Polymers and the Environment* 22: 449–459. https://doi.org/10.1007/s10924-014-0684-5.

162 Rivas, L.F., Casarin, S.A., Nepomuceno, N.C. et al. (2017). Reprocessability of PHB in extrusion: ATR-FTIR, tensile tests and thermal studies. *Polímeros* 27: 122–128. http://dx.doi.org/10.1590/0104-1428.2406.

163 Ferreira, J.A. and Åkesson, D. (2020). Aerobic and anaerobic degradation pathways of PHA. In: *The Handbook of Polyhydroxyalkanoates - Postsynthetic Treatment, Processing and Application*, vol. III (ed. M. Koller), 317–338. CRC press Taylor & Francis Group ISBN 9780367275594.

164 Morikawa, H. and Marchessault, R.H. (1981). Pyrolysis of bacterial polyalkanoates. *Canadian Journal of Chemistry* 59: 2306–2313. https://doi.org/10.1139/v81-334.

165 Sato, S., Ishii, N., Hamada, Y. et al. (2012). Utilization of 2-alkenoic acids for biosynthesis of medium-chain length polyhydroxyalkanoates in metabolically engineered *Escherichia coli* to construct a novel chemical recycling system. *Polymer Degradation and Stability* 97: 329–336. https://doi.org/10.1016/j.polymdegradstab.2011.12.007.

166 Yang, X., Odelius, K., and Hakkarainen, M. (2014). Microwave-assisted reaction in green solvents recycles PHB to functional chemicals. *ACS Sustainable Chemistry and Engineering* 2: 2198–2203. https://doi.org/10.1021/sc500397h.

167 Niaounakis, M. (2013). *Biopolymers: Reuse, Recycling, and Disposal*, PDL Handbook Series. Elsevier.

168 Van Walsem, J., Licata, J., Anderson, E. A., Sparks, K. A., et al. (2014). U.S. Patent Application No. 14/237, 808.

169 Yañez, L., Conejeros, R., Vergara-Fernández, A., and Scott, F. (2020). Beyond intracellular accumulation of polyhydroxyalkanoates: chiral hydroxyalkanoic acids and polymer secretion. *Frontiers in Bioengineering and Biotechnology* 8: 248. https://doi.org/10.3389/fbioe.2020.00248.

170 Jendrossek, D. (1998). Microbial degradation of polyesters: a review on extracellular poly(hydroxyalkanoic acid) depolymerases. *Polymer Degradation and Stability* 59: 317–325. https://doi.org/10.1016/S0141-3910(97)00190-0.

171 Bano, K., Kuddus, M., and Zaheer, M.R. (2018). A novel, thermotolerant, extracellular PHB depolymerase producer *Paenibacillus alvei* PHB28 for bioremediation of biodegradable plastics. *Turkish Journal of Biochemistry* 44 (3): 344–353. https://doi.org/10.1515/tjb-2018-0207.

172 Gao, D., Maehara, A., Yamane, T., and Ueda, S. (2001). Identification of the intracellular polyhydroxyalkanoate depolymerase gene of *Paracoccus denitrificans* and some properties of the gene product. *FEMS Microbiology Letters* 196 (2): 159–164. https://doi.org/10.1111/j.1574-6968.2001.tb10558.x.

173 Grigore, M.E., Grigorescu, R.M., Iancu, L. et al. (2019). Methods of synthesis, properties and biomedical applications of polyhydroxyalkanoates: a review. *Journal of Biomaterials Science, Polymer Edition* 30: 695–712. https://doi.org/10.1080/09205063.2019.1605866.

174 Ong, S.Y., Chee, J.Y., and Sudesh, K. (2017). Degradation of polyhydroxyalkanoate (PHA): a review. *Journal of Siberian Federal University. Biology* 10: 211–225. http://elib.sfu-kras.ru/handle/2311/33371.

175 Urbanek, A.K., Mirończuk, A.M., García-Martín, A. et al. (2020). Biochemical properties and biotechnological applications of microbial enzymes involved in the degradation of polyester-type plastics. *Biochimica et Biophysica Acta (BBA), Proteins and Proteomics* 1868 (2): 140315. https://doi.org/10.1016/j.bbapap.2019.140315.

176 Jendrossek, D., Schirmer, A., and Schlegel, H.G. (1996). Biodegradation of polyhydroxyalkanoic acids. *Applied Microbiology and Biotechnology* 46: 451–463. https://doi.org/10.1007/s002530050844.

177 Knoll, M., Hamm, T.M., Wagner, F. et al. (2009). The PHA depolymerase engineering database: a systematic analysis tool for the diverse family of polyhydroxyalkanoate (PHA) depolymerases. *BMC Bioinformatics* 10 (1): 89. https://doi.org/10.1186/1471-2105-10-89.

178 Shah, A.A., Hasan, F., Hameed, A., and Ahmed, S. (2008). Biological degradation of plastics: a comprehensive review. *Biotechnology Advances* 26: 246–265. https://doi.org/10.1016/j.biotechadv.2007.12.005.

179 Aneja, P. and Charles, T.C. (1999). Poly-3-hydroxybutyrate degradation in *Rhizobium (Sinorhizobium) meliloti*: isolation and characterization of a gene encoding 3-hydroxybutyrate dehydrogenase. *Journal of Bacteriology* 181: 849–857. https://doi.org/10.1128/JB.181.3.849-857.1999.

180 Lee, S.Y. and Choi, J.I. (1999). Production and degradation of polyhydroxyalkanoates in waste environment. *Waste Management* 19: 133–139. https://doi.org/10.1016/S0956-053X(99)00005-7.

181 Doi, Y., Segawa, A., Kawaguchi, Y., and Kunioka, M. (1990). Cyclic nature of poly(3-hydroxyalkanoate) metabolism in *Alcaligenes eutrophus. FEMS Microbiology Letters* 67: 165–169. https://doi.org/10.1111/j.1574-6968.1990.tb13856.x.

182 Kusaka, S., Iwata, T., and Doi, Y. (1999). Properties and biodegradability of ultra-high-molecular-weight poly[(*R*)-3-hydroxybutyrate] produced by a recombinant *Escherichia coli. International Journal of Biological Macromolecules* 25: 87–94. https://doi.org/10.1016/S0141-8130(99)00019-7.

183 Adaya, L., Millán, M., Peña, C. et al. (2018). Inactivation of an intracellular poly-3-hydroxybutyrate depolymerase of *Azotobacter vinelandii* allows to obtain a polymer of uniform high molecular mass. *Applied Microbiology and Biotechnology* 102: 2693–2707. https://doi.org/10.1007/s00253-018-8806-y.

184 Choi, G.G., Kim, H.W., and Rhee, Y.H. (2004). Enzymatic and nonenzymatic degradation of poly (3-hydroxybutyrate-*co*-3-hydroxyvalerate) copolyesters produced by *Alcaligenes* sp. MT-16. *Journal of Microbiology* 42: 346–352. PubMed ID: 15650693.

185 Steinbüchel, A. (2005). Non-biodegradable biopolymers from renewable resources: perspectives and impacts. *Current Opinion in Biotechnology* 16: 607–613. https://doi.org/10.1016/j.copbio.2005.10.011.

186 Kourilova, X., Pernicova, I., Sedlar, K. et al. (2020). Production of polyhydroxyalkanoates (PHA) by a thermophilic strain of *Schlegelella thermodepolymerans* from xylose rich substrates. *Bioresource Technology* 315: 123885. https://doi.org/10.1016/j.biortech.2020.123885.

187 Koller, M. (2014). Poly(hydroxyalkanoates) for food packaging: application and attempts towards implementation. *Applied Food Biotechnology* 1 (1): 3–15. https://doi.org/10.22037/afb.v1i1.7127.

188 Kovalcik, A., Machovsky, M., Kozakova, Z., and Koller, M. (2015). Designing packaging materials with viscoelastic and gas barrier properties by optimized processing of poly(3-hydroxybutyrate-*co*-3-hydroxyvalerate) with lignin. *Reactive and Functional Polymers* 94: 25–34. https://doi.org/10.1016/j.reactfunctpolym.2015.07.001.

189 Plackett, D. and Siró, I. (2011). Polyhydroxyalkanoates (PHAs) for food packaging. In: *Multifunctional and Nanoreinforced Polymers for Food Packaging* (ed. J.M. Lagarón), 498–526. Woodhead Publishing https://doi.org/10.1533/9780857092786.4.498.

190 Haugaard, V.K., Udsen, A.M., Hoegh, L. et al. (2001). Potential food applications of biobased materials – an EU-concerted action project. *Starch* 53: 189–2000. https://doi.org/10.1002/1521-379X(200105)53:53.0.CO;2-3.

191 https://www.bacardilimited.com/media/news-archive/bacardi-first-in-fight-against-plastic-pollution-with-100-biodegradable-spirits-bottle (accessed 11 November 2021).

192 Triantafillopoulos, N. and Koukoulas, A.A. (2020). The future of single-use paper coffee cups: current progress and outlook. *BioResources* 15 (3): 7260–7287. https://doi.org/10.15376/biores.15.3.

193 Wu, B., Zhou, M., Song, L. et al. (2021). Mechanism insights into polyhydroxyalkanoate-regulated denitrification from the perspective of peri-cytoplasmic nitrate reductase expression. *Science of the Total Environment* 754: 142083. https://doi.org/10.1016/j.scitotenv.2020.142083.

194 Cyras, V.P., Commisso, M.S., Mauri, A.N., and Vázquez, A. (2007). Biodegradable double-layer films based on biological resources: polyhydroxybutyrate and cellulose. *Journal of Applied Polymer Science* 106: 749–756. https://doi.org/10.1002/app.26663.

195 Rastogi, V.K. and Samyn, P. (2020). Compression molding of polyhydroxybutyrate nano-composite films as coating on paper substrates. *Proceedings of The 2nd Coatings and Interfaces Web Conference (CIWC 2020)*. MDPI: Basel, Switzerland. 7 p.

196 Online resource: Kemira Announces Partnership with Danimer Scientific to Develop Biodegradable Coating for Paper and Board Industry, Business Wire (7.1.21)

197 Amelia, T.S.M., Govindasamy, S., Tamothran, A.M. et al. (2019). Applications of PHA in agriculture. In: *Biotechnological Applications of Polyhydroxyalkanoates* (ed. V. Kalia), 347–361. Singapore: Springer.

198 Chen, L. (2009). Next generation of electrospun textiles for chemical and biological protection and air filtration. Massachusetts Inst of Tech Cambridge.

199 Online resource NAFIGATE Corporation Natural polymer Vyvíjíme technologie pro lepší svět.https://www.nafigate.com/ (accessed 11 November 2021).

200 Online resource. Bio-on | Turn Off Pollution (bio-on.it). http://www.bio-on.it/?lin=inglese (accessed 11 November 2021).

201 Rodriguez-Contreras, A. (2019). Recent advances in the use of polyhydroyalkanoates in biomedicine. *Bioengineering* 6 (3): 82. https://doi.org/10.3390/bioengineering6030082.

202 Rodríguez-Contreras, A., Calafell-Monfort, M., and Marqués-Calvo, M.S. (2012). Enzymatic degradation of poly(3-hydroxybutyrate) by a commercial lipase. *Polymer Degradation and Stability* 97: 2473–2476. https://doi.org/10.1016/j.polymdegradstab.2012.07.042.

203 Wu, L.P., Wang, D., Parhamifar, L. et al. (2014). Poly(3-hydroxybutyrate-*co-R*-3-hydroxyhexanoate) nanoparticles with polyethylenimine coat as simple, safe, and versatile vehicles for cell targeting: population characteristics, cell uptake, and intracellular trafficking. *Advanced Healthcare Materials* 3: 817–824. https://doi.org/10.1002/adhm.201300533.

204 Mukheem, A., Muthoosamy, K., Manickam, S. et al. (2018). Fabrication and characterization of an electrospun PHA/graphene silver nanocomposite scaffold for antibacterial applications. *Materials* 11 (9): 1673. https://doi.org/10.3390/ma11091673.

205 Chai, J.M., Amelia, T.S.M., Mouriya, G.K. et al. (2021). Surface-modified highly biocompatible bacterial-poly (3-hydroxybutyrate-*co*-4-hydroxybutyrate): a review on the promising next-generation biomaterial. *Polymers* 13 (1): 51. https://doi.org/10.3390/polym13010051.

206 Pereira, J.R., Araújo, D., Marques, A.C. et al. (2019). Demonstration of the adhesive properties of the medium-chain-length polyhydroxyalkanoate produced by *Pseudomonas chlororaphis* subsp. *aurantiaca* from glycerol. *International Journal of Biological Macromolecules* 122: 1144–1151. https://doi.org/10.1016/j.ijbiomac.2018.09.064.

207 Bhardwaj, A., Alam, T., Sharma, V. et al. (2020). Lignocellulosic agricultural biomass as a biodegradable and eco-friendly alternative for polymer-based food packaging. *Journal of Packaging Technology and Research* 4: 205–216. https://doi.org/10.1007/s41783-020-00089-7.

208 Giubilini, A., Siqueira, G., Clemens, F.J. et al. (2020). 3D-printing nanocellulose-poly(3-hydroxybutyrate-*co*-3-hydroxyhexanoate) biodegradable composites by fused deposition modeling. *ACS Sustainable Chemistry & Engineering* 8 (27): 10292–10302. https://doi.org/10.1021/acssuschemeng.0c03385.

209 Meischel, M., Eichler, J., Martinelli, E. et al. (2016). Adhesive strength of bone-implant interfaces and *in-vivo* degradation of PHB composites for load-bearing applications. *Journal of the Mechanical Behavior of Biomedical Materials* 53: 104–118. https://doi.org/10.1016/j.jmbbm.2015.08.004.

210 Elmowafy, E., Abdal-Hay, A., Skouras, A. et al. (2019). Polyhydroxyalkanoate (PHA): applications in drug delivery and tissue engineering. *Expert Review of Medical Devices* 16 (6): 467–482. https://doi.org/10.1080/17434440.2019.1615439.

211 Sosa-Hernández, J.E., Villalba-Rodríguez, A.M., Romero-Castillo, K.D. et al. (2020). Poly-3-hydroxybutyrate-based constructs with novel characteristics for drug delivery and tissue engineering applications – a review. *Polymer Engineering & Science* 60 (8): 1760–1772. https://doi.org/10.1002/pen.25470.

212 Wu, Q., Wang, Y., and Chen, G.Q. (2009). Medical application of microbial biopolyesters polyhydroxyalkanoates. *Artificial Cells, Blood Substitutes, and Biotechnology* 37 (1): 1–12. https://doi.org/10.1080/10731190802664429.

213 Puppi, D., Pecorini, G., and Chiellini, F. (2019). Biomedical processing of polyhydroxyalkanoates. *Bioengineering* 6 (4): 108. https://doi.org/10.3390/bioengineering6040108.

214 Hablot, E., Dharmalingam, S., Hayes, D.G. et al. (2014). Effect of simulated weathering on physicochemical properties and inherent biodegradation of PLA/PHA nonwoven mulches. *Journal of Polymers and the Environment.* 22 (4): 417–429. https://doi.org/10.1007/s10924-014-0697-0.

215 Online resource. http://www.bio-on.it/minerv-supertoys.php (accessed 11 November 2021).

216 Online resource Restore Foodware – Products (accessed 11 November 2021).

217 Online resource Covalent. covalentfashion.com (accessed 11 November 2021).

218 Zhang, X., Luo, R., Wang, Z. et al. (2009). Application of (*R*)-3-hydroxyalkanoate methyl esters derived from microbial polyhydroxyalkanoates as novel biofuels. *Biomacromolecules* 10 (4): 707–711. https://doi.org/10.1021/bm801424e.

219 Wang, S.Y., Wang, Z., Liu, M.M. et al. (2010). Properties of a new gasoline oxygenate blend component: 3-hydroxybutyrate methyl ester produced from bacterial poly-3-hydroxybutyrate. *Biomass and Bioenergy* 34 (8): 1216–1222. https://doi.org/10.1016/j.biombioe.2010.03.02.

7

Ring-Opening Polymerization Strategies for Degradable Polyesters

An Sofie Narmon[1], *Liliana M. Jenisch*[1], *Louis M. Pitet*[2], *and Michiel Dusselier*[1]

[1] *KU Leuven, Center for Sustainable Catalysis and Engineering (CSCE), Faculty of Bioscience Engineering, Celestijnenlaan 200F, 3001 Leuven, Belgium*
[2] *Universiteit Hasselt, Institute for Material Research (IMO), Advanced Functional Polymers Group, Department of Chemistry, Martelarenlaan 42, 3500 Hasselt, Belgium*

7.1 Introduction

Polymer synthesis by ring-opening polymerization (ROP) dates back to the early 1900s when Leuchs (1906) described the synthesis of *N*-carboxyanhydride, which could be polymerized by ROP to prepare polypeptides [1]. Later (1918), ROP was applied in the synthesis of polysaccharides starting from anhydro sugars [2]. In 1932, Carothers et al. [3] described the first ROP of lactide (LD) to obtain what is now one of the most prominent polyester bioplastics on the market, poly(lactic acid) (PLA). In 1954, this method was patented by Du Pont [4] and, until the late 1970s, mainly used in the context of biomedical applications, due to its particularly costly production at the time [5]. In addition to the synthesis of PLA and other polyesters such as poly(ε-caprolactone) (PCL) and poly(glycolic acid) (PGA), contemporary ROP is used to supply industry with a number of other essential polymer materials, including polyethers (such as poly(oxy methylene), poly(ethylene glycol), or poly(tetrahydrofuran)), polysiloxanes, polyphosphazenes, poly(cyclooctene), poly(norbornene), poly(ethylene imines) made from aziridine or oxazoline monomers and several polylactams, such as Nylon 6 [6, 7].

ROP is a chain-growth polymerization reaction in which a cyclic monomer is added to a growing polymer chain by ring-opening the monomer via a reaction with the active terminal group of this polymer (Figure 7.1a). The type of cyclic monomer, as well as the catalyst/initiator system used, will determine the nature of the resulting active end-group of the growing chain. The nature of the resulting end-group then determines the type of mechanism through which the polymerization reaction will take place. The most important ROP mechanisms include radical, ionic (cationic or anionic), coordination–insertion, metathetic, and enzymatic [8].

A wide variety of cyclic molecules can react via one or more ROP mechanisms. Some general structures that are amenable to ROP include cyclic alkanes and alkenes as well as molecules containing heteroatoms in the ring, such as oxygen

Biodegradable Polymers in the Circular Plastics Economy, First Edition.
Edited by Michiel Dusselier and Jean-Paul Lange.

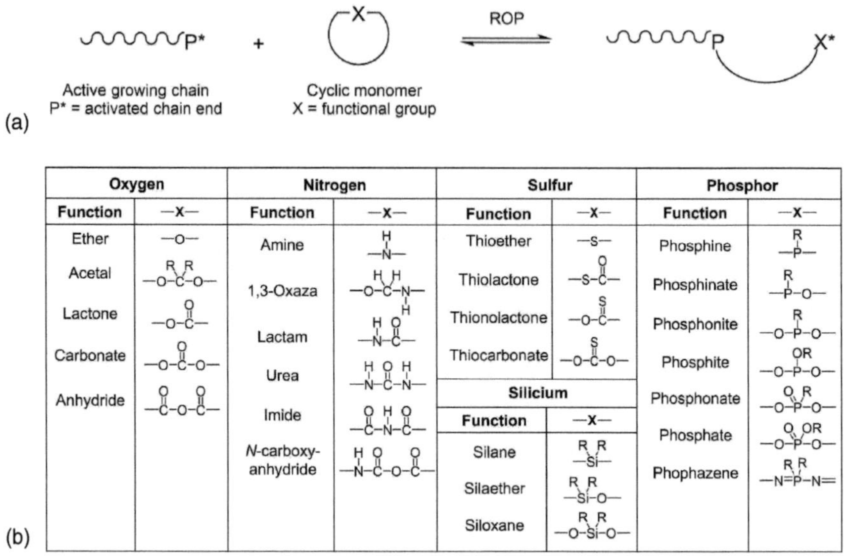

Figure 7.1 (a) The general mechanism of ring-opening polymerization, (b) Functional groups (–X–) of cyclic monomers used in ROP.

(ethers, acetals, esters [lactones and carbonates], and anhydrides), nitrogen (amines, amides [lactams], urea, imides, *N*-carboxyanhydrides, and 1,3-oxaza derivatives), sulfur (thioethers, thioesters [thiolactone, thionolactone], and thiocarbonates), phosphorus (phosphines, phosphinates, phosphonites, phosphites, phosphonates, phosphates, and phosphazenes) [9], or silicon (siloxanes, silaethers, and silanes) (Figure 7.1b) [10].

In contrast to some prominent industrial polymerizations (e.g. polyolefins), ROP enables the incorporation of heteroatoms within the polymer backbone, leading to materials with some unique properties, such as (bio)degradability. Moreover, the reversible nature of ROP reactions permits in certain cases for easy depolymerization and thus chemical recycling. This chapter primarily focuses on ROP for the synthesis of polyesters, containing ester linkages in their chains. At certain conditions, these ester linkages can make a polymer material amenable to (bio)degradation (e.g. via (a)biotic hydrolysis), offering the potential to create environmentally benign plastics. Although various polyesters can be obtained through condensation polymerization of diols with diacids or of hydroxy acids, ROP holds several advantages over this polymerization method. Compared to addition polymerizations, such as ROP, condensation polymerization involves repeated condensation reactions between two complementary functional groups (step-growth polymerization) and results in the production of by-products (i.e. condensates). When by-products are not efficiently removed from the reaction, they may promote depolymerization (equilibria), resulting in limited chain lengths and poor mechanical properties as a consequence. In addition, condensation polymerizations are often carried out at high temperatures (>200 °C) and are generally more difficult to control. Moreover, they only lead to high degrees of polymerization at very high conversions (>99%), whereas dispersity (i.e. M_w (weight-average molar mass)/M_n (number-average molar mass), i.e. a measure for the uniformity of chain lengths of a polymer) is often broad [11]. Nevertheless, many factors may have an impact on the course of a ROP, such as ring structure (size, number and position of substituents, functional groups), reaction conditions (e.g. temperature, solvent, monomer concentration), the catalyst system, and side reactions due to transesterification (i.e. chain transfer, backbiting, end-to-end biting). Therefore, ROP can be highly complex and sensitive to external factors [12]. In addition, it has to be mentioned that polymer structures obtained via polycondensation of hydroxy acids can often also be obtained via ROP of lactones, while the chain structures created by polycondensation of diacids with diols cannot be synthesized via ROP.

This chapter will first discuss several key ROP mechanisms and the associated thermodynamics and kinetics, followed by an overview of the most important degradable (or potentially degradable) polyesters obtained by ROP. A classification of the different polyesters will be made based on the ring size of their corresponding cyclic monomers.

7.2 Ring-Opening Polymerization Mechanisms

In the following sections only the most essential ROP mechanisms for the synthesis of aliphatic polyesters will be elaborated (cationic, anionic, coordination–insertion, and enzymatic). Although radical ROP can give rise to certain (bio)degradable polyester structures, research is relatively limited and will not be further discussed here. For more information on radical ROP the reader is referred to some excellent reviews [13, 14]. Ring-opening metathesis polymerization (ROMP) converts an unsaturated cyclic monomer into a polymer chain by means of metathesis. However, ROMP for the synthesis of degradable polyesters has only been described through polymerization of multifunctional macrocyclic monomers [15, 16] or macrocyclic unsaturated lactones [17, 18]. As such, further discussions on ROMP are not included in this chapter.

7.2.1 Cationic Ring-Opening Polymerization

Cationic ring-opening polymerization (cROP) occurs through the formation of a positively charged intermediate. Positive charges typically originate from addition of electrophilic/acidic initiators, such as Brønsted acids, carbenium ions, or onium ions. Typical Brønsted acids include HCl, H_2SO_4, $HClO_4$, or heteropolyacids ($H_nX_mM_zO_y$ with X = P, Si, and M = Mo, W, V) of which $H_3PMo_{12}O_{40}$ and $H_3PW_{12}O_{40}$ are extensively described in literature. In addition, more recently described derivatives of phosphoric acid [19, 20] (e.g. diphenylphosphate [DPP]) and triflic acid (CF_3SO_3H) have become more prominent. Lewis acids, of which the most important ones in cROP include BF_3 and PF_5, typically require a co-initiator such as water or alkyl halides, resulting in Brønsted acidic catalyst systems. Examples of cROP-initiating carbenium ions include $(C_6H_5)_3C^+$, $(C_6H_5)_2CH^+$, $(C_6H_5)CH_2{}^+$,

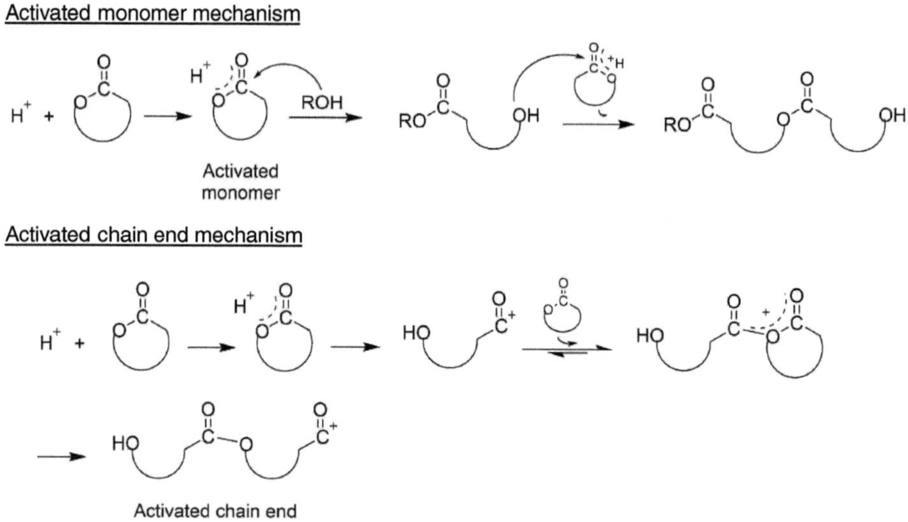

Figure 7.2 The activated monomer (AM) and activated chain end (ACE) mechanisms for cROP of lactone monomers. Source: Figure adapted from Kaluzynski et al. [23].

$H_2C{=}CH{-}CH_2^+$, $(CH_3)_3C^+$, etc., while trialkyl oxonium is an often-used onium ion initiator [6, 21, 22].

cROP can proceed via two main mechanisms: an activated chain-end (ACE) mechanism or an activated monomer (AM) mechanism, as depicted in Figure 7.2 for lactone monomers [12, 23, 24]. Typical monomers that can be polymerized via a cationic mechanism include cyclic ethers, sulfides, imines, amines, acetals, lactones, siloxanes, phosphazenes, oxazolines, etc.

The ACE mechanism is most common in cROP, while the number of monomers polymerizing via the AM mechanism is rather limited [25]. The ACE mechanism proceeds via a nucleophilic attack of the cyclic monomer on the positively charged onium ion at the growing chain-end of the polymer. The AM mechanism proceeds via a nucleophilic attack of the growing polymer chain on the positively charged, AM molecule [6, 12, 24, 25].

cROPs can undergo various termination reactions. In certain cases, living polymerization can be conducted, in which termination is limited. Since the ACE mechanism proceeds via a nucleophilic attack of the monomer on the ACE, other nucleophiles present in the reaction mixture could interact with the growing polymer chains and terminate the reaction. Typical terminating nucleophiles are counter-anions of the initiator or heteroatoms present in the growing chains. Counter-anions can transform the active chain-end into an uncharged system. Depending on the nucleophilicity of the initiating system, this covalent interaction can be permanent, reversible, or even negligible. Heteroatoms present in a polymer chain can interact with the active chain end of their own polymer molecule or of another chain. Both reactions can be either reversible or irreversible. In the former case, branched structures are created, while in the second case cyclic structures can be obtained. If in the latter case termination is reversible, cyclic oligomeric molecules can arise [24].

7.2.2 Anionic Ring-Opening Polymerization

Anionic ring-opening polymerization (aROP) proceeds via a nucleophilic initiator cleaving a cyclic monomer with high electrophilicity. While during cROP a protonated ring structure is opened, aROP proceeds through the cleavage of an uncharged cyclic monomer. Since the latter reaction requires more energy than the first, the number of monomers that can be polymerized through aROP is more limited compared to cROP [12]. Cleaving uncharged bonds is not evident and therefore needs activating agents, typically nucleophilic compounds. Examples of some common catalysts capable of aROP initiation include alkyl metals (alkyl lithium, alkyl magnesium bromide [Grignard reagent], alkyl aluminum, alkyl zinc, etc.), metal amides, alkoxides, phosphines, amines, thiols, alcohols, and water.

Monomers that can efficiently be polymerized via an aROP mechanism include (thio)esters, (thio)carbonates, amides, urethanes, phosphates, etc. Less electrophilic cyclic molecules, such as (thio)ethers or amines, can only be polymerized via aROP when the ring-strain in the molecules is sufficiently high such as in three-membered rings (epoxides, episulfides, aziridines, etc.) [12, 26].

The general (ACE) mechanism of aROP is depicted in Figure 7.3. Through ring-opening, a negatively charged structure is formed which represents a new nucleophilic species. This new nucleophile can attack a second molecule to form a dimer. The nucleophilic attack can, depending on the monomer, both occur on the carboxyl carbon (leading to acyl–oxygen bond cleavage) and on the alkoxide carbon (leading to alkyl–oxygen bond cleavage) (Figure 7.3) [6]. This repeating procedure will eventually provide a polymer chain. As is the case for cROP, aROP can occur via both an ACE and/or an AM mechanism. However, in aROP the AM mechanism

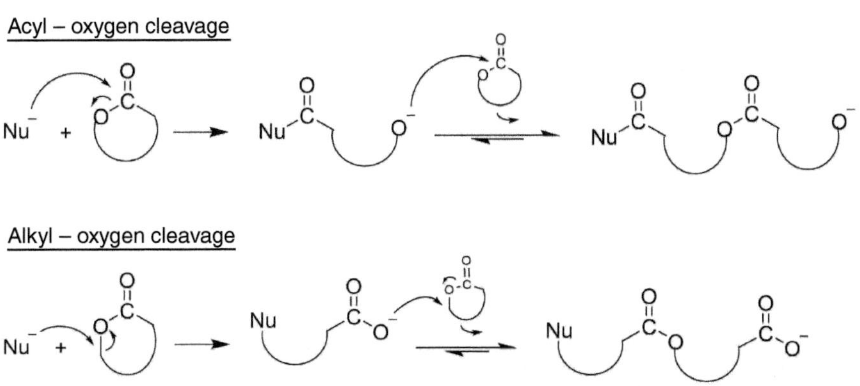

Figure 7.3 General mechanism of aROP: the monomer ring breaks via a nucleophilic attack of the initiator on the carboxyl carbon (acyl–oxygen cleavage) or the alkoxide carbon (alkyl–oxygen cleavage).

is only achieved when deprotonation of the functional group in the monomer by the nucleophile is possible, such as in the case of lactam polymerization [25].

Termination of aROP can occur by similar termination reactions as described for cROP through the presence of electrophilic compounds.

In addition to the catalytic systems for ionic ROP mentioned above, organic catalysts have gained more attention lately. These type of catalysts include amines (triethylamine, 4-dimethylaminopyridine [DMAP], 4-pyrrolidinopyridine [PPY], etc.), phosphines, N-heterocyclic carbenes [NHCs], amidines (1,8-diazabicyclo[5.4.0] undec-7-ene [DBU]), guanidines (1,5,7-triazabicyclo[4.4.0]dec-5-ene [TBD]), etc. For more information on organocatalysts for ROP the reader is referred to the review by A. Dove [27].

7.2.3 Coordination–Insertion Ring-Opening Polymerization

Coordination–insertion-based ring-opening polymerization (ciROP) starts with the interaction of a cyclic monomer with a catalytic system, without the formation of charges or radicals. The polymer chain is built up by successive insertion of monomers at the activated catalyst-bound chain. The mechanism of these types of ROP depends on the cyclic monomer and the specific catalytic system. Most catalytic systems are metal-based complexes.

ciROP is widely used in the polymerization of lactones. A generic ROP of lactones by a coordination–insertion mechanism is depicted in Figure 7.4. In this mechanism, the organometallic species (typically a metal alkoxide) initiates the reaction by coordinating with the carbonyl group of the monomer. This step is followed by cleavage of the acyl–oxygen bond to open the ring structure and simultaneous insertion of the monomer into the metal alkoxide bond [28, 29].

The two most widely used metal complexes for ROP of cyclic esters (also used in polycondensation) are Sn(II) 2-ethylhexanoate ($Sn(Oct)_2$) and aluminum alkoxides such as aluminum isopropoxide (Figure 7.5) due to their high selectivity and high control over the final molar mass (MM) of the polyesters. By reaction with alcohols

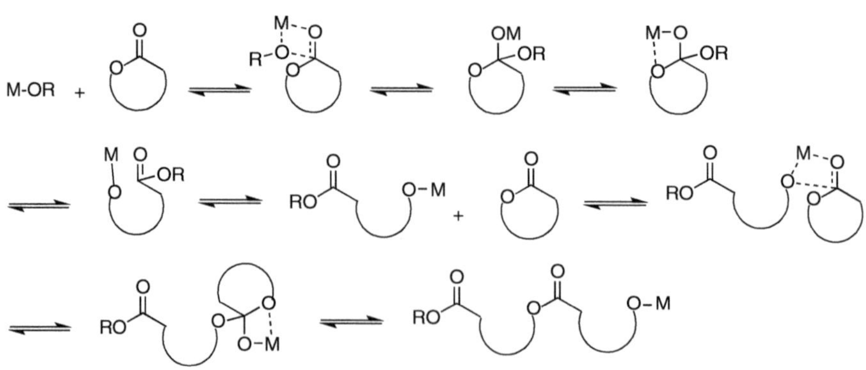

Figure 7.4 Ring-opening polymerization of lactones by coordination–insertions with metal alkoxides (M-OR). Source: Figure adapted from Jérôme and Lecomte [28].

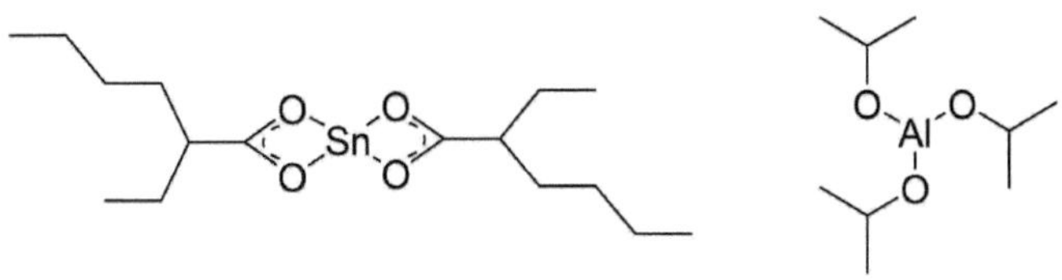

Figure 7.5 Structures of Sn(II) ethylhexanoate (left) and Al isopropoxide (right).

or other protic impurities, $Sn(Oct)_2$ is converted into the corresponding tin alkoxide, which acts as the active initiator. Every active initiator will theoretically create one growing polymer chain (when side reactions, such as transesterification, are neglected). As a consequence, by controlling the amount of alcohol added to the ROP reaction, the molar mass of the final polymer can be predicted.

An extremely wide range of other metals and complex structures have been investigated in the field of cyclic ester polymerization. Some extensively studied metal complexes and derivatives mainly contain rare earth metals (mainly La, Ln, Y, Sc, Nd, Sm, etc.) [30], alkali metals (Li, Na, K) [31–33], alkaline earth metals (Mg, Ca, Sr, Ba, Zn) [32–34], group 4 metals (Ti, Zr, Hf) [35–37], group 13 metals (Al [38], Ga, In) [33, 39], Sn, etc.

7.2.4 Enzymatic Ring-Opening Polymerization

Enzymatic ring-opening polymerization (eROP) can be defined as a ROP reaction induced by enzymatic activity. The most widely studied lactones in eROP include ε-caprolactone (ε-CL), ω-pentadecanolide, and LD, while lipases are the main class of enzymatic catalysts [40]. For a complete overview of all lactones and derivatives polymerizable by eROP the reader is referred to the review by Engel et al. [40]. Although eROP is much less discussed than the other ROP methods described in this chapter, enzymes have some important advantages over chemocatalysts. Enzymes can exhibit very high selectivities, while being active in mild reaction conditions and potentially also in greener reaction media, such as water, ionic liquids, or supercritical fluids. In addition, no toxic compounds remain after polymerization, making it especially suitable for biomedical applications. However, the main challenge in eROP is the rather low molar masses that are currently obtained [40].

7.3 ROP-Based Polyesters

7.3.1 Lactones

ROP to create polyesters most often takes place starting from lactone monomers. Lactones are ring-shaped molecules containing an ester function as part of their ring structure. Both monolactones and dilactones exist in ROP in which the ring contains one or two ester groups, respectively. Monolactones are usually not called by their IUPAC name, but the rings are based on the name of their hydroxy acid precursor, in which the Greek letter refers to the position of the hydroxy group relative to the acid function (e.g. β-propionic acid → β-propiolactone [β-PL]). In the name of the lactone, the letter thus often indicates the ring size of the molecule (e.g. β-lactone = four-membered ring and γ-lactone = five-membered ring) [11]. For the sake of consistency, this nomenclature will be used throughout this chapter. Naturally occurring lactones are mainly γ- and δ-lactones (five- and six-membered lactones), due to their high stability resulting from their low ring strain. However, more stable rings translate to more challenging polymerizations. Whether a lactone can undergo ROP is determined by polymerization thermodynamics and kinetics. The following section will give a basic introduction to the thermodynamics and kinetics of ROP.

7.3.2 Thermodynamics and Kinetics

The general ROP reaction can be written as a reaction between a cyclic monomer (M) and a growing polymer chain consisting of n monomer (M_n^*) units (with * representing the activated end-group) resulting in a chain with $n+1$ units (M_{n+1}^*) (Eq. (7.1)).

$$M_n^* + M \underset{k_\mathrm{d}}{\overset{k_\mathrm{p}}{\rightleftharpoons}} M_{n+1}^* \tag{7.1}$$

$$r_\mathrm{p} = k_\mathrm{p}\left[M_n^*\right][M] \qquad r_\mathrm{d} = k_\mathrm{d}\left[M_{n+1}^*\right]$$

$$K = \frac{k_p}{k_d} = \frac{\left[M_{n+1}^*\right]}{\left[M_n^*\right][M]} \tag{7.2}$$

The parameters k_p and k_d are the rate constants of polymerization and depolymerization, respectively. The rates of polymerization (r_p) and depolymerization (r_d) are both dependent on the rate constants and concentrations of monomer ([M]) and growing polymer ($\left[M_n^*\right]$). At equilibrium r_p and r_d are equal, which is expressed by the equilibrium constant K, as shown in Eq. (7.2).

Polymerization can only take place if the Gibbs free energy of polymerization (ΔG_p) is negative, i.e. when the polymer has a lower free energy than the monomer. ΔG_p is determined by the enthalpy (ΔH_p) and entropy (ΔS_p) of polymerization (Eq. (7.3)) with T being the absolute temperature.

$$\begin{aligned}\Delta G_\mathrm{p} &= G_\mathrm{polymer} - G_\mathrm{monomer} \\ &= H_\mathrm{polymer} - H_\mathrm{monomer} - T\left(S_\mathrm{polymer} - S_\mathrm{monomer}\right) \\ &= \Delta H_\mathrm{p} - T\Delta S_\mathrm{p}\end{aligned} \tag{7.3}$$

Since polymerization results in the enchainment of a large number of monomers in macromolecular chains, the degrees of freedom decrease, and therefore ROP (and other polymerization types) is most often entropically unfavored (i.e. $\Delta S_\mathrm{p} < 0$). As such, ROP is often an enthalpy-driven process ($\Delta H_\mathrm{p} < 0$) in which ΔH_p is strongly dependent on ring strain. Ring strain is dependent on the conformation, repulsion, and angle strain of the molecule, and therefore ring size and degree of substitution play a critical role in ROP thermodynamics. Small lactone rings (four to seven membered rings) often exhibit sufficient strain to enthalpically drive ROP ($\Delta S_p < 0$, $\Delta H_p < 0$), whereas the ROP of larger rings (nine membered rings or larger) can become entropically driven ($\Delta S_p > 0$, $\Delta H_p > 0$) [41, 42]. Since this chapter will focus on small lactone rings, the rest of this section will focus on enthalpically driven ROP.

Although entropic changes are generally limited compared to enthalpic changes, the positive entropic term in Eq. (7.3) ($T\Delta S_p$) becomes more dominant when the temperature (T) at which ROP is performed is high. This indicates, from a purely thermodynamic point of view, that ROP is preferably performed at lower temperatures, particularly when $\Delta S_p < 0$.

ΔG_p can also be written as the sum of the standard free energy of polymerization (ΔG_p^0) and a term related to the equilibrium constant K with R the gas constant (Eq. (7.4)). As shown in Eq. (7.2), K is related to the concentrations of monomer and growing polymers, which results in Eq. (7.5).

$$\Delta G_p = \Delta G_p^0 + RT \ln K \tag{7.4}$$

$$\Delta G_p = \Delta G_p^0 + RT \ln \frac{[M_{n+1}^*]}{[M_n^*]\,[M]} \tag{7.5}$$

When the polymer chains are sufficiently long, one can assume that the reactivity of an active center on the polymer molecule does not depend on the degree of polymerization. Following this assumption, Eq. (7.5) can be written according to Eqs. (7.6) and (7.7) [10, 43].

$$\Delta G_p = \Delta H_p^0 - T\Delta S_p^0 - RT \ln [M] = \Delta H_p^0 - T\left(\Delta S_p^0 + R \ln [M]\right) \tag{7.6}$$

$$T = \frac{\Delta H_p^0 - \Delta G_p}{\Delta S_p^0 + R \ln [M]_{eq}} \tag{7.7}$$

$$T_c = \frac{\Delta H_p^0}{\Delta S_p^0 + R \ln [M]_0} \tag{7.8}$$

Equation (7.8) shows that at equilibrium (when $\Delta G_p = 0$) a critical temperature exists at which no net polymerization takes place ($r_p = r_d$). This temperature is called the ceiling temperature (T_c). According to IUPAC, T_c is defined as "*the temperature at or above which the concentration of monomer in equilibrium with its polymer becomes essentially equal to the initial monomer concentration*" [44]. Therefore, Eq. (7.8) can be written with the starting monomer concentration ($[M]_0$) instead of $[M]_{eq}$. This essentially means that to promote polymerization the reaction should be performed at a temperature below T_c, while above T_c depolymerization will take place.

Since T_c is strongly dependent on the initial monomer concentration, this value is typically determined at either a monomer concentration of 1 M or in neat monomer (bulk). It is noteworthy that different monomers have intrinsically different bulk concentrations due to variations in MM and density of the monomers [42, 44].

In a lot of literature, it has been described that changes in concentration or solvent only result in changes in entropy, assuming enthalpy and thus ring strain is not affected by concentration or solvent. Nevertheless, this is not completely true,

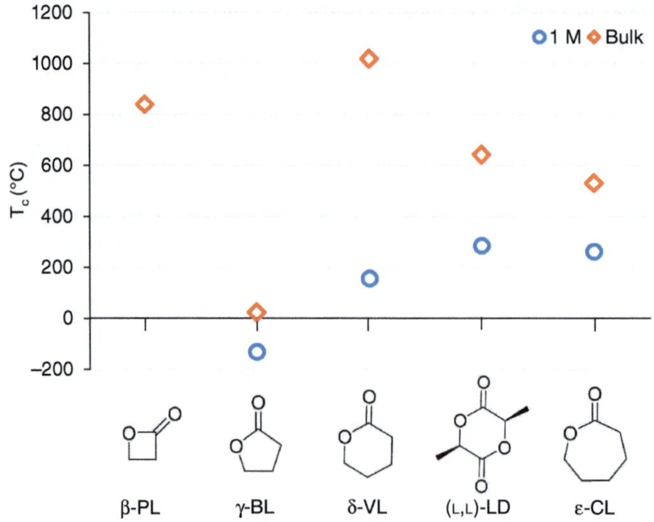

Figure 7.6 Ceiling temperature (T_c, °C) for different lactones at an initial monomer concentration ($[M]_0$) of 1 M or in bulk (10–13 M). Source: Data obtained from Duda and Kowalski [10] and Olsén et al. [42].

since ring strain can be affected by interactions of the monomer with the solvent depending on, among others, concentration, and polarity [42]. Finally, it should be stressed that enthalpy and entropy of polymerization are affected by the aggregation states of both monomer and polymer during polymerization, resulting from the corresponding phase transition (e.g. melting and crystallization) enthalpies and entropies. Therefore, one should be careful when comparing the values of enthalpy and entropy in literature [10].

The thermodynamic properties of different lactones, as summarized by the T_c value, are shown in Figure 7.6. For lactones of small ring size, thermodynamics of ROP are most favorable for four-membered (β-PL), six-membered (monolactone: δ-valerolactone [δ-VL], dilactone: LD), and seven-membered (ε-CL) rings, mainly resulting from the high ring strain of these molecules. Five-membered rings (γ-butyrolactone [γ-BL]), however, exhibit very stable ring structures with low ring strains, reflected in the exceptionally low T_c values of ROP.

7.3.3 Functionalization

The polyesters obtained from the ROP of simple lactones only have an ester bond which is susceptible to hydrolysis and biodegradation. However, certain applications require further tuning of degradability through, for example, increasing hydrophilicity and decreasing crystallinity. Introducing additional functional groups is an essential strategy to increase (bio)degradation capabilities. Attributes such as hydrophobicity can severely limit biodegradation (see Chapter 2). One way to increase the hydrophilicity is the direct ROP of ether-lactones, thus adding ether bonds into the polymer backbone. Some prominent examples of poly(ether-lactones) are mentioned in further detail in Section 7.3.5.4, 7.3.6.4, 7.3.7.3.

Additionally, functionalities can also be introduced in the polymer backbone in the form of pendant functional groups. In this way, the hydrophilicity can be increased by incorporating polar functional pendants, including small groups such as $-NH_2$, $-COOH$, $-OH$, and $-C{=}O$ [45, 46]. These functionalities can be introduced either through ROP of substituted, functional lactone monomers or through post-polymerization modification and grafting onto the polyester backbone.

7.3.3.1 ROP of Functional Lactones

There are a multitude of examples in which ROP has been performed on functionalized lactones, including halide, ester, and acetyl functional groups. Direct ROP of cyclic lactones bearing, e.g. hydroxyl groups, needs to be bypassed because these groups can undergo undesirable transesterification reactions and, unprotected, those lactones would act as inimers, i.e. act as both monomers and initiators alike, resulting in hyperbranched structures [47, 48]. Therefore, protective groups are used or the lactones are first polymerized with a suitable functional group and replaced post-polymerization by the desired group.

The benefit of using functionalized lactones is the possibility of purification before polymerization. Possible disadvantages include multiple steps for monomer synthesis as well as difficulties related to the choice of the protective group. If too labile the polymerization is impeded, but stronger groups (e.g. $-OSiEt_3$) require more stringent conditions for deprotection, risking degradation of the polymer [28]. Dilactones such as benzoyloxymethyl glycolide can be polymerized using conventional ROP catalysts and readily deprotected to yield the corresponding polyester with pendant OH-groups [28, 49]. An additional reason for the protection of carboxylic acid groups in the form of esters is their incompatibility with common metallic ROP catalysts [48].

7.3.3.2 Post-polymerization Functionalization

ROP of certain functionalized lactones takes place forming a precursor polymer which is subsequently modified to obtain the desired functionality. Halogenated lactones are an interesting group of such precursor monomers as the resulting functionalized polyesters can be derivatized in a number of ways, while the halogen groups do not disrupt the polymerization. This is demonstrated here by the derivatization of poly(α-Cl-ε-CL) from the ROP of α-chloro-ε-caprolactone (α-Cl-ε-CL) in Figure 7.7.

After polymerization, the chlorinated groups can be quantitatively replaced by azido pendants, which can then undergo so-called click reactions with alkynes to graft the desired functional groups onto the polymer [28, 47]. The advantage of click reactions compared to deprotection reactions is that they can readily take place under mild conditions, ensuring that degradation is minimized. For certain applications, the toxicological drawbacks of copper catalysts for these reactions make them unsuitable [50].

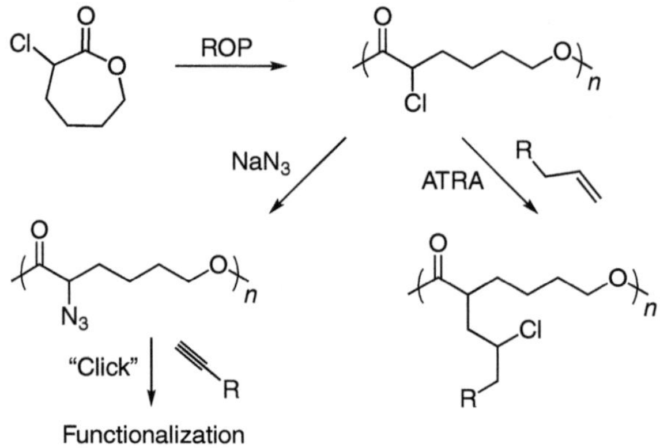

Figure 7.7 Derivatization pathways of chlorinated PCL from ROP of α-Cl-ε-CL: (left) derivatization to azido moiety with sodium azide followed by click reaction with an alkyne, (right) ATRA reaction with a functional alkene.

A second possibility is via so-called atom transfer radical addition (ATRA), see Figure 7.7 (right). The halogen atom is activated by the adjacent ester function and can react with functional alkenes (e.g. 3-buten-1-ol) in the presence of organometallic catalysts. Here, the copper catalyst required can be of concern for toxicological reasons, especially since it is typically used in stoichiometric amounts and some cases require elevated temperatures leading to polymer degradation [28].

Polyesters with pendant allyl groups or unsaturated backbones are another group of possible intermediates for the creation of more hydrophilic polymers. Reported post-polymerization functionalizations for such unsaturated polyesters include (di)hydroxylation, halogenation, epoxidation, hydrosilylation, thiol-ene reaction, and crosslinking and cross-metathesis reactions with functionalized alkenes [51–55]. Examples of polyesters with halogenated or unsaturated functional groups and their post-polymerization modifications can be found in Sections 7.3.4.4, 7.3.5.2, 7.3.6.2, and 7.3.7.2.

7.3.3.3 Grafting

In grafting, side chains or groups are added onto preformed polymer chains. An example of grafting with functional groups onto a polyester backbone can be seen in Figure 7.8. A precursor is formed via an anionic reaction using lithium diisopropylamide (LDA). The benefit of this strategy is that this precursor acts as a template for several different functionalized polyesters, as is illustrated here with some examples [28].

7.3.4 Four-Membered Lactones

Four-membered lactones or β-lactones are widely described as useful monomers for the ROP of some interesting, biodegradable polyesters.

The two structurally most simple β-lactones are β-PL and β-butyrolactone (β-BL), which are industrially synthesized through cycloaddition of ethenone with an aldehyde (formaldehyde for β-PL, acetaldehyde for β-BL) (Figure 7.9) and are therefore most often non-renewable [56]. Another synthesis method includes the carbonylation of epoxides [57, 58].

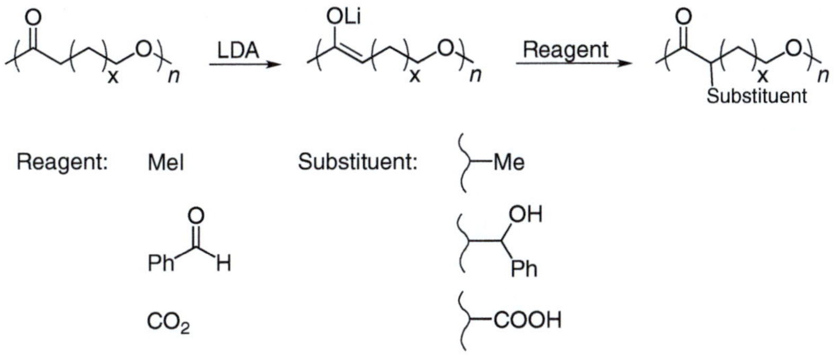

Figure 7.8 Grafting reaction onto polyesters with lithium diisopropylamide (LDA) and selection of reagents and substituents.

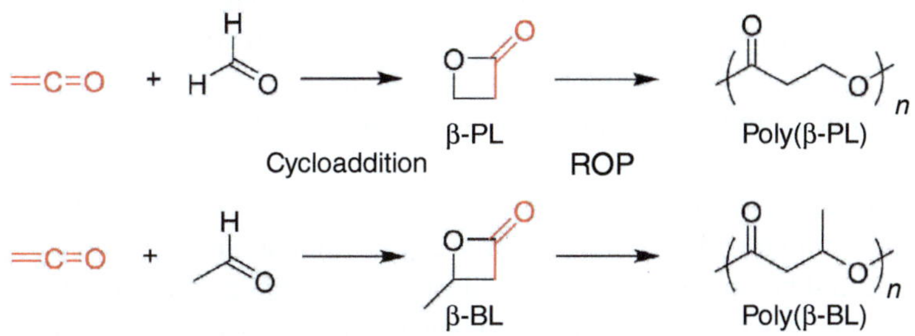

Figure 7.9 Industrial synthesis method of β-propiolactone (β-PL) and β-butyrolactone (β-BL) and their corresponding ROP-based polyesters.

Due to their very high internal ring strain and high polarity, β-lactones exhibit notably different polymerization behavior than their larger counterparts. Whereas larger lactone structures most often proceed through acyl-cleavage, creating alkoxide end groups, β-lactones regularly undergo both acyl- and alkyl-cleavage (see Section 7.2.2 on ROP mechanisms). While acyl-cleavage will ensure the preservation of the monomer stereochemistry, alkyl-cleavage will invert the stereochemistry. Depending on the reaction conditions, monomer, and catalytic system, one mechanism may dominate over the other, while in some cases both mechanisms may occur simultaneously [12, 59]. Alkyl-cleaved β-lactones are susceptible to a widely described and often unwanted side reaction, leading to acrylate-type side products. These species can also serve as initiating sites leading to polymers with unsaturated end-groups. This side reaction limits the control over molar mass and influences the dispersity. It is important to take these possible effects into account when choosing an appropriate ROP protocol [12].

Although ROP of β-PL results in materials with a T_g (glass transition temperature) around −20 °C, a T_m (melting temperature) around 80 °C [60–63], and excellent degradability [64–67], industrial applications of this polymer are limited. β-PL is highly toxic and by the International Agency for Research on Cancer "*reasonably anticipated to be a human carcinogen.*" Therefore, research on ROP of β-lactones has mainly been shifted to the less harmful methyl-substituted counterpart of β-PL, namely β-BL, as an alternative route to produce poly(3-hydroxybutyrate) (PHB) (or poly(β-BL)).

7.3.4.1 β-Butyrolactone

PHB is a biopolymer belonging to the class of poly(hydroxyalkanoates) (PHAs). PHAs have gained increasing attention in the field of bioplastics, due to their exceptional biocompatibility and biodegradability in compost, soil, and even marine environments [59]. PHB is the most common naturally available PHA. (R)-PHB (isotactic) is a highly crystalline polymer with a T_m around 175–180 °C and a T_g around −4 °C. Due to its high crystallinity, PHB homopolymers exhibit a high tensile strength, Young's modulus, and impact strength, while elongation at break is low. Moreover, PHB has a decomposition temperature close to its melting point, which makes processing very challenging.

PHAs are naturally produced by microorganisms as intracellular granules to serve as carbon and energy reserve. For more details on PHAs, the reader is referred to Chapter 6 in this book. However, the biotechnological synthesis of PHAs is currently very time consuming, complex, and expensive. In addition, PHAs made by microorganisms are strictly isotactic, exclusively exhibiting (*R*)-configurations without the possibility to vary the stereochemistry. As a consequence, the performance of these polymers is rather limited and with that, to some degree, their applicability [68–70]. These problems could be overcome by ROP of β-BL and other β-lactones. ROP does not only provide the possibility to change stereochemistry but also enables copolymerization and functionalization to tune properties to meet specific requirements [59, 71].

The ROP of β-BL has mostly been described in the presence of metal-based salts or metal complexes. Metal-based salts mainly include potassium salts of carboxylic acids (potassium acetate, benzoate, crotonate, etc.) [72–74] and potassium alkoxides (MeOK, t-BuOK, etc.) [72, 74–76]. Most of these anionic polymerization reactions (aROP) lead (partly) to acyl-cleavage and resultant crotonate side product formation. However, the introduction of macrocyclic ligands such as crown ethers (e.g. 18-crown-6) is described to promote acyl-cleavage mechanisms and lead to living polymerizations [76, 77]. Metal complexes for ciROP, in general, enable better control over the molar mass, dispersity, and stereochemistry of PHB. Typical metal complexes contain Cr [78], Sn [79, 80], Zn [81–84], Al [85], In [86–88], or rare earths (Y, La) [51, 89–94]. More recently, organocatalysts have also been studied, with particular promise shown by DBU, 2-*tert*-butylimino-2-diethylamino-1,3-dimethylperhydro-1,3,2-diazaphosphorine (BEMP), and TBD [71, 95–97].

7.3.4.2 Acid-Substituted β-Lactones (β-Malolactonate)

Although PHAs are one of the best degradable polyesters described, an important shortcoming of PHB is its hydrophobic character that often limits its use or (bio)degradability in certain conditions. This disadvantage can be overcome by introducing hydrophilic groups on the polymer backbone. An important PHA-type polymer described in this field is poly(malic acid) (PMLA), which contains a

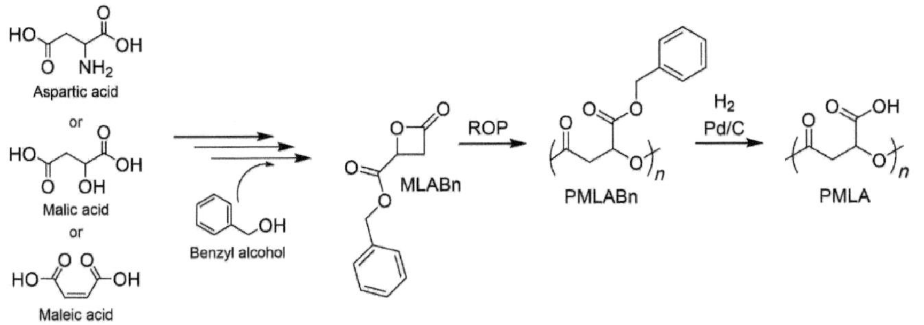

Figure 7.10 Synthesis and ROP of MLABn to PMLABn and removal of the benzyl group via catalytic hydrogenolysis to form PMLA.

carboxylic acid side group on each repeating unit (Figure 7.10). PMLA can be produced both chemically and naturally via bacterial fermentation. Due to some shortcomings of the current fermentation process (limited control over MM, low productivity, limited availability of sources, lengthy procedure, etc.) interest in the chemical ROP route via substituted lactones has risen [98]. The synthesis procedure of PMLA starts with the ROP of benzyl-β-malolactonate (MLABn) to poly(benzyl-β-malolactonate) (PMLABn). PMLA can then be obtained by removing the benzyl groups via catalytic hydrogenolysis with Pd/C to create a carboxylic acid side group (Figure 7.10). This side group leads to water solubility as well as provides a convenient handle for further modification or functionalization. Moreover, PMLA can degrade *in vitro* via hydrolysis to create malic acid, an intermediate of the Krebs cycle [99, 100]. As a consequence, this bioresorbable polymer is extensively studied in the biomedical field as a drug-carrier system.

MLABn is a biobased monomer that can be obtained from either malic acid, aspartic acid, or maleic acid, three biobased platform molecules. Different multi-step (four to six steps) synthesis methods have been described in the presence of benzyl alcohol (BnOH) with yields ranging between 3% and 11%, of which synthesis from aspartic acid is most efficient [98, 101, 102]. For more details and a clear overview of all synthesis options the reader is referred to the review by Jaffredo and Guillaume [98]. Enantiopure MLABn can be obtained starting from enantiomerically pure aspartic acid [103, 104]. Substituting benzyl alcohol by other alcohols provides access to various other β-malolactonates and thus various polyesters with diverging properties [105, 106].

7.3.4.3 Alkoxy-Substituted β-Lactones

The ROP of 4-alkoxy-methylene-β-propiolactones (BPLORs) (Figure 7.11) was first described in 2017 by Carpentier and coworkers [107]. The lactones were synthesized in high yields by carbonylation of their corresponding racemic epoxides in the presence of CO [58]. These racemic alkoxy-substituted (with R = allyl (All), methyl (Me), benzyl (Bn), or *tert*-butyl-dimethyl-silane (TBDMS)) β-lactones were enantio-selectively polymerized in the presence of Y-based catalyst complexes

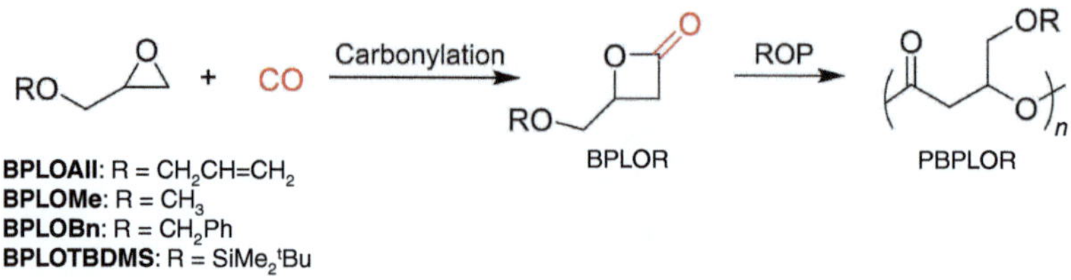

Figure 7.11 Synthesis and ROP of BPLOR with R being an allyl, methyl, or benzyl group.

resulting in either pure syndiotactic or isotactic chains depending on the ligands of the catalyst [107, 108].

The same research group recently demonstrated different ring-opening copolymerizations (ROCOPs) of these monomers in different combinations of functional groups and chirality resulting in copolymers with high degrees of alternation in both functional groups and tacticities [109]. Yet another study described the successful ROP of these substituted β-lactones aided by organocatalysts BEMP, DBU, and TBD [71]. These types of functionalized β-lactones can be interesting in the context of potential post-polymerization functionalization (Section 7.3.3.2) or biodegradation.

7.3.4.4 Alkene-Substituted β-Lactones

Alkene-substituted β-lactones are interesting monomers due to the possibility of creating polyesters with unsaturated pendant groups for post-polymerization functionalization. Two important monomers described in this context are β-6-heptenolactone (BHL) (Figure 7.13) and α-methylene-β-butyrolactone (MBL) (Figure 7.12).

The homopolymerization of MBL (synthesized from CO_2 and 2-butyne) (Figure 7.12) was recently described in the presence of Al salen complexes [110]. Later, the same research group synthesized different block copolymers with LD, ε-CL, and ω-pentadecalactone (ω-PDL) and random copolymers with β-BL aided by Al or Y complexes. Hydrolysis tests in the presence of a Lewis base showed that PMBL has significantly improved degradability compared to PHB. Tuning the amount of MBL units in random copolymers with β-BL enabled control over the polymer degradability. In addition, the unsaturation both enabled intra- and inter-cross-linking reactions between copolymers with MBL as well as post-polymerization functionalization through thiol-ene click reactions [111].

The ROP of BHL was first described in 2009 aided by a Y-based catalyst [51]. Both homopolymers of BHL and copolymers with β-BL (up to 50 mol% BHL) were synthesized with high MMs (M_n up to 58 kg/mol) and low polydispersities.

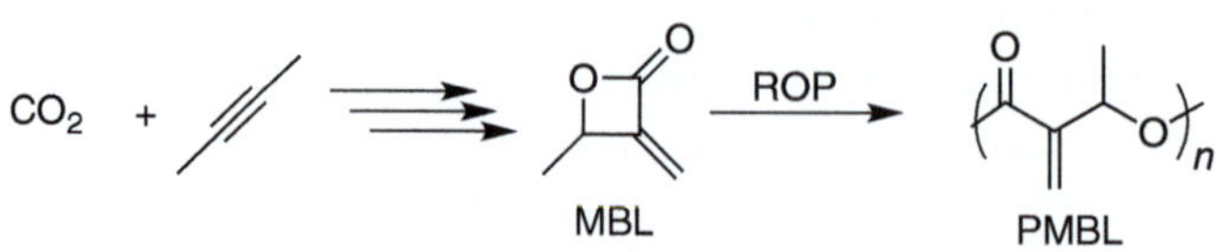

Figure 7.12 Synthesis of MBL from CO_2 and its ROP to produce PMBL.

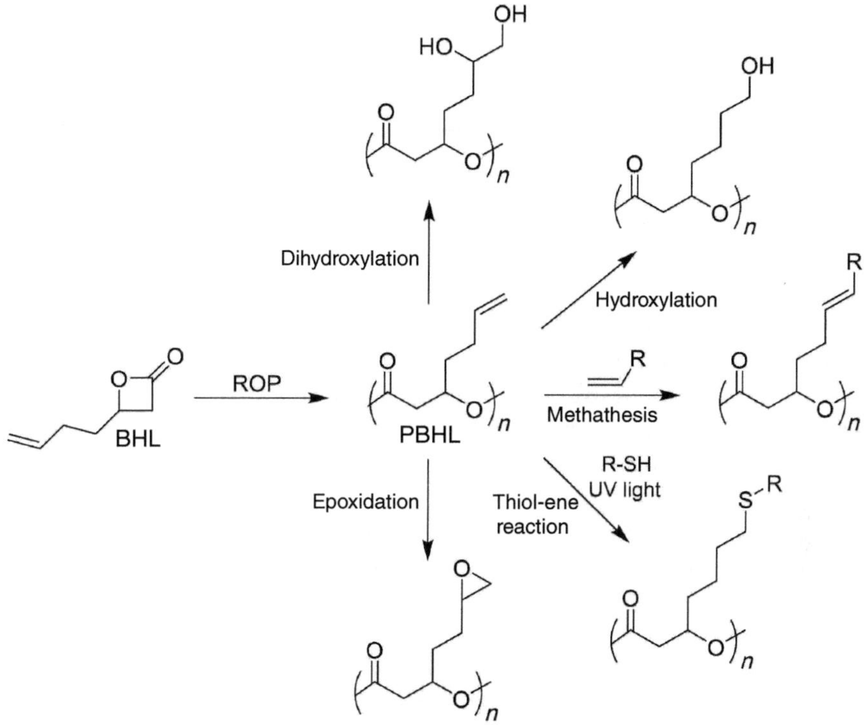

Figure 7.13 ROP of BHL and possible post-polymerization functionalizations.

In addition, the authors performed post-polymerization functionalization of the copolymers, whereby three possibilities were tested: dihydroxylation, hydroxylation via hydroboration/oxidation, and epoxidation (Figure 7.13) [51, 52]. More recent research describes BHL homopolymerization and copolymerization with LD, followed by post-polymerization cross-metathesis functionalization with alkenes bearing a wide range of functional groups (Figure 7.13) [53]. Yet another study reports the synthesis of block copolymers of ethylene oxide with BHL after which the alkene pending groups were modified using thiol-ene reactions under UV light (Figure 7.13) [112].

7.3.5 Five-Membered Lactones

7.3.5.1 γ-Butyrolactone

Carothers et al. [3] stated in 1932 that five-membered lactones do not show any tendency to form chain structures compared to their six-membered counterparts. As such, these ring structures, of which γ-BL (Figure 7.14) is the simplest, were long thought to be impossible to polymerize. This phenomenon was later explained as a thermodynamic problem resulting from the limited ring strain and high stability of five-membered lactones, leading to a positive Gibbs free energy of polymerization under normal conditions (e.g. see very low T_c in Figure 7.6) [113, 114].

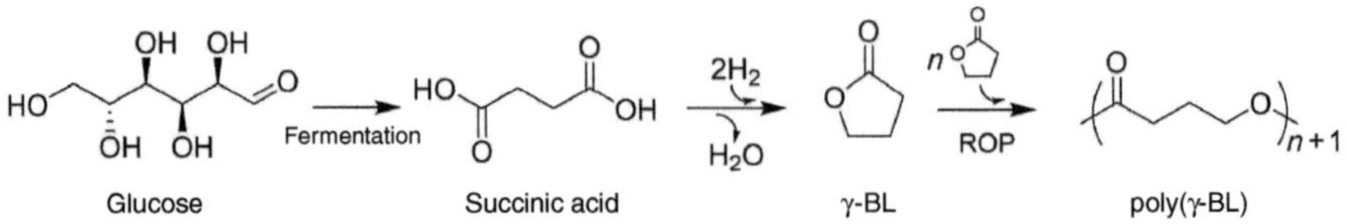

Figure 7.14 Synthesis of γ-BL from glucose via the platform molecule succinic acid and the ROP of γ-BL toward poly(γ-BL).

Despite its theoretically implausible ROP, researchers have devoted substantial effort to γ-BL due to the promising properties of this polyester (poly(γ-butyrolactone) [poly(γ-BL)]), better known as poly(4-hydroxybutyrate) (see also Chapter 6 about microbial PHA). First, γ-BL can easily be synthesized from succinic acid (Figure 7.14), a key bio-derived platform molecule [115]. Second, this polymer exhibits desirable properties: poly(γ-BL) is not only strong, but also flexible, a property which is often lacking in typically brittle bioplastics such as PHB, PGA, or PLA. Furthermore, poly(γ-BL) degrades faster *in vivo* than PLA or PCL, showing its (potential) applicability in the medical field [116–118].

Despite the thermodynamic limitations, the first successful homopolymerization of γ-BL was described in 1966 at extreme reaction conditions (20 000 atm, 160 °C, 4 hours), leading to a polymer with a low MM ranging from 1.20 to 3.35 kg/mol [119]. In its wake, various attempts have been made to improve the ROP of γ-BL under very high pressures, but the reported molar masses remained rather low using either lipase or acid catalysts [120–126]. However, more recently, the ROP of γ-BL under ambient pressure has been described. The application of temperatures below zero (−40 °C), high monomer concentrations (10 M), and the presence of a suitable La- or Y-based catalyst, resulted in high monomer conversions up to 90% and molar masses up to 30 kg/mol, creating both linear and cyclic polymer chains. By working at very low temperatures (i.e. below T_c) and high monomer concentrations, entropic barriers were overcome. In addition, controlling the reaction conditions (solvent, concentration) so that polymer chains precipitated upon formation further drove the equilibrium of the reaction toward polymerization. Moreover, the polymer structures are described to be easily recyclable to their monomers after catalyst removal without formation of by-products by simply heating the materials at 200–300 °C. [127, 128] Following this breakthrough, other successful catalytic systems have been discovered for the ROP of γ-BL at low temperatures, such as organocatalysts. Strong bases (e.g. phosphazenes) exhibited interesting results [129, 130]. However, these catalytic systems afforded both linear and cyclic polymer chains. The addition of (thio)urea to the phosphazene bases led to a selective formation of linear polymer chains of high MM, whereas a combination of urea and alkoxides proved to be successful as well [131–134]. Copolymerization of γ-BL with high ring strain monomers, such as β-PL, β-BL, G, LD, δ-VL, and ε-CL have been described as well [135–144]. The use of a high ring-strain comonomer enables ROP at temperatures above the T_c of γ-BL. In addition, copolymerization with γ-BL can potentially facilitate the recycling or degradation of other polyesters [134]. For a complete overview of the literature on

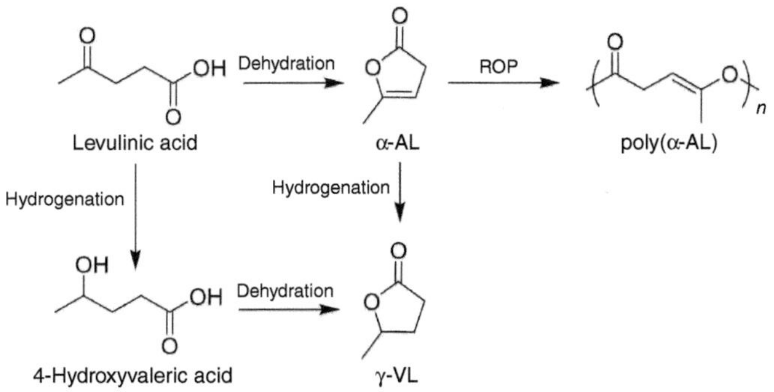

Figure 7.15 Synthesis of α-AL and γ-VL from levulinic acid and ROP of α-AL to poly(α-AL).

copolymerization of γ-BL the reader is referred to the recent reviews by Song et al. [145] and Liu et al. [134].

Due to the rather recent breakthroughs in ROP of five-membered lactones, the amount of γ-BL analogs used as monomers in ROP is limited. Only the ones interesting in the context of (bio)degradation will be further elaborated.

7.3.5.2 α-Angelicalactone

α-Angelicalactone (α-AL) is an unsaturated γ-BL analog with a methyl group at the γ-position relative to the ester carbonyl (Figure 7.15). α-AL has received increasing attention lately, due to its easy synthesis from levulinic acid through intramolecular condensation followed by dehydration. Both levulinic acid, which can be produced from lignocellulosic biomass and sugars, and α-AL are considered important biobased platform molecules that can be converted to a broad range of industrially important chemicals [146].

In 2011, the successful ROP of α-AL has been described in the presence of $Sn(Oct)_2$ at 130 °C for 30 hours in toluene [147]. The polymers, of up to 29 kg/mol, contain unsaturations in their backbone, providing a handle for further chemical modification. In addition, these double bonds are sensitive to light or heat and can generate radicals to promote facile degradation. A weight loss of 34.8% was described after exposing the polymers to daylight for 49 days, whereas placing them in an aqueous acidic or basic solution for the same time resulted in a weight loss of 63.3% and 52.0%, respectively. The rather simple ROP of α-AL compared to γ-BL was tentatively explained by a higher strain energy present in α-AL due to smaller bond angles for C—O—(C=O) and O—(C=O)—C resulting from the unsaturated bond [147–149].

The homopolymerization of γ-valerolactone (γ-VL), which can easily be made by hydrogenation of α-AL or by dehydration of 4-hydroxyvaleric acid (Figure 7.15), has not been described yet. Although copolymerization of this monomer with ε-CL or β-BL has been studied, molar masses and extent of incorporation remain rather low [150–152].

7.3.5.3 α-Methylene-γ-Butyrolactone

α-Methylene-γ-butyrolactone (MGBL) or tulipalin A is a γ-BL derivative with an unsaturated methylene group at the α-position relative to the carbonyl

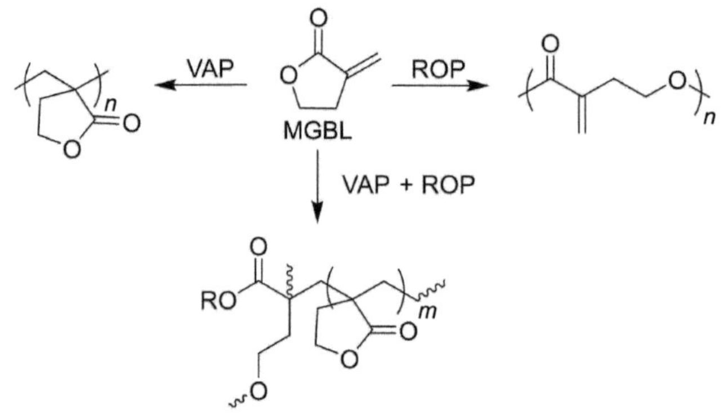

Figure 7.16 Polymerization of MGBL via vinyl-addition polymerization (VAP), ring-opening polymerization (ROP), or a combination of both. Source: Adapted from ref. Liu et al. [134].

group (Figure 7.16). This lactone is naturally present in tulips or can be chemically synthesized from biomass [153, 154]. MGBL, together with its methyl-substituted counterparts β-methyl-α-methylene-γ-butyrolactone and γ-methyl-α-methylene-γ-butyrolactone, which are derived from the biobased platform molecules itaconic acid and levulinic acid, respectively, is mainly studied with respect to its vinyl-addition polymerization (VAP) in absence of ring opening [155–157]. The ROP of MGBL is thus challenging for two reasons: (i) the high stability and low ring strain of the γ-BL ring suppresses ring opening ($\Delta H^0{}_p$ = −5.9 kJ/mol, $\Delta S^0{}_p$ = −40.1 J/(mol K), T_c = −126 °C (1 M)) and (ii) the highly reactive unsaturated bond tends to polymerize via vinyl addition.

One way to overcome these problems is via copolymerization of MGBL with a high ring-strain lactone, such as ε-CL [158–160]. However, recently the exclusive homopolymerization of MGBL was described in the presence of a $La[N(SiMe_3)_2]_3$ catalyst in combination with an alcohol initiator at low temperatures (−60 °C) and high monomer concentration (5 M) [161]. Linear polymer chains were obtained with M_n up to 21.0 kg/mol and low T_g (−46 to −35 °C) and T_m (39–44 °C) and could be depolymerized back to MGBL through chemolysis [134, 161]. Decreasing the amount of alcohol initiator, increasing the temperature, or decreasing the monomer concentration led to a combination of vinyl addition and ROP and in some cases to a crosslinked polymer network (Figure 7.16). This was explained by a calculation of T_c = −51 °C at a monomer concentration of 5 M. More recent research describes this polymerization aided by a cyclic trimetric phosphazene superbase catalyst. Both catalyst systems described here were also applied in copolymerization of MGBL with ε-CL or δ-VL, where decreasing temperature resulted in higher incorporation of the MGBL, lower melting points, and lower crystallinities [159, 160, 162]. The exomethylene groups on the polymer chains provide the possibility of post-polymerization functionalization (e.g. via thiol-ene-based click chemistry [160, 161]) of the polyester chains, among others to tune degradability properties.

Other α-substituted-γ-butyrolactones, such as α-bromo-γ-BL and α-acetyl-γ-BL, are interesting monomers in the context of degradability owing to their functionalities enabling potential post-polymerization modification, without hindering ring opening. α-Bromo-γ-BL has been described in the context of copolymerizations with ε-CL, trimethylene carbonates, and L-LD ((L,L)-lactide) [163–165]. Olsén et al. [164] performed the copolymerization of α-bromo-γ-BL with ε-CL and L-LD in the presence of $Sn(Oct)_2$ at 110 °C in bulk for 30 hours. Up to 28% of α-bromo-γ-BL for ε-CL and 35% for L-LD were built-in in the polymer chains, obtaining M_n values between 17.0 and 54.8 kg/mol. The same group synthesized block copolymers with trimethylene carbonates, ε-CL, and α-bromo-γ-BL using DPP as a catalyst [163]. Both studies describe the post-polymerization modifications of the Br side group with various acrylate monomers, indicating the grafting capabilities of this functionality.

7.3.5.4 Ether γ-Lactones

Another group of five-membered lactones, more recently investigated in ROP, are the ether γ-lactones. More specifically, research has focused on the polymerization of various 1,3-dioxolane-4-one (DOX) monomers. DOXs are five-rings with both an ester and ether functional group as depicted in Figure 7.17 and can easily be synthesized by, among others, condensation of different, (often) renewable α-hydroxy acids and an aldehyde or ketone [166].

ROP of DOX monomers is an interesting alternative route to produce a wide range of divergent polyesters (Figure 7.17), being driven by the release of formaldehyde or acetone. Chain structures that are rather hard to obtain through classical cyclic diester polymerization, due to challenging monomer synthesis, can occasionally be more easily obtained via ROP of the corresponding DOX.

However, only one study describes the polymerization of the most simple ether γ-lactone, DOX ($R_1 = R_2 = R_3 = H$), and its copolymerization with LD, maintaining the acetal (O–C–O) functionality in the final polymer structure to generate polyester-acetals (up to 36 mol% DOX) (Figure 7.17 (**1**)) [167]. Incorporation of acetal groups resulted in an increase in T_g in the range of 4–19 mol% and decreased T_m of PLA. In addition, facile degradation of the copolymers was observed under acidic, neutral, and slightly basic conditions. At pH 7 (distilled water) the copolymer (4 mol% DOX) degraded with nearly 2% mass loss and more than 15% loss in molar mass (M_n) over 45 days, while no degradation was observed in PLA. Compared to separate ether functions, actals and ketals are even more labile and hydrolytically sensitive functionalities, enabling very fast degradation.

A more recent study describes the cationic ROCOP of different alkyl-substituted DOXs with two different oxiranes (Figure 7.17) using strong Lewis acid catalysts, such as $B(C_6F_5)_3$ in DCM at −78 °C. Although molar masses were mediocre (up to 18.8 kg/mol) polyesteracetal structures were obtained in which the DOX groups (9–25 mol%) were always interrupted by at least one comonomer group. The copolymers were easily hydrolyzed under acidic conditions (0.5 M HCl in 1,2-dimethoxyethane, 1 wt% polymer) at room temperature, resulting in more than 90% losses in molar mass after three hours. It was noticed that the acid-labile acetal functions disappeared most quickly from the nuclear magnetic resonance (NMR) spectra during hydrolysis [168].

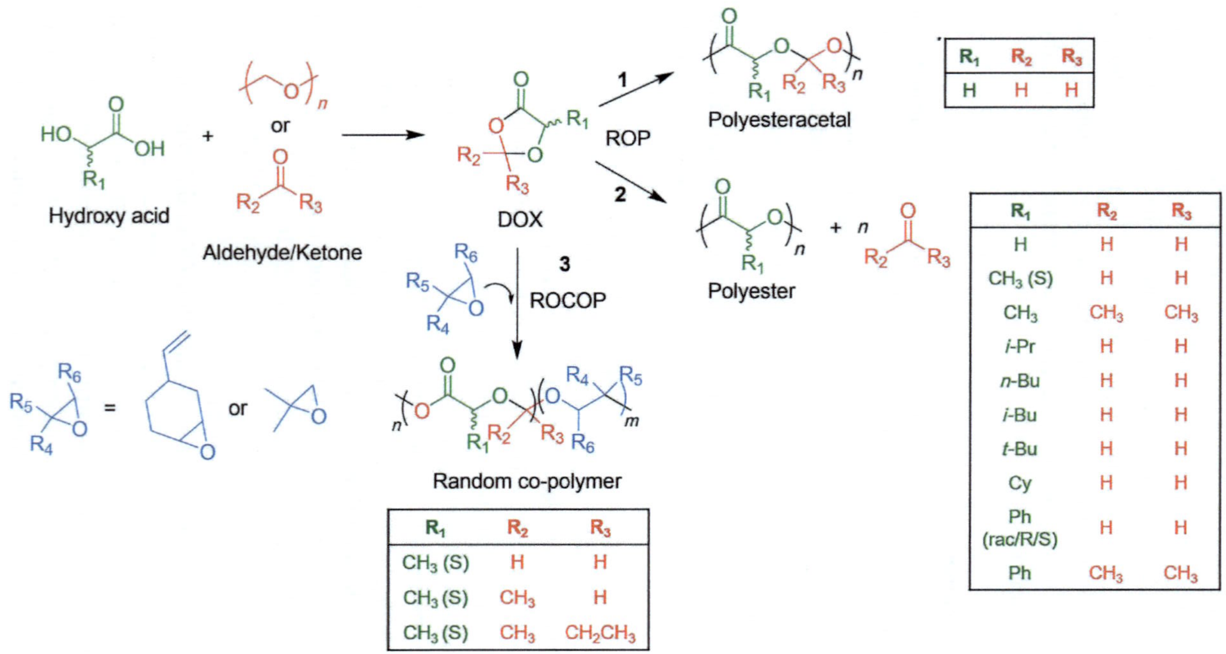

R_1	R_2	R_3
H	H	H

R_1	R_2	R_3
H	H	H
CH_3 (S)	H	H
CH_3	CH_3	CH_3
i-Pr	H	H
n-Bu	H	H
i-Bu	H	H
t-Bu	H	H
Cy	H	H
Ph (rac/R/S)	H	H
Ph	CH_3	CH_3

R_1	R_2	R_3
CH_3 (S)	H	H
CH_3 (S)	CH_3	H
CH_3 (S)	CH_3	CH_2CH_3

Figure 7.17 Synthesis and ROP of 1,3-dioxolan-4-ones (DOXs) toward (**1**) polyesteracetals, (**2**) polyesters, or (**3**) substituted polyesteracetals via ROCOP with oxiranes.

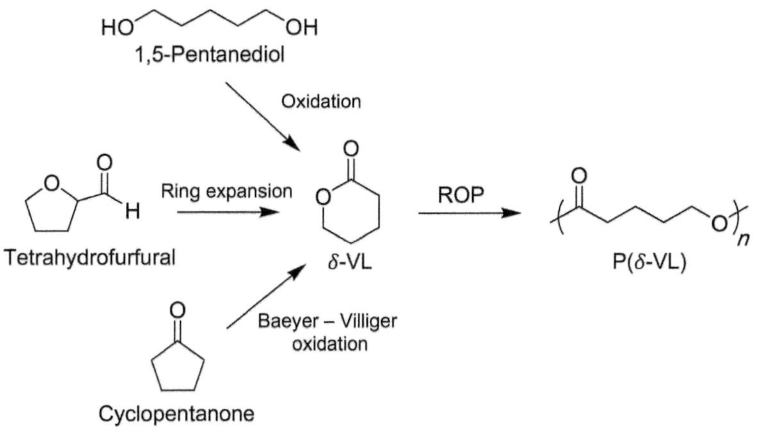

Figure 7.18 Synthesis methods and ROP of δ-VL.

7.3.6 Six-Membered Lactones

7.3.6.1 δ-Valerolactone

The simplest, unsubstituted six-membered monolactone is δ-VL. δ-VL can be synthesized in different ways such as via oxidation of 1,5-pentanediol, through Baeyer–Villiger oxidation of cyclopentanone or by ring expansion of tetrahydrofurfural (Figure 7.18) [169–171].

The first polymerization of δ-VL was described by Carothers et al. in 1932 after 29 days at room temperature [3]. Since then a tremendous amount of research has been devoted to the polymerization and copolymerization of δ-VL with the usage of a broad variety of catalysts (mainly metal-based or organocatalysts) and reaction conditions, which will not be further elaborated in detail. Poly(δ-valerolactone) (poly(δ-VL)) (Figure 7.18) is a semi-crystalline polymer with a T_g of −60 °C and a T_m around 58 °C. It is a rather strong material with a high Young's modulus of 570 MPa and a tensile strength of 12.5 MPa, but it is less flexible compared to poly(γ-BL) with an elongation at break of 150–200% [117, 172–174]. δ-Lactones in general exhibit a low-to-moderate conversion at conventional ROP conditions, and this conversion is strongly dependent on the presence and type of substituents on the six-membered ring [42].

Polymerization of analogs of δ-VL has received limited attention, certainly when compared to ε-CL or G analogs. The following sections will discuss a selection of δ-VL derivatives described in ROP, interesting in the context of (potentially) improved degradability.

7.3.6.2 Unsaturated δ-Lactones

As discussed above for four- and five-membered lactones, creating unsaturated (pending) structures on lactones or polyester backbones can enable functionalization of the lactone (before ROP) or polyester (post-polymerization modification).

5,6-dihydro-2H-pyran-2-one (DHP) is an α,β-unsaturated δ-VL (Figure 7.19a) that enables the generation of polymerizable, functional δ-lactones by a one-step Michael

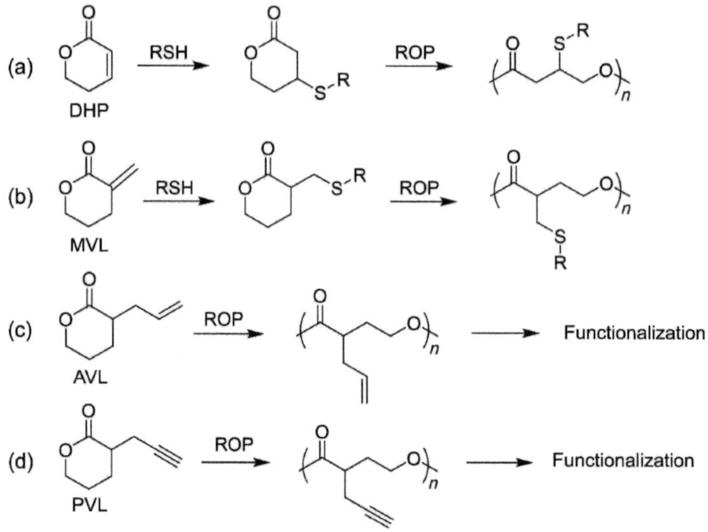

Figure 7.19 ROP and functionalization of unsaturated δ-lactones: (a) 5,6-dihydro-2H-pyran-2-one, (b) α-methylene-δ-valerolactone, (c) α-allyl-δ-valerolactone, and (d) α-propargyl-δ-valerolactone.

addition of thiols bearing various functionalities. ROP of these functionalized lactones has been described both in the presence of TBD and aided by methyllithium (aROP) [175, 176].

α-Methylene-δ-valerolactone (MVL), which can be synthesized from δ-VL, is another unsaturated δ-lactone in which the unsaturation is present as a pending methylene group at the α-position relative to the carbonyl group (Figure 7.19b). ROP of MVL was achieved with $Sn(Oct)_2$, TBD, DPP, or NHC catalysts. Functionalization, mainly performed by addition of thiols, has been described before executing ROP [177–181].

Finally, ROP of α-allyl-δ-valerolactone (AVL) (Figure 7.19c) and α-propargyl-δ-valerolactone (PVL) (Figure 7.19d) have been successfully executed as well. Various polymerizations and copolymerization have been described in which functionalization of the polyester chains always happened after polymer synthesis [182–188].

7.3.6.3 Ester-Substituted δ-Lactones

4-Carboalkoxyvalerolactone

Very recently, the ROP of ester-substituted δ-VLs has been published in which the ester function is located at the γ-position relative to the carbonyl group of the lactone. These 4-carboalkoxyvalerolactones (CAVLs) can be synthesized in three steps starting from malic acid (Figure 7.20). Malic acid is a biobased platform molecule and an intermediate in the Krebs cycle, widely available through microbial fermentation of biomass. In the first step, malic acid is treated with sulfuric acid (H_2SO_4) to create coumalic acid, which is subsequently esterified in the presence of an alcohol (methanol, ethanol, isopropanol, *n*-butanol, or *tert*-butanol) and hydrogenated over

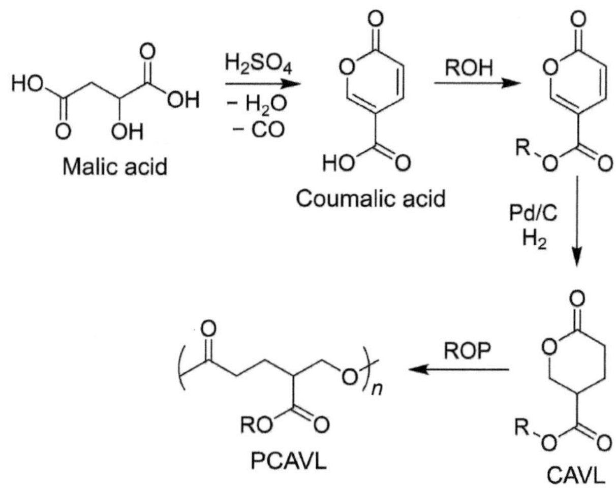

Figure 7.20 Three-step synthesis and ROP of CAVL to produce PCAVL. Source: Adapted from Fahnhorst and Hoye [189].

Pd/C in the presence of hydrogen gas (H_2) to create carboalkoxy-substituted δ-VLs (37–59% yields) [189–191].

These monomers could undergo bulk ROP in the presence of DPP at room temperature to produce linear polyesters with M_n values ranging between 8.0 and 10.1 kg/mol. Furthermore, bulk ROP at increased temperature (70 °C) allowed for the synthesis of high MM (M_n = c. 100 kg/mol) polymers containing small amounts of branched chain structures. Despite the monomers typically being racemic mixtures, the poly(carboalkoxy valerolactones) (PCAVLs) are semicrystalline, in which T_g, T_m, crystallinity, mechanical properties, and hydrolytic degradation rates vary with the type of alkoxy groups on the pending esters. For more details, the reader is referred to the research by Fahnhorst et al. [189–191].

Carbohydrate δ-Valerolactone

In 2009 Williams and coworkers [192] successfully polymerized a δ-VL monomer (acetic acid 5-acetoxy-6-oxotetrahydropyran-2-yl methyl ester [AOME]), synthesized from the carbohydrate derived molecule D-gluconolactone in two high yield steps (Figure 7.21).

In the first step, D-gluconolactone is reacted with acetic anhydride in the presence of a base and converted quantitatively to compound **2**. Subsequently, hydrogenolysis at 50 bar H_2 aided by Pd/C yielded AOME in 95% yield. The hydrogenolysis proceeded to selectively afford the syn-enantiomer (Figure 7.21). ROP was achieved in solution in the presence of $Sn(OBu)_2$ at 80 °C resulting in cyclic polyesters (PAOME) with M_n values of 1.8–7.3 kg/mol. Although amorphous polymers with a T_g of 18 °C were obtained, the materials were thermally stable showing thermal degradation at 250 °C [192]. Copolymerizations of AOME with LD [193, 194] or ε-CL [195] provided for high MM materials. Polymerization with LD resulted in random copolymers (25 mol% AOME) in which hydrolysis was observed to be four times faster than homopolymers of LD.

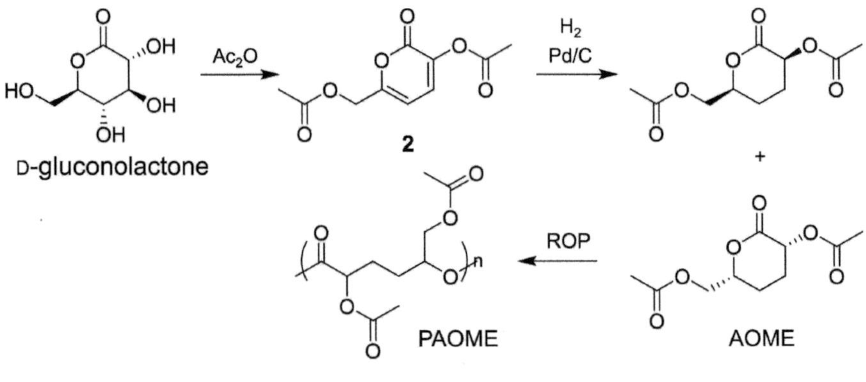

Figure 7.21 Two-step synthesis and ROP of carbohydrate-derived AOME.

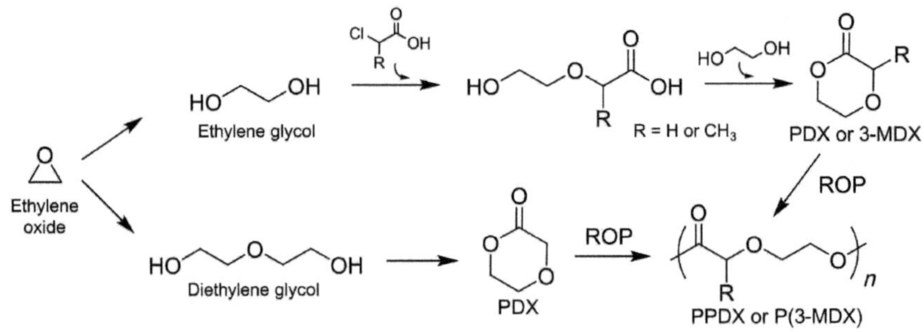

Figure 7.22 Synthesis and polymerization of p-dioxanone from ethylene glycol or diethylene glycol.

7.3.6.4 Ether δ-Lactones

Ether-δ-lactones are δ-lactone monomers containing an additional ether oxygen in their ring structure. The most well-known ether-δ-lactone in the field of ROP is *p*-dioxanone or 1,4-dioxane-2-one (PDX). PDX can be synthesized from ethylene glycol or diethylene glycol, which can both be obtained from petroleum-based feedstock as well as from biomass (Figure 7.22, R = H).

The addition of an ether functional group in a lactone ring changes the reaction properties of ROP as well as the final polymer features (Table 7.1). Poly(*p*-dioxanone) (PPDX) is a biodegradable polymer with a higher T_g and T_m than poly(δ-VL). Furthermore, it is slightly stronger and due to the oxygen atoms in the backbone, PPDX is more flexible. PPDX is most often used in degradable sutures (e.g. PDS® by Ethicon) and to a lesser extent in other applications in the medical field [198, 199]. The ring strain of PDX is comparable to the one of δ-lactones. However, the entropic change during ROP is more pronounced for PDX than for poly(δ-VL), which could be explained by increased flexibility due to the presence of the ether group in the lactone ring.

Only a few dioxanone analogs have been described in literature with regard to ROP. D,L-3-methyl-1,4-dioxanone (3-MDX) (Figure 7.22, R = CH_3), a PDX containing a racemic methyl group at the α-position relative to the carbonyl functional

Table 7.1 Polymerization thermodynamic features and polymer properties of PDX and analogs.

Monomer		ROP thermodynamics			Polymer properties				
	$[M]_0$ (M)	ΔH^0_p (kJ/mol)	ΔS^0_p (J (K/mol))	T_c (°C)	T_g (°C)	T_m (°C)	Young's modulus (MPa)	Tensile strength (MPa)	Elongation at break (%)
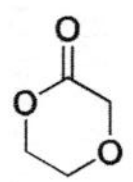 PDX	Bulk (~12 M)	−15.7	−28.9	271	−30 to −1	110	620	50	500–600
	Solution (1 M)	−15.7	−49.7	43					
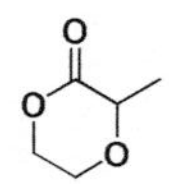 3-MDX	Bulk (~10 M)	−9.9	−20.1	216	−24	Amorphous	—	—	—
	Solution (1 M)	−9.9	−41	−32					
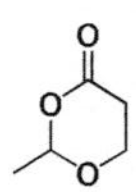 2-MDX	Bulk (~10 M)	−10.9	−29.6	95	−32 to −25	Amorphous	—	—	—
	Solution (1 M)	−10.9	−41	−57					

Source: Data obtained from Refs. [42, 196, 197].

group (Table 7.1), was synthesized and polymerized for the first time in 2001 [196, 200]. The synthesis method is based on the synthesis of PDX from ethylene glycol but in the presence of D,L-chloropropionic acid instead of chloroacetic acid (Figure 7.22). 3-MDX has a reduced ring strain compared to PDX. This can be explained by the emergence of a more preferable angle at the ether bond [42]. The random copolymers [201] and block-copolymers [202] of 3-MDX with PDX have been studied, as well as the block copolymerization of poly(PDX-*co*-3-MDX) with ε-CL [203]. Another PDX analog, the cyclic esteracetal 2-methyl-1,3-dioxane-4-one (2-MDX) (Table 7.1), was synthesized via Baeyer–Villiger oxidation of the commercial product 2-methyltetrahydrofuran-3-one [197]. ROP of 2-MDX was performed in the presence of diethylzinc ($ZnEt_2$) as a catalyst and benzyl alcohol as initiator. While only low $ZnEt_2$ loadings led to poly(2-MDX) of up to 30 kg/mol, high catalyst loadings led to a loss of acetaldehyde and the formation of poly(β-PL). Degradation of poly(2-MDX) at room temperature proceeded significantly faster than degradation of poly(β-PL) due to the presence of the labile acetal function in the polymer backbone. The cationic polymerization of 2-MDX has also been described in the presence of DPP [204].

7.3.6.5 Dilactones

Dilactones are cyclic molecules containing two lactone functional groups instead of one. This additional functional group affects both polymerization characteristics and the properties of the resulting polymers. Their chains contain shorter repeated ester units (less C-atoms) compared to polymers obtained from monolactones, often making the materials less hydrophobic and less flexible, while increasing the T_g.

The most important six-membered dilactones are without any doubt glycolide (G) (Figure 7.23, R = H) and LD (Figure 7.23, R = CH_3). Polyglycolide or poly(glycolic acid) (PGA) and poly(lactide) or poly(lactic acid) (PLA) can either be synthesized by condensation polymerization (polycondensation) of the corresponding α-hydroxy acids glycolic acid (GA) and lactic acid (LA) or via ROP of G and LD, respectively (Figure 7.23). However, as mentioned previously, due to the nature of polycondensation, reaching high MM is challenging [205–207]. While measures exist to remove water from the equilibrium (e.g. azeotropic polycondensation [208, 209] and solid-state polycondensation [210]) or increase molar masses (e.g. chain extension [211, 212]), these approaches will not be further discussed here. On the other hand, ROP of G and LD circumvents the challenge with equilibrium in polycondensation and is accordingly the preferred route in industry to obtain high MM polyesters.

Glycolide and Lactide

The most commonly described and industrially applied method to synthesize G and LD is a two-step process (Figure 7.23). The first step consists of a condensation polymerization of the corresponding α-hydroxy acid to produce a low MM pre-polymer. Subsequently, this pre-polymer is broken down in the presence of a catalyst (e.g. ZnO, $Sn(Oct)_2$, and SnO) to form dilactones via a backbiting depolymerization reaction. This reaction is typically performed at high temperatures and under reduced

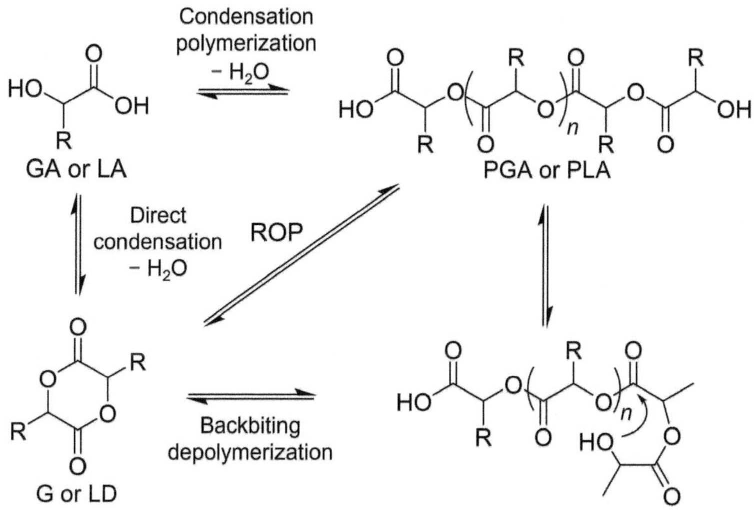

Figure 7.23 Condensation polymerization of GA (R = H) or LA (R = CH_3) to PGA or PLA, backbiting depolymerization of PGA or PLA to create G or LD and ROP of G or LD toward PGA or PLA.

pressure to constantly remove the dilactones and drive equilibrium toward depolymerization [213–215]. Although widely described and applied, this two-step process is time- and energy-intensive, whereby often only mediocre dilactone yields are obtained with, in the case of LD, loss of control over the stereochemistry [213, 216]. Another widely described method to synthesize G and LD is through cyclization of the salts of halogen-derived hydroxycarboxylic acids [217]. More recently different one-step synthesis methods have been developed directly starting from GA and LA or their corresponding esters [218, 219]. For a more detailed overview of the most prominent synthesis methods the reader is referred to some interesting reviews [217, 220].

Compared to six-membered monolactones, G and LD are thermodynamically much easier to polymerize. The reason for this can be found in the higher ring strain of dilactones compared to monolactones resulting from the presence of two ester groups (instead of one) with a planar conformation [221, 222]. In addition, the entropy change for ROP is much lower for δ-VL compared to LD (see Figure 7.6), facilitating PLA synthesis.

Poly(Glycolic Acid)

GA, which is often the main starting compound to produce G and PGA (Figure 7.23), is industrially obtained, among others, by hydrolysis of monochloroacetic acid with sodium hydroxide or by treating formaldehyde with carbon monoxide and water (main method) [56]. GA is widely used in the food industry as a preservative and flavor agent, in the cosmetic and pharmaceutical industry as a skin care product, in the textile industry as a tanning and dyeing compound, etc. [223]. Nevertheless, GA is a naturally occurring compound that can be biochemically obtained as well through enzymatic processes [223]. Due to its biological origin, GA can be assimilated by living organisms and be eliminated via the Krebs cycle as water and carbon dioxide [222]. The good availability and biocompatibility of GA thus make it an attractive monomer for biodegradable (co-)polymers.

PGA is the structurally simplest aliphatic polyester. It is a semicrystalline, biodegradable, and biocompatible thermoplastic material first described and patented by DuPont in 1954 [4]. PGA has a T_g value of 35–40 °C, a T_m around 225–230 °C, and a high degree of crystallinity of 45–55% [224]. Compared to most other degradable polyesters, PGA is a very strong material with high tensile strength (115 MPa) and Young's modulus (7 GPa) resulting from its high crystallinity [225]. In 1962 the American company Cyanamid Co. developed PGA as the first synthetic adsorbable suture under the tradename Dexon®, which has been commercially available since the early 1970s [226–228]. This invention has led to the evaluation of PGA in other biomedical applications, such as in implants, orthopedics, drug-delivery systems, and tissue engineering [229, 230]. Nevertheless, PGA generally degrades quite fast *in vitro* and high levels of GA in the body can result in undesired inflammatory responses. Therefore, copolymers of LD and G, poly(lactic acid-*co*-glycolic acid) (PLGA), have been widely investigated in the context of biomedical applications and have been used in sutures (Vicryl®, Panacryl®, Polysorb®, Purasorb®, etc.), tissue engineering and drug delivery (Lupron®, Zoladex®, etc.) [225, 230, 231]. Changing compositions of G and LD allows tailoring of properties toward desired mechanical features and degradation rates for specific applications.

Poly(Lactic Acid)

PLA is a biodegradable, biorenewable, and biocompatible aliphatic thermoplastic polyester. LA has various natural sources; it is produced from pyruvate in the presence of the enzyme lactate dehydrogenase in metabolic processes in various living organisms.

Due to the presence of a methyl group at the α position of the hydroxy acid, LA consists of two different stereoisomers: L-lactic acid (L-LA) and D-lactic acid (D-LA) (Figure 7.24). Today, LA is industrially synthesized by microbial fermentation of carbohydrates (from corn, sugar beet, sugar cane, etc.), in which micro-organisms mainly produce the L-form [206, 232, 233]. LA can also be synthesized chemically from acetaldehyde and hydrogen cyanide to form lactonitrile, which is further hydrolyzed to LA resulting in a 50 : 50 mixture of L- and D-LA [56]. LA and its salts or esters are used in the food industry as flavoring agents, emulsifiers, or to inhibit bacterial spoilage. In addition, similar to GA, LA can be used as a tanning agent in the textile industry or is widely applied in pharmaceuticals or cosmetics [233].

PLA is top of the league in today's growing synthetic bioplastics market. In 2020, 2.11 million tons of bioplastics were produced globally, of which 13.9% was PLA [234]. PLA owes its popularity to the wide availability and promising cost structure of LA, as well as its excellent rigidity and tensile strength, comparable with commodity plastics such as poly(ethylene terephthalate) (PET) and polystyrene (PS) [235, 236].

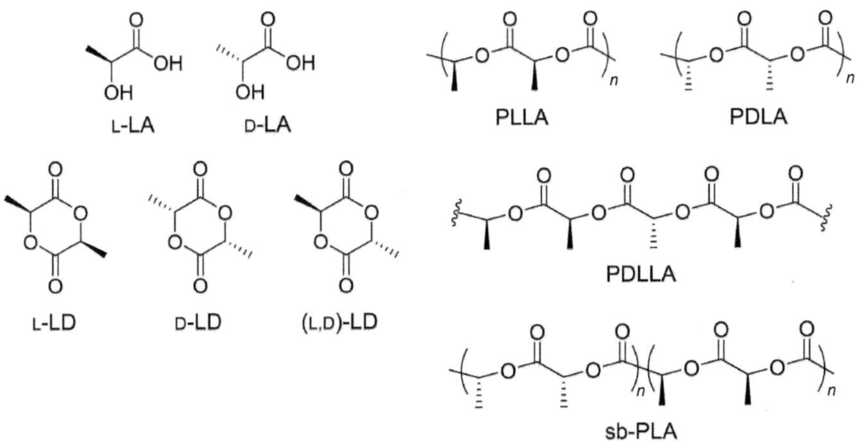

Figure 7.24 Stereochemistry of lactic acid (LA), lactide (LD), and PLA.

PLA is currently used primarily in rigid and flexible packaging, textiles, consumer goods, and medical applications (sutures, implants, drug-delivery systems, etc.).

Due to the stereochemistry of LA, LD and PLA can be produced in different isomeric forms. In contrast with G, LD consists of three stereoisomers: (L,L)-LD (L-LD), (D,D)-LD (D-LD), and (D,L)-LD or meso-LD (Figure 7.24). The properties of PLA are strongly dependent on the stereochemical configuration of the chains. The amount and sequence of L and D chiral centers will have an influence on mechanical and thermal properties as well as crystallinity and degradation rate. PLA with pure L (PLLA) or pure D (PDLA) rotating chiral centers, also called isotactic PLA (Figure 7.24), results in materials with a high crystallinity due to the possibility of the chains to easily stack into ordered structures. The high crystallinity of isotactic PLA results in excellent mechanical properties and a high T_m value (Table 7.2). When optical purity decreases by addition of for example randomly distributed D units in PLLA (atactic PLA, PDLLA [poly((D,L)-lactic acid)], (Figure 7.24)), the crystallinity, T_m and T_g values will slowly drop until the polymer becomes completely amorphous with addition of 12–15% D units. Atactic, amorphous PLA has no melting point. In addition, amorphous PLA is less strong than the semi-crystalline counterparts, showing a reduced tensile strength and Young's modulus [215, 237]. Another interesting feature of PLA lies in the possibility to create stereocomplexes. Mixing of PLLA and PDLA chains or block copolymers of PLLA and PDLA (stereoblock PLA [sb-PLA], (Figure 7.24)) can result in packing of L-based chains with D-based chains to form different crystal structures, called stereocomplex crystals. Stereocomplex PLA exhibits even better thermal stability and mechanical strength than isotactic PLA resulting in materials with a very high T_m (Table 7.2).

Although PLA is often proposed as the bioplastic with the highest potential to drive the biobased plastics industry, it exhibits some drawbacks which put a limit on the sustainability and applicability of this polyester. Although PLA is strong, it also is a brittle material, exhibiting a low-impact strength and elongation at break

Table 7.2 Thermal and mechanical properties of PLA materials with different stereochemistry.

Polymer	T_g (°C)	T_m (°C)	Tensile strength (GPa)	Young's modulus (GPa)	Elongation at break (%)
PGA	35–40	225–230	0.08–0.98	3.9–14	30–40
PLLA/PDLA	50–65	170–190	0.12–2.3	6.9–9.8	12–26
PDLLA	50–60	–	0.04–0.05	1.5–1.9	5–10
Sc-PLA	65–72 [224]	220–240	0.88	8.6	30

Source: Data based on Tsuji [224] and Hirata and Kimura [237].

(Table 7.2). In addition, this bioplastic has a poor melt strength and low melt viscosity and elasticity which limits its processability in strong elongational-flow-dominated processes, such as film blowing, blow molding, and foaming [238–241]. Although PLA is described as biodegradable, its degradability is rather limited and only occurs at a sufficient rate under industrial composting conditions. Although its degradability is described as limited, for certain applications PLA is too hydrolytically unstable. Besides copolymerization, polymer blending, or the preparation of polymer composites, functionalization of PLA can provide solutions for some of these deficiencies. For an overview of some interesting functionalized G and LD monomers used in ROP the reader is referred to some excellent reviews [49, 242, 243].

7.3.7 Seven-Membered Lactones

In general, seven-membered rings have a higher ring strain compared to their five- and six-membered counterparts, which results in high equilibrium conversions during ROP. They are less commonly found in nature and are thus mainly obtained from fossil-based sources, although more sustainable routes are emerging [244].

7.3.7.1 ε-Caprolactone

The simplest seven-membered lactone, ε-CL is used to make the well-established biodegradable polymer, PCL. The main synthetic route makes use of petroleum-derived cyclohexanone (Figure 7.25). This industrial process (i.e. Baeyer–Villiger oxidation of cyclohexanone) uses peracetic acid and is far from being "green," entailing ecological and toxicological concerns [245, 246]. Hence, efforts are ongoing to find a competitive biobased route. Novel strategies suggest a synthesis starting from the biobased platform chemical 5-hydroxymethylfufural, in which ε-CL is obtained from the intermediary product 1,6-hexanediol and in some cases 6-hydroxyhexanoic acid (Figure 7.25) [246, 247].

A myriad of catalysts can be used for the ROP of ε-CL, which can proceed accordingly via anionic, (monomer-activated) cationic, enzymatic, or coordination–insertion ROP. For an extensive review on the matter of catalysts and reaction conditions, the reader is referred to the work of Labet and Thielemans [245].

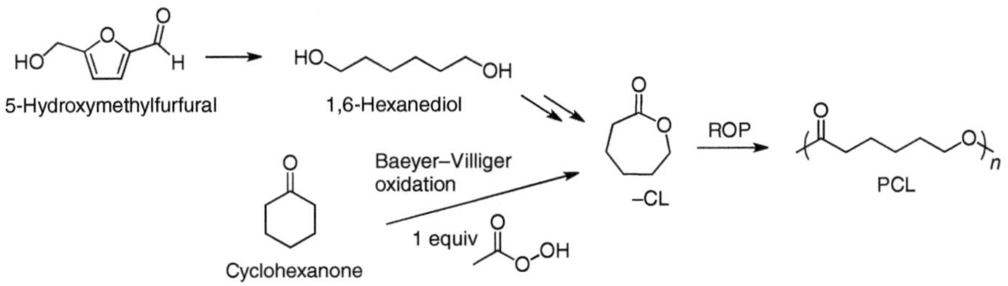

Figure 7.25 ROP and synthesis routes toward ε-CL via the biobased platform chemical 5-hydroxymethylfurfural and industrial process from cyclohexanone via Baeyer–Villiger oxidation.

The first mention of PCL dates back to the ROP reported by Carothers in the 1930s. PCL is semicrystalline, with a crystallinity content ranging from 33% for higher MMs to 70% for lower MMs [248]. It has a low T_g (−65 to −60 °C) and a low melting range of 56–65 °C and shows elastic behavior at room temperature [245, 248, 249]. Mechanical properties also depend strongly on MMs; tensile strength ranges from 4 to 785 MPa, elongation at break from 20% to 1000% with a Young's modulus of 0.21–0.44 GPa [245]. The low T_m facilitates thermal processing, for example in 3D printing (e.g. Facilan™ filament), compression molding, and heat-deformable prosthetics (X-lite™) [250]. After warming it in hot water it is used as a modeling compound for crafting purposes (e.g. Shapelock™ and Friendly Plastic™). Because of its ease of miscibility with many other polymers, PCL is also a popular plastic used for polymer blends with (non)-degradable plastics [251]. PCL is sold under the tradename Capa™ and Capromer™ for applications ranging from compostable food packaging to polyols as precursor components for the development of polyurethane elastomers.

PCL is both biocompatible and biodegradable, although the time scale of degradation largely depends on the environment. In humans, the degradation of PCL is very slow (two to four years), because of the shortfall of appropriate enzymes and inherent hydrophobicity [245, 249, 252, 253]. Its applications in the biomedical field were therefore limited to drug-delivery systems and sutures that require longer active periods (e.g. Capronor™ capsules). With the advent of tissue engineering, PCL regained traction because its material properties enable facile shaping into a multitude of scaffolds. Sutures made from copolymers of ε-CL with G (Monocryl™, Monocryl Plus™, and Suruglyde™) and LD (Neurolac™) are widely available. A review by Woodruff and Hutmacher describes in detail the applications in the biomedical field [249].

Nevertheless, PCL is biodegradable by a large population of fungi and bacteria [254]. These include both anaerobic and aerobic bacterial species from diverse habitats such as a heat-resistant strain of *Aspergillus* sp. capable of degrading PCL to completion after six days at 50 °C [251, 254]. In the presence of suitable bacteria and enzymes, the degradation time is significantly reduced to an average of several months. Enzymes that assist the degradation of PCL include esterases, lipases, and cutinases. The latter is believed to be effective because the polyester bears sufficient structural resemblance to cutin, a natural waxy polymer found in plant cuticles

[251, 255, 256]. The greater abundance and population of species capable of degrading this plastic could indicate that PCL is more prone to degradation in nature than other biodegradable plastics such as PLA [251]. Toxicological tests showed that the degradation product ε-caproic acid is less harmful to living organisms than the degradation product of PLA, LA [257]. The major drawbacks of PCL that limit its degradability and applications are its high crystallinity, hydrophobic nature, and lack of functionality [54, 252, 253].

7.3.7.2 Substituted and Functionalized ε-Caprolactone

As is the case with other lactones, properties influencing the degradability can be fine-tuned by substituting the ε-CL monomer structure with alkyl- and/or functional groups.

Alkyl Substitutions

Whereas PCL is generally semicrystalline, the addition of alkyl-groups disrupts crystallite formations by hindering stacking of the backbone, rendering the polyesters fully amorphous with low T_gs [258]. Alkyl-substituted PCLs bring added value as soft segments in block copolymers with more crystalline polyesters (e.g. PLA or PCL) and widen the scope of their applications to where ductility and elasticity are required, i.e. as thermoplastic (bio)elastomers [259, 260].

Alkyl substituents also reduce the propagation rate of polymerization relative to that of ε-CL, the extent of which depends on the nature and position of the substituent and the ROP mechanism. The addition of alkyl groups also introduces chirality into the monomers, which can result in different selectivity and propagation rates when using enantioselective catalysts [261–264]. Similarly, the rate of enzymatic hydrolysis is also influenced by the length of the alkyl pendant groups, albeit to a lesser extent than the rate of polymerization [265]. The choice of the length of the alkyl chain is always a trade-off, as it increases hydrophobicity but also the amorphous content of the resulting polymer [244].

A promising alkyl-substituted analog to ε-CL is menthide (Figure 7.28), due to its biorenewable nature. It can be synthesized via Baeyer–Villiger oxidation of (−)-menthone, a ketone derivative of (−)-menthol, a plant extract used in large quantities for a multitude of purposes across different industries [266]. The ring strain (i.e. ΔH_p) of menthide is comparable to that of ε-CL. The ΔS_p value, on the other hand, is significantly lower as, in accordance with the Thorpe–Ingold effects, the ring is more stable due to its bulky side-groups. Unsurprisingly, the polymerization rate is significantly lower than that of ε-CL because of the steric hindrance of the ester group by the isopropyl substituent. Nonetheless, ROP yields amorphous poly(menthide) with a low T_g (−25 °C) and a M_n up to 90 kg/mol [266, 267].

The steric hindrance of the alkyl groups at the ε-position can inhibit or slow down transesterification reactions. These side reactions are usually undesirable as they lead to random copolymers with large dispersities and lower MM for one-pot syntheses. Wilson et al. proved this by copolymerizing menthide with higher-membered lactones (≥ 8), which have a slower polymerization rate, and used this for controlled one-pot synthesis of the copolymers [267].

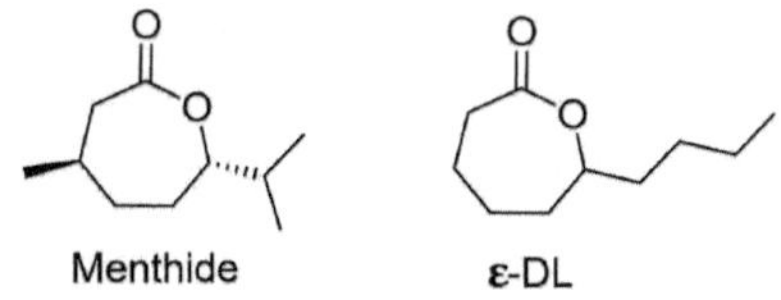

Figure 7.26 Chemical structures of menthide (right) and ε-decalactone (left).

Another biobased seven-membered lactone is ε-decalactone (ε-DL), which is derived from natural castor oil (Figure 7.26) [260]. The T_c of poly(ε-decalactone) is sufficiently high for the ROP to take place at elevated temperatures to obtain high conversion (96% at 150 °C), despite the long alkyl chain at the ε-position slowing down the polymerization [268]. The fully amorphous polymer has a T_g of −58 °C [260].

The physical crosslinking via multiphase separation of crystalline (PLLA) segments and soft segments (poly(menthide) or poly(ε-decalactone)) gives the block copolymers improved and tunable mechanical properties. These thermoplastic bioelastomers showed excellent elongations at break of 530–960% and 723–1060%, respectively [260, 268, 269].

Pendant Functional Groups

As described previously for other lactones, attaching functional groups to the backbone of the polyesters can be a powerful way to influence degradability. A crucial drawback of PCL in this regard is its hydrophobicity, which can be countered by incorporating polar functionalities such as hydroxyl groups. As mentioned before, suitable protective groups are necessary for the ROP of lactones. A prominent example in the category of protected lactones is 1,4,8-trioxaspiro[4.6]-9-undecanone (TOSUO), which is readily homo- and copolymerized (Figure 7.27).

The ketal group protects the ketone functionality and is formed by reaction with a diol. Deprotection forms the ketone group in the backbone, which can then be reduced to a hydroxyl group. Alternatively, this poly(ketone-ester) can be obtained more directly by the ROP of oxepane-1,5-dione (Figure 7.27), provided

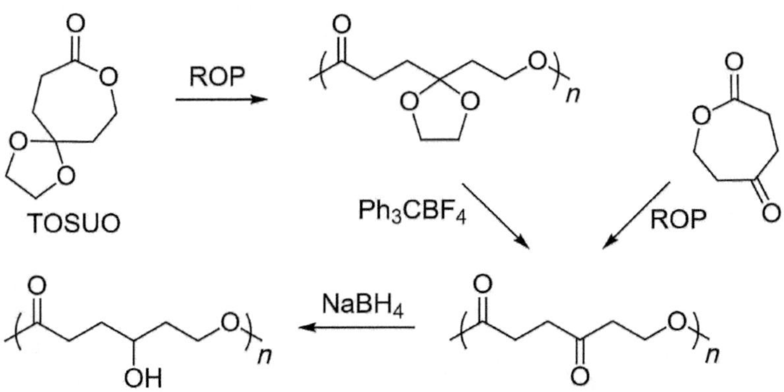

Figure 7.27 Route to poly(4-hydroxy-ε-caprolactone) from the reduction of poly(ketone ester) which is obtained from ROP of TOSUO and subsequent deprotection vs. direct ROP of oxepane-1,5-dione.

a catalyst incapable of undergoing side reactions with the ketone groups is used, such as tin-based alkoxides [48]. The cyclic ketal groups themselves can also bring added value as they decrease the crystallinity, render the polymer more flexible, and can reduce the T_m to room temperature or body temperature depending on the anticipated application of the soft material (e.g. in biocompatible physical-sensing platforms). Values for elongations of up to 1500% were observed for poly(CL-*co*-TOSUO) [270].

Obtaining PCL with amino pendant groups can be accomplished by polymerization of γ-(carbamic acid benzyl ester)-ε-CL followed by deprotection [271]. The poly(ketone ester) can also be derivatized to create an amine by reacting it with 2-hydroxyethyl hydrazine [50], or it can be used to create a graft copolymer by reacting it with an aminooxy-terminated poly(ethylene oxide) [272].

As previously mentioned, halide pendant groups offer interesting post-polymerization modification options. Yao et al., for instance, substituted the chloro-groups from ROP of α-Cl-ε-CL with azido functions to subsequently graft biocompatible rosin ester moieties onto the polyester. The resulting polymer was capable of fast microbial degradation, despite decreased hydrophilicity and higher T_g [273]. Other halogen-functionalized PCL homo- and copolymers are reported in literature, including that of α-fluoro-ε-CL [274], α-iodo-ε-CL [275], α-bromo-ε-CL [276], γ-(2-bromo-2-methylpropionate)-ε-CL, and γ-bromo-ε-CL [48, 277, 278]. The reader is referred to the reviews by Jérôme and Lecomte for further information on the matter [28, 48]. Elimination of halogen atoms attached to larger pendant groups or directly attached to the PCL backbone has also been reported. This dehydrohalogenation results in either pendant groups with allyl functions or PCL with the unsaturated function in the polymer backbone which can be further functionalized [55, 278].

7-Allyl-1-oxa-cycloheptan-2-one (Figure 7.28a) is an example of ε-CL substituted with a pendant allyl group forming poly(6-allyl ε-caprolactone) upon ROP, which could be quantitatively transformed via bromination, epoxidation, and silylation [279]. Similarly, acrylate-functionalized ε-CL, e.g. γ-acrylic-ε-CL, can also undergo ROP and can then be derivatized via Michael addition of various thiols without the necessity of an organometallic catalyst. However, some loss to cross-linking and backbiting reactions is inevitable [28]. Alkyne-substituted ε-CL such as α-propargyl-ε-CL (Figure 7.28b) can also undergo ROP and subsequently a click reaction with azide-capped compounds for post-polymerization modification [280, 281].

Recently, Wu et al. reported the ROP of ε-CL substituted with a phenylseleno group at the α-position (Figure 7.28c). Through an oxidative elimination, the pendant group is replaced by a double bond in the backbone structure, similar to the Cl-elimination reaction seen in Figure 7.7, but with the added benefit that the phenylseleno groups can be recovered (up to 90%) and reused. The authors found that both the poly(α-phenylseleno-ε-caprolactone) and unsaturated polyester, resulting from its oxidation, hydrolyze more slowly than PCL in acidic conditions. However, derivatization to a suitable functional group could increase the degradability [54]. Interestingly, this route led to a trans-(E) configuration at

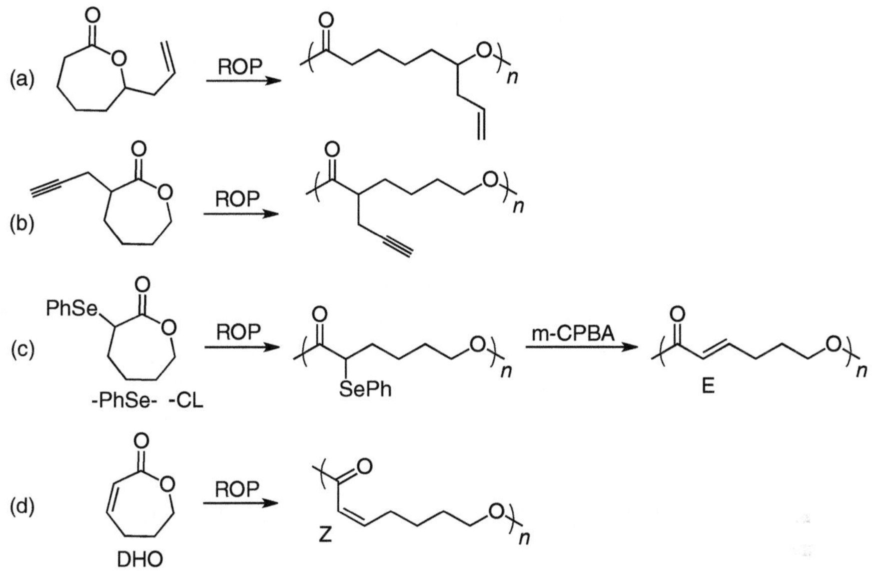

Figure 7.28 Selection of ROP reactions to obtain PCL-derived unsaturated polymers: (a) 7-allyl-1-oxa-cycloheptan-2-one, (b) α-propargyl-ε-caprolactone, (c) α-phenylseleno-ε-caprolactone, (d) 6,7-dihydro-2-(5H)-oxepinone.

the double bond, whereas the direct ROP of 6,7-dihydro-2-(5*H*)-oxepinone (DHO) (Figure 7.28d) results in a cis-(Z) configuration. Because of the trans-configuration, the polymer becomes stiffer and the T_g is higher (−28 °C) compared to that of the polymer of DHO (−50 °C) [54, 282].

7.3.7.3 Ether-ε-Lactones

Ether groups can be incorporated into the polymer backbone to effectively increase the hydrophilicity of PCL. Research on polymerizations of ether-ε-caprolactones has focused on 1,5-dioxepan-2-one (DXO), which can be synthesized from 4-oxotetrahydropyran via Baeyer–Villiger oxidation (Figure 7.29) [283].

ROP proceeds readily using numerous catalysts of metallic, organic, and enzymatic nature to yield high MM poly(1,5-dioxepan-2-one) (PDXO). PDXO is an amorphous, hydrophilic poly(ether ester) with a low T_g of −39 °C, which readily hydrolyzes [284].

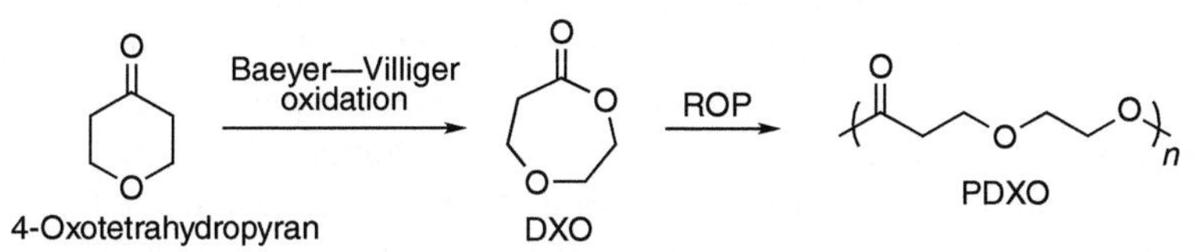

Figure 7.29 Synthesis of DXO from 4-oxotetrahydropyran and ROP forming the poly(ether ester) PDXO.

These properties greatly limit the applications of PDXO as homopolymers but make it attractive for uses in copolymers, which widens the scope of the resulting plastic to biomedical applications. These copolymers combine the flexibility and hydrophilicity of PDXO with the stiffness of, among others, PLLA and PCL [285–287].

The hydrophilicity, as measured by the contact angle, is improved by increasing the ratio of DXO to the more hydrophobic block comonomer. Similarly, PDXO blocks are more readily hydrolyzed compared to the more hydrophobic PCL and PLLA blocks [285, 288, 289]. After 23 weeks in a buffer solution at 37 °C, 70% of the DXO monomers and 10–20% of the L-LA units were hydrolyzed [289]. Despite the instant hydrolysis, the block copolymers of DXO and L-LA could retain good tensile properties (elongation of 600–800% and stress at break of 8–20 MPa) for several months [285].

7.4 Relations Between ROP Polymers and Degradability

The biodegradability of a polymer does not necessarily coincide with that of the (biobased) feedstock used to synthesize the monomer or polymer. The degree to which a polymer can (bio)degrade depends on the chemical composition of its chains and their behavior at macromolecular level, determining the possibility for water and/or enzymes to break down the chains and for micro-organisms to assimilate and convert the resulting oligomers and/or monomers to water, carbon dioxide or methane, and biomass. More details on the mechanisms of (bio)degradation can be found in Chapter 2 of this book.

In general, due to their main chain ester linkages, (aliphatic) polyesters are inherently susceptible to (bio)degradation. Nevertheless, the rate and ease of (bio)degradation do not exclusively depend on the presence of ester bonds. Characteristics such as crystallinity, molar mass, and hydrophilicity also affect a polyester's sensitivity to degradation [290]. Hence, substitution and functionalization of lactones and polyesters aid in changing these characteristics, and thus, substantially influence (bio)degradability. In this section, we will briefly discuss a few interesting examples of such relations in order of ring size, focusing on hydrolysis and not complete biodegradation.

The group of Lu [111] performed hydrolysis tests in alkaline solutions on homopolymers of α-methylene-β-butyrolactone (MBL) compared to poly(β-BL) and random copolymers of (β-BL) and MBL (see also Section 7.3.4.4). The polymers were dissolved in toluene or tetrahydrofuran in the presence of an amine (triethylamine or diisopropylethylamine) and were either heated (80 °C) or kept at room temperature while stirring. The presence of the vinylidene pendant groups in the polymer backbone resulted in faster hydrolysis rates compared to poly(β-BL), even at room temperature. Significant changes in MM were observed for PMBL (homopolymer of MBL) (88% loss) compared to poly(β-BL) (10% loss) at 80 °C after 12 hours. Increases in dispersity were linked to the chain scission mechanism of degradation taking place. Adjusting the amount of MBL units in the copolymers

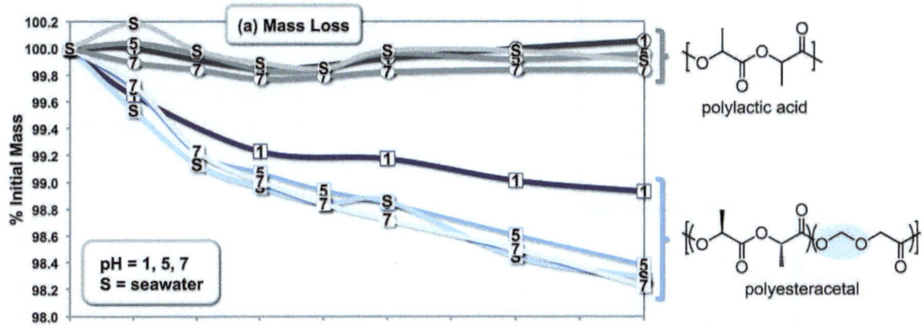

Figure 7.30 Whereas poly(DL-lactide) (PDLLA) is not affected, the polyesteracetal (4% DOX) shows considerable molar mass loss upon exposure to aqueous environments of pH 1, pH 5, pH 7 (distilled water), and seawater, over 45 days. Source: Reproduced from Martin et al. [167] with permission from the Royal Society of Chemistry.

(9–46% of MBL) could enable tuning of the degradability of the final polymer chains.

Miller and coworkers [167] described an improved hydrolytic degradability of PLA by copolymerization of (L,L)-LD with 1,3-dioxolan-4-one (DOX) to create polyesteracetals (also see Section 7.3.5.2). The acetal functional group in the main chain results in a more hydrophilic backbone with weaker linkages. Hydrolysis tests were performed on thin films (0.25 mm thickness) immersed and stirred in aqueous acidic solutions of pH 1, 5, and 7 (distilled water) and seawater (Atlantic Ocean, pH = 7.5) at room temperature for 45 days.

The mass of the films as well as the molar mass (gel permeation chromatography) were monitored over time after rinsing and drying of the films (Figure 7.30, only showing mass loss). A polyesteracetal copolymer with 4% DOX was compared to a commercial, rather amorphous Ingeo™ PLA (Natureworks), approximated as PDLLA. Figure 7.30 shows that the polyesteracetal copolymer lost more than 1% of its mass (and up to 35% of its molar mass) over 45 days, while PDLLA did not show noticeable degradation under all applied conditions. In neutral conditions (distilled water, pH 7) the polyesteracetal copolymer degraded most rapidly by mass, while molar mass loss was most pronounced under acidic conditions. Although the mechanism of degradation was not studied in this work, the authors presumed the hydrolytic cleavage to occur at the less sterically hindered acetal groups or glycolic ester groups instead of at the lactic ester groups, with formaldehyde being the final by-product [167].

As mentioned in Section 7.3.6.3, Williams and coworkers [192] performed ROCOP of (L,L)-lactide (monomer of PLLA) with a new carbohydrate-based six-membered lactone containing two methyl ester side groups (AOME), resulting in random copolymers with 1 (RP1), 6 (RP2), 11 (RP3), and 25 wt% (RP4) AOME. Incorporation of AOME led to copolymers with a lower degree of crystallinity and a lower T_g. While RP2 and RP3 showed splitting of the melting peak due to microphase separation of both crystalline regions, RP4 was completely amorphous. Degradation tests were performed on spin-coated films of the copolymers. Static

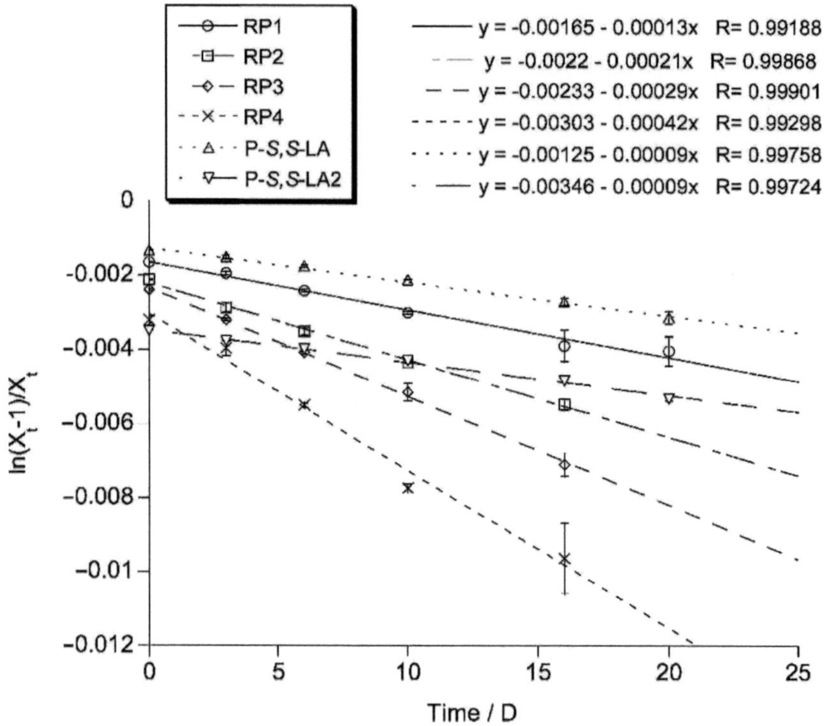

Figure 7.31 Plots of $\ln((X_t - 1)/X_t)$ vs. time for the copolymers (RP1 (88 kg/mol), RP2 (68 kg/mol), RP3 (61 kg/mol), RP4 (44 kg/mol), P((L,L)-LD) (P–S,S–LA 108 kg/mol. P–S,S–LA2 41 kg/mol). Each degradation experiment was conducted three times, and the errors for each measurement are included. Source: Reprinted with permission from Tang et al. [193]. 2010 American Chemical Society.

water contact angle measurements on the films showed a decrease in contact angle (RP4: 7° lower than P–S,S–LA (= PLLA)) with more incorporation of AOME, proving the higher hydrophilicity of AOME compared to LD.

Degradation experiments were performed in $CHCl_3$ (2 mg polymer/ml) (to dissolve the polymers) and phosphate-buffered saline (pH 7.4) at 25 °C. Figure 7.31 shows the loss in molar mass (M_n) over time with X_t being the degree of polymerization ($X_t = M_{n,t}/M_0$ with $M_{n,t}$ the number average MM at time t and M_0 the MM of the monomer). While PLLA showed a constant rate of degradation independent of initial MM, copolymers of AOME and (L,L)-LD exhibited a higher rate of degradation, while the rate increased with higher AOME contents. This higher rate of degradation was attributed to the greater hydrophilicity of the copolymers compared to PLLA.

Li et al. [291] studied the hydrolysis of copolymers of LA and GA (ratio 1 : 1, not made via ROP), PLGA, in phosphate-buffered solutions with pH 7.4. The crucial difference lies in the sequencing of the monomer units. They compared the degradation of PLGA with alternating sequences (named polyLG with MMs of 16 and 26 kDa) to that of L-PLGA (copolymer of L-LA and GA) and PLGA (copolymer of racemic LA and GA) with random sequencing (named R-SAP and R-ROP with MMs of 31

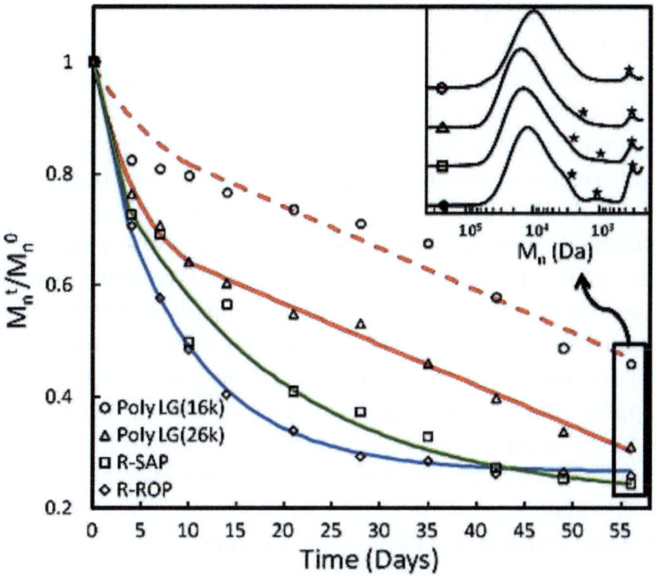

Figure 7.32 Hydrolysis profiles of PLGA copolymers with alternating (o 16 kDa, △ 26 kDa) and random sequencing (□ L-PLGA (R-SAP), ◇PLGA (R-ROP)) normalized relative to their original M_n. Source: Reprinted with permission from Li et al. [291]. Copyright 2011 American Chemical Society.

and 32 kDa, respectively). Figure 7.32 shows the degradation profiles with marked differences. The typical randomly sequenced PLGA follows an exponential decaying trendline with a fast initial degradation followed by a stagnant degradation phase and can be problematic for the biomedical field where this polymer is often applied. The random sequencing results statistically in short block-type sequences, some of which will contain purely PGA units prone to fast hydrolysis, while other segments comprise exclusively PLA, which are less susceptible and persist longer. In contrast, the profile of the alternating copolymer shows much better control over degradation by circumventing the formation of such blocks. After an initial rapid decline in MM (by 20% and 35%), the degradation rate slows down and follows a linear trendline, which is beneficial to for instance avoid the formation of small slow degrading remnants in the body.

Hakkarainen et al. [288] investigated the hydrolysis behavior of homo- and block copolymers of ε-CL and the ether lactone analog thereof, DXO, in water at 37 °C and analyzed the degradation products via mass spectrometry (MS). They observed larger oligomers resulting from the hydrolysis of PDXO homopolymer than from PCL (up to heptadecamers of DXO versus only hexamers of ε-CL), proving that the polyester ether has a higher degree of hydrophilicity. As a result, the PDXO is preferentially hydrolyzed, which was confirmed by the degradation of copolymers. MS spectra of copolymer degradation products strongly resembled those of PDXO, while DXO signals in the NMR spectra of copolymer remnants decreased.

As can be seen from the examples shown above, functionalization and/or ROCOP of lactones (or polyesters) can provide better or more controlled degradability when hydrophilicity is increased and/or crystallinity is reduced. Nevertheless, degradability studies are performed under strongly varying conditions, making comparing different materials over different studies ambiguous at best. Hence, the development of standard methods for (bio)degradation studies of polyester, or polymers in general, could provide for higher levels of insights in degradation–structure relations of plastics. The examples highlighted above show degradation studies that were performed through hydrolysis and often in the dissolved state, a method that is regularly favored in polymer synthesis due to the ease of application. While comparisons in such conditions are surely valuable, it is not sure how these relations survive in biological and often dry real-world degradation environments. Nevertheless, other degradation assessments are also applied and thus of great importance. These will often show different results due to the different conditions. Other relevant methods in this field include hydrolysis in presence of electrolytes and/or enzymes and microorganisms, photodegradation with UV-light (especially relevant for polymers with unsaturated backbones), and composting. Further detail is found in Chapter 2 of this book.

7.5 Conclusion

ROP is, compared to some important industrial polymerization procedures (e.g. polyolefins), an interesting strategy to incorporate multiple types of heteroatoms in polymer backbones. ROP of lactone monomers enables the synthesis of aliphatic polyesters in which the ester units are inherently susceptible to (bio)degradation in certain conditions. Even though condensation polymerization of diacids with di-alcohols or of hydroxy acids also results in polyesters, ROP often has some pronounced benefits over polycondensation. In a lot of cases, ROP reactions need less stringent reaction conditions. Moreover, higher molar masses and lower polydispersities are often obtained due to the absence of equilibrium shifting side-products formed during polycondensation. On the other hand, ROP reactions are not always straightforward and can be challenging due to the sensitivity of the process to various parameters. Furthermore, certain polymer structures can only be obtained via polycondensation (di-acids with di-alcohols) and not via ROP.

This chapter presents an overview of the most prominent biodegradable ROP-based polyesters classified by ring size (four to seven membered rings) of their lactone monomers. These polyesters and some of their thermal and mechanical properties are summarized in Table 7.3. PGA, PLLA, and (R)-PHB (containing 1–2 CH_2 groups between the ester units) are strong, but brittle materials exhibiting high tensile strengths and Young's moduli, but low elongations at break. In addition, these materials have high T_ms and relatively high T_gs. However, aliphatic polyesters with chains consisting of longer C-containing units in their backbone (containing >3 CH_2 groups between ester units), such as poly(γ-BL), poly(δ-VL), and PCL, exhibit more elastomeric behaviors, showing low T_ms, negative T_gs, and high elongations at break.

Table 7.3 Most prominent ROP-based polyesters (isotactic homopolymers) and some of their thermal and mechanical properties.

Polymer	Acronym	T_g (°C)	T_m (°C)	Tensile strength (MPa)	Young's modulus (MPa)	Elongation at break (%)	References
Poly(glycolic acid)	PGA	35–40	225–230	80–98	3900–3914 000	30–40	[224, 237]
Poly(L-lactic acid)	PLLA	50–65	170–190	120–2300	6900–9800	12–26	[224, 237]
Poly(β-butyrolactone)	(R)-PHB	−4 to 1	175–180	40	3500–4000	3.0–8.0	[68–70]
Poly(γ-butyrolactone)	Poly(γ-BL)	−51 to −48	53–60	50	70	1000	[116, 117]
Poly(δ-valerolactone)	Poly(δ-VL)	−67 to −60	57–60	12.5	570	150–200	[117, 172–174]
Poly(ε-caprolactone)	PCL	−65 to −60	56–65	4–785	210–440	20–1000	[245]

Although potentially very promising, ROP-based polyesters still have a very limited presence in industry. PLA is by far the most abundant type accounting for 13.9% (≈ 293 kt) of the bioplastics market (2019) [234]. Due to its excellent strength and rigidity, PLA has similarities with commodity petroleum-based plastics such as PET and polystyrene (PS), and is mainly used in both rigid and flexible packaging, consumer goods, textiles, and medical applications. [235, 236] However, the brittle character (mainly isotactic PLA) and rather low T_g still limit its replacement potential in a broad range of other applications. Although currently microbially synthesized, PHAs (including PHB) are the second most important plastics in this group accounting for 1.2% of the bioplastics market (2020), which is expected to rise to 11.5% by 2025 [234]. Due to their good barrier properties, high strength, and good heat resistance, PHAs are interesting potential substitutes (see Chapter 6 However, current microbial production processes are still very expensive and of limited scale, and an ROP process could be envisioned if a more efficient route to the corresponding monomeric lactone could be developed [292]. PCL and PGA are also industrially available at the moment, but only to a very limited extent, and are predominantly used in medical applications.

Even though these commodity bioplastics are defined as biodegradable, their degradation capacity is often limited by the conditions at which degradation takes place. While most of these aliphatic polyesters are readily degradable under industrial composting conditions, degradation in natural environments at lower temperatures and/or lower microbial activity (soil, rivers, oceans, etc.) is often limited. Introducing additional functional groups, either pendant or within the polymer backbone, can significantly improve the hydrophilicity of a plastic and with that often also its degradation capacity. Functionalities can be introduced by either direct ROP of functionalized lactone monomers or via grafting on the polymer backbone post-polymerization. Various techniques and examples of functionalization have been described in this chapter. Functionalization of polyesters constitutes a powerful method to increase degradation rates by tuning properties such as hydrophilicity. Nevertheless, it remains a challenging undertaking for numerous reasons. For instance, the use of protective groups is often unavoidable, which requires extra reaction steps and optimization efforts. Moreover, substituted groups on lactones often hinder opening of the rings, impeding polymerization. Grafting also requires specific functionalities, such as halogens or unsaturated bonds, which are not always readily obtained, while the grafting reaction itself also demands optimization. However, examples in literature have been described in which smart and efficient chemistry and catalysis can help overcome these difficulties, making the synthesis processes less costly and time consuming. Examples of this include the use of very reactive organocatalysts, ROP of ether-lactones containing functional groups in the ring structure itself that do not interact with ROP catalysts, changing reaction conditions in a smart way (low temperatures, high concentrations, etc.), etc.

Multi-step syntheses, challenging purifications, and price-intensive processes limit the scalability of functionalized polyesters. As a consequence, a gap is frequently noticed between research on polymer chemistry and analysis of the material properties (thermal, mechanical, rheological, degradation, etc.). The gap is often a matter of scale as new polymers made in sub-10 g amounts are hard to assess mechanically for instance. Novel synthesized polyesters are also quickly described as "biodegradable" simply due to the structural resemblance to classic polyesters such as PLA and PCL, without properly analyzing their degradation potential. However, small changes in structure can greatly influence degradability. Furthermore, there is a lack of consensus on how to analyze degradability in different environments. The great variety in how degradation is assessed makes it near-impossible to compare results among different polymers, resulting in a lack of generally accepted polymer chemistry–degradability relationships.

Finally, it is important to mention that degradation should be mainly prioritized in applications in which the degradability itself has an additional value (e.g. medical applications) or in applications in which the plastic comes in close contact with the environment and leakages are hard to avoid (e.g. agriculture and fishing) [293]. Although degradation is important in these contexts, a trade-off between the plastic's degradability and performance should always be taken into account.

7.6 Outlook and Recommendations

Although ROP can be a very challenging polymerization method, certainly when applied in the context of functionalized polymers, it exhibits important advantages which can make it an industrially viable process once optimized.

A first important outlook in polyester research via ROP should be to close the gap between lab-scale polymer chemistry and material analysis. Focus should be laid on smart and efficient chemistry and catalysis to decrease synthesis times and costs, and improve scalability. It is suggested that new polymers are aimed at a scale of at least 100 g, which allows some mechanical and rheology tests, but also enables shaping for degradation experiments of solid objects. Additionally, a closer collaboration between scientists focusing on monomer synthesis, polymer chemistry, and material science (engineering, mechanical properties, and plastics processing) is recommended to improve overall efficiency and quickly spot industrially viable polymers.

Besides improving scalability, it is important to get better insights into the exact relationships between the chemical structures of polymers and their (bio)degradation behavior under specific conditions. An important step that should be taken in this context is the development of universal and straightforward degradation testing procedures, mimicking (bio)degradation in various conditions (e.g. hydrolysis and composting).

List of Relevant Abbreviations

Abbreviation	Full description
α-AL	α-Angelicalactone
ACE	Activated chain-end
α-Cl-ε-CL	α-chloro-ε-caprolactone
AM	Activated monomer
aROP	Anionic ring-opening polymerization
β-BL	β-butyrolactone
BEMP	2-*tert*-butylimino-2-diethylamino-1,3-dimethylperhydro-1,3,2-diazaphosphorine
β-PL	β-propiolactone
ciROP	Coordination–insertion-based ring-opening polymerization
cROP	Cationic ring-opening polymerization
DBU	1,8-diazabicyclo[5.4.0]undec-7-ene
ΔG_p	Gibbs free energy of polymerization
ΔG_p^0	Standard Gibbs free energy of polymerization
ΔH_p	Enthalpy of polymerization
D-LA	D-lactic acid
D-LD	(D,D)-lactide
DMAP	4-dimethylaminopyridine
DOX	1,3-dioxolane-4-one
DPP	Diphenyl phosphate
ΔS_p	Entropy of polymerization
δ-VL	δ-valerolactone
DXO	1,5-dioxepan-2-one
ε-CL	ε-caprolactone
eROP	Enzymatic ring-opening polymerization
G	Glycolide
GA	Glycolic acid
γ-BL	γ-butyrolactone
γ-VL	γ-valerolactone
LA	Lactic acid
LD	Lactide
L-LA	L-lactic acid
L-LD	(L,L)-lactide

Abbreviation	Full description
$[M]$	Monomer concentration
$[M]_0$	Initial monomer concentration
$[M]_{eq}$	Equilibrium monomer concentration
Meso-LD = (D,L)-LD	(D,L)-lactide
MM	Molar mass
M_n	Number-average molar mass
M_w	Weight-average molar mass
NHC	N-heterocyclic carbene
PCL	Poly(ε-caprolactone)
PDLA	Poly(D-lactic acid)
PDLLA	Poly((D,L)-lactic acid)
PDX	*p*-dioxanone or 1,4-dioxane-2-one
PGA	Poly(glycolic acid)
PHA	Poly(hydroxyalkanoate)
PHB	Poly(3-hydroxybutyrate)
PLA	Poly(lactic acid)
PLGA	Poly(lactic acid-*co*-glycolic acid)
PLLA	Poly(L-lactic acid)
Poly(β-BL)	Poly(β-butyrolactone)
Poly(β-PL)	Poly(β-propiolactone)
Poly(γ-BL)	Poly(γ-butyrolactone)
Poly(δ-VL)	Poly(δ-valerolactone)
PPY	4-pyrrolidinopyridine
ROCOP	Ring-opening copolymerization
ROP	Ring-opening polymerization
(*R*)-PHB	Poly((*R*)-3-hydroxybutyrate)
$Sn(Oct)_2$	Sn(II) 2-ethylhexanoate
TBD	1,5,7-triazabicyclo[4.4.0]dec-5-ene
T_c	Critical temperature
T_g	Glass transition temperature
T_m	Melting temperature

References

1 Leuchs, H. (1906). Ueber die Glycin-carbonsäure. *Berichte der Deutschen Chemischen Gesellschaft* 39 (1): 857–861.

2 Pictet, A. (1918). Sur la transformation de la lévoglucosane en dextrine. *Helvetica Chimica Acta* 1 (1): 226–230.

3 Carothers, W.H., Dorough, G.L., and Natta, F.J. (1932). Studies of polymerization and ring formation. X. The reversible polymerization of six-membered cyclic esters. *Journal of the American Chemical Society* 54 (2): 761–772.

4 Lowe, C.E. (1954). Preparation of high molecular weight polyhydroxyacetic ester. U.S. Patent 2, 668, 162, issued 1954.

5 Masutani, K. and Kimura, Y. (2014). PLA synthesis and polymerization. In: *Poly(Lactic Acid) Science and Technology: Processing, Properties, Additives and Applications* (ed. A. Jiménez, M. Peltzer and R. Ruseckaite), 1–36. RSC Publishing.

6 Nuyken, O. and Pask, S.D. (2013). Ring-opening polymerization – an introductory review. *Polymers* 5 (2): 361–403.

7 Su, W.-F. (2013). Ring-opening polymerization. In: *Principles of Polymer Design and Synthesis* (ed. W.-F. Su), 267–299. Berlin/Heidelberg: Springer-Verlag.

8 Takeshi, E. (2009). General mechanisms in ring-opening polymerization. In: *Handbook of Ring-Opening Polymerization*, 53–64.

9 Lapienis, G. (2012). *Ring-Opening Polymerization of Cyclic Phosphorus Monomers*. Elsevier B.V.

10 Duda, A. and Kowalski, A. (2009). Thermodynamics and kinetics of ring-opening polymerization. In: *Handbook of Ring-Opening Polymerization* (ed. P. Dubois, O. Coulembier and J.-M. Raquez), 1–51. Wiley.

11 Lecomte, P. and Jérôme, C. (2012). Recent developments in ring-opening polymerization of lactones. *Synthetic Biodegradable Polymers* 187–217.

12 Dubois, P., Coulembier, O., and Raquez, J. (2009). *Handbook of Ring-Opening Polymerization*. Wiley.

13 Tardy, A., Nicolas, J., Gigmes, D. et al. (2017). Radical ring-opening polymerization: scope, limitations, and application to (bio)degradable materials. *Chemical Reviews* 117 (3): 1319–1406.

14 Pesenti, T. and Nicolas, J. (2020). Degradable polymers from radical ring-opening polymerization: latest advances, new directions, and ongoing challenges. *ACS Macro Letters* 9 (12): 1812–1835.

15 Nowalk, J.A., Fang, C., Short, A.L. et al. (2019). Sequence-controlled polymers through entropy-driven ring-opening metathesis polymerization: theory, molecular weight control, and monomer design. *Journal of the American Chemical Society* 141 (14): 5741–5752.

16 Gutekunst, W.R. and Hawker, C.J. (2015). A general approach to sequence-controlled polymers using macrocyclic ROMP. *Journal of the American Chemical Society* 137 (25): 8038–8041.

17 Boadi, F.O., Zhang, J., Yu, X. et al. (2020). Alternating ring-opening metathesis polymerization provides easy access to functional and fully degradable polymers. *Macromolecules* 53 (14): 5857–5868.

18 Araceli, M., Tlenkopatchev, M.A., and Gutiérrez, S. (2018). The unsaturated polyester via ring-opening metathesis polymerization (ROMP) of ω-6-hexadecenelactone. *Current Organic Synthesis* 15 (4): 566–571.

19 Makiguchi, K., Satoh, T., and Kakuchi, T. (2011). Diphenyl phosphate as an efficient cationic organocatalyst for controlled/living ring-opening polymerization of δ-valerolactone and ε-caprolactone. *Macromolecules* 44 (7): 1999–2005.

20 Delcroix, D., Couffin, A., Susperregui, N. et al. (2011). Phosphoric and phosphoramidic acids as bifunctional catalysts for the ring-opening polymerization of ε-caprolactone: a combined experimental and theoretical study. *Polymer Chemistry* 2 (10): 2249–2256.

21 Bednarek, M., Brzezińska, K., Stasiński, J. et al. (1989). Heteropolyacids – new efficient initiators of cationic polymerization. *Die Makromolekulare Chemie* 190 (5): 929–938.

22 Fèvre, M., Vignolle, J., Gnanou, Y., and Taton, D. (2012). Organocatalyzed ring-opening polymerizations. In: *Polymer Science: A Comprehensive Reference*, vol. 4 (ed. K. Matyjaszewski and M. Möller), 67–115. Elsevier B.V.

23 Kaluzynski, K., Lewinski, P., Pretula, J. et al. (2019). ε-Caprolactone polymerization catalyzed by heteropolyacid. Derivation of the kinetic equation for activated monomer propagation and determination of the rate constants of propagation. *Macromolecules* 52: 6405–6415.

24 Goethals, E.J. and Penczek, S. (1989). Cationic ring-opening polymerization: introduction and general aspects. *Comprehensive Polymer Science: Supplement* 3 (1989): 711–717.

25 Penczek, S., Duda, A., Kubisa, P., and Slomkowski, S. (2007). Ionic and coordination ring-opening polymerization. In: *Macromolecular Engineering. Precise Synthesis, Materials Properties, Applications* (ed. K. Matyjaszewski, Y. Gnanou and L. Leibler), 103–159. Wiley.

26 Sudo, A. (2013). Anionic ring-opening polymerization. In: *Encyclopedia of Polymeric Nanomaterials* (ed. S. Kobayashi and K. Müllen), 33–43. Berlin/Heidelberg: Springer.

27 Dove, A.P. (2012). Organic catalysis for ring-opening polymerization. *ACS Macro Letters* 1 (12): 1409–1412.

28 Jérôme, C. and Lecomte, P. (2008). Recent advances in the synthesis of aliphatic polyesters by ring-opening polymerization. *Advanced Drug Delivery Reviews* 60 (9): 1056–1076.

29 Platel, R.H., Hodgson, L.M., and Williams, C.K. (2008). Biocompatible initiators for lactide polymerization. *Polymer Reviews* 48 (1): 11–63.

30 Lyubov, D.M., Tolpygin, A.O., and Trifonov, A.A. (2019). Rare-earth metal complexes as catalysts for ring-opening polymerization of cyclic esters. *Coordination Chemistry Reviews* 392: 83–145.

31 Sutar, A.K., Maharana, T., Dutta, S. et al. (2010). Ring-opening polymerization by lithium catalysts: an overview. *Chemical Society Reviews* 39 (5): 1724–1746.

32 Wu, J., Yu, T.L., Chen, C.T., and Lin, C.C. (2006). Recent developments in main group metal complexes catalyzed/initiated polymerization of lactides and related cyclic esters. *Coordination Chemistry Reviews* 250 (5, 6): 602–626.

33 Gao, J., Zhu, D., Zhang, W. et al. (2019). Recent progress in the application of group 1, 2 & 13 metal complexes as catalysts for the ring opening polymerization of cyclic esters. *Inorganic Chemistry Frontiers* 6 (10): 2619–2652.

34 Harinath, A., Bhattacharjee, J., Sarkar, A. et al. (2018). Ring opening polymerization and copolymerization of cyclic esters catalyzed by group 2 metal complexes supported by functionalized P–N ligands. *Inorganic Chemistry* 57 (5): 2503–2516.

35 He, J.X., Duan, Y.L., Kou, X. et al. (2015). Dinuclear group 4 alkoxides: excellent initiators for ring-opening polymerization of cyclic esters. *Inorganic Chemistry Communications* 61: 144–148.

36 Chmura, A.J., Davidson, M.G., Jones, M.D. et al. (2006). Group 4 complexes with aminebisphenolate ligands and their application for the ring opening polymerization of cyclic esters. *Macromolecules* 39 (21): 7250–7257.

37 Sauer, A., Kapelski, A., Fliedel, C. et al. (2013). Structurally well-defined group 4 metal complexes as initiators for the ring-opening polymerization of lactide monomers. *Dalton Transactions* 42 (25): 9007–9023.

38 Jianming, R., Anguo, X., Hongwei, W., and Hailin, Y. (2014). Review – recent development of ring-opening polymerization of cyclic esters using aluminum complexes. *Designed Monomers and Polymers* 17 (4): 345–355.

39 Dagorne, S., Normand, M., Kirillov, E., and Carpentier, J.F. (2013). Gallium and indium complexes for ring-opening polymerization of cyclic ethers, esters and carbonates. *Coordination Chemistry Reviews* 257 (11, 12): 1869–1886.

40 Engel, J., Cordellier, A., Huang, L., and Kara, S. (2019). Enzymatic ring-opening polymerization of lactones: traditional approaches and alternative strategies. *ChemCatChem* 11 (20): 4983–4997.

41 Duda, A., Kowalski, A., Penczek, S. et al. (2002). Kinetics of the ring-opening polymerization of 6-, 7-, 9-, 12-, 13-, 16-, and 17-membered lactones. Comparison of chemical and enzymatic polymerizations. *Macromolecules* 35: 4266–4270.

42 Olsén, P., Odelius, K., and Albertsson, A.C. (2016). Thermodynamic presynthetic considerations for ring-opening polymerization. *Biomacromolecules* 17 (3): 699–709.

43 Sawada, H. (1969). Thermodynamics of polymerization I. *Journal of Molecular Science Part C Polymer Reviews* 3 (2): 313–338.

44 Penczek, S. (2002). Terminology of kinetics, thermodynamics, and mechanisms of polymerization. *Journal of Polymer Science Part A: Polymer Chemistry* 40: 1665–1676.

45 Chen, M., Zhang, Y., Zhou, Y. et al. (2015). Pendant small functional groups on poly(ε-caprolactone) substrate modulate adhesion, proliferation and differentiation of human mesenchymal stem cells. *Colloids and Surfaces B: Biointerfaces* 134: 322–331.

46 Zhang, Y., Li, J., Du, Z., and Lang, M. (2014). Synthesis and pH-responsive assembly of methoxy poly(ethylene glycol)-*b*-poly(ε-caprolactone) with pendant

carboxyl groups. *Journal of Polymer Science Part A: Polymer Chemistry* 52 (2): 188–199.

47 Riva, R., Schmeits, S., Jérôme, C. et al. (2007). Combination of ring-opening polymerization and "click chemistry": toward functionalization and grafting of poly(ε-caprolactone). *Macromolecules* 40 (4): 796–803.

48 Lecomte, P. and Jerome, C. (2009). Recent advances in the functionalization of aliphatic polyesters by ring-opening polymerization. In: *New Smart Materials Via Metal Mediated Macromolecular Engineering* (ed. E. Khosravi, Y. Yagci and Y. Savelyev), 343–360. Springer.

49 Gerhardt, W.W., Noga, D.E., Hardcastle, K.I. et al. (2006). Functional lactide monomers: methodology and polymerization. *Biomacromolecules* 7 (6): 1735–1742.

50 Prime, E.L., Cooper-White, J.J., and Qiao, G.G. (2006). Coupling hydrophilic amine-containing molecules to the backbone of poly(ε-caprolactone). *Australian Journal of Chemistry* 59 (8): 534–538.

51 Ajellal, N., Thomas, C.M., and Carpentier, J.-F. (2009). Functional syndiotactic poly(β-hyoxyalkanoate)s via stereoselective ring-opening copolymerization of rac-β-butyrolactone and rac-allyl-β-buryolactone. *Journal of Polymer Science Part A: Polymer Chemistry* 47 (12): 3177–3189.

52 Guillaume, C., Ajellal, N., Carpentier, J.-F., and Guillaume, S.M. (2011). Boron-functionalized poly (3-hydroxybutyrate)s from hydroboration of poly(allyl-*b*-hydroxyalkanoate)s: synthesis and insights into the microstructure. *Journal of Polymer Science Part A: Polymer Chemistry* 49: 907–917.

53 Sinclair, F., Chen, L., Greenland, B.W., and Shaver, M.P. (2016). Installing multiple functional groups on biodegradable polyesters via post-polymerization olefin cross-metathesis. *Macromolecules* 49: 6826–6834.

54 Wu, J., Ding, C., Xing, D. et al. (2019). The functionalization of poly(ε-caprolactone) as a versatile platform using ε-(α-phenylseleno) caprolactone as a monomer. *Polymer Chemistry* 10 (28): 3851–3858.

55 Lou, X., Detrembleur, C., and Jérôme, R. (2003). Novel aliphatic polyesters based on functional cyclic (di)esters. *Macromolecular Rapid Communications* 24 (2): 161–172.

56 Miltenberger, K. (2012). Hydroxycarboxylic acids, aliphatic. In: *Ullmann's Encyclopedia of Industrial Chemistry* (ed. B. Elvers and C. Ley), 481–491. Wiley.

57 Getzler, Y.D.Y.L., Mahadevan, V., Lobkovsky, E.B., and Coates, G.W. (2002). Synthesis of β-lactones: a highly active and selective catalyst for epoxide carbonylation. *Journal of the American Chemical Society* 124 (7): 1174–1175.

58 Kramer, J.W., Lobkovsky, E.B., and Coates, G.W. (2006). Practical β-lactone synthesis: epoxide carbonylation at 1 atm. *Organic Letters* 8 (17): 3709–3712.

59 Reichardt, R. and Rieger, B. (2012). Poly(3-hydroxybutyrate) from carbon monoxide. *Synthetic Biodegradable Polymers* 245, 245: 49–90.

60 Watanabe, M., Togo, M., Sanui, K. et al. (1984). Ionic conductivity of polymer complexes formed by poly(β-propiolactone) and lithium perchlorate. *Macromolecules* 17 (12): 2908–2912.

61 Kumagai, Y. and Doi, Y. (1992). Enzymatic degradation of binary blends of microbial poly(3-hydroxybutyrate) with enzymatically active polymers. *Polymer Degradation and Stability* 37 (3): 253–256.

62 Crescenzi, V., Manzini, G., Calzolari, G., and Borri, C. (1972). Thermodynamics of fusion of poly-β-propiolactone and poly-ε-caprolactone. Comparative analysis of the melting of aliphatic polylactone and polyester chains. *European Polymer Journal* 8 (3): 449–463.

63 Shimamura, E., Scandola, M., and Doi, Y. (1994). Microbial synthesis and characterization of poly(3-hydroxybutyrate-*co*-3-hydroxypropionate). *Macromolecules* 27 (16): 4429–4435.

64 Asano, M., Yoshida, M., Fukuzaki, H. et al. (1990). Effect of crystallinity on the in vivo degradation of poly(β-propiolactone) prepared by radiation-induced solid polymerization in organic solvent system at low temperature. *European Polymer Journal* 26 (1): 29–33.

65 Mathisen, T., Lewis, M., and Albertsson, A. (1991). Hydrolytic degradation of nonoriented poly(β-propiolactone). *Journal of Applied Polymer Science* 42 (8): 2365–2370.

66 Furuhashi, Y., Iwata, T., Kimura, Y., and Doi, Y. (2003). Structural characterization and enzymatic degradation of α-, β-, and γ-crystalline forms for poly(β-propiolactone). *Macromolecular Bioscience* 3 (9): 462–470.

67 Cortizo, S.M., Molinuevo, S.M., and Cortizo, A.M. (2008). Biocompatibility and biodegradation of polyester and polyfumarate based-scaffolds for bone tissue engineering. *Journal of Tissue Engineering and Regenerative Medicine* 2 (1): 33–42.

68 Sudesh, K., Abe, H., and Doi, Y. (2000). Synthesis, structure and properties of polyhydroxyalkanoates: biological polyesters. *Progress in Polymer Science* 25 (10): 1503–1555.

69 Lenz, R.W. and Marchessault, R.H. (2005). Bacterial polyesters: biosynthesis, biodegradable plastics and biotechnology. *Biomacromolecules* 6 (1): 1–8.

70 Strong, P., Laycock, B., Mahamud, S. et al. (2016). The opportunity for high-performance biomaterials from methane. *Microorganisms* 4 (1): 11.

71 Shakaroun, R.M., Jéhan, P., Alaaeddine, A. et al. (2020). Organocatalyzed ring-opening polymerization (ROP) of functional β-lactones: new insights into the ROP mechanism and poly(hydroxyalkanoate)s (PHAs) macromolecular structure. *Polymer Chemistry* 10: 2640–2652.

72 Kricheldorf, H.R. and Scharnagl, N. (1989). Polylactones. 17. Anionic polymerization of β-D,L-butyrolactone. *Journal of Macromolecular Science: Part A – Chemistry* 26 (7): 951–968.

73 Duda, A. (1992). Anionic polymerization of 4-methyl-2-oxetanone. *Journal of Polymer Science Part A: Polymer Chemistry* 30: 21–29.

74 Hofman, A., Słomkowski, S., and Penczek, S. (1984). Structure of active centers and mechanism of the anionic polymerization of lactones. *Die Makromolekulare Chemie* 185 (1): 91–101.

75 Grobelny, Z., Golba, S., and Jurek-Suliga, J. (2019). Ring-opening polymerization of β-butyrolactone in the presence of alkali metal salts: investigation of initiation course and determination of polymers structure by MALDI-TOF mass spectrometry. *Polymer Bulletin* 76 (10): 4951–4966.

76 Jedlinski, Z., Kowalczuk, M., and Kurcok, P. (1991). What is the real mechanism of anionic polymerization of β-lactones by potassium alkoxides? A critical approach. *Macromolecules* 24 (5): 1218–1219.

77 Slomokowski, S. and Penczek, S. (1976). Influence of dibenzo-18-crown-6 ether on the kinetics of anionic polymerization of β-propiolactone. *Macromolecules* 9 (2): 367–369.

78 Zintl, M., Molnar, F., Urban, T. et al. (2008). Variably isotactic poly(hydroxybutyrate) from racemic β-butyrolactone: microstructure control by achiral chromium(III) salophen complexes. *Angewandte Chemie International Edition* 47 (18): 3458–3460.

79 Kemnitzer, J.E., McCarthy, S.P., and Gross, R.A. (1993). Syndiospecific ring-opening polymerization of β-butyrolactone to form predominantly syndiotactic poly(β-hydroxybutyrate) using tin(IV) catalysts. *Macromolecules* 26 (23): 6143–6150.

80 Kricheldorf, H.R. and Eggerstedt, S. (1997). Polylactones. 41. Polymerizations of β-D,L-butyrolactone with dialkyltinoxides as initiators. *Macromolecules* 30 (19): 5693–5697.

81 Kramer, J.W. and Coates, G.W. (2008). Fluorinated β-lactones and poly(β-hydroxyalkanoate)s: synthesis via epoxide carbonylation and ring-opening polymerization. *Tetrahedron* 64 (29): 6973–6978.

82 Guillaume, C., Carpentier, J.F., and Guillaume, S.M. (2009). Immortal ring-opening polymerization of β-butyrolactone with zinc catalysts: catalytic approach to poly(3-hydroxyalkanoate). *Polymer* 50 (25): 5909–5917.

83 Jaffredo, C.G., Carpentier, J.F., and Guillaume, S.M. (2013). Poly(hydroxyalkanoate) block or random copolymers of β-butyrolactone and benzyl β-malolactone: a matter of catalytic tuning. *Macromolecules* 46 (17): 6765–6776.

84 Ebrahimi, T., Aluthge, D.C., Hatzikiriakos, S.G., and Mehrkhodavandi, P. (2016). Highly active chiral zinc catalysts for immortal polymerization of β-butyrolactone form melt processable syndio-rich poly(hydroxybutyrate). *Macromolecules* 49 (23): 8812–8824.

85 García-Valle, F.M., Tabernero, V., Cuenca, T. et al. (2018). Biodegradable PHB from rac-β-butyrolactone: highly controlled ROP mediated by a pentacoordinated aluminum complex. *Organometallics* 37 (6): 837–840.

86 Xu, C., Yu, I., and Mehrkhodavandi, P. (2012). Highly controlled immortal polymerization of β-butyrolactone by a dinuclear indium catalyst. *Chemical Communications* 48 (54): 6806–6808.

87 Quan, S.M. and Diaconescu, P.L. (2015). High activity of an indium alkoxide complex toward ring opening polymerization of cyclic esters. *Chemical Communications* 51 (47): 9643–9646.

88 Osten, K.M. and Mehrkhodavandi, P. (2017). Indium catalysts for ring opening polymerization: exploring the importance of catalyst aggregation. *Accounts of Chemical Research* 50 (11): 2861–2869.

89 Ajellal, N., Bouyahyi, M., Amgoune, A. et al. (2009). Syndiotactic-enriched poly(3-hydroxybutyrate)s via stereoselective ring-opening polymerization of racemic β-butyrolactone with discrete yttrium catalysts. *Macromolecules* 42 (4): 987–993.

90 Bouyahyi, M., Ajellal, N., Kirillov, E. et al. (2011). Exploring electronic versus steric effects in stereoselective ring-opening polymerization of lactide and β-butyrolactone with amino-alkoxy-bis(phenolate)-yttrium complexes. *Chemistry A European Journal* 17 (6): 1872–1883.

91 Fang, J., Tschan, M.J.L., Roisnel, T. et al. (2013). Yttrium catalysts for syndioselective β-butyrolactone polymerization: on the origin of ligand-induced stereoselectivity. *Polymer Chemistry* 4 (2): 360–367.

92 Kiriratnikom, J., Robert, C., Guérineau, V. et al. (2019). Stereoselective ring-opening (co)polymerization of β-butyrolactone and ε-decalactone using an yttrium bis(phenolate) catalytic system. *Frontiers in Chemistry* 7 (May): 1–9.

93 Ajellal, N., Durieux, G., Delevoye, L. et al. (2010). Polymerization of racemic β-butyrolactone using supported catalysts: a simple access to isotactic polymers. *Chemical Communications* 46 (7): 1032–1034.

94 Zhuo, Z., Zhang, C., Luo, Y. et al. (2018). Stereo-selectivity switchable ROP of: rac-β-butyrolactone initiated by salan-ligated rare-earth metal amide complexes: the key role of the substituents on ligand frameworks. *Chemical Communications* 54 (85): 11998–12001.

95 Jaffredo, C.G., Carpentier, J.F., and Guillaume, S.M. (2013). Organocatalyzed controlled ROP of β-lactones towards poly(hydroxyalkanoate)s: from β-butyrolactone to benzyl β-malolactone polymers. *Polymer Chemistry* 4 (13): 3837–3850.

96 Barouti, G., Jarnouen, K., Cammas-Marion, S. et al. (2015). Polyhydroxyalkanoate-based amphiphilic diblock copolymers as original biocompatible nanovectors. *Polymer Chemistry* 6 (30): 5414–5429.

97 Khalil, A., Cammas-Marion, S., and Coulembier, O. (2019). Organocatalysis applied to the ring-opening polymerization of β-lactones: a brief overview. *Journal of Polymer Science Part A: Polymer Chemistry* 57 (6): 657–672.

98 Jaffredo, C.G. and Guillaume, S.M. (2014). Benzyl β-malolactonate polymers: a long story with recent advances. *Polymer Chemistry* 5 (14): 4168–4194.

99 Braud, C., Bunel, C., and Vert, M. (1985). Poly(β-malic acid) : a new polymeric drug-carrier - evidence for degradation in vitro. *Polymer Bulletin* 13 (4): 293–299.

100 Braud, C. and Vert, M. (1992). Degradation of poly(β-malic acid) – monitoring of oligomers formation by aqueous SEC and HPCE. *Polymer Bulletin* 29 (1, 2): 177–183.

101 Vert, M. and Lenz, R.W. (1979). Preparation and properties of poly-β-malic acid: a functional polyester of potential biomedical importance. *American Chemical Society, Division of Polymer Chemistry Preprints* 20 (1): 608–611.

102 Johns, D.B. and Lenz, R.W. (1986). Poly(malic acid) Part I – Preparation and polymerization of benzyl malolactonate. *Journal of Bioactive and Compatible Polymers* 1: 47–60.

103 Guerin, P., Vert, M., Braud, C., and Lenz, R.W. (1985). Optically active poly (β-malic-acid). *Polymer Bulletin* 14 (2): 187–192.

104 Guérin, P., Francilette, J., Braud, C., and Vert, M. (1986). Benzyl esters of optically active malic acid stereocopolymers as obtained by ring-opening polymerization of (R)-(+) and (S)-(−)-benzyl malolactonates. *Die Makromolekulare Chemie* 6 (1): 305–314.

105 Arnold, S.C. and Lenz, R.W. (1986). Synthesis of stereoregular poly(alkyl malolactonates). *Die Makromolekulare Chemie* 6: 285–303.

106 Gammas-Marion, S. and Guérin, P. (2000). 4-Alkyloxycarbonyl-2-oxetanones and 3-alkyl-4-alkyloxycarbonyl-2-oxetanones as versatile chiral precursors in the design of functionalized polyesters with a controlled architecture. *Designed Monomers and Polymers* 3 (1): 77–93.

107 Ligny, R., Hänninen, M.M., Guillaume, S., and Carpentier, J.-F. (2017). Highly syndiotactic to isotactic polyhydroxyalkanoates via ligand-controlled yttrium-catalyzed stereoselective ring-opening polymerization of functional racemic β-lactones. *Angewandte Chemie International Edition* 56: 10388–10393.

108 Ligny, R., Hänninen, M.M., Guillaume, S.M., and Carpentier, J.F. (2018). Steric vs. electronic stereocontrol in syndio- or iso-selective ROP of functional chiral β-lactones mediated by achiral yttrium-bisphenolate complexes. *Chemical Communications* 54 (58): 8024–8031.

109 Ligny, R., Guillaume, S.M., and Carpentier, J.F. (2019). Yttrium-mediated ring-opening copolymerization of oppositely configurated 4-alkoxymethylene-β-propiolactones: effective access to highly alternated isotactic functional PHAs. *Chemistry A European Journal* 25 (25): 6412–6424.

110 Xu, Y.-C., Zhou, H., Sun, X.-Y. et al. (2016). Crystalline polyesters from CO_2 and 2-butyne via α-methylene-β-butyrolactone intermediate. *Macromolecules* 49: 5782–5787.

111 Xu, Y., Ren, W., Zhou, H. et al. (2017). Functionalized polyesters with tunable degradability prepared by controlled ring-opening (co)polymerization of lactones. *Macromolecules* 50: 3131–3142.

112 Raycraft, B.M., MacDonald, J.P., McIntosh, J.T. et al. (2017). Post-polymerization functionalization of poly(ethylene oxide)-poly(b-6-heptenolactone) diblock copolymers to tune properties and self-assembly. *Polymer Chemistry* 8: 557–567.

113 Lebedev, B.V. (1996). Thermodynamics of polylactones. *Russian Chemical Reviews* 65 (12): 1063–1082.

114 Saiyasombat, W., Molloy, R., Nicholson, T.M. et al. (1998). Ring strain and polymerizability of cyclic esters. *Polymer* 39 (23): 5581–5585.

115 Bozell, J.J. and Petersen, G.R. (2010). Technology development for the production of biobased products from biorefinery carbohydrates – The US Department of Energy's "Top 10" revisited. *Green Chemistry* 12 (4): 539–554.

116 Martin, D.P. and Williams, S.F. (2003). Medical applications of poly-4-hydroxybutyrate: a strong flexible absorbable biomaterial. *Biochemical Engineering Journal* 16 (2): 97–105.

117 Moore, T., Adhikari, R., and Gunatillake, P. (2005). Chemosynthesis of bioresorbable poly(γ-butyrolactone) by ring-opening polymerisation: a review. *Biomaterials* 26 (18): 3771–3782.

118 Utsunomia, C., Ren, Q., and Zinn, M. (2020). Poly(4-hydroxybutyrate): current state and perspectives. *Frontiers in Bioengineering and Biotechnology* 8 (April): 1–18.

119 Korte, F. and Glet, W. (1966). Hochdruckreaktionen. II. Die polymerisation von γ-butyrolacton und δ- valerolactam bei hohen drücken. *Journal of Polymer Science Part B: Polymer Letters* 4 (10): 685–689.

120 Nobes, G.A.R., Kazlauskas, R.J., and Marchessault, R.H. (1996). Lipase-catalyzed ring-opening polymerization of lactones: a novel route to poly(hydroxyalkanoate)s. *Macromolecules* 29 (14): 4829–4833.

121 Dong, H., Da Wang, H., Cao, S.G., and Shen, J.C. (1998). Lipase-catalyzed polymerization of lactones and linear hydroxyesters. *Biotechnology Letters* 20 (10): 905–908.

122 Kadokawa, J., Iwasaki, Y., and Tagaya, H. (2002). Ring-opening polymerization of lactones catalyzed by ion-exchanged clay montmorillonite. *Green Chemistry* 4 (1): 14–16.

123 Oishi, A., Taguchi, Y., Fujita, K., et al. (2000). Production of poly-γ-butyrolactone. JP2000281767, issued 2000.

124 Oishi, A., Taguchi, Y., and Fujita, K. (2003). Production method of poly(γ-butyrolactone) using metal complex catalyst. JP2003252968, issued 2003.

125 Yamashita, K., Yamamoto, K., and Kadokawa, J.I. (2014). Acid-catalyzed ring-opening polymerization of γ-butyrolactone under high-pressure conditions. *Chemistry Letters* 43 (2): 213–215.

126 Gorke, J.T., Okrasa, K., Louwagie, A. et al. (2007). Enzymatic synthesis of poly(hydroxyalkanoates) in ionic liquids. *Journal of Biotechnology* 132 (3): 306–313.

127 Hong, M. and Chen, E.Y.X. (2016). Completely recyclable biopolymers with linear and cyclic topologies via ring-opening polymerization of γ-butyrolactone. *Nature Chemistry* 8 (1): 42–49.

128 Myers, D., Cyriac, A., and Williams, C.K. (2016). To react the impossible ring. *Nature Chemistry* 8 (1): 3–4.

129 Hong, M. and Chen, E.Y.X. (2016). Towards truly sustainable polymers: a metal-free recyclable polyester from biorenewable non-strained γ-butyrolactone. *Angewandte Chemie International Edition* 55 (13): 4188–4193.

130 Zhao, N., Ren, C., Li, H. et al. (2017). Selective ring-opening polymerization of non-strained γ-butyrolactone catalyzed by a cyclic trimeric phosphazene base. *Angewandte Chemie International Edition* 56 (42): 12987–12990.

131 Zhang, C.J., Hu, L.F., Wu, H.L. et al. (2018). Dual Organocatalysts for highly active and selective synthesis of linear poly(γ-butyrolactone)s with high molecular weights. *Macromolecules* 51 (21): 8705–8711.

132 Lin, L., Han, D., Qin, J. et al. (2018). Nonstrained γ-butyrolactone to high-molecular-weight poly(γ-butyrolactone): facile bulk polymerization using economical ureas/alkoxides. *Macromolecules* 51 (22): 9317–9322.

133 Shen, Y., Zhao, Z., Li, Y. et al. (2019). A facile method to prepare high molecular weight bio-renewable poly(γ-butyrolactone) using a strong base/urea binary synergistic catalytic system. *Polymer Chemistry* 10 (10): 1231–1237.

134 Liu, Y., Wu, J., Hu, X. et al. (2021). Advances, challenges, and opportunities of poly(γ-butyrolactone)-based recyclable polymers. *ACS Macro Letters* 10: 284–296.

135 Shen, Y., Zhang, J., Zhao, N. et al. (2018). Preparation of biorenewable poly(γ-butyrolactone)-*b*-poly(L-lactide) diblock copolyesters via one-pot sequential metal-free ring-opening polymerization. *Polymer Chemistry* 9 (21): 2936–2941.

136 Hong, M., Tang, X., Newell, B.S., and Chen, E.Y.X. (2017). "Nonstrained" γ-butyrolactone-based copolyesters: copolymerization characteristics and composition-dependent (thermal, eutectic, cocrystallization, and degradation) properties. *Macromolecules* 50 (21): 8469–8479.

137 Qiu, H., Shah, M.I., Bai, T. et al. (2020). Zwitterionic copolymerization of γ-butyrolactone with 3,3-bis(chloromethyl) oxacyclobutane catalyzed by scandium triflates. *Polymer Chemistry* 11 (11): 1845–1851.

138 Tada, K., Numata, Y., Saegusa, T., and Furukawa, J. (1964). Copolymerization of γ-butyrolactone and β-propiolactone. *Die Makromolekulare Chemie* 77 (1): 220–228.

139 Duda, A., Biela, T., Libiszowski, J. et al. (1998). Block and random copolymers of ε-caprolactone. *Polymer Degradation and Stability* 59 (1–3): 215–222.

140 Nakayama, A., Kawasaki, N., Aiba, S. et al. (1998). Synthesis and biodegradability of novel copolyesters containing γ-butyrolactone units. *Polymer* 39 (5): 1213–1222.

141 Nishiura, M., Hou, Z., Koizumi, T.A. et al. (1999). Ring-opening polymerization and copolymerization of lactones by samarium(II) aryloxide complexes. *Macromolecules* 32 (25): 8245–8251.

142 Hori, Y., Yamaguchi, A., and Hagiwara, T. (1995). Chemical synthesis of high molecular weight poly(3-hydroxybutyrate-*co*-4-hydroxybutyrate). *Polymer* 36 (24): 4703–4705.

143 Duda, A., Penczek, S., Dubois, P. et al. (1996). Oligomerization and copolymerization of γ-butyrolactone - a monomer known as unable to homopolymerize, 1: copolymerization with ε-caprolactone. *Macromolecular Chemistry and Physics* 197 (4): 1273–1283.

144 Shen, Y., Zhang, J., Zhao, Z. et al. (2019). Preparation of amphiphilic poly(ethylene glycol)-*b*-poly(γ-butyrolactone) diblock copolymer via ring opening polymerization catalyzed by a cyclic trimeric phosphazene base or alkali alkoxide. *Biomacromolecules* 20 (1): 141–148.

145 Song, Q., Pascouau, C., Zhao, J. et al. (2020). Ring-opening polymerization of γ-lactones and copolymerization with other cyclic monomers. *Progress in Polymer Science* 110.

146 Lima, C.G.S., Monteiro, J.L., de Melo Lima, T. et al. (2018). Angelica lactones: from biomass-derived platform chemicals to value-added products. *ChemSusChem* 11 (1): 25–47.

147 Chen, T., Qin, Z., Qi, Y. et al. (2011). Degradable polymers from ring-opening polymerization of α-angelica lactone, a five-membered unsaturated lactone. *Polymer Chemistry* 2 (5): 1190–1194.

148 Gümüştaş, S., Balcan, M., and Kinal, A. (2018). Computational determination of ring opening polymerization reaction mechanism of α-angelica lactone. *Computational & Theoretical Chemistry* 1142: 1–8.

149 Kaygorodov, K.L., Tarabanko, V.E., and Tarabanko, N. (2018). Thermodynamics of α-angelicalactone polymerization. *Cogent Chemistry* 4 (1): 1–10.

150 Lee, C.W., Urakawa, R., and Kimura, Y. (1997). Copolymerization of γ-valerolactone and β-butyrolactone. *Macromolecular Chemistry and Physics* 198 (4): 1109–1120.

151 Gagliardi, M., Di Michele, F., Mazzolai, B., and Bifone, A. (2015). Chemical synthesis of a biodegradable PEGylated copolymer from ε-caprolactone and γ-valerolactone: evaluation of reaction and functional properties. *Journal of Polymer Research* 22 (2): 6–12.

152 Gagliardi, M. and Bifone, A. (2018). Ring-opening copolymerization thermodynamics and kinetics of γ-valerolactone/ε-caprolactone. *PLoS One* 13 (6): 1–15.

153 Hoffmann, H.M.R. and Rabe, J. (1985). Synthesis and biological activity of α-methylene-γ-butyrolactones. *Angewandte Chemie International Edition English* 24 (2): 94–110.

154 Kitson, R.R.A., Millemaggi, A., and Taylor, R.J.K. (2009). The renaissance of α-methylene-γ-butyrolactones: new synthetic approaches. *Angewandte Chemie International Edition* 48 (50): 9426–9451.

155 Agarwal, S., Jin, Q., and Maji, S. (2012). Biobased polymers from plant-derived Tulipalin A. *ACS Symposium Series* 1105: 197–212.

156 Gowda, R.R. and Chen, E.Y.-X. (2014). Sustainable polymers from biomass-derived α-methylene-γ-butyrolactones. *Encyclopedia of Polymer Science and Technology* 8: 235–271.

157 Gowda, R.R. and Chen, E.Y.X. (2016). Organocatalytic and chemoselective polymerization of multivinyl-functionalized γ-butyrolactones. *ACS Macro Letters* 5 (6): 772–776.

158 Zhou, J., Schmidt, A.M., and Ritter, H. (2010). Bicomponent transparent polyester networks with shape memory effect. *Macromolecules* 43 (2): 939–942.

159 Hong, M. and Chen, E.Y.X. (2014). Coordination ring-opening copolymerization of naturally renewable α-methylene-γ-butyrolactone into unsaturated polyesters. *Macromolecules* 47 (11): 3614–3624.

160 Danko, M., Basko, M., Ďurkáčová, S. et al. (2018). Functional polyesters with pendant double bonds prepared by coordination-insertion and cationic ring-opening copolymerizations of ε-caprolactone with renewable Tulipalin A. *Macromolecules* 51 (10): 3582–3596.

161 Tang, X., Hong, M., Falivene, L. et al. (2016). The quest for converting biorenewable bifunctional α-methylene-γ-butyrolactone into degradable and recyclable polyester: controlling vinyl-addition/ring-opening/cross-linking pathways. *Journal of the American Chemical Society* 138 (43): 14326–14337.

162 Zhao, Z., Shen, Y., Kou, X. et al. (2020). Organocatalytic ring-opening copolymerization of biorenewable α-methylene-γ-butyrolactone toward functional copolyesters: preparation and composition dependent thermal properties. *Macromolecules* 53 (9): 3380–3389.

163 Undin, J., Olsén, P., Godfrey, J. et al. (2016). Controlled copolymerization of the functional 5-membered lactone monomer, α-bromo-γ-butyrolactone, via selective organocatalysis. *Polymer* 87: 17–25.

164 Olsén, P., Undin, J., Odelius, K., and Albertsson, A.C. (2014). Establishing α-bromo-γ-butyrolactone as a platform for synthesis of functional aliphatic polyesters-bridging the gap between ROP and SET-LRP. *Polymer Chemistry* 5 (12): 3847–3854.

165 Gao, C., Tsou, C.H., Zeng, C.Y. et al. (2018). Organocatalyzed ring-opening copolymerization of α-bromo-γ-butyrolactone with ε-caprolactone for the synthesis of functional aliphatic polyesters – pre-polymers for graft copolymerization. *Designed Monomers and Polymers* 21 (1): 193–201.

166 Cairns, S.A., Schultheiss, A., and Shaver, M.P. (2017). A broad scope of aliphatic polyesters prepared by elimination of small molecules from sustainable 1,3-dioxolan-4-ones. *Polymer Chemistry* 8 (19): 2990–2996.

167 Martin, R.T., Camargo, L.P., and Miller, S.A. (2014). Marine-degradable polylactic acid. *Green Chemistry* 16 (4): 1768–1773.

168 Hyoi, K., Kanazawa, A., and Aoshima, S. (2019). Cationic ring-opening co- and terpolymerizations of lactic acid-derived 1,3-dioxolan-4-ones with oxiranes and vinyl ethers: nonhomopolymerizable monomer for degradable co- and terpolymers. *ACS Macro Letters* 8 (2): 128–133.

169 Wang, L., Thai, K., and Gravel, M. (2009). NHC-catalyzed ring expansion of oxacycloalkane-2-carboxaldehydes: a versatile synthesis of lactones. *Organic Letters* 11 (4): 891–893.

170 Caretto, A., Noè, M., Selva, M., and Perosa, A. (2014). Upgrading of biobased lactones with dialkylcarbonates. *ACS Sustainable Chemistry & Engineering* 2 (9): 2131–2141.

171 Hara, T., Hatakeyama, M., Kim, A. et al. (2012). Preparation of clay-supported Sn catalysts and application to Baeyer–Villiger oxidation. *Green Chemistry* 14 (3): 771–777.

172 Aubin, M. and Prud'homme, R.E. (1981). Preparation and properties of poly(valerolactone). *Polymer* 22 (9): 1223–1226.

173 Kasuya, K.I., Ohura, T., Masuda, K., and Doi, Y. (1999). Substrate and binding specificities of bacterial polyhydroxybutyrate depolymerases. *International Journal of Biological Macromolecules* 24 (4): 329–336.

174 Kasuya, K.I., Inoue, Y., and Doi, Y. (1996). Adsorption kinetics of bacterial PHB depolymerase on the surface of polyhydroxyalkanoate films. *International Journal of Biological Macromolecules* 19 (1): 35–40.

175 Kim, H., Olsson, J.V., Hedrick, J.L., and Waymouth, R.M. (2012). Facile synthesis of functionalized lactones and organocatalytic ring-opening polymerization. *ACS Macro Letters* 1 (7): 845–847.

176 Hao, J., Kos, P., Zhou, K. et al. (2015). Rapid synthesis of a lipocationic polyester library via ring-opening polymerization of functional valerolactones for efficacious siRNA delivery. *Journal of the American Chemical Society* 137 (29): 9206–9209.

177 Stöhr, O. and Ritter, H. (2014). Networks based on poly(α-methylene-δ-valerolactone-*co*-δ-valerolactone): crosslinking through free-radical vinyl copolymerization. *Macromolecular Chemistry and Physics* 215: 426–430.

178 Stöhr, O. and Ritter, H. (2015). Hyperbranched polyesters based on hydroxyalkyl-lactones via thiol-ene click reaction. *Polymer International* 64: 37–41.

179 Li, X., Li, H., Zhao, Y. et al. (2015). Facile synthesis of well-defined hydrophilic polyesters as degradable poly(ethylene glycol)-like biomaterials. *Polymer Chemistry* 6: 6452–6456.

180 Song, L., Ding, A.-X., Zhang, K.-X. et al. (2017). Degradable polyesters via ring-opening polymerization of functional valerolactones for efficient gene delivery. *Organic & Biomolecular Chemistry* 15: 6567–6574.

181 Xu, T.Q., Yu, Z.Q., and Zhang, X.M. (2019). Recyclable vinyl-functionalized polyesters via chemoselective organopolymerization of bifunctional α-methylene-δ-valerolactone. *Macromolecular Chemistry and Physics* 220 (12): 1–9.

182 Silvers, A.L., Chang, C., and Emrick, T. (2012). Functional aliphatic polyesters and nanoparticles prepared by organocatalysis and orthogonal grafting chemistry. *Journal of Polymer Science Part A: Polymer Chemistry* 50: 3517–3529.

183 Parrish, B., Quansah, J.K., and Emrick, T. (2002). Functional polyesters prepared by polymerization of α-allyl (valerolactone) and its copolymerization with ε-caprolactone and δ-valerolactone. *Journal of Polymer Science Part A: Polymer Chemistry* 40: 1983–1990.

184 Yiu, A., Simchuk, D., and Hao, J. (2020). Facile synthesis of novel thermo-responsive polyvalerolactones with tunable LCSTs. *Macromolecular Chemistry and Physics* 221 (2000136): 1–6.

185 van der Ende, A.E., Harrell, J., Sathiyakumar, V. et al. (2010). "Click" reactions: novel chemistries for forming well-defined polyester nanoparticles. *Macromolecules* 43: 5665–5671.

186 Parrish, B. and Emrick, T. (2004). Aliphatic polyesters with pendant cyclopentene groups: controlled synthesis and conversion to polyester-*graft*-PEG copolymers. *Macromolecules* 37: 5863–5865.

187 Stevens, D.M., Watson, H.A., Leblanc, M.A. et al. (2013). Practical polymerization of functionalized lactones and carbonates with $Sn(OTf)_2$ in metal catalysed ring-opening polymerization methods. *Polymer Chemistry* 4 (8): 2470–2474.

188 Le Devedec, F., Boucher, H., Dubins, D., and Allen, C. (2018). Factors controlling drug release in cross-linked poly(valerolactone) based matrices. *Molecular Pharmaceutics* 15: 1565–1577.

189 Fahnhorst, G.W. and Hoye, T.R. (2018). A carbomethoxylated polyvalerolactone from malic acid: synthesis and divergent chemical recycling. *ACS Macro Letters* 7 (2): 143–147.

190 Fahnhorst, G.W., Stasiw, D.E., Tolman, W.B., and Hoye, T.R. (2018). Isomerization of linear to hyperbranched polymers: two isomeric lactones converge via metastable isostructural polyesters to a highly branched analogue. *ACS Macro Letters* 7 (10): 1144–1148.

191 Fahnhorst, G.W., De Hoe, G.X., Hillmyer, M.A., and Hoye, T.R. (2020). 4-Carboalkoxylated polyvalerolactones from malic acid: tough and degradable polyesters. *Macromolecules* 53: 3194–3201.

192 Tang, M., White, A.J.P., Stevens, M.M., and Williams, C.K. (2009). Biomaterials from sugars: ring-opening polymerization of a carbohydrate lactone. *Chemical Communications* 3 (8): 941–943.

193 Tang, M., Dong, Y., Stevens, M.M., and Williams, C.K. (2010). Tailoring polylactide degradation: copolymerization of a carbohydrate lactone and S,S-lactide. *Macromolecules* 43 (18): 7556–7564.

194 Lee, K.Y., Tang, M., Williams, C.K., and Bismarck, A. (2012). Carbohydrate derived copoly(lactide) as the compatibilizer for bacterial cellulose reinforced polylactide nanocomposites. *Composites Science and Technology* 72 (14): 1646–1650.

195 Tang, M., Purcell, M., Steele, J.A.M. et al. (2013). Porous copolymers of ε-caprolactone as scaffolds for tissue engineering. *Macromolecules* 46 (20): 8136–8143.

196 Bechtold, K., Hillmyer, M.A., and Tolman, W.B. (2001). Perfectly alternating copolymer of lactic acid and ethylene oxide as a plasticizing agent for polylactide. *Macromolecules* 34 (25): 8641–8648.

197 Neitzel, A.E., Petersen, M.A., Kokkoli, E., and Hillmyer, M.A. (2014). Divergent mechanistic avenues to an aliphatic polyesteracetal or polyester from a single cyclic esteracetal. *ACS Macro Letters* 3 (11): 1156–1160.

198 Goonoo, N., Jeetah, R., Bhaw-Luximon, A., and Jhurry, D. (2015). Polydioxanone-based bio-materials for tissue engineering and drug/gene delivery applications. *European Journal of Pharmaceutics and Biopharmaceutics* 97: 371–391.

199 Martins, J.A., Lach, A.A., Morris, H.L. et al. (2020). Polydioxanone implants: a systematic review on safety and performance in patients. *Journal of Biomaterials Applications* 34 (7): 902–916.

200 Lochee, Y., Jhurry, D., Bhaw-Luximon, A., and Kalangos, A. (2010). Biodegradable poly(ester-ether)s: ring-opening polymerization of D,L-3-methyl-1,4-dioxan-2-one using various initiator systems. *Polymer International* 59 (9): 1310–1318.

201 Lochee, Y., Bhaw-Luximon, A., Jhurry, D., and Kalangos, A. (2009). Novel biodegradable poly(ester-ether)s: copolymers from 1,4-dioxan-2-one and D,L-3-methyl-1,4-dioxan-2-one. *Macromolecules* 42 (19): 7285–7291.

202 Jeetah, R., Bhaw-Luximon, A., and Jhurry, D. (2012). New amphiphilic PEG-*b*-P(ester–ether) micelles as potential drug nanocarriers. *Journal of Nanoparticle Research* 14 (10).

203 Goonoo, N., Bhaw-Luximon, A., Bowlin, G.L., and Jhurry, D. (2012). Diblock poly(ester)-poly(ester–ether) copolymers: I. Synthesis, thermal properties, and degradation kinetics. *Industrial and Engineering Chemistry Research* 51 (37): 12031–12040.

204 Neitzel, A.E., Haversang, T.J., and Hillmyer, M.A. (2016). Organocatalytic cationic ring-opening polymerization of a cyclic hemiacetal ester. *Industrial and Engineering Chemistry Research* 55 (45): 11747–11755.

205 Garlotta, D. (2001). A literature review of poly(lactic acid). *Journal of Polymers and the Environment* 9 (2): 63–84.

206 Auras, R., Lim, L.-T., Susan, E.M., and Selke, H.T. (2010). *Poly(Lactic Acid): Synthesis, Structures, Properties, Processing, and Applications*. New York: Wiley.

207 Inkinen, S., Hakkarainen, M., Albertsson, A.-C., and Södergård, A. (2011). From lactic acid to poly(lactic acid) (PLA): characterization and analysis of PLA and its precursors. *Biomacromolecules* 12 (3): 523–532.

208 Kim, K.W. and Woo, S.I. (2002). Synthesis of high-molecular-weight poly (L-lactic acid) by direct polycondensation. *Macromolecular Chemistry and Physics* 203 (15): 2245–2250.

209 Chen, G.-X., Kim, H.-S., Kim, E.-S., and Yoon, J.-S. (2006). Synthesis of high-molecular-weight poly(L-lactic acid) through the direct condensation polymerization of L-lactic acid in bulk state. *European Polymer Journal* 42 (2): 468–472.

210 Maharana, T., Mohanty, B., and Negi, Y.S. (2009). Melt-solid polycondensation of lactic acid and its biodegradability. *Progress in Polymer Science* 34 (1): 99–124.

211 Tuna, B. and Ozkoc, G. (2017). Effects of diisocyanate and polymeric epoxidized chain extenders on the properties of recycled poly(lactic acid). *Journal of Polymers and the Environment* 25 (4): 983–993.

212 Khankrua, R., Pivsa-Art, S., Hiroyuki, H., and Suttiruengwong, S. (2014). Effect of chain extenders on thermal and mechanical properties of poly(lactic acid) at high processing temperatures: potential application in PLA/polyamide 6 blend. *Polymer Degradation and Stability* 108: 232–240.

213 Yoo, D.K. and Kim, D. (2006). Synthesis of lactide from oligomeric PLA: effects of temperature, pressure, and catalyst. *Macromolecular Research* 14 (5): 510–516.

214 De Clercq, R., Dusselier, M., and Sels, B.F. (2017). Heterogeneous catalysis for bio-based polyester monomers from cellulosic biomass: advances, challenges and prospects. *Green Chemistry* 19 (21): 5012–5040.

215 Groot, W., Van Krieken, J., Sliekersl, O., and De Vos, S. (2010). Production and purification of lactic acid and lactide. In: *Poly(Lactic Acid): Synthesis, Structure, Properties, Processing and Applications* (ed. R. Auras, L.-T. Lim, S.E.M. Selke and H. Tsuji), 3–18. Wiley.

216 Gruber, P.R., Hall, E.S., Kolstad, J.J., et al. (1992). Continuous process for manufacture of lactide polymers with controlled optical purity. 5,142,023, issued 1992.

217 Botvin, V., Karaseva, S., Salikova, D., and Dusselier, M. (2021). Syntheses and chemical transformations of glycolide and lactide as monomers for biodegradable polymers. *Polymer Degradation and Stability* 183: 109427–109478.

218 Dusselier, M., Van Wouwe, P., Dewaele, A. et al. (2015). Shape-selective zeolite catalysis for bioplastics production. *Science* 349 (6243): 79–81.

219 De Clercq, R., Dusselier, M., Makshina, E., and Sels, B.F. (2018). Catalytic gas-phase production of lactide from renewable alkyl lactates. *Angewandte Chemie International Edition* 57 (12): 3074–3078.

220 Van Wouwe, P., Dusselier, M., Vanleeuw, E., and Sels, B. (2016). Lactide synthesis and chirality control for polylactic acid production. *ChemSusChem* 9 (9): 907–921.

221 Duda, A. and Penezek, S. (1990). Thermodynamics of L-lactide polymerization. Equilibrium monomer concentration. *Macromolecules* 23 (6): 1636–1639.

222 Dechy-Cabaret, O., Martin-Vaca, B., and Bourissou, D. (2004). Controlled ring-opening polymerization of lactide and glycolide. *Chemical Reviews* 104 (12): 6147–6176.

223 Salusjärvi, L., Havukainen, S., Koivistoinen, O., and Toivari, M. (2019). Biotechnological production of glycolic acid and ethylene glycol: current state and perspectives. *Applied Microbiology and Biotechnology* 103 (6): 2525–2535.

224 Tsuji, H. (2005). Poly(lactide) stereocomplexes: formation, structure, properties, degradation, and applications. *Macromolecular Bioscience* 5 (7): 569–597.

225 Gentile, P., Chiono, V., Carmagnola, I., and Hatton, P.V. (2014). An overview of poly(lactic-*co*-glycolic) acid (PLGA)-based biomaterials for bone tissue engineering. *International Journal of Molecular Sciences* 15 (3): 3640–3659.

226 Schmitt, E.E. (1967). Surgical Sutures. U.S. Patent 3,297,033, issued 1967.

227 Gilding, D.K. and Reed, A.M. (1979). Biodegradable polymers for use in surgery polyglycolic/poly(lactic acid) homo- and copolymers: 1. *Polymer* 20 (12): 1459–1464.

228 Frazza, E.J. and Schmitt, E.E. (1971). A new absorbable suture. *Journal of Biomedical Materials Research* 1: 43–58.

229 Griffith, L.G. and Naughton, G. (2002). Tissue engineering – current challenges and expanding opportunities. *Science* 295 (5557): 1009–1014.

230 Ulery, B.D., Nair, L.S., and Laurencin, C.T. (2011). Biomedical applications of biodegradable polymers. *Journal of Polymer Science Part B: Polymer Physics* 49 (12): 832–864.

231 Essa, D., Kondiah, P.P.D., Choonara, Y.E., and Pillay, V. (2020). The design of poly(lactide-*co*-glycolide) nanocarriers for medical applications. *Frontiers in Bioengineering and Biotechnology* 8 (February): 1–20.

232 Jem, K.J., van der Pol, J.F., and de Vos, S. (2010). Microbial lactic acid, its polymer poly(lactic acid), and their industrial applications. In: *Plastics from Bacteria*, 323–346.

233 Dusselier, M., Van Wouwe, P., Dewaele, A. et al. (2013). Lactic acid as a platform chemical in the biobased economy: the role of chemocatalysis. *Energy & Environmental Science* 6 (5): 1415.

234 European Bioplastics, and nova-Institute (2019). Bioplastics market data 2019 – Global production capacities of bioplastics 2019–2024.

235 Farah, S., Anderson, D.G., and Langer, R. (2016). Physical and mechanical properties of PLA, and their functions in widespread applications – a comprehensive review. *Advanced Drug Delivery Reviews* 107: 367–392.

236 Drumright, R.E., Gruber, P.R., and Henton, D.E. (2000). Polylactic acid technology. *Advanced Materials* 12 (23): 1841–1846.

237 Hirata, M. and Kimura, Y. (2010). Structure and properties of stereocomplex-type poly(lactic acid). In: *Poly(Lactic Acid): Synthesis, Structure, Properties, Processing and Applications* (ed. R. Auras, L.-T. Lim, S.E.M. Selke and H. Tsuji), 59–65. Wiley.

238 Lim, L.T., Auras, R., and Rubino, M. (2008). Processing technologies for poly(lactic acid). *Progress in Polymer Science* 33 (8): 820–852.

239 Dorgan, J. (2010). Rheology of poly(lactic acid). In: *Poly(Lactic Acid): Synthesis, Structure, Properties, Processing and Applications* (ed. R. Auras, L.-T. Lim, E.M. Susan and H.T. Selke), 125–139. New York: Wiley.

240 Nofar, M. and Park, C.B. (2014). Poly (lactic acid) foaming. *Progress in Polymer Science* 39 (10): 1721–1741.

241 Narmon, A.S., Dewaele, A., Bruyninckx, K. et al. (2021). Boosting PLA melt strength by controlling the chirality of co-monomer incorporation. *Chemical Science* 12 (15): 5672–5681.

242 Pounder, R.J. and Dove, A.P. (2010). Towards poly(ester) nanoparticles: recent advances in the synthesis of functional poly(ester)s by ring-opening polymerization. *Polymer Chemistry* 1 (3): 260–271.

243 Becker, G. and Wurm, F.R. (2018). Functional biodegradable polymers: via ring-opening polymerization of monomers without protective groups. *Chemical Society Reviews* 47 (20): 7739–7782.

244 Schneiderman, D.K. and Hillmyer, M.A. (2016). Aliphatic polyester block polymer design. *Macromolecules* 49 (7): 2419–2428.

245 Labet, M. and Thielemans, W. (2009). Synthesis of polycaprolactone: a review. *Chemical Society Reviews* 38 (12): 3484–3504.

246 Pyo, S.H., Park, J.H., Srebny, V., and Hatti-Kaul, R. (2020). A sustainable synthetic route for biobased 6-hydroxyhexanoic acid, adipic acid and ε-caprolactone by integrating bio- and chemical catalysis. *Green Chemistry* 22 (14): 4450–4455.

247 Buntara, T., Noel, S., Phua, P.H. et al. (2011). Caprolactam from renewable resources: catalytic conversion of 5-hydroxymethylfurfural into caprolactone. *Angewandte Chemie International Edition* 123 (31): 7221–7225.

248 Sisson, A.L., Ekinci, D., and Lendlein, A. (2013). The contemporary role of ε-caprolactone chemistry to create advanced polymer architectures. *Polymer* 54 (17): 4333–4350.

249 Woodruff, M.A. and Hutmacher, D.W. (2010). The return of a forgotten polymer – polycaprolactone in the 21st century. *Progress in Polymer Science* 35 (10): 1217–1256.

250 Arakawa, C.K. and DeForest, C.A. (2017). Polymer design and development. In: *Biology and Engineering of Stem Cell Niches*, 295–314. Elsevier Inc.

251 Tokiwa, Y., Calabia, B.P., Ugwu, C.U., and Aiba, S. (2009). Biodegradability of plastics. *International Journal of Molecular Sciences* 10 (9): 3722–3742.

252 Pitt, C.G., Chasalow, F.I., Hibionada, Y.M. et al. (1981). Aliphatic polyesters. I. The degradation of poly(ε-caprolactone) in vivo. *Journal of Applied Polymer Science* 26 (11): 3779–3787.

253 Pitt, G.G., Gratzl, M.M., Kimmel, G.L. et al. (1981). Aliphatic polyesters II. The degradation of poly (DL-lactide), poly (ε-caprolactone), and their copolymers in vivo. *Biomaterials* 2 (4): 215–220.

254 Goldberg, D. (1995). A review of the biodegradability and utility of poly(caprolactone). *Journal of Environmental Polymer Degradation* 3 (2): 61–67.

255 Suzuki, M., Tachibana, Y., and Kasuya, K. (2021). Biodegradability of poly(3-hydroxyalkanoate) and poly(ε-caprolactone) via biological carbon cycles in marine environments. *Polymer Journal* 53 (1): 47–66.

256 Tokiwa, Y. and Suzuki, T. (1977). Hydrolysis of polyesters by lipases. *Nature* 270 (5632): 76–78.

257 Taylor, M.S., Daniels, A.U., Andriano, K.P., and Heller, J. (1994). Six bioabsorbable polymers: in vitro acute toxicity of accumulated degradation products. *Journal of Applied Biomaterials* 5 (2): 151–157.

258 Lee, R.-S., Huang, Y.-T., and Chen, W.-H. (2010). Synthesis and characterization of temperature-sensitive block copolymers from poly(*N*-isopropylacrylamide) and 4-methyl-ε-caprolactone or 4-phenyl-ε-caprolactone. *Journal of Applied Polymer Science* 118 (3): 1634–1642.

259 Wanamaker, C.L., O'Leary, L.E., Lynd, N.A. et al. (2007). Renewable-resource thermoplastic elastomers based on polylactide and polymenthide. *Biomacromolecules* 8 (11): 3634–3640.

260 Lin, J.O., Chen, W., Shen, Z., and Ling, J. (2013). Homo- and block copolymerizations of ε-decalactone with L-lactide catalyzed by lanthanum compounds. *Macromolecules* 46 (19): 7769–7776.

261 Ten Breteler, M.R., Zhong, Z., Dijkstra, P.J. et al. (2007). Ring-opening polymerization of substituted ε-caprolactones with a chiral (salen) AlOiPr complex. *Journal of Polymer Science Part A: Polymer Chemistry* 45 (3): 429–436.

262 Peeters, J., Palmans, A.R.A., Veld, M. et al. (2004). Cascade synthesis of chiral block copolymers combining lipase catalyzed ring opening polymerization and atom transfer radical polymerization. *Biomacromolecules* 5 (5): 1862–1868.

263 Küllmer, K., Kikuchi, H., Uyama, H., and Kobayashi, S. (1998). Lipase-catalyzed ring-opening polymerization of α-methyl-δ-valerolactone and α-methyl-ε-caprolactone. *Macromolecular Rapid Communications* 19 (2): 127–130.

264 Kikuchi, H., Uyama, H., and Kobayashi, S. (2002). Lipase-catalyzed ring-opening polymerization of substituted lactones. *Polymer Journal* 34 (11): 835–840.

265 Peeters, J.W., Van Leeuwen, O., Palmans, A.R.A., and Meijer, E.W. (2005). Lipase-catalyzed ring-opening polymerizations of 4-substituted ε-caprolactones: mechanistic considerations. *Macromolecules* 38 (13): 5587–5592.

266 Zhang, D., Hillmyer, M.A., and Tolman, W.B. (2005). Catalytic polymerization of a cyclic ester derived from a "cool" natural precursor. *Biomacromolecules* 6 (4): 2091–2095.

267 Wilson, J.A., Hopkins, S.A., Wright, P.M., and Dove, A.P. (2016). Dependence of copolymer sequencing based on lactone ring size and ε-substitution. *ACS Macro Letters* 5 (3): 346–350.

268 Olsén, P., Borke, T., Odelius, K., and Albertsson, A.C. (2013). ε-Decalactone: a thermoresilient and toughening comonomer to poly(L-lactide). *Biomacromolecules* 14 (8): 2883–2890.

269 Liu, Q., Jiang, L., Shi, R., and Zhang, L. (2012). Synthesis, preparation, in vitro degradation, and application of novel degradable bioelastomers – a review. *Progress in Polymer Science* 37 (5): 715–765.

270 Ye, C., Wang, J., Xu, Y. et al. (2019). De novo synthesis and assembly of flexible and biocompatible physical sensing platforms. *Advanced Materials Technologies* 4 (1): 1–9.

271 Chen, T., Cai, T., Jin, Q., and Ji, J. (2015). Design and fabrication of functional polycaprolactone. *E-Polymers* 15 (1): 3–13.

272 Lecomte, P., Riva, R., Schmeits, S. et al. (2006). New prospects for the grafting of functional groups onto aliphatic polyesters. Ring-opening polymerization of α- or γ-substituted ε-caprolactone followed by chemical derivatization of the substituents. *Macromolecular Symposia* 240: 157–165.

273 Yao, K., Wang, J., Zhang, W. et al. (2011). Degradable rosin-ester-caprolactone graft copolymers. *Biomacromolecules* 12 (6): 2171–2177.

274 Al-Azemi, T.F. and Mohamod, A.A. (2011). Fluorinated ε-caprolactone: synthesis and ring-opening polymerization of new α-fluoro-ε-caprolactone monomer. *Polymer* 52 (24): 5431–5438.

275 El Habnouni, S., Darcos, V., and Coudane, J. (2009). Synthesis and ring opening polymerization of a new functional lactone, α-iodo- ε-caprolactone: a novel route to functionalized aliphatic polyesters. *Macromolecular Rapid Communications* 30 (3): 165–169.

276 Massoumi, B., Abdollahi, M., Fathi, M. et al. (2013). Synthesis of novel thermoresponsive micelles by graft copolymerization of *N*-isopropylacrylamide on poly(ε-caprolactone-*co*-α-bromo-ε-caprolactone) as macroinitiator via ATRP. *Journal of Polymer Research* 20 (2): 1–8.

277 Detrembleur, C., Mazza, M., Halleux, O. et al. (2000). Ring-opening polymerization of γ-bromo-ε-caprolactone: a novel route to functionalized aliphatic polyesters. *Macromolecules* 33 (1): 14–18.

278 Detrembleur, C., Mazza, M., Lou, X. et al. (2000). New functional aliphatic polyesters by chemical modification of copolymers of ε-caprolactone with γ-(2-bromo-2-methylpropionate)-ε-caprolactone, γ-bromo-ε-caprolactone, and a mixture of β- and γ-ene-ε-caprolactone. *Macromolecules* 33 (21): 7751–7760.

279 Mecerreyes, D., Miller, R.D., Hedrick, J.L. et al. (2000). Ring-opening polymerization of 6-hydroxynon-8-enoic acid lactone. *Journal of Polymer Science Part A: Polymer Chemistry* 38: 870–875.

280 Darcos, V., El Habnouni, S., Nottelet, B. et al. (2010). Well-defined PCL-graft-PDMAEMA prepared by ring-opening polymerisation and click chemistry. *Polymer Chemistry* 1 (3): 280–282.

281 Jazkewitsch, O., Mondrzyk, A., Staffel, R., and Ritter, H. (2011). Cyclodextrin-modified polyesters from lactones and from bacteria: an approach to new drug carrier systems. *Macromolecules* 44 (6): 1365–1371.

282 Lou, X., Detrembleur, C., Lecomte, P., and Jérôme, R. (2001). Living ring-opening (co)polymerization of 6,7-dihydro-2(5*H*)-oxepinone into unsaturated aliphatic polyesters. *Macromolecules* 34 (17): 5806–5811.

283 Hassannia, A., Piercy, G., and Merbouh, N. (2009). A simple and efficient synthesis of 1,4- and 1,5-dioxepan-2-one. *Letters in Organic Chemistry* 6 (6): 478–480.

284 Löfgren, A. and Albertsson, A.C. (1995). Synthesis and characterization of high molecular weight poly(1,5-dioxepan-2-one) with narrow molecular weight distribution. *Polymer* 36 (19): 3753–3759.

285 Ryner, M. and Albertsson, A.C. (2002). Resorbable and highly elastic block copolymers from 1,5-dioxepan-2-one and L-lactide with controlled tensile properties and hydrophilicity. *Biomacromolecules* 3 (3): 601–608.

286 Odelius, K., Plikk, P., and Albertsson, A.C. (2005). Elastomeric hydrolyzable porous scaffolds: copolymers of aliphatic polyesters and a polyether-ester. *Biomacromolecules* 6 (5): 2718–2725.

287 Löfgren, A., Renstad, R., and Albertsson, A.-C. (1995). Synthesis and characterization of a new degradable thermoplastic elastomer based on 1,5-dioxepan-2-one and ε-caprolactone. *Journal of Applied Polymer Science* 55 (11): 1589–1600.

288 Hakkarainen, M., Adamus, G., Hoglund, A. et al. (2008). ESI-MS reveals the influence of hydrophilicity and architecture on the water-soluble degradation product patterns of biodegradable homo- and copolyesters of 1,5-dioxepan-2-one and ε-caprolactone. *Macromolecules* 41 (10): 3547–3554.

289 Stridsberg, K. and Albertsson, A.C. (2000). Changes in chemical and thermal properties of the tri-block copolymer poly(L-lactide-*b*-1,5-dioxepan-2-one-*b*-L-lactide) during hydrolytic degradation. *Polymer* 41 (20): 7321–7330.

290 Woodard, L.N. and Grunlan, M.A. (2018). Hydrolytic degradation and erosion of polyester biomaterials. *ACS Macro Letters* 7 (8): 976–982.

291 Li, J., Stayshich, R.M., and Meyer, T.Y. (2011). Exploiting sequence to control the hydrolysis behavior of biodegradable PLGA copolymers. *Journal of the American Chemical Society* 133 (18): 6910–6913.

292 Poltronieri, P. and Kumar, P. (2017). Polyhydroxyalkanoates (PHAs) in industrial applications. In: *Handbook of Ecomaterials* (ed. L. Martínez, O. Kharissova and B. Kharisov), 1–30. Cham: Springer.

293 Ellen MacArthur Foundation (2016). The New Plastics Economy – rethinking the future of plastics.

8

Recent Developments in Biodegradable Cellulose-Based Plastics

Karin Molenveld and Ted M. Slaghek

Wageningen Food and Biobased Research, Bornse Weilanden 9, 6708 WG Wageningen, The Netherlands

8.1 General Introduction

The abundance of carbohydrates like cellulose and starch, combined with their specific properties, makes them very suited for the development of bio-based and biodegradable plastics. This has been recognized in the past as cellulose esters are among the oldest plastics found in industrial applications. With the development of environmentally benign plastics, including plastics from bio-based origin and plastics that have a lower potential to accumulate in the natural environment, there is a renewed interest in carbohydrate-based plastics. A requirement for plastic behavior of polymers is that the melting temperature be lower than the degradation temperature [1]. Pure (dry), unmodified cellulose does not melt and cellulose needs extensive chemical modification to allow thermal processing. Despite the abundant availability of functional groups, cellulose cannot be easily modified. Cellulose is insoluble in water and most organic solvents. The crystalline structure of cellulose limits accessibility of solvents and reagents, and highly pure cellulose (dissolving pulp) is needed as a raw material [2]. The most commonly used cellulose plastic is cellulose acetate (CA). Its main application is in cigarette filters, and moreover it is used in durable applications where the excellent optical properties, hypoallergenic nature, and high toughness are beneficial. Examples include spectacle frames, tool handles, and liquid crystal displays (LCDs). Cellophane films are used in packaging applications, but these regenerated cellulose materials (hydrolyzed version of a CS_2-modified cellulose) are not true plastics as they do not melt or flow and require different conversion processes. This book chapter summarizes the latest developments in biodegradable cellulose-based plastics with an emphasis on developments from 2015 onwards. The main focus is on applications that apply cellulose as a matrix material and as fiber reinforcement rather than as a filler. Cellulose plastics discussed are typically cellulose esters and cellophane as these are most commonly used in plastic applications.

Biodegradable Polymers in the Circular Plastics Economy, First Edition.
Edited by Michiel Dusselier and Jean-Paul Lange.

8.2 Cellulose

Cellulose is the most abundant polymer on earth. It is estimated that annually 10^{11}–10^{12} tonne cellulose is produced via photosynthesis. It is a structural component of the primary cell wall of plants. It is also found in various algae and produced by some species of bacteria and fungi. Cellulose for industrial use is most commonly obtained from wood pulp and cotton. Plant-derived cellulose is found in a composite structure with hemicellulose, lignin, pectin, and other substances [3]. The cellulose content of cotton fiber is about 90%, whereas that of wood is typically 40–50% [4].

Cellulose is a linear polymer where the glucose units are linked together via an β-1,4 linkage [5], see Figure 8.1. The typical length of cellulose from wood pulp ranges between 300 and 1700 glucose units, whereas cotton displays a degree of polymerization (DP) ranging from 800 to 10 000. Cellulose consists of fibrils with crystalline and amorphous regions. A high crystallinity together with extensive hydrogen bonding explains its stiffness, high tensile strength, and insolubility in water and most organic solvents.

The main industrial uses of cellulose are paperboard, paper, and textiles. The global annual production volume of cellulose pulp has exceeded 400 million tonnes [6]. Only about 2% is used for the production of cellulose derivatives like cellophane, viscose, and cellulose esters [5]. More recent is the development of industrial processes to convert cellulose into biofuels like ethanol, and this boosts cellulose utilization.

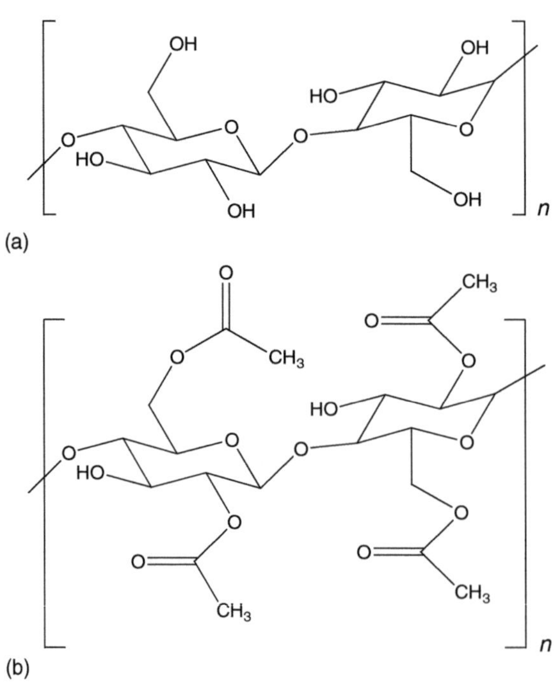

Figure 8.1 Chemical structure of (a) cellulose, (b) cellulose acetate with a degree of substitution of 2.

8.3 The Development of Cellulose Plastics

Cellulose-based plastics are of historical importance, as the development of the plastics industry started with the discovery and production of bio-based plastics like cellulose nitrate, celluloid, viscose, and cellulose acetate [7]. These plastics were used to replace expensive and scarce materials like ivory, and allowed the development and production of cars and aircrafts [8]. The production of cellulose-based plastics originates in the nineteenth century. Cellulose nitrate was first used as guncotton and in 1862 cellulose mononitrate film was patented under the tradename Parkesine. It was produced by treating cellulose with nitric acid and a solvent. A few years later it was discovered that cellulose nitrate could be plasticized with camphor, an oil extracted from wood, to produce celluloid. Celluloid is a highly flammable material that was used in photographic films and lacquers. The most widely used process to manufacture regenerated cellulose (cellophane film and viscose fibers) originates from 1891 when it was discovered that treatment of cellulose with alkali and carbon disulfide generated a soluble cellulose derivative that could be regenerated to cellulose by immersion in a bath containing dilute acid. Regenerated cellulose is chemically identical to cellulose and combines the advantages of natural cellulose (biodegradable, high strength, and stiffness) with the uniformity of man-made materials. The production of cellophane film was patented in 1912 by Brandenberger, and in the same year it was introduced as candy wrapping [9]. Cellulose acetate films, lacquers, and yarns were developed in the early twentieth century. Reacting cellulose with acetic anhydride resulted in the production of cellulose triacetate (CTA) and, via selective partial hydrolysis, cellulose diacetate was produced with an improved solubility in organic solvents which extended application opportunities. The development of new fossil-based plastics produced from building blocks like ethylene, propylene, butadiene, benzene, and toluene started just before the World War II, and cheaper fossil-based plastics replaced most bio-based plastics [10] (see Chapter 4, tutorial on polymers). Still, cellulose-based plastics regained a fair market share based on their specific properties. The increasing demand for bio-based and biodegradable plastics has led to a renewed interest in cellulose-based plastics. An asset of cellulose-based plastics is that they are produced from feedstocks that are not in direct competition with food.

Important topics related to the development and production of cellulose-based plastics include the cellulose source and cellulose purification, the derivatization process, and the extent and type of derivatization. Originally cotton was used as feedstock for the production of cellulose derivatives, the development of new purification and bleaching processes enabled the use of soft wood as the main source. New purification and derivatization processes also aim to reduce the environmental footprint of regenerated celluloses and cellulose derivatives. A well-known example is the development of the Lyocell process for the production of textile fibers [11]. In this process regenerated cellulose fibers are spun from a solution of cellulose in *N*-methyl morpholine *N*-oxide (NMMO). Lyocell fibers are stronger than viscose fibers, also when they are wet due to an improved orientation of cellulose.

8.3.1 Cellulose Feedstock and Dissolving Pulp

At present wood is the most important source for the recovery of cellulose. As wood contains about 15–35% lignin and 20–30% hemicellulose, extensive purification is needed to extract clean cellulose fibers. Most effective are chemical pulping processes that rely on chemical reactants and heat to dissolve lignin, and on subsequent mechanical refining to separate the cellulose fibers. The annual production volume of paper-grade pulp amounted to 178 million tonnes in 2018. Additional purification is needed to produce "dissolving-grade" cellulose pulps that are needed as raw material for the manufacture of cellulose derivatives. The cellulose content of dissolving-grade cellulose pulp needs to be higher than 90% [12]. Moreover, a low hemicellulose content is required as it is chemically similar to cellulose and can interfere during derivatization. The lignin content in dissolving pulp is generally very low, around 0.05%. The global annual production volume of dissolving pulp in 2020 is about 8.5 million tonnes of which 80% is used to produce textile fibers (viscose) [11].

Typical processes used to produce dissolving pulp are the sulfite process and the prehydrolysis kraft pulp process. In the sulfite process wood chips are treated with an aqueous solution of hydrogen sulfite and sulfur dioxide at elevated pressure and temperature. The bonds between cellulose and lignin components are cleaved, and lignin is converted to lignosulfonates that are soluble and can be removed. Prehydrolysis kraft pulping involves treatment of wood chips at acidic conditions to remove hemicelluloses and a subsequent treatment with an aqueous solution of sodium hydroxide and sodium sulfide to remove lignin [4]. A drawback of this process is the required pH switch that contributes to the chemical use and costs.

Cotton linters are an alternative feedstock to produce dissolving pulp. Cotton linters are a by-product of the production of staple cotton for textile applications. They can be used to manufacture high-quality dissolving pulps with a high cellulose content, high degree of polymerization (DP), and are free of lignin [13].

There is an increasing interest in the recovery of cellulose from non-wood plant sources [14], agricultural residues, and waste streams [15]. Non-wood plant sources like sisal, hemp, and flax have the advantage of a higher cellulose content and lower lignin content as compared to wood, and thus require less chemicals during pulping. Non-wood fibers require pre-treatments to prevent deterioration during storage. During the production of dissolving pulp from non-wood fibers typically a highly viscous black liquor is formed that is difficult to remove. This is an important drawback of non-wood fibers [16].

Agricultural residues like rice straw, wheat straw, sugarcane bagasse, and corn cobs are widely available because they are associated with the increasing production of sugar for biofuels [17]. Commonly agricultural residues have a low cellulose content and a high ash content, and this complicates cellulose recovery.

8.3.2 Cellulose Derivatization

The large number of free hydroxyl groups along the polymer chain (3 per glucose unit, see Figure 8.1) offers plenty of opportunity to derivatize cellulose and change the properties of the material. These hydroxyl groups can react with alcohols and acids to form ether and esters, and can be oxidized to ketones, aldehydes, and carboxylic groups [18]. The maximum number of substituent groups per glucose is referred to as the degree of substitution (DS) and is equal to the amount of hydroxyl groups per glucose unit. Typically, cellulose ethers are water soluble and commonly used as rheology modifiers or thickeners, adhesives, film formers, and water-retention agents. Cellulose esters are soluble in organic solvents and used in cigarette filter tow, membranes, coatings, and plastics. The enhanced solubility and thermoplastic behavior can be explained by the disruption of the crystalline structure of cellulose and the inhibition of hydrogen bond formation.

Both heterogeneous (cellulose in a solid phase) and homogeneous (cellulose completely dissolved) reaction paths can be applied to modify cellulose. Dissolution of cellulose in DMAc/LiCl allows a homogeneous reaction with for example anhydrides or acyl chloride. Industrial production of cellulose acetate is initially a heterogenous process and cellulose is present in a solid phase. Cellulose is treated with a mixture of acetic acid, acetic anhydride, and mineral acid catalyst in the presence of an organic solvent. Successive layers of cellulose react and the produced CTA is "peeled of" as it is soluble in the reaction mixture. At the end of the reaction, the reaction mixture is homogeneous and free of undissolved cellulose [19]. Cellulose esters with a DS lower than 2.9 cannot be produced directly as this will lead to an inhomogeneous product that is not evenly modified. The DS in amorphous regions will be higher than in crystalline regions, and this will lead to issues like gel formation. The required lower DS is obtained by partial hydrolysis of the fully substituted product.

Important research and development topics in cellulose derivatization are:

- ✓ Activation of cellulose to make it more susceptible for dissolution and derivatization [19–21]
- ✓ Controlling the molecular mass (depolymerization) during dissolution and derivatization [22–24]
- ✓ Novel derivatization agents and derivatives [25, 26]
- ✓ The properties of derivatives and the influence of the DS and regularity of substitution [27–29]

8.3.3 Cellulose Acetate and Cellulose Esters

The most important group of cellulose derivatives in thermoplastic applications are cellulose esters. Cellulose acetate (CA, see Figure 8.1) is the most commonly used cellulose ester and is known from its application in Lego bricks until 1967, when it was replaced by poly acrylonitrile butadiene styrene (ABS). Less common cellulose esters include cellulose acetate butyrate (CAB) and cellulose propionate [30].

The synthesis of cellulose esters is a three-step process: activation of cellulose by solvents or heat, functionalization and dissolution of cellulose, and hydrolysis to obtain the desired DS [31]. The pre-treatment process aims to accelerate the acetylation process and the most used method is exposure to glacial acetic acid [30]. The acetylation mixture contains three components: the acetylation agent, a catalyst, and a diluent. A common diluent is acetic acid that can be used in combination with methylene dichloride. The most common catalyst is sulfuric acid. Anhydrides are used as reactants. Hydrolysis with water and dilute acid can be performed to partially reverse the acetylation reaction without isolation of CTA. During the production of both dissolving pulp and cellulose acetates the molar mass of cellulose is reduced. Usually cellulose acetates have a degree of polymerization (number of glucose units) of 100–360 [19].

The DS determines the solubility and thermal properties of cellulose acetate and the desired DS depends on the application [30]:

- Cellulose acetate with a DS of 2.2–2.3 is soluble in acetone and suitable for injection molding
- Cellulose acetate with a DS of 2.3–2.4 is used in films
- Cellulose acetate with a DS of 2.4–2.6 is used in lacquers
- Cellulose triacetate with a DS of 2.8–3.0 is soluble in chloroform and used in triacetate film or fibers.

The T_g of cellulose acetates decreases with an increasing DS [1]. CTA tends to crystallize because of its highly symmetric structure. This behavior is shown by Teramoto [1] by measuring the T_g of cellulose acetates with a DS ranging from 1.8 to 2.95 (See Figure 8.2).

Commonly, plasticizers are used to lower the melting temperature and simultaneously reduce the melt viscosity. The most frequently used plasticizers are diethyl phthalate and triacetin (glycerol triacetate). Properties like the heat distortion temperature (HDT), stiffness, and strength can be tuned by varying the plasticizer content. Cellulose acetate propionate and cellulose butyrate have longer side chains, and this results in a lower stiffness, lower HDT, lower density, and improved processability (flow).

Cellulose esters are used for their excellent toughness and appearance (high gloss and clarity). These properties are beneficial in injection-molded articles, e.g. tool

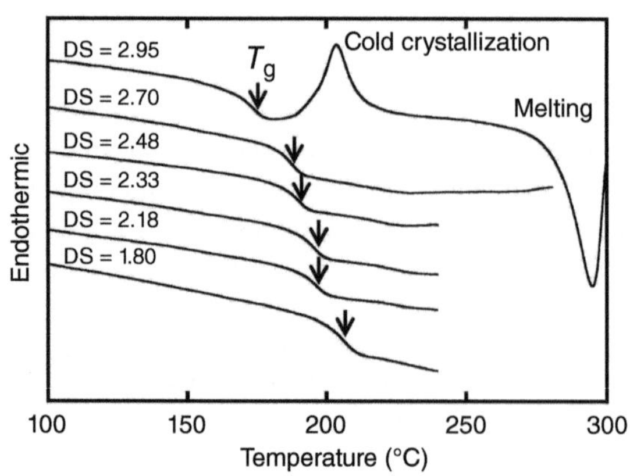

Figure 8.2 Differential scanning calorimetry (DSC) thermograms showing the T_g as a function of the DS. DSC thermograms obtained during the second heating scan. Arrows indicate Tg positions taken as midpoints of heat flow discontinuity. Source: Teramoto [1] / MDPI./ CC BY 4.0.

handles, spectacle frames, and ophthalmics (also see Chapter 9 on microbial polysaccharides and their applications). Moreover, cellulose acetate films are used in adhesive tapes, lamination films, and labels.

Since 1990 there is an increased interest in biodegradable cellulose acetate compounds for the production of packaging and disposables. Typically cellulose acetates with a DS below 2.5 are biodegradable [32–34]. Biodegradation is enhanced by the presence of plasticizers and is faster for cellulose acetates with lower DS. Cellulose propionates are biodegradable when the DS is below 1.85.

Besides the potential biodegradability and the bio-based origin, important advantages of cellulose acetates in packaging applications are the excellent water resistances, high transparency and gloss, and the high HDT. Comparative life-cycle assessments that include cellulose acetates are not available [35]. Based on the high energy and chemical use during the production of dissolving pulp and cellulose acetates, it can be expected that the environmental footprint is high compared to other bio-based plastics. Strategies to lower this footprint include heat recovery and reduction of reaction times as well as recovery of solvents and reduction of the water usage.

8.3.4 Cellophane

Although regenerated cellulose films are not true plastics (they do not melt upon heating) they are considered an important class of cellulose-based bioplastics because of their excellent biodegradability in all natural environments [36]. Most commonly, cellophane is produced via the xanthate process. Cellulose is first activated with alkali and subsequently treated with carbon disulfide to produce sodium cellulose xanthate by converting a few –OH groups (DS ~0.5) to $-O-C(S)S^{(-)}$ [30]. Next the solution is dispersed into an aqueous caustic soda solution to ripen. The solution is then extruded through a slit into a bath of dilute sulfuric acid and sodium sulfate. This results in the complete hydrolysis of the xanthate and regeneration into a transparent cellulose film [36]. The brittleness of the cellulose film is reduced by blending ethylene glycol or glycerol that act as a plasticizer. Novel, more environment-friendly production routes are reported for regenerated cellulose fibers but not yet for cellophane films [30].

The cellulose films can be provided with lacquers or can be laminated with other thermoplastic films to improve the moisture barrier or to allow sealing. The use of the lacquers or other thermoplastic films can compromise the biodegradability of the film. Cellulose films are used in packaging applications typically for food (candy) and tobacco. Limited information is available on the environmental footprint of cellophane production. Starting from wood, extensive purification and modification is required. Most concerns are related to the use of highly toxic CS_2. Through the development of new production processes, the recovery of chemicals including CS_2 and the reduction of energy use the environmental footprint of regenerated celluloses and toxicity issues can be reduced [37]. CS_2 can be recovered to a great extent or alternatively converted to sulfuric acid. Futamura, one of the main producers of cellophane films, reports significant reductions of the environmental footprint of cellophane films over the past years. It is indicated that at present the global warming potential of cellophane is slightly higher compared to polyethylene terephthalate (PET) [38].

8.3.5 Cellulose Fibers in Thermoplastic Formulations

Cellulose fibers obtained from wood or annual plants can be used to produce biodegradable thermoplastic composites. Cellulose fibers are added to fossil-based, bio-based, and biodegradable plastics to improve mechanical properties like stiffness and strength, reduce costs but also to enhance the biodegradability or rather the disintegration rate [39–41]. Cellulose nanofibers are used as a nucleating agent, enhancing crystallization. Interaction between the polymer matrix and the cellulose fiber can be improved by chemical modification, and this generally results in improved mechanical properties.

Besides wood fibers a broad range of cellulose fibers from various parts of plants can be used including bast fibers (hemp, flax, kenaf), seed fibers (cotton, kapok), fruit fibers (coir, oil palm), leaf fibers (sisal, abaca, banana), and grass fibers (bamboo, bagasse, wheat, rice). Composites with a range of biodegradable and bio-based plastics are reported in the literature including polyhydroxyalkanoates (PHAs), polylactic acid (PLA), polybutylene succinate (PBS), and starch-based plastics. "All cellulose composites" (ACCs) are reported in which both the matrix and the fibers are produced from cellulose [42]. Typically cellulose (nano-)fibers are then dispersed in a regenerated cellulose film.

Typical applications include packaging and disposables, automotive (interior parts), building, and furniture [43]. Advantages of natural fibers over glass fibers are low costs, low weight, low abrasion, and high a specific stiffness. Moreover, at end-of-life incineration of natural-fiber-reinforced composites are more advantageous compared to glass-containing composites.

The development of nanocellulose has boosted research related to cellulose fibers and composites [44, 45]. A common classification of nanocellulosic material includes cellulose nanocrystals (CNCs), cellulose nanofibrils (CNFs), and bacterial nanocellulose (BNC) [46]. CNCs are rod-like structure that can be acquired via acid hydrolysis of lignocellulosic fibers, removing the amorphous regions of the cellulosic source. CNFs are produced via a combination of mechanical and chemical techniques and have a complex fiber structure with thin fibrils in a web-like structure. Target applications range from composites, paper applications, membranes to biomedical applications [47].

8.4 Recent Developments in Thermoplastic Cellulose Derivatives

This section lists the developments related to thermoplastic cellulose derivatives from 2015 onwards. Many developments are driven by the aim to improve sustainability of cellulose derivatives and to reduce costs. Developments are related

Table 8.1 Summary of the research and development topics.

Topic	Development
Feedstock use and dissolving pulp production	Characterization of biomass feedstock Alternative feedstocks
Derivatization processes	Novel solvents Alternative processes
New derivatives and formulations	Plasticizers Mixed esters Blends New derivatives
Processing and properties and applications	Foaming Flame retardancy Anti-microbial properties
Products and applications	Biosensors Drug delivery All cellulose composites Nano-fibers
End-of-life	Biodegradability

to the feedstock and use of alternative biomass feedstocks, the development of more environment-friendly production processes up to materials that are more sustainable at end-of-life. The research topics are summarized in Table 8.1.

8.4.1 Characterization Methods for Lignocellulosic Biomass

Incentives for using and producing biofuels and bioenergy has resulted in a renewed interest in lignocellulosic biomass. New deconstruction strategies are developed to maximize yields and improve recovery of cellulose from biomass [48]. Improved and advanced analysis techniques are needed to characterize lignocellulosic biomass and allow the use of alternative biomass sources.

Lupoi et al. reviewed the opportunities of Raman spectroscopy to analyze lignocellulosic biomass and its derivatives, and for downstream process monitoring [49]. As Raman imaging techniques allow the real-time evaluation of deconstruction strategies and provide fundamental insights into how specific reagents affect the morphology of the biomass, the technique is valuable for future developments in cellulose chemistry and production of cellulose plastics for instance for the development of dissolving celluloses. As shown in Figure 8.3 Raman spectroscopy can show the distribution of cellulose and lignin in cellulose fiber bundles [50].

Another valuable new technique to analyze the recalcitrant nature of lignocellulosic biomass is infrared photo-induced force microscopy (IR PiFM) [51]. IR PiFM has a spatial resolution comparable to transmission electron microscopy (TEM) and

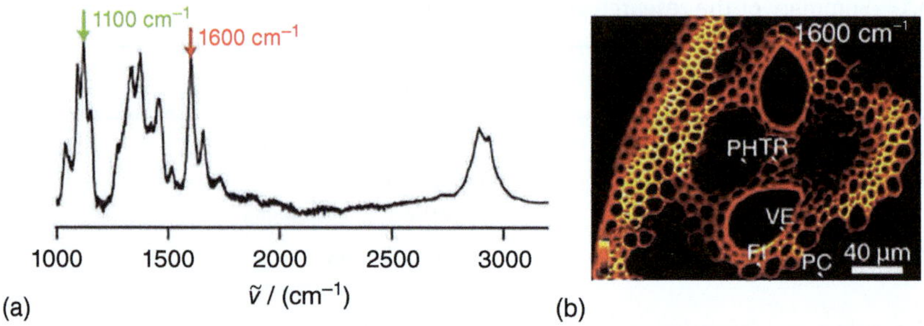

Figure 8.3 Raman spectroscopy and Stimulated Raman Scattering (SRS) imaging of corn stover. (a) Raman spectrum of raw corn stover. The peak at 1600 cm^{-1} (red arrow) corresponds to the lignin distribution, and the peak at 1100 cm^{-1} (green arrow) corresponds to cellulose. (b) SRS image of the vascular bundle including the edge of the stem in raw corn stover at 1600 cm^{-1}, showing the lignin distribution. Labeled structures are discussed in the text: parenchyma (PC), phloem (PH), vessel (VE), tracheid (TR), fiber (FI). Source: Reproduced with permission from Saar et al. [50].

more than 50 times higher than Raman imaging. An additional benefit of IR PiFM is that it can resolve structures based on their chemical composition. This allows nanoscale analysis of the physical and chemical structure of plant cell walls.

A third technique that needs to be mentioned is scanning near-field optical microscopy (SNOM) [52]. SNOM images provide chemical information on the spatial distribution of the cell wall components in secondary cell walls at a resolution similar to atomic force microscopy (AFM) and TEM.

8.4.2 Alternative Feedstocks for Dissolving Pulp and Production Routes

Whereas wood and cotton remain the most important sources for dissolving pulp and cellulose derivatives, various studies are found using alternative feedstocks typically from waste streams (sawmill) or non-wood plants like bamboo [23, 53, 54]. High-quality dissolving pulps were obtained from *Eulaliopsis bininata* (grass) feedstock, the *Acacia farnesiana* L. Willd. plant (huizache), jute, banana plant stem, palm empty fruit bunch, and bamboo willow [55–60]. Other developments with respect to dissolving pulp production include the conversion of paper grade pulp to dissolving pulp, and the valorization of by-products from the pulping process (like hemicellulose) to value-added products [54].

Developing dedicated processes to produce dissolving pulp from lignocellulose waste is important to achieve high purities. Using a pre-hydrolysis potassium hydroxide pulping process, dissolving pulp suitable for the production of viscose fibers could be produced from rice and wheat straws, corn stalk, dhaincha, and jute stick [61]. Steam explosion at 190 °C for 15 minutes combined with oxygen-assisted alkaline delignification was used to simultaneously produce oxidized lignin and cellulosic pulp from sugarcane bagasse [62]. Oxidation using oxygen and ozone followed by an alkaline extraction proved to be an effective additional purification process to produce dissolving grade pulp from sawdust and sugarcane bagasse [63].

8.4.3 Ionic Liquids and Deep Eutectic Solvents for Cellulose Regeneration and Modification

Dissolution of cellulose and accessibility of cellulose for derivatization in ionic liquids is a rather new topic. Ionic liquids are a combination of bulky organic cations with mostly inorganic anions, resulting in salts that are liquids at temperatures below 100 °C. The cellulose dissolution in ionic liquids is controlled by the anion selection, since its size and geometry plays an important role in cellulose dissolution. The chlorine ion is one of the most promising anions, but other halides, carboxylates, or phosphates anions can also be used for cellulose dissolution. Other parameters that affect cellulose dissolution in ionic liquids include the presence of water (that negatively influences cellulose dissolution), the dissolution temperature, and the cellulose source. Advantages of ionic liquids include the relatively mild dissolution conditions, easy processability, low volatility, and the environment-friendly nature. This makes them promising solvents to apply in industrial processes for biomass processing [23, 64, 65]. A review by Yuan et al. aims to provide an improved understanding of the mechanism of cellulose dissolution in ionic liquids [66]. To allow cellulose dissolution, the native hydrogen-bonding network in cellulose needs to be disturbed or destroyed and ionic liquids have this ability. The authors conclude that there are three interacting modules that determine the mechanisms of cellulose dissolution in ionic liquids: (i) the structure of elementary fibrils; (ii) the solvation of cellulose in ionic liquids by interaction of cation, anion, and cellulose on different length scales, and (iii) solution structure of cellulose solubilized in ionic liquids.

The practical application of ionic liquids for the manufacture of regenerated cellulose fibers is reviewed by Mendes et al. [23]. Advantages in fiber spinning are the ability to dissolve cellulose in high concentrations and to recover the solvent after the regeneration of the cellulose film or fiber. Moreover, water pollution is moderate and sometimes a higher fibrillation resistance is obtained. Drawbacks include the slow rate of dissolution, high costs of the solvents, difficulties to reuse or recycle the solvents, and some ionic liquids are toxic and non-biodegradable. Imidazolium ionic liquids with chloride anions are suitable for cellulose spinning, but can induce cellulose degradation at temperatures above 90 °C [67]. Regenerated cellulose fibers with a strength comparable to Lyocell could be produced using 1,5-diazabicyclo[4.3.0]non-5-enium acetate [22]. This solvent allowed high cellulose concentrations at moderate temperatures (60–100 °C) while cellulose degradation was limited.

Pang et al. report the properties of regenerated cellulose films fabricated from different cellulose feedstocks using ionic liquids [68]. Feedstocks studied included pine, cotton, and bamboo and the ionic liquid used was 1-ethyl-3-methylimidazolium acetate. Using X-ray diffraction (XRD) it was confirmed that the transition of cellulose I to cellulose II had occurred due to dissolution and regeneration. The films prepared from cotton linter and pine exhibited excellent thermal stability and mechanical properties.

The use of ionic liquids allows the homogenous production of CTA [69]. CTA with a DS of 2.98 was prepared by transesterification of lignocellulosic biomass with isopropenyl acetate in 1-ethyl-3-methylimidazolium acetate. Homogenous esterification of cellulose pulp was also achieved using 1,5-diazabicyclo[4.3.0]non-5-ene acetate as a solvent. A recyclable system with low boiling by-products was demonstrated [70].

With an increased use of ionic liquids, their regeneration, recovery, and recycling becomes even more important [71]. Filtration and rotary evaporation to recover 3-butyl-1-methylimidazolium chloride (BMIM Cl) from corn stover hydrolysate enabled at least 10 reuse cycles. Also, recovery using vacuum distillation was reported with an efficiency higher than 95%. Alternative methods include electrodialysis and in some cases this ionic liquid can be recovered by washing with water. Although various procedures are proposed in literature there are still a very limited number of methods for the direct reuse or regeneration of ionic liquids. Extensive studies on procedures regarding the reuse of ionic liquids are needed.

Another class of promising solvents that are currently studied for cellulosic biomass dissolution and biomass conversions are deep eutectic solvent (DESs). These solvents are formed by a hydrogen bond acceptor (commonly an ammonium salt) and a hydrogen bond donor (e.g. amines and carboxylic acids), linked by hydrogen bonds at a certain molar ratio. Advantages of DES in biomass pre-treatment are lower production costs, higher biodegradability, low vapor pressure, safety, and innocuous nature [23, 24]. Cellulose solubility in DES is limited as compared to ionic liquids, due to the strong hydrogen bond network in DES that results in a decrease of interactions between cellulose and the DES solvents. A review on the use of DES for lignocellulosic biomass fractionation and subsequent conversion to bio-based products is prepared by Tan et al. [24]. Studies indicate a DES based on choline chloride and urea is an effective plasticizer for cellulose leading to soft materials having enhanced elongation [72].

8.4.4 New Derivatization Routes

Apart from the use of new solvents (ionic liquids) there is little progress in the development of new derivatization routes. A new strategy to produce thermoplastic cellulose oleate involved ball milling [73]. Cellulose oleates with a high DS (1.29–2.55) were produced using a low dosage of oleic acid (6 mol per anhydro glucose unit) and no additional solvents. Materials with a high DS (2.34–2.55) could be processed into homogeneous flexible transparent films via compression molding at 160 °C.

8.4.5 Plasticizers

Several publications study the interactions of cellulose acetates with plasticizers [74–76]. Glycerine, triethyl citrate, and dioctyl phthalate were found to significantly enhance crystallization of CTA. Dioctyl phthalate was the most effective plasticizer leading to a T_g depression of 46 °C at an addition level of 25 wt%. Related to this effect, the compounds with dioctyl phthalate had the lowest E-modulus [74]. Bao studied the miscibility of cellulose diacetate with diethyl phthalate. At a diethyl phthalate content higher than 25% phase separation occurred [75]. Alternative plasticizers include essential oils (EO) [77]. Essential oils containing limonene and pinenes (e.g. from lime and nutmeg) gave the strongest plasticizing effect. Essentials oils containing fatty acids (e.g. from coffee) were the weakest plasticizers. When essential oils were added to cellulose acetate the water barrier was improved. Most of the essential oils decreased the transparency of the films. Eugenol, pimento berry, and anise were notable exceptions. In an earlier study essential oils (EO) from lemongrass, rosemary pepper, and basil were found to plasticize cellulose acetate, cellulose acetate propionate, and CAB [78]. Li studied the thermal behavior of cellulose diacetate melts, using ionic liquids as a plasticizer aiming to develop a melt-spinning process. At higher shear rates and at higher concentrations the thermal stability was found to decrease [76].

8.4.6 Mixed Cellulose Esters

A detailed analysis of the thermoplastic behavior of cellulose derivatives is published by Teramoto [1] (Figure 8.2). The review includes the effect of single-substituent derivatization, derivatization with multi-substituents, blending of simple derivatives with synthetic polymers, and graft copolymerization.

Multi-substitution offers the opportunity to develop new cellulose-based thermoplastic polymers that combine excellent thermal processability with tunable mechanical performance. It allowed the development of the first melt-spun cellulosic fiber by Toray Industries. Derivatization of cellulose diacetates with modified cardanol (phenolic compound extracted from cashew nutshell liquid) was presented as an alternative to the use of plasticizers [79]. Mixed cellulose acetate oleate esters demonstrate promising properties including an improve oxygen and water vapor barrier. In this mixed ester the oleate functions as an internal plasticizer [80]. A wide range of cellulose mixed esters was synthesized through homogeneous acylation of cellulose diacetate (DS 2.45), by using a series of commercially available aliphatic acid chloride modifiers (fatty chains with different length, ramified, bulky, aliphatic, or aromatic rings disposed more or less near to cellulose backbone) [81]. The covalent introduction of the acyl moieties leads to a significant lowering of T_g, as a consequence of the enhancement of the overall mobility of the macromolecular chains. Acylating reagents composed of bulky moieties at the extremity of a soft linear chain led to an efficient plasticization. The water transport properties of casted films were evaluated and were decreased by the presence of apolar side groups. The effects of long acyl chain structures on mechanical and thermal properties are reported by Tanaka et al. [82]. The study focused on two types of renewable long acyl chains for mixed cellulose esters: the aromatic 3-pentadecylphenoxy acetyl (PA) group derived from cardanol extracted from cashew nutshells and the aliphatic stearoyl (St) group made from vegetable oils. Mixed substitution of long and short acyl chains (acetate) prevented the crystallization of the long acyl chain moieties in mixed cellulose esters. Materials with the PA group had a slightly higher bending strength and glass transition temperature than those with the St group but showed

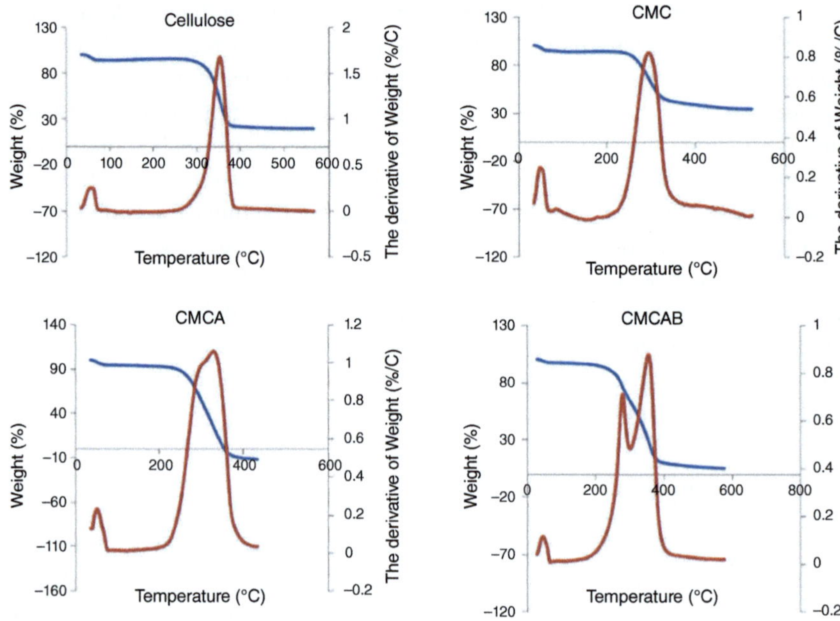

Figure 8.4 Thermogravimetric analysis (TGA) and differential thermal analysis (DTA) curves of cellulose, CMC, CMCA, and CMCAB [83]. Source: Reproduced by permission of Romanian Academy Publishing House, the owner of the publishing rights.

low impact strength. Materials with the St group showed remarkably improved impact strength with sufficient balanced mechanical properties for use in durable products.

Grafting caprolactone and lactic acid onto cellulose diacetate resulted in a transparent thermoplastic material that could be processed at 200 °C without the use of an additional plasticizer [72].

Carboxymethyl cellulose acetate (CMCA) and carboxymethyl cellulose acetate butyrate (CMCAB) are glassy thermoplastic materials with T_g of 138.35 °C and T_m of 200 °C [83]. From Figure 8.4 (weight loss in blue, differential weight loss in red) it can be seen that the CMCAB demonstrates an increased thermal stability compared to those of cellulose, carboxymethyl cellulose (CMC), and CMCA.

New thermoplastic cellulose stearate and laurate esters were prepared by Krasnou et al. [84]. The authors indicate that more development is needed to improve the properties of these cellulose esters.

Synthesis of cellulose acetate propionate with controllable contents and distributions of acetyl and propionyl in the C2, C3, and C6 positions was studied by Meng et al. [85]. The reactivity as well as selective hydrolysis of the different OH-positions of the anhydro glucose unit was determined.

8.4.7 Cellulose–Polymer Blends

Various studies focus on blending as a route to avoid the use of plasticizers. Polyvinyl pyrrolidone forms a miscible blend with cellulose acetate [86]. Of specific interest however are biodegradable blends with aliphatic polyesters like polycaprolactone (PCL, see Chapter 7). The miscibility of cellulose acetates with PCL depends on the type of cellulose derivative and the DS, and the most commonly used cellulose acetates are immiscible with PCL [87]. CAB and PBS can form miscible blends. It was found that PBS acts as a plasticizer for CAB, lowering the E-modulus and enhancing elongation [88]. Cross-linked blends of cellulose acetate and starch acetate were claimed to combine the advantages of these two materials while reducing their disadvantages [89]. The films displayed improved mechanical properties as compared to fragile starch acetate films, and the strength and strain at break increased at higher cellulose acetate contents. Specifically films with higher starch acetate were more readily biodegradable in soil as compared to cellulose acetate. Blends of plasticized CAB and PLA were studied by Kunthadong et al. [90]. Limited compatibility of the materials negatively influenced the mechanical properties of the blend.

8.4.8 (New) Properties and Processing Routes

Extrusion foaming of cellulose palmitate with a DS of 1.0 was studied by Rokkonen et al. [91]. The blowing agents isobutane, carbon dioxide, and nitrogen all enable foam formation. Foams with the lowest density were obtained using isobutane; nitrogen resulted in the most uniform cell morphology. Further studies are needed to improve the quality of the foams. Extrusion foaming of cellulose acetate with various co-blowing agents was studied by Hendriks et al. [92]. It was found that the effect of the co-blowing agent depended on its solubility in the melt. Ethyl acetate gave optimal results at low addition levels, while isopropanol and acetone gave improved foaming at higher addition levels. Further research is needed to study the effect of co-blowing agents on cell nucleation.

Development of flame-retardant cellulose acetate foams was not successful [93]. The flame retardants had a strong effect on the processing behavior, especially on the rheological properties. The higher viscosity of the flame-retardant compounds negatively influences processing and foaming.

Cellulose acetate with antimicrobial activity was achieved by embedding molybdenum trioxide particles [94].

8.4.9 New Applications

Development of new cellulose and cellulose acetate products for biomedical applications is frequently reported [95–98]. Cellulose acetate fibers produced via electrospinning are biocompatible and biodegradable and, hence, suitable for medical applications like tissue engineering, drug delivery, and wound dressings [99]. Wound dressings based on cellulose acetate/PCL blends are prepared via electrospinning [96]. Electrospun cellulose acetate nanofibers are studied for drug-delivery systems [97]. These electrospun fibers are also very suitable for application in biosensors including optical/colorimetric, and electrochemical-based sensing systems. The use of electrospinning to produce 3D nano-fiber stacks of cellulose acetate for biomedical and pharmaceutical applications such as tissue engineering, cell cultures, immobilization of bioactive substances, and drug delivery is discussed by Cheng et al. [100]. Cellulose-based hydrogels have promising properties for application in various biomedical applications including drug delivery, tissue engineering, wound dressing, bioimaging, and wearable sensors [101]. The use of cellulose-based materials in sodium-ion batteries and capacitors is presented in an extensive review showing these have excellent opportunities in these applications [102].

Various publications can be found on the application of cellulose fibers in composite materials. These include review articles on the opportunities of natural composites, publications on specific fibers like bamboo, coir, abaca, and curaua, but also articles focusing on isolation and modification of cellulosic fibers and the properties of cellulose-fiber-based composites [103–109]. Various articles discuss "ACCs" [110–112]. These ACCs consist of a matrix of dissolved and regenerated cellulose reinforced by undissolved or partly dissolved cellulose. Two routes are applied to produce ACCs; the complete dissolution of a cellulose material followed by conventional impregnation methods of fibrous cellulose material, or partial dissolution of a cellulose fabric to form a matrix phase that bonds undissolved cellulose fiber cores together. This last method offers more opportunities for industrial upscaling. ACCs can be used in a wide range of applications including structural materials, biomedical engineering, photoelectric devices, electro-active paper, sensors, electrical displays, filtration materials, biodegradable food packaging materials, and mulching films for agriculture. Other biodegradable natural-fiber-reinforced composites are based on biodegradable polymers like starch, PBS, PCL, or polylactic acid [106, 113–118].

A clear trend is the development and use of cellulosic nano-fibers because of their outstanding mechanical properties. Commonly chemo-mechanical processes are used to produce nano-fibers [119]. Alternatives routes include biological synthesis of nano fibers via controlled hydrolysis of cotton fibers by purified cellulase enzymes [120]. Electrospinning is reported as a technique to convert both cellulose and derivatives like cellulose acetates into nanofibers [121]. Apart from composites for engineering applications, potential uses of cellulose nano-fibers include thickening agents, tissue engineering scaffolds, nucleation agent, dry-strength additive for paper and adsorbents [122, 123]. Another promising application field for cellulosic materials are liquid crystal polymers. This includes the use of cellulose nanocrystals that can form ordered structures leading to lyotropic liquid crystals [124].

8.5 Biodegradation of Cellulose Derivatives

Whereas natural cellulose and regenerated cellulose are readily biodegradable in all natural environments, there is still little fundamental understanding regarding the biodegradation mechanisms of cellulose esters. This is illustrated by contradictory literature and two review papers tried to clarify the biodegradation behavior of cellulose acetates [125, 126]. The aerobic biodegradation of cellulose acetates in soil is discussed by Polman, whereas Yadav reviewed the complicated interplay between structure, environment, and degradation. Both studies confirm that chemical modification negatively affects biodegradability of natural polymers and that the biodegradability of cellulose acetate largely depends on the degree of acetylation. At a DS <1.8, acetyl esterase enzymes are able to remove acetyl groups and this has a significant effect on the biodegradation rate, as deacetylation is the rate-determining step during biodegradation of cellulose acetate. Regio-selectively substituted cellulose acetate degrades faster than randomly substituted cellulose acetate with a similar DS. Other material modifications (like plasticizers and other additives) can enhance or further limit the degradability. The biodegradation rate is highly influenced by the actual degradation environment, and often a long lag phase is reported before the biodegradation process starts. Several biodegradable cellulose acetate grades are commercially available, and these are certified biodegradable at industrial composting conditions or even in the marine environment.

8.6 Conclusions

Although they are a fairly old class of plastics, there are still developments that increase the applicability and improve the environmental footprint of cellulose plastics like cellophane and cellulose acetates. Reducing energy use and chemical loads, for example by using novel solvents and new recovery processes, is demonstrated in textile applications and is under development for plastic applications. Optimized processing could also reduce the costs of cellulose plastics. This can be envisaged using new plasticizers that are less volatile or do not migrate from the plastic matrix, by developing new blends that do not need plasticizers or via the production of mixed esters with internal plasticization. Historically, cellulose plastics are used in durable applications. The applicability of these new plasticizer systems depends on the development of cost-efficient production routes. In recent literature new opportunities of cellulose plastics in biomedical applications are explored (drug delivery, biosensors). Moreover new cellulose acetate grades are developed that are biodegradable in the natural environment. Typically, these grades have a relatively low DS and additives including plasticizers are used to enhance biodegradation. These new materials can be of interest in agricultural applications and in fishing nets where littering cannot be avoided and where the excellent mechanical properties of cellulose plastics are beneficial. Biodegradability is also an important asset of cellulose based (nano-)fibers that are studied for a wide range of applications.

As compared to other bio-based plastics, cellulose plastics have a rather large (historical) market share. Still, they are rather expensive and this limits applicability in bulk applications like packaging. Exploiting the biodegradability in the natural environment could help to obtain market shares in new applications.

References

1 Teramoto, Y. (2015). Functional thermoplastic materials from derivatives of cellulose and related structural polysaccharides. *Molecules* 20 (4): 5487–5527.

2 Kennedy, J.F. and Hasamudin, W.H.W. (1995). Cellulose: structure, accessibility and reactivity, edited by H. A. Krassig. Gordon & Breach, Pennsylvania, 1993. Pp. xvi + 376, price £169.00, US$260.00. ISBN 2-88124-798-9. *Polymer International* 36 (1): 101–101.

3 Abe, M.M., Martins, J.R., and Sanvezzo, P.B. (2021). Advantages and disadvantages of bioplastics production from starch and lignocellulosic components. *Polymers* 13 (15): 2484.

4 Heinze, T., El Seoud, O.A., and Koschella, A. (2018). Production and characteristics of cellulose from different sources. In: *Cellulose Derivatives: Synthesis, Structure, and Properties* (ed. T. Heinze, O.A. El Seoud and A. Koschella), 1–38. Cham: Springer International Publishing.

5 Ganster, J. and Fink, H.-P. (2013). Cellulose and cellulose acetate. In: *Bio-based Plastics* (ed. S. Kabasci), 35–62. Hoboken, NJ, USA: Wiley Online Library.

6 Chauhan, S. and Meena, B.L. (2021). Introduction to pulp and paper industry: global scenario. *Physical Sciences Reviews* 6 (5): 81–109.

7 Rasmussen, S.C. (2021). From Parkesine to celluloid: the birth of organic plastics. *Angewandte Chemie International Edition* 60 (15): 8012–8016.

8 Zhibitsky, J.T. (1943). Plastics and their application to aircraft construction. *Journal of the Roy Aeronautical Society* 47 (389): 138–143.

9 Carlisle, R.P. (2004). *Scientific American Inventions and Discoveries : All the Milestones in Ingenuity From the Discovery of Fire to the Invention of the Microwave Oven*. Hoboken: Wiley.

10 Geyer, R. (2020). A brief history of plastics. In: *Mare Plasticum – The Plastic Sea: Combatting Plastic Pollution Through Science and Art* (ed. M. Streit-Bianchi, M. Cimadevila and W. Trettnak), 31–47. Cham: Springer International Publishing.

11 Kallio, A.M.I. (2021). Wood-based textile fibre market as part of the global forest-based bioeconomy. *Forest Policy and Economics* 123: 102364.

12 Sixta, H. (2006). Pulp properties and applications. In: *Handbook of Pulp* (ed. H. Sixta), 1009–1067. Weinheim, Germany: Wiley.

13 Sczostak, A. (2010). Cotton linters: an alternative cellulosic raw material. *Macromolecular Symposia* 280: 45–53.

14 Kamoga, O.L.M., Byaruhanga, J.K., and Kirabira, J.B. (2013). A review on pulp manufacture from non wood plant materials. *International Journal of Chemical Engineering and Applications* 4: 144–148.

15 Nuruddin, M., Chowdhury, A., and Haque, S.A. (2011). Extraction and characterization of cellulose microfibrils from agricultural wastes in an integrated biorefinery initiative. *Cellulose Chemistry and Technology* 45 (5, 6): 347–354.

16 Hammett, A.L., Youngs, R.L., and Sun, X. (2001). Non-wood fiber as an alternative to wood fiber in Chinas pulp and paper industry. *Holzforschung* 55 (2): 219–224.

17 Ali, M., Saleem, M., and Khan, Z. (2019). The use of crop residues for biofuel production, Chapter 16. In: *Biomass, Biopolymer-Based Materials, and Bioenergy* (ed. D. Verma, E. Fortunati and S. Jain), 369–395. Woodhead Publishing.
18 Heinze, T., Koschella, A., and Liebert, T. (2012). Cellulose: chemistry of cellulose derivatization. In: *The European Polysaccharide Network of Excellence (EPNOE): Research Initiatives and Results* (ed. P. Navard), 283–327. Vienna: Springer Vienna.
19 Heinze, T., El Seoud, O.A., and Koschella, A. (2018). Principles of cellulose derivatization. In: *Cellulose Derivatives: Synthesis, Structure, and Properties* (ed. T. Heinze), 259–292. Cham: Springer International Publishing.
20 Yan, L., Li, W., and Qi, Z. (2006). Solvent-free synthesis of cellulose acetate by solid superacid catalysis. *Journal of Polymer Research* 13 (5): 375–378.
21 Matsumoto, T., Tatsumi, D., and Tamai, N. (2001). Solution properties of celluloses from different biological origins in LiCl DMAc. *Cellulose* 8 (4): 275–282.
22 Asaadi, S., Hummel, M., and Ahvenainen, P. (2018). Structural analysis of ioncell-F fibres from birch wood. *Carbohydrate Polymers* 181: 893–901.
23 Mendes, I.S.F., Prates, A., and Evtuguin, D.V. (2021). Production of rayon fibres from cellulosic pulps: state of the art and current developments. *Carbohydrate Polymers* 273: 118466.
24 Tan, Y.T., Chua, A.S.M., and Ngoh, G.C. (2020). Deep eutectic solvent for lignocellulosic biomass fractionation and the subsequent conversion to bio-based products – a review. *Bioresource Technology* 297: 122522.
25 Aly, A.A. and El-Bisi, M.K. (2018). Grafting of polysaccharides: recent advances. In: *Biopolymer Grafting: Synthesis and Properties* (ed. V.K. Thakur), 469–519. Elsevier Inc.
26 Hon, D.N.S. (2017). Cellulose and its derivatives: structures, reactions, and medical uses. In: *Polysaccharides in Medicinal Applications* (ed. S. Dumitriu), 87–105. CRC Press.
27 Suzuki, S., Hikita, H., and Hernandez, S.C. (2021). Direct conversion of sugarcane bagasse into an injection-moldable cellulose-based thermoplastic via homogeneous esterification with mixed acyl groups. *ACS Sustainable Chemistry & Engineering* 9 (17): 5933–5941.
28 Battisti, R., Hafemann, E., and Claumann, C.A. (2019). Synthesis and characterization of cellulose acetate from royal palm tree agroindustrial waste. *Polymer Engineering and Science* 59 (5): 891–898.
29 Niu, X., Liu, Y., and King, A.W.T. (2019). Plasticized cellulosic films by partial esterification and welding in low-concentration ionic liquid electrolyte. *Biomacromolecules* 20 (5): 2105–2114.
30 Gilbert, M. (2017). Cellulose plastics, Chapter 22. In: *Brydson's Plastics Materials*, 8e (ed. M. Gilbert), 617–630. Butterworth-Heinemann.
31 El Seoud, O.A. and Heinze, T. (2005). Organic esters of cellulose: new perspectives for old polymers. In: *Polysaccharides I*, Advances in Polymer Science (ed. T. Heinze), 103–149. Berlin/Heidelberg, Germany: Springer.

32 Buchanan, C.M., Dorschel, D., and Gardner, R.M. (1996). The influence of degree of substitution on blend miscibility and biodegradation of cellulose acetate blends. *Journal of Environmental Polymer Degradation* 4 (3): 179–195.

33 Edgar, K.J., Buchanan, C.M., and Debenham, J.S. (2001). Advances in cellulose ester performance and application. *Progress in Polymer Science* 26 (9): 1605–1688.

34 Sakai, K., Yamauchi, T., and Nakasu, F. (1996). Biodegradation of cellulose acetate by *Neisseria sicca*. *Bioscience, Biotechnology, and Biochemistry* 60 (10): 1617–1622.

35 Foroughi, F., Rezvani Ghomi, E., and Morshedi Dehaghi, F. (2021). A review on the life cycle assessment of cellulose: from properties to the potential of making it a low carbon material. *Materials* 14 (4): 714.

36 McKeen, L.W. (2017). Environmentally friendly polymers, Chapter 13. In: *Permeability Properties of Plastics and Elastomers*, 4e (ed. L.W. McKeen), 305–323. William Andrew Publishing.

37 Shen, L., Worrell, E., and Patel, M.K. (2010). Environmental impact assessment of man-made cellulose fibres. *Resources, Conservation and Recycling* 55 (2): 260–274.

38 Futamura. *Carbon Footprint*. 2021 [cited 2021 October 25th]. http://www.futamuragroup.com/jp/sustainability/carbon-footprint.

39 Mu, B., Tang, W., and Liu, T. (2021). Comparative study of high-density polyethylene-based biocomposites reinforced with various agricultural residue fibers. *Industrial Crops and Products* 172: 114053.

40 Morin, S., Dumoulin, L., and Delahaye, L. (2021). Green composites based on thermoplastic starches and various natural plant fibers: impacting parameters of the mechanical properties using machine-learning. *Polymer Composites* 42 (7): 3458–3467.

41 Rajeshkumar, G., Arvindh Seshadri, S., and Devnani, G.L. (2021). Environment friendly, renewable and sustainable poly lactic acid (PLA) based natural fiber reinforced composites – a comprehensive review. *Journal of Cleaner Production* 310: 127483.

42 Nishino, T. and Peijs, T. (2013). All-cellulose composites. In: *Handbook of Green Materials* (ed. K. Oksman, A.P. Mathew and A. Bismarck), 201–216. World Scientific.

43 Peças, P., Carvalho, H., and Salman, H. (2018). Natural fibre composites and their applications: a review. *Journal of Composites Science* 2 (4): 66.

44 Khalid, M.Y., Al Rashid, A., and Arif, Z.U. (2021). Recent advances in nanocellulose-based different biomaterials: types, properties, and emerging applications. *Journal of Materials Research and Technology* 14: 2601–2623.

45 Zhang, L., Guo, L., and Wei, G. (2021). Recent advances in the fabrication and environmental science applications of cellulose nanofibril-based functional materials. *Materials* 14 (18): 5390.

46 Zinge, C. and Kandasubramanian, B. (2020). Nanocellulose based biodegradable polymers. *European Polymer Journal* 133: 109758.

47 Sharma, A., Thakur, M., and Bhattacharya, M. (2019). Commercial application of cellulose nano-composites – a review. *Biotechnology Reports* 21: e00316.

48 Liao, Y., Op de Beeck, B., and Thielemans, K. (2020). The role of pretreatment in the catalytic valorization of cellulose. *Molecular Catalysis* 487: 110883.

49 Lupoi, J.S., Gjersing, E., and Davis, M.F. (2015). Evaluating lignocellulosic biomass, its derivatives, and downstream products with Raman spectroscopy. *Frontiers in Bioengineering and Biotechnology* 3: 50.

50 Saar, B.G., Zeng, Y., and Freudiger, C.W. (2010). Label-free, real-time monitoring of biomass processing with stimulated Raman scattering microscopy. *Angewandte Chemie International Edition* 49 (32): 5476–5479.

51 Kesari, K.K., O'Reilly, P., and Seitsonen, J. (2021). Infrared photo-induced force microscopy unveils nanoscale features of Norway spruce fibre wall. *Cellulose* 28 (11): 7295–7309.

52 Keplinger, T., Konnerth, J., and Aguié-Béghin, V. (2014). A zoom into the nanoscale texture of secondary cell walls. *Plant Methods* 10 (1): 1.

53 Choudhury, M.J. and Khan, G.M.A. (2015). Utilization of sawmill by-product for making cellulose and its valuable derivatives. In: *Biomass and Bioenergy: Applications* (ed. K.R. Hakeem, M. Jawaid and U. Rashid), 165–186. Springer International Publishing.

54 Kumar, H. and Christopher, L.P. (2017). Recent trends and developments in dissolving pulp production and application. *Cellulose* 24 (6): 2347–2365.

55 Harsono, H., Putra, A.S., and Maryana, R. (2016). Preparation of dissolving pulp from oil palm empty fruit bunch by prehydrolysis soda-anthraquinone cooking method. *Journal of Wood Science* 62 (1): 65–73.

56 Das, A.K., Nakagawa-Izumi, A., and Ohi, H. (2016). Quality evaluation of dissolving pulp fabricated from banana plant stem and its potential for biorefinery. *Carbohydrate Polymers* 147: 133–138.

57 Nayeem, J., Sarkar, M., and Quadery, A.H. (2017). High purity dissolving pulp from jute. *Nordic Pulp and Paper Research Journal* 32 (4): 623–629.

58 Wu, C.J., Zhang, J.C., and Yu, D.M. (2018). Dissolving pulp from bamboo-willow. *Cellulose* 25 (1): 777–785.

59 Casillas, R.R., López, M.C.L., and Aguilar, B.B. (2019). Obtaining dissolving grade cellulose from the huizache (*Acacia farnesiana* L. Willd.) plant. *BioResources* 14 (2): 3301–3318.

60 Wu, G., Zhang, Y., and Yuan, B. (2019). Production of dissolving pulp from *Eulaliopsis binata* with the concept of integrated biorefinery. *Cellulose* 26 (3): 2087–2097.

61 Sarkar, A.M., Nayeem, J., and Rahaman, M.M. (2021). Dissolving pulp from non-wood plants by prehydrolysis potassium hydroxide process. *Cellulose Chemistry and Technology* 55 (1, 2): 117–124.

62 Rocha, G.J.M., Andrade, L.P., and Martín, C. (2020). Simultaneous obtaining of oxidized lignin and cellulosic pulp from steam-exploded sugarcane bagasse. *Industrial Crops and Products* 147: 11227.

63 Ehman, N.V., Rodriguez Rivero, G.E., and Area, M.C. (2017). Dissolving pulps by oxidation of the cellulosic fraction of lignocellulosic waste. *Cellulose Chemistry and Technology* 51 (9, 10): 863–870.

64 Cao, Y., Zhang, R., and Cheng, T. (2017). Imidazolium-based ionic liquids for cellulose pretreatment: recent progresses and future perspectives. *Applied Microbiology and Biotechnology* 101 (2): 521–532.

65 Michud, A., Hummel, M., and Sixta, H. (2016). Influence of process parameters on the structure formation of man-made cellulosic fibers from ionic liquid solution. *Journal of Applied Polymer Science* 133 (30): 43718.

66 Yuan, X. and Cheng, G. (2015). From cellulose fibrils to single chains: understanding cellulose dissolution in ionic liquids. *Physical Chemistry Chemical Physics* 17 (47): 31592–31607.

67 Michud, A., Tanttu, M., and Asaadi, S. (2015). Ioncell-F: ionic liquid-based cellulosic textile fibers as an alternative to viscose and lyocell. *Textile Research Journal* 86 (5): 543–552.

68 Pang, J., Wu, M., and Zhang, Q. (2015). Comparison of physical properties of regenerated cellulose films fabricated with different cellulose feedstocks in ionic liquid. *Carbohydrate Polymers* 121: 71–78.

69 Suzuki, S., Shibata, Y., and Hirose, D. (2018). Cellulose triacetate synthesis via one-pot organocatalytic transesterification and delignification of pretreated bagasse. *RSC Advances* 8 (39): 21768–21776.

70 Kakko, T., King, A.W.T., and Kilpeläinen, I. (2017). Homogenous esterification of cellulose pulp in [DBNH][OAc]. *Cellulose* 24 (12): 5341–5354.

71 Dołżonek, J., Kowalska, D., and Maculewicz, J. (2020). Regeneration, recovery, and removal of ionic liquids. In: *Encyclopedia of Ionic Liquids* (ed. S. Zhang), 1–9. Singapore: Springer Singapore.

72 Müller, K., Zollfrank, C., and Schmid, M. (2019). Natural polymers from biomass resources as feedstocks for thermoplastic materials. *Macromolecular Materials and Engineering* 304 (5): 1800760.

73 Hou, D.-F., Li, M.-L., and Yan, C. (2021). Mechanochemical preparation of thermoplastic cellulose oleate by ball milling. *Green Chemistry* 23 (5): 2069–2078.

74 Azadimanesh, F. and Mohammadi, N. (2015). A plasticizer index to universally correlate the normalized work of fracture and elastic modulus of plasticized cellulose triacetates. *Carbohydrate Polymers* 130: 316–324.

75 Bao, C.Y., Long, D.R., and Vergelati, C. (2015). Miscibility and dynamical properties of cellulose acetate/plasticizer systems. *Carbohydrate Polymers* 116: 95–102.

76 Li, Z., Liu, N., and Yao, Y. (2015). Thermal behavior of cellulose diacetate melt using ionic liquids as plasticizers. *RSC Advances* 5 (2): 901–907.

77 Biswas, A., do Socorro Rocha Bastos, M., and Furtado, R.F. (2020). Evaluation of the properties of cellulose ester films that incorporate essential oils. *International Journal of Polymer Science* 2020: 1–8.

78 do Socorro Rocha Bastos, M., Da Silva Laurentino, L., and Canuto, K.M. (2016). Physical and mechanical testing of essential oil-embedded cellulose ester films. *Polymer Testing* 49: 156–161.

79 Iji, M., Moon, S., and Tanaka, S. (2011). Hydrophobic, mechanical and thermal characteristics of thermoplastic cellulose diacetate bonded with cardanol from cashew nutshell. *Polymer Journal* 43 (8): 738–741.

80 Tedeschi, G., Guzman-Puyol, S., and Paul, U.C. (2018). Thermoplastic cellulose acetate oleate films with high barrier properties and ductile behaviour. *Chemical Engineering Journal* 348: 840–849.

81 Boulven, M., Quintard, G., and Cottaz, A. (2019). Homogeneous acylation of cellulose diacetate: towards bioplastics with tuneable thermal and water transport properties. *Carbohydrate Polymers* 206: 674–684.

82 Tanaka, S., Iwata, T., and Iji, M. (2017). Long/short chain mixed cellulose esters: effects of long acyl chain structures on mechanical and thermal properties. *ACS Sustainable Chemistry & Engineering* 5 (2): 1485–1493.

83 El-Sakhawy, M., Tohamy, H.A.S., and Salama, A. (2019). Thermal properties of carboxymethyl cellulose acetate butyrate. *Cellulose Chemistry and Technology* 53 (7–8): 667–675.

84 Krasnou, I., Tarasova, E., and Märtson, T. (2015). Thermoplastic cellulose stearate and cellulose laurate: melt rheology, processing and application potential. *International Polymer Processing* 30 (2): 210–216.

85 Meng, C., Cao, G.P., and Yan, Y.Z. (2017). Synthesis of cellulose acetate propionate with controllable contents and distributions of acetyl and propionyl in the C2, C3 and C6 positions. *Reaction Kinetics, Mechanisms and Catalysis* 122 (2): 1031–1047.

86 Ohno, T. and Nishio, Y. (2007). Estimation of miscibility and interaction for cellulose acetate and butyrate blends with *N*-vinylpyrrolidone copolymers. *Macromolecular Chemistry and Physics* 208 (6): 622–634.

87 Kusumi, R., Inoue, Y., and Shirakawa, M. (2008). Cellulose alkyl ester/poly(ε-caprolactone) blends: characterization of miscibility and crystallization behaviour. *Cellulose* 15 (1): 1–16.

88 Wang, X., Huang, S., and Wang, Y. (2017). Eco-friendly cellulose acetate butyrate/poly(butylene succinate) blends: crystallization, miscibility, thermostability, rheological and mechanical properties. *Journal of Polymer Research* 24 (2): 16.

89 Fei, Z., Huang, S., and Yin, J. (2015). Preparation and characterization of bio-based degradable plastic films composed of cellulose acetate and starch acetate. *Journal of Polymers and the Environment* 23 (3): 383–391.

90 Kunthadong, P., Molloy, R., and Worajittiphon, P. (2015). Biodegradable plasticized blends of poly(L-lactide) and cellulose acetate butyrate: from blend preparation to biodegradability in real composting conditions. *Journal of Polymers and the Environment* 23 (1): 107–113.

91 Rokkonen, T., Willberg-Keyriläinen, P., and Ropponen, J. (2021). Foamability of cellulose palmitate using various physical blowing agents in the extrusion process. *Polymers* 13 (15): 2416.

92 Breuer, R., Zhang, Y., and Erdmann, R. (2018). Extrusion foaming of thermoplastic cellulose acetate sheets with HFO-1234ze and co-blowing agents. *Polymer Engineering & Science* 58 (S1): E182–E188.

93 Breuer, R., Zhang, Y., and Erdmann, R. (2020). Development and processing of flame retardant cellulose acetate compounds for foaming applications. *Journal of Applied Polymer Science* 137 (28): 48863.

94 Shafaei, S., Dörrstein, J., and Guggenbichler, J.P. (2017). Cellulose acetate-based composites with antimicrobial properties from embedded molybdenum trioxide particles. *Letters in Applied Microbiology* 64 (1): 43–50.

95 Liu, J., Willför, S., and Xu, C. (2015). A review of bioactive plant polysaccharides: biological activities, functionalization, and biomedical applications. *Bioactive Carbohydrates and Dietary Fibre* 5 (1): 31–61.

96 Suteris, N.N., Misnon, I.I., and Roslan, R. (2020). Synthesis and characterization of polycaprolactone/cellulose acetate by electrospinning for wound dressing applications, Chapter 3. In: *Materials Science Forum* (ed. A.G.E. Sutjipto, F.H. Zulkifli and I.I. Misnon), 291–295. Trans Tech Publications Ltd.

97 Khoshnevisan, K., Maleki, H., and Samadian, H. (2018). Cellulose acetate electrospun nanofibers for drug delivery systems: applications and recent advances. *Carbohydrate Polymers* 198: 131–141.

98 Baharifar, H., Honarvarfard, E., and Malek-Kheili, M.H. (2017). The potentials and applications of cellulose acetate in biosensor technology. *Nanomedicine Research Journal* 2 (4): 216–223.

99 Suwantong, O. and Supaphol, P. (2015). Applications of cellulose acetate nanofiber mats. In: *Handbook of Polymer Nanocomposites. Processing, Performance and Application: Volume C: Polymer Nanocomposites of Cellulose Nanoparticles* (ed. J.K. Pandey, H. Takagi, A.N. Nakagaito and H.-J. Kim), 355–368. Berlin/Heidelberg: Springer.

100 Cheng, M., Qin, Z., and Hu, S. (2017). Use of electrospinning to directly fabricate three-dimensional nanofiber stacks of cellulose acetate under high relative humidity condition. *Cellulose* 24 (1): 219–229.

101 Fu, L.H., Qi, C., and Ma, M.G. (2019). Multifunctional cellulose-based hydrogels for biomedical applications. *Journal of Materials Chemistry B* 7 (10): 1541–1562.

102 Zhang, L., Guo, L., and Wei, G. (2018). Recent advances of cellulose-based materials and their promising application in sodium-ion batteries and capacitors. *Small* 14 (47): e1802444.

103 Abdul Khalil, H.P.S., Alwani, M.S., and Islam, M.N. (2015). The use of bamboo fibres as reinforcements in composites. In: *Biofiber Reinforcements in Composite Materials* (ed. O. Faruk and M. Sain), 488–524. Elsevier Inc.

104 Agarwal, J., Sahoo, S., and Mohanty, S. (2020). Progress of novel techniques for lightweight automobile applications through innovative eco-friendly composite materials: a review. *Journal of Thermoplastic Composite Materials* 33 (7): 978–1013.

105 Ferreira, D.P., Cruz, J., and Fangueiro, R. (2018). Surface modification of natural fibers in polymer composites. In: *Green Composites for Automotive Applications* (ed. G. Koronis and A. Silva), 3–41. Elsevier.

106 Gutiérrez, T.J. and Alvarez, V.A. (2017). Cellulosic materials as natural fillers in starch-containing matrix-based films: a review. *Polymer Bulletin* 74 (6): 2401–2430.

107 Panthapulakkal, S., Raghunanan, L., and Sain, M. (2017). Natural fiber and hybrid fiber thermoplastic composites: advancements in lightweighting applications. In: *Green Composites: Waste and Nature-based Materials for a Sustainable Future*, 2e (ed. C. Baillie and R. Jayasinghe), 39–72. Elsevier Inc.

108 Ramamoorthy, S.K., Skrifvars, M., and Persson, A. (2015). A review of natural fibers used in biocomposites: plant, animal and regenerated cellulose fibers. *Polymer Reviews* 55 (1): 107–162.

109 Souza, S.F., Ferreira, M., and Sain, M. (2015). The use of curaua fibers as reinforcements in composites. In: *Biofiber Reinforcements in Composite Materials* (ed. O. Faruk and M. Sain), 700–720. Elsevier Inc.

110 Sarasini, F. (2017). Fully biodegradable all-cellulose composites. In: *Handbook of Composites From Renewable Materials* (ed. V.K. Thakur, M.K. Thakur and M.R. Kessler), 303–322. Wiley.

111 Baghaei, B. and Skrifvars, M. (2020). All-cellulose composites: a review of recent studies on structure, properties and applications. *Molecules* 25 (12): 2836.

112 Fujisawa, S., Saito, T., and Isogai, A. (2018). All-cellulose (cellulose-cellulose) green composites. In: *Advanced Green Composites* (ed. A. Netravali), 111–133. Wiley.

113 de Oliveira Sousa Neto, V. and Do Nascimento, R.F. (2017). Recent progress in biocomposite of biodegradable polymer. In: *Handbook of Composites from Renewable Materials* (ed. V.K. Thakur, M.K. Thakur and M.R. Kessler), 61–94. Wiley.

114 Dufresne, A. and Castaño, J. (2017). Polysaccharide nanomaterial reinforced starch nanocomposites: a review. *Starch/Staerke* 69 (1, 2): 1500307.

115 Espinach, F.X., Méndez, J.A., and Granda, L.A. (2017). Bleached kraft softwood fibers reinforced polylactic acid composites, tensile and flexural strengths. In: *Natural Fiber-Reinforced Biodegradable and Bioresorbable Polymer Composites* (ed. A.K. Lau and A.P.-Y. Hung), 73–90. Elsevier Inc.

116 Khosravi, A., Fereidoon, A., and Khorasani, M.M. (2020). Soft and hard sections from cellulose-reinforced poly(lactic acid)-based food packaging films: a critical review. *Food Packaging and Shelf Life* 23: 100429.

117 Mokhena, T.C., Sefadi, J.S., and Sadiku, E.R. (2018). Thermoplastic processing of PLA/cellulose nanomaterials composites. *Polymers* 10 (12): 1363.

118 Nazrin, A., Sapuan, S.M., and Zuhri, M.Y.M. (2020). Nanocellulose reinforced thermoplastic starch (TPS), polylactic acid (PLA), and polybutylene succinate (PBS) for food packaging applications. *Frontiers in Chemistry* 8: 213.

119 Jonoobi, M., Oladi, R., and Davoudpour, Y. (2015). Different preparation methods and properties of nanostructured cellulose from various natural resources and residues: a review. *Cellulose* 22 (2): 935–969.

120 Vigneshwaran, N., Satyamurthy, P., and Jain, P. (2015). Biological synthesis of nanocrystalline cellulose by controlled hydrolysis of cotton fibers and linters. In: *Handbook of Polymer Nanocomposites. Processing, Performance and Application: Volume C: Polymer Nanocomposites of Cellulose Nanoparticles* (ed. J.K. Pandey, H. Takagi, A.N. Nakagaito and H.-J. Kim), 27–36. Berlin/Heidelberg: Springer.

121 Abdul Khalil, H.P., Davoudpour, Y., and Bhat, A.H. (2015). Electrospun cellulose composite nanofibers. In: *Handbook of Polymer Nanocomposites. Processing, Performance and Application: Volume C: Polymer Nanocomposites of Cellulose Nanoparticles* (ed. J.K. Pandey, H. Takagi, A.N. Nakagaito and H.-J. Kim), 191–228. Berlin/Heidelberg: Springer.

122 Boufi, S. (2015). Nanofibrillated cellulose: sustainable nanofiller with broad potentials use. In: *Biomass and Bioenergy: Applications* (ed. K.R. Hakeem, M. Jawaid and U. Rashid), 267–305. Springer International Publishing.

123 Jorfi, M. and Foster, E.J. (2015). Recent advances in nanocellulose for biomedical applications. *Journal of Applied Polymer Science* 132 (14): 41719.

124 Joseph, K., Sandhya, K.Y., and Saritha, A. (2015). Liquid crystalline polymers from renewable resources: synthesis and properties. In: *Liquid Crystalline Polymers: Volume 1-Structure and Chemistry* (ed. V.K. Thakur and M.R. Kessler), 273–306. Springer International Publishing.

125 Polman, E.M.N., Gruter, G.-J.M., and Parsons, J.R. (2021). Comparison of the aerobic biodegradation of biopolymers and the corresponding bioplastics: a review. *Science of the Total Environment* 753: 141953.

126 Yadav, N. and Hakkarainen, M. (2021). Degradable or not? Cellulose acetate as a model for complicated interplay between structure, environment and degradation. *Chemosphere* 265: 128731.

9

Ester Derivatives of Microbial Synthetic Polysaccharides

Hakyong Lee, Hongyi Gan, Azusa Togo, Yuya Fukata, and Tadahisa Iwata

The University of Tokyo, Graduate School of Agricultural and Life Sciences, Department of Biomaterial Sciences, Science of Polymeric Materials, 1-1-1, Yayoi, Bunkyo-ku, Tokyo, 113-8657 Japan

9.1 Introduction

9.1.1 Background of Bio-Based Plastics

Plastics manufactured from petroleum-based polymers are among the most conspicuous products in modern life. Light, cheap plastics that are easily processed and shaped have improved the comfort and convenience of modern life through applications ranging from automobiles to medical devices and food packaging. However, the constant increase in plastic fabrication has resulted in serious environmental problems, such as global warming, waste management and treatment, and marine microplastics pollution. With a lower environmental impact, bio-based plastics derived from sustainable and renewable natural resources, and, preferably, the biodegradable ones that degrade into carbon dioxide and water under the action of microorganisms, have attracted attention as potential alternatives to petroleum-based non-biodegradable plastics. For the definitions of bio-based plastics, we refer to the introductory chapter of this book. When it comes to bio-based plastics, these are expected to help achieving carbon neutrality. Furthermore, biodegradable plastics are considered key polymers for alleviating part of the problem of plastic waste as marine microplastics pollution.

Bio-based plastics are generally classified into three categories [1]:

1) Those with identical chemical structures to petroleum-based plastics, such as bio-polyethylene (bio-PE) derived from bio-ethanol and bio-poly(ethylene terephthalate) (bio-PET) produced from bio-based ethylene glycol.
2) Bio-based plastics that are biodegradable and have similar physical properties to petroleum-based plastics, such as poly(lactic acid) (PLA) and polyhydroxyalkanoates (PHAs).
3) Plastics with unique structures and properties derived only from biomass such as polysaccharide esters and polymers derived from aromatic plants. There are also lignin-based plastics, which fall in this category.

Biodegradable Polymers in the Circular Plastics Economy, First Edition.
Edited by Michiel Dusselier and Jean-Paul Lange.

Among bio-based plastics, PLA and PHAs are the most widely studied. PLA is an aliphatic polyester derived from fermented plant starch, such as corn, sugarcane, or sugar beet pulp. PLA has a glass transition temperature (T_g) of 60 °C and a melting temperature (T_m) of 175 °C [2], and films, fibers, and injection molding of PLA have been developed. PLA degrades under composting conditions (temperature >65 °C and humidity >60%) through hydrolysis but degrades slowly under lower temperature conditions. PHAs are produced in nature by numerous microorganisms from sugars or lipids [3]. More than 150 types of PHAs have been reported and can be combined to produce materials with different properties. The physical properties of poly[(*R*)-3-hydroxybutyrate] (P(3HB)) are similar to those of polypropylene (PP) with a T_g of −4 °C and a T_m of 170 °C. However, P(3HB) is relatively brittle, while P(3HB) copolymers with different second monomer units, such as 3-hydroxyvalerate, 3-hydroxyhexanoate, and 4-hydroxybutyrate, provide ductility. PHAs have high biocompatibility, which is promising for their widespread use in the medical field.

9.1.2 Polysaccharides

Polysaccharides are the most abundant natural polymer resources on Earth and are mainly composed of glucose units bonded by two types of glycosidic linkages (α and β). Figure 9.1 shows typical polysaccharides and their chemical structures, including cellulose contained in plant cell walls and plant fibers, glucomannan isolated from the tubers of *Amorphophallus konjac* plants, chitin or chitosan obtained from crustacean shells, pullulan produced by the strains of the fungus *Aureobasidium pullulans*, and paramylon derived from microalgae. Polysaccharides with different glycosidic linkages and molecular structures have varying physical properties. As the most abundant natural polymer resource, polysaccharides could potentially be used as raw materials to produce many useful bioplastics. As polysaccharides are already polymerized, they do not need to be synthesized from monomers. However, their thermal processability and solubility in organic solvents are restricted by strong intermolecular and intramolecular hydrogen bonds between the hydroxyl groups of monosaccharide units. To gain solubility and thermoplasticity, hydrogen bonds are suppressed by substituting hydroxyl groups with functional ester groups through esterification. Usually, esterification of polysaccharides is performed using carboxylic acids and trifluoroacetic anhydride (TFAA) as a catalyst at 50 °C for one hour. Then, the reaction solution is poured into methanol or water to obtain polysaccharide esters as powder samples.

Cellulose found in plant cell walls and plant fibers is the most abundant and commonly used polysaccharide with a molecular structure comprising β-1,4-glucan without branches. Cellulose triacetate (CTA) is an important polymer that is mainly used in optical films, cigarette filters, and fiber materials [4]. However, despite its excellent properties, a large amount of plasticizer is required for heat molding owing to poor thermal fluidity [5]. In addition to cellulose, many types of natural polysaccharides with various characteristic structures can be found in nature, including hemicellulose containing xylan, chitin, and chitosan, which are

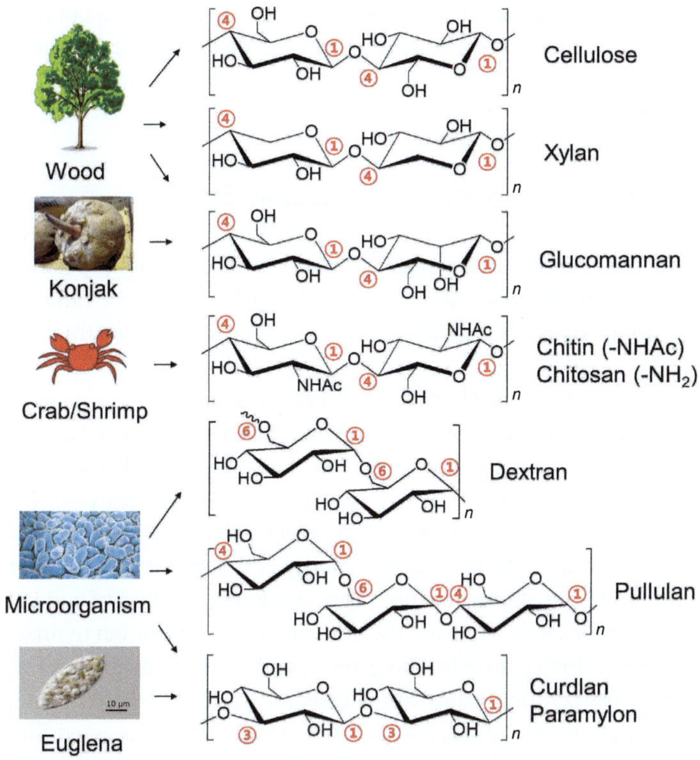

Figure 9.1 Representative polysaccharides existed in nature. Source: Tim/Adobe Stock Photos.

obtained from crabs and prawns, and curdlan and pullulan, which are synthesized by microorganisms. Although most of these materials are used as food additives and paint thickeners, there are few reports on their use as plastics.

In recent years, we have developed polysaccharide esters fabricated from various polysaccharides. As a result, some unique properties have been developed using techniques of esterification processes. In particular, pullulan esters consisting of α-1,3- and α-1,4-linkages show outstanding transparency that is insensitive to light polarization (zero birefringence) and, thereby, deliver sharp transparency [6, 7], dextran esters consisting of α-1,6-glucan show excellent adhesive property [8], and β-1,3-glucan derived from paramylon or curdlan shows high thermal fluidity after esterification [9–11]. An organic solvent is generally needed to extract polysaccharides from natural resources, resulting in a decrease in molecular weight. To solve the problem, we have synthesized a non-natural polysaccharide with a new α-1,3-glucan structure *in vitro* using enzymatic polymerization, followed by esterification [12, 13]. Furthermore, to improve the physical properties, ester groups with branches were adopted in cellulose, curdlan, and α-1,3-glucan derivatives [14–16]. In the case of enzymatic synthesis, regioselective esterification of dextrin was achieved [17].

This chapter introduces the excellent thermal and mechanical properties obtained from various polysaccharide linear and branched esters and discusses the enzymatic polymerization and esterification of polysaccharides.

9.2 Zero Birefringence Property of Pullulan Esters

Various α-glucans are found in nature depending on the glycosidic linkage position. The most representative α-glucan is a starch composed of α-1,4-glycosidic linkages. In this section, other α-glucan esters are introduced, namely, pullulan, dextran, and α-1,3-glucan.

Pullulan is a natural linear polysaccharide produced by strains of fungus *Aureobasidium pullulans*. This polysaccharide consists of α-1,6-linked maltotriose units, which comprise three α-1,4-linked glucose units [18]. The commercial production of pullulan at Hayashibara is at approximately 300 metric tons per year [18]. It is applied as a food additive and a biomedical material owing to its valuable properties, such as biodegradability, water solubility, non-toxicity, and film-forming abilities [19, 20]. However, this valuable glucan has not been previously applied as a plastic material. To date, some studies have reported the derivatization of pullulan hydroxyl groups with substituents including sulfate, cholesterol, and carboxylate groups [21]. Among these various derivatization studies, much research has focused on the esterification of pullulan (Figure 9.2) [6, 7, 22, 23]. Solution-cast and melt-pressed films were successfully prepared from pullulan esters with different carbon numbers (n), from acetic acid ($n = 2$) to myristic acid ($n = 14$) [6]. The characteristics of pullulan ester films are shown in Table 9.1. The tensile strength of these solvent-cast films is in the range of 1.4–24 MPa, while the elongation at break is in the range of 8–1120%. The higher the carbon number of the acyl groups, the lower the strength, but the higher the elasticity. Pullulan esters have no melting point, suggesting that they are amorphous polymers [6]. However, the T_g of pullulan acetate is 174 °C, which is higher than that of polycarbonate (145 °C), poly(methyl methacrylate) (115 °C), and polystyrene (100 °C). This result indicates that pullulan acetate has higher thermal stability than oil-based amorphous polymers. Transparent melt-pressed films of pullulan esters can be easily prepared above the T_g.

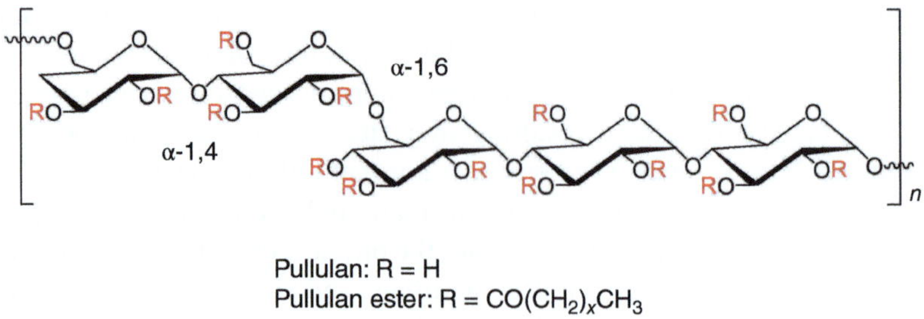

Figure 9.2 Chemical structure of pullulan esters.

Table 9.1 Characteristics of α-glucan.

Name	Substituent	M_w (×10^5)	M_n (×10^5)	Polydispersity index (PDI)	T_g (°C)	T_m (°C)	$T_{d5\%}$ (°C)	Tensile strength (MPa)	Elongation strain (%)	Young's modulus (MPa)
Pullulan	Acetate ($n = 2$)	2.1	0.92	2.3	174	—	332	23.4	8	772
	Propionate ($n = 3$)	1.5	0.67	2.2	127	—	331	23.5	14	650
	Butyrate ($n = 4$)	4.1	1.9	2.1	99	—	332	15.6	9	571
	Valerate ($n = 5$)	5.9	2.3	2.4	67	—	334	9.5	37	274
	Hexanoate ($n = 6$)	5.9	2.5	2.4	46	—	349	2.1	410	23
	Octanoate ($n = 8$)	7.1	3.4	2.3	42	—	354	2.4	1120	8
	Decanoate ($n = 10$)	8.2	3.1	2.4	35	—	351	1.4	830	9
	Laurate ($n = 12$)	9.0	3.6	2.5	38	—	348	1.8	1000	16
	Myristate ($n = 14$)	8.1	3.1	2.6	38	—	365	1.4	1090	7
Dextran	Acetate ($n = 2$)	—	—	—	—	—	346	—	—	—
	Propionate ($n = 3$)	2.1	0.74	2.8	—	—	364	—	—	—
	Butyrate ($n = 4$)	2.2	0.55	4.1	75	—	360	7.9	14	71
	Valerate ($n = 5$)	1.5	0.61	2.4	18	—	285	1.4	84	10
	Hexanoate ($n = 6$)	1.6	0.63	2.5	−1.6	—	302	0.76	35	10
	Octanoate ($n = 8$)	2.3	0.60	3.8	2.4	—	330	0.44	20	5.5
	Decanoate ($n = 10$)	3.0	0.72	4.2	9.4	—	273	0.48	33	4.7
	Laurate ($n = 12$)	3.2	0.79	4.0	—	—	269	2.2	27	24
α(1 → 3) Glucan	Acetate ($n = 2$)	1.6	0.71	2.3	177	339	350	41	7	840
	Propionate ($n = 3$)	2.1	0.89	2.3	150	299	355	37	8	710
	Butyrate ($n = 4$)	1.8	0.84	2.2	110	242	354	15	12	360
	Valerate ($n = 5$)	1.6	0.77	2.1	97	210	353	8.2	32	220
	Hexanoate ($n = 6$)	1.7	0.84	2.0	91	142	350	7.7	90	150
	Octanoate ($n = 8$)	2.2	0.90	2.4	28	—	351	4.8	212	70

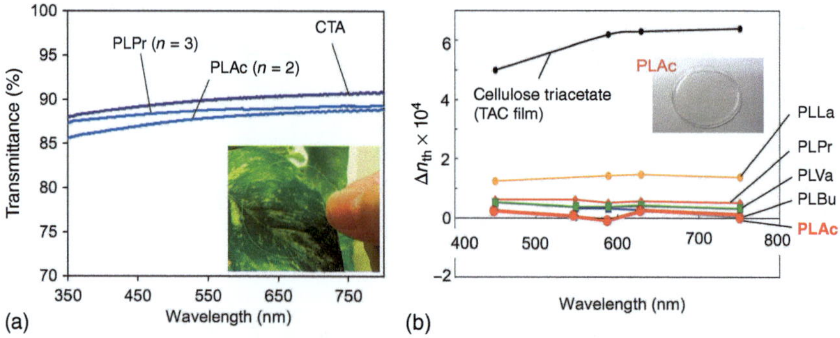

Figure 9.3 (a) Optical transmittance of pullulan acetate (PLAc), propionate (PLPr), and cellulose triacetate (CTA) and (b) out-of-plane birefringence of pullulan ester films. Source: Danjo et al. [7]/Springer Nature/CC BY 4.0.

As shown in Figure 9.3a, the optical transmittance percentage of these films was within the range of 83–88% in the visible-light wavelength range (380–750 nm) [7]. These values are similar to those of cellulose triacetate (CTA, 87–90%), which is applied as an liquid crystal display (LCD) protective film. In particular, Figure 9.3b shows a lower out-of-plane birefringence for pullulan ester films than for CTA films, regardless of the acyl carbon number [7]. The birefringence of the CTA film is known to decrease with the introduction of some additives. However, surprisingly, the out-of-plane birefringence of pullulan acetate was almost "zero" throughout the visible-light wavelength region without additives. This result suggests that pullulan acetate has an excellent zero birefringence property with high thermal stability without any additive, which has great potential for applications in optical films.

9.3 Bio-Based Adhesives from Dextran (α-1,6-Glucan)

Similar to pullulan, dextran is a water-soluble polysaccharide. However, it shows unique properties different from those of pullulan. Dextran is a natural polysaccharide comprising glucose linked by α-1,6-glycocidic linkages (Figure 9.1) [24]. This α-1,6-glucan is biologically synthesized from sucrose by lactic acid bacteria, such as *Leuconostoc mesenteroides*. The commercial production of dextran is estimated to be c. 2000 t/yr [25]. It has been widely applied as a medical material and a cosmetic binding agent owing to its water solubility and biocompatibility [24, 26–28]. However, dextran derivatives can also be used as adhesives, including hot-melt adhesives and safety solution adhesives, by esterifying the hydroxyl groups of dextran with linear carboxylic acids [8]. As shown in Tables 9.1 and 9.2, these dextran esters have different physical properties and solubilities depending on their acyl carbon number (*n*) and degree of substitution (DS), which is the number of hydroxyl groups substituted among the three hydroxyl groups in glucose units [8, 29]. For example, the solubility of dextran esters can reportedly be controlled by changing the DS. Table 9.2 shows the solubility of dextran esters in common solvents. Dextran butyrate (DexBu,

Table 9.2 Solubility of dextran butyrate with different DS in common solvents.

Name	DS	Water	Methanol	Ethanol	Acetone	Chloroform
Dextran-neat	0.0	○	×	×	×	×
DexBu-0.7	0.7	○	○	×	×	×
DexBu-1.1	1.1	×	○	○	○	×
DexBu-1.7	1.7	×	○	○	○	○
DexBu-2.2	2.2	×	○	○	○	○
DexBu-2.5	2.5	×	○	○	○	○
DexBu-3.0	3.0	×	×	×	○	○

○; soluble, ×; insoluble.

$n = 4$) with a low DS dissolves in hydrophilic solvents, such as alcohol or water, while DexBu with a high DS does not dissolve in hydrophilic solvents such as water [29]. Specifically, the water resistance of dextran was expected to be improved by esterification. Notably, DexBu with a DS of 1.1–2.5 dissolves in ethanol. Figure 9.4 shows some typical examples of adhesion with dextran ester adhesives using ethanol as a

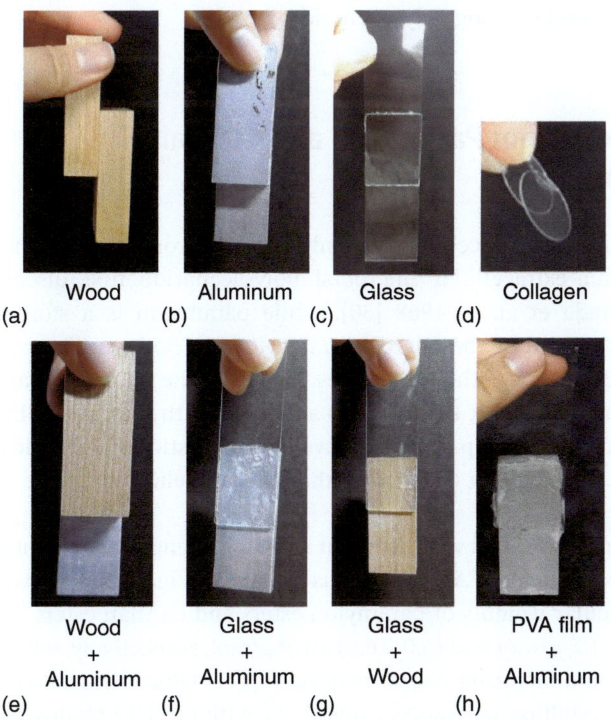

Figure 9.4 Adhesive property of dextran esters with different materials. (a) wood + wood, (b) aluminum + aluminum, (c) glass + glass, (d) collagen + collagen, (e) wood + aluminum, (f) glass + aluminum, (g) glass + wood, (h) PVA film + aluminum.

solvent. Ethanol solutions of dextran esters have shown adhesive performance with a wide range of materials, including wood, glass, glass coated by a collagen film, and gladius of squid [8, 29]. The tensile strength with wood is comparable to that of commercial polyvinyl alcohol (PVA)–water adhesives [29]. Both the water resistance and drying speed have been reported to be significantly improved by using ethanol as a solvent compared with water-based adhesives, such as PVA–water and dextran–water adhesives, although the latter leads to increased emission of volatile organic compounds. For example, in a water resistance test, neat dextran–water adhesives and commercial PVA–water adhesives completely lost adhesive strength after immersion in water for 24 hours, while DexBu ($n = 4$, DS = 1.1) maintained 51% of the adhesive strength (1.0 MPa) measured before immersion in water [29]. Another advantage of dextran ester adhesives is applicability to medical adhesives. As shown in Figure 9.4d, ethanol-born dextran esters show adhesive properties with medical materials, such as collagen films and gladius of squid.

Finally, some dextran esters, such as dextran valerate, have also shown adhesive performance as hot-melt adhesives by heating above the T_g [29]. This is a meaningful property because hot-melt adhesives are more eco-friendly because no solvent is required. Therefore, dextran ester adhesives can be applied using two eco-friendly approaches: hot-melt and ethanol as a solvent. These good adhesive properties with various materials suggest that dextran esters have high functionality as an adhesive. This valuable property is unique among polysaccharides.

9.4 Films and Fibers from Paramylon and Curdlan (β-1,3-Glucan) Esters

β-1,3-Glucan is a crystalline polysaccharide mainly derived from curdlan and paramylon. Curdlan is an extracellular microbial polysaccharide first discovered and studied by Harada et al. in 1968 [30], while paramylon is a storage polysaccharide photosynthesized by the microalga *Euglena* [31]. Both curdlan and paramylon are linear β-1,3-glucans without branches. Normally, the weight-average molecular weight of paramylon (200×10^3 g/mol) is about one-fifth that of curdlan (1000×10^3 g/mol). Both curdlan and paramylon have characteristic sixfold helical main chain structures [32–35], which differ from the twofold helical structure of cellulose.

Fully substituted β-1,3-glucan esters with different side-chain lengths can be prepared by a simple carboxylic acid/TFAA system, as shown in Figure 9.5 [9, 10]. The weight-average molecular weights of paramylon esters and curdlan esters are in the range of $(3.0–9.6) \times 10^5$ g/mol and $(6.0–16.0) \times 10^5$ g/mol, respectively, which are adequate values for manufacturing plastic materials. β-1,3-Glucan esters with C2–C6 alkyl chains are crystalline polymers, while those with C8–C12 chains are amorphous polymers, as summarized in Table 9.3. Furthermore, despite the molecular weight differences between curdlan and paramylon, their esters with the same substituted ester groups show the same thermal properties.

Table 9.3 Characteristics of β-1,3-glucan.

Name	Substituent	M_w (×10^5)	M_n (×10^5)	PDI	T_g (°C)	T_m (°C)	$T_{d5\%}$ (°C)	Tensile strength (MPa)	Elongation strain (%)	Young's modulus (MPa)
Curdlan	Acetate ($n = 2$)				171	287	318	21.2 ± 4.0	12 ± 3	448 ± 33
	Propionate ($n = 3$)	6.0	3.3	1.8	110	213	364	23.7 ± 4.2	46 ± 20	260 ± 99
	Butyrate ($n = 4$)	8.9	4.6	1.9	74	190	351	15.4 ± 1.9	26 ± 5	204 ± 12
	Valerate ($n = 5$)	13.0	6.4	2.0	51	168	296	12.4 ± 2.3	79 ± 9	149 ± 35
	Hexanoate ($n = 6$)	14.0	7.1	2.0	69	167	274	14.7 ± 1.0	150 ± 28	92 ± 10
	Octanoate ($n = 8$)	14.0	6.7	2.1	50		281	6.0 ± 2.0	112 ± 43	46 ± 3
	Decanoate ($n = 10$)	16.0	7.4	2.1	55		292	4.6 ± 0.4	144 ± 11	19 ± 1
	Laurate ($n = 12$)	16.0	7.0	2.3	47		293	1.7 ± 0.3	172 ± 35	12 ± 1
Paramylon	Acetate ($n = 2$)				168	281	327			
	Propionate ($n = 3$)	3.4	1.9	1.8	112	221	340	31.4 ± 3.4	62 ± 6	597 ± 52
	Butyrate ($n = 4$)	5.8	3.2	1.8	76	207	337	20.6 ± 1.2	87 ± 25	352 ± 54
	Valerate ($n = 5$)	8.1	3.9	2.1		196	330	7.9 ± 0.7	61 ± 28	133 ± 19
	Hexanoate ($n = 6$)	8.5	4.0	2.1		114	328	7.5 ± 0.6	446 ± 18	47 ± 3
	Octanoate ($n = 8$)	8.9	4.5	1.9			330	2.7 ± 0.3	415 ± 66	3 ± 1
	Decanoate ($n = 10$)	9.2	4.8	1.9			324	2.8 ± 0.6	328 ± 48	12 ± 1
	Laurate ($n = 12$)	9.6	5.4	1.8			333	2.7 ± 0.5	529 ± 108	5 ± 1

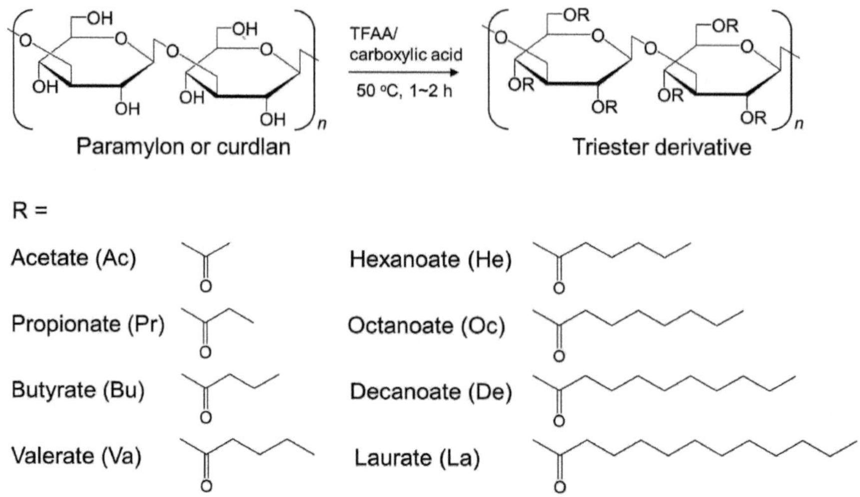

Figure 9.5 Synthetic scheme of β-1,3-glucan esters.

Figure 9.6a shows a comparison of the thermal properties of β-1,3-glucan esters with those of commonly used polymers. By adjusting the carbon number of the ester groups, esters with T_m values of 114–281 °C can be obtained. This range covers the T_m values of commonly used plastics, such as polyethylene (PE, 120 °C), polypropylene (PP, 170 °C), and polyethylene terephthalate (PET, 280 °C). The T_g values of β-1,3-glucan esters are between 47 and 168 °C, with those of β-1,3-glucan acetate (168 °C), β-1,3-glucan propionate (112 °C), and β-1,3-glucan butyrate (76 °C) higher than that of PET (70 °C). Furthermore, the 5% mass reduction temperatures evaluated by thermogravimetric analysis (TGA) are all above 300 °C, indicating that β-1,3-glucan esters, except acetate derivatives, have a wide window of thermal process temperatures.

All β-1,3-glucan esters can form self-standing films using the solvent-cast method, while those with C2–C6 alkyl chains can also make melt-quench films. With increasing alkyl side-chain length, the tensile strength and Young's modulus decrease, while the elongation at break tends to increase. Figure 9.6b shows a comparison of the tensile properties of β-1,3-glucan esters with those of commonly used polymers. By adjusting the side-chain length, the tensile properties of β-1,3-glucan esters can be tuned to be similar to those of PP and PE. Regarding the thermal and mechanical properties, they can have similar mechanical properties, and much higher thermal stabilities, compared with those of popular petroleum-based plastics. These properties show that β-1,3-glucan esters are prospective biomass plastics with excellent thermal and mechanical properties.

In addition to films, melt-spun fibers can be manufactured from β-1,3-glucan propionate, butyrate, and valerate derivatives owing to their favorable thermal fluidity at the melt-processing temperature indicated by differential scanning calorimetry (DSC) and melt flow test results [11]. The image of β-1,3-glucan propionate fiber is

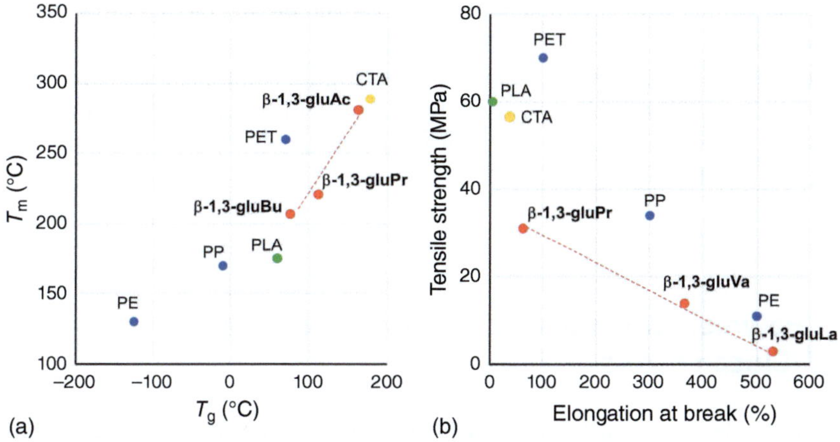

Figure 9.6 (a) Thermal and (b) mechanical properties of β-1,3-glucan esters compared with those of general petroleum- or bio-based plastics.

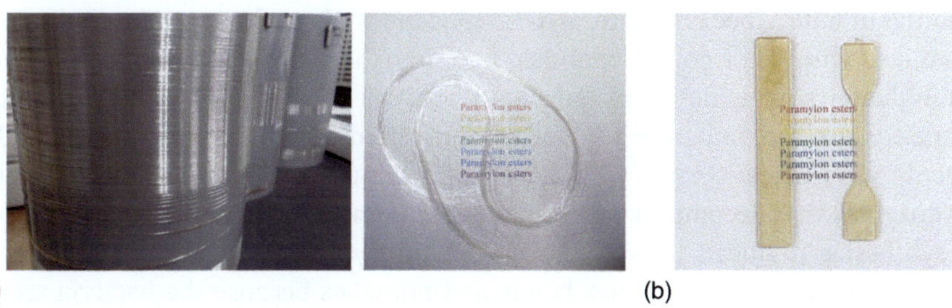

Figure 9.7 (a) Melt-spun fiber and (b) injection-molded specimens of β-1,3-glucan propionate.

shown in Figure 9.7a as a representative one. Each of the β-1,3-glucan esters was successively melt-spun at various take-up rates, without the need for additives, and the molecular weight did not change significantly. By adjusting the take-up rate, crystalline or amorphous melt-spun fibers could be produced. The fiber strength could be improved by increasing the take-up rate or annealing treatment, both of which promoted the crystallization of molecular chains. For β-1,3-glucan propionate, in addition to fibers, a bulk specimen could also be fabricated in different shapes by injection molding, as shown in Figure 9.7b. The specimens are yellowish because of the heating procedure during thermal processing, which might be prevented by adding an antioxidant.

In addition to homoesters, Shibakami et al. prepared paramylon esters with mixed short and long alkyl side chains, aiming to control the physical properties precisely by adjusting carboxylic acid types and ratios [36–38]. Among injection-molded specimens of the mixed esters, some showed bending strength and bending elastic modulus comparable to those of petroleum-based acrylonitrile–butadiene–styrene (ABS) resin and cellulose acetate stearate mixed ester. Meanwhile, the impact strength of mixed paramylon esters was comparable to that of PLA, indicating that mixed paramylon esters have potential uses as thermoplastic materials in practical applications.

In summary, β-1,3-glucan esters show a wide range of controllable physical properties with changing side-chain lengths. By changing the ester group from valerate to acetate, the glucan ester shows thermal and mechanical properties that vary from PE-like to PET-like properties.

9.5 Polymerization of α-1,3-Glucan and Films of α-1,3-Glucan Esters

α-1,3-Glucan is a polysaccharide in which glucose is linked by α-1,3-glycosidic linkages. This glucan is a structural polysaccharide in the cell walls of fungi and yeast [39–42] and is also known as a major component of plaque produced by *Streptococcus* in the mouth [43, 44]. However, no pure and complete linear α-1,3-glucan is found in nature because natural α-1,3-glucans are branched by other linkages or are mixed with other polysaccharides. Recently, the synthesis of completely linear α-1,3-glucan was achieved by enzymatic polymerization [12]. As such, glucan sucrase is used to prepare α-1,3-glucan from sucrose, a disaccharide composed of glucose and fructose (Figure 9.8). Glycosyltransferase (Gtf) is among the extracellular enzymes that decompose sucrose and simultaneously polymerize only glucose to synthesize α-glucan [45–48]. In nature, the synthesized polysaccharide is an α-glucan network with complex bonds and branches because the bacteria secrete multiple types of glucan sucrase. Therefore, *in vitro* enzymatic polymerization with the GtfJ enzyme cloned from *Streptococcus salivarius* (ATCC25975) was adopted in their research. The GtfJ enzyme was produced by culturing *Escherichia coli* expressing GtfJ. After the produced enzyme was added to sucrose aqueous solution adjusted to pH 5.5, a large amount of white precipitate was observed within a few days (Figure 9.9). As shown in Figure 9.9d, this white precipitate was found to have a crystalline fiber structure with a width of 10–20 mm and length of about 1 μm. The synthesized linear α-1,3-glucan was readily collected by filtration or centrifugation because it was insoluble in water. This enzymatic polymerization is an environmentally friendly one-pot water-based reaction without organic solvent.

The yield and molecular weight of the synthesized glucan are important factors in its application in plastic materials because they directly affect production cost and the physical properties of the materials, respectively. By conducting an enzymatic reaction at 30 °C, α-1,3-glucan with a M_w of 2.0×10^5 g/mol and a polydispersity of 2.1 can be reportedly synthesized at about 50 g for the 1 l system in four days. Although the optimum temperature for this enzyme is around 37 °C, similar to the human oral temperature, the glucan molecular weight increased to over 7.0×10^5 when conducting the enzymatic reaction at 15 °C. This shows that the synthesized molecular weight can be controlled by adjusting the reaction conditions of this enzymatic polymerization.

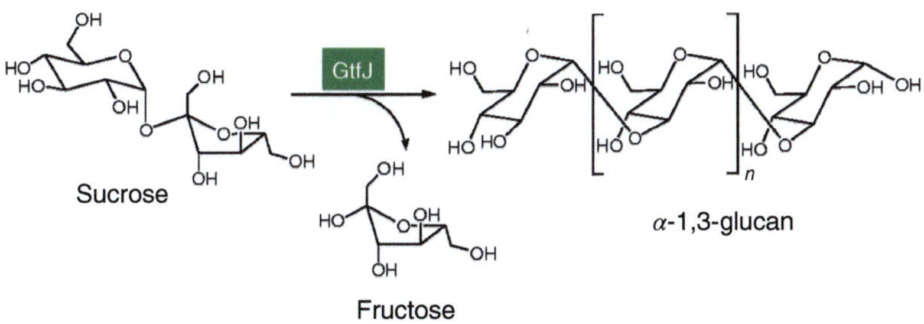

Figure 9.8 Enzymatic polymerization of α-1,3-glucan.

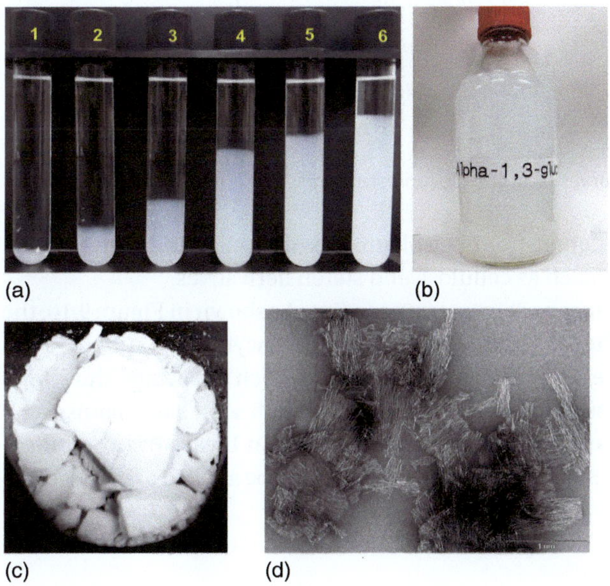

Figure 9.9 (a) *In vitro*-synthesized α-1,3-glucan; (b) synthesis of α-1,3-glucan in large scale; (c) filtered and dried α-1,3-glucan; and (d) transmission electron microscopy (TEM) image of synthesized α-1,3-glucan.

The designed synthesis of high-molecular-weight α-1,3-glucan is indispensable for improving the mechanical properties and processability into fibers. The establishment of high-yielding and high-molecular-weight enzyme reaction conditions is required. Like other polysaccharides, this synthesized α-1,3-glucan does not exhibit thermal plasticity and solubility in common organic solvents because of hydrogen bonding derived from its hydroxyl groups. However, esterification of the hydroxyl groups imparts the glucans with thermal plasticity and solubility in organic solvent [12]. α-1,3-Glucan esters have much higher T_m values than other polysaccharide esters and common petroleum-based plastics, such as CTA and PET. For example, the T_m of α-1,3-glucan acetate is 340 °C. Furthermore, the T_g of α-1,3-glucan esters is in the range of 120–180 °C, which is higher than that of nylon-6 (55 °C) and PET (70 °C). This suggests that novel bio-based plastics derived from α-1,3-glucan have

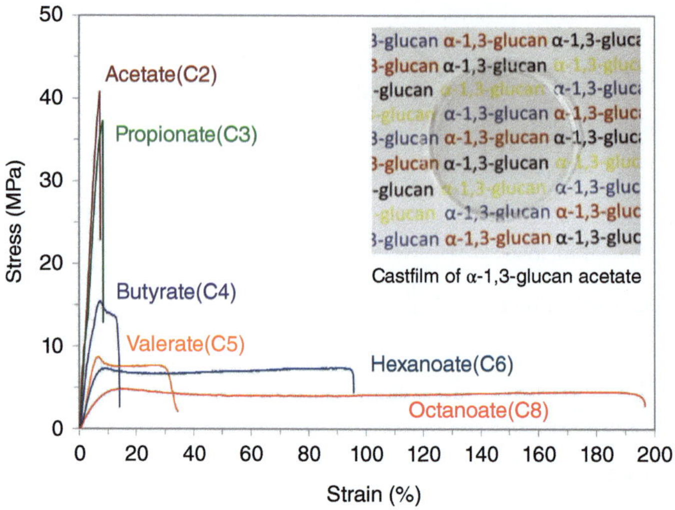

Figure 9.10 Stress–strain curves of α-1,3-glucan ester films.

new thermal properties different from petroleum-based plastics and conventional polysaccharide derivatives, such as cellulose and starch derivatives.

The films of these α-1,3-glucan esters are transparent. As shown in Figure 9.10, the mechanical properties of these films are controlled by the acyl carbon number [13]. For example, α-1,3-glucan esters with short side chains, such as acetate, show high maximum stress and low strain at break, while those with long side chains, such as octanoate, show low maximum stress and high strain at break. Based on these mechanical properties, α-1,3-glucan esters are expected to be applied as engineering plastics in the future.

9.6 High-Performance Polysaccharide-Branched Esters

In addition to the length, or carbon number, the end structure of side chain ester groups plays a prominent role in the physical properties of polysaccharide esters as thermoplastic resins. In particular, ester groups with terminally branched shapes (*iso*-branched or *tert*-branched ones such as isovalerate or pivalate, respectively, as shown in Figure 9.11) give unique thermal properties, as characterized by high melting temperatures and high glass transition temperatures that cannot be attained by linear esters. Table 9.4 shows the characteristics of polysaccharide-branched esters compared with those of cellulose, curdlan, and α-1,3-glucan.

9.6.1 Cellulose-Branched Esters [14]

Iso-branched acyl groups can be introduced into cellulose by a heterogeneous reaction with the corresponding carboxylic acids and TFAA, which is the same method used to introduce linear acyl groups. Fully substituted cellulose *iso*-branched esters

Table 9.4 Characteristics of polysaccharide-branched esters.

Name	Substituent	M_w (×10^5)	M_n (×10^5)	PDI	T_g (°C)	T_m (°C)	$T_{d5\%}$ (°C)	Tensile strength (MPa)	Elongation strain (%)	Young's modulus (MPa)
Cellulose	Isobutyrate ($n = 4$)	4.4	2.3	1.9	115	248	300	28 ± 1	25 ± 4	350 ± 20
	Isovalerate ($n = 5$)	6.3	3.1	2.1	79	141	294	25 ± 1	44 ± 20	330 ± 40
	Isohexanoate ($n = 6$)	9.3	4.7	2.0	70	106	330	19 ± 2	208 ± 41	300 ± 50
	Isoheptanoate ($n = 7$)	6.3	2.6	2.4	61	104	331	16 ± 2	403 ± 27	150 ± 20
Curdlan	Isobutyrate ($n = 4$)	8.7	4.0	2.2	110	339	360	28.6 ± 2.0	68 ± 28	470 ± 170
	Isovalerate ($n = 5$)	20.6	8.0	2.6	91	250	368	15.5 ± 1.4	52 ± 20	250 ± 60
	Isohexanoate ($n = 6$)	21.2	6.8	3.1	64	199	369	18.6 ± 2.1	180 ± 21	280 ± 50
	Isoheptanoate ($n = 7$)	17.7	6.1	2.9	41	126	368	5.0 ± 0.6	349 ± 64	50 ± 10
	Pivalate ($n = 5$)	14.3	6.8	2.1	173	337	360	35.7 ± 1.3	29 ± 5	580 ± 60
α-1,3-Glucan	Isobutyrate ($n = 4$)	1.9	0.9	2.1	206	251	385	22.9 ± 3.4	8 ± 1	500 ± 70
	Isovalerate ($n = 5$)	2.1	1.0	2.2	132	292	387	14.5 ± 1.3	9 ± 2	410 ± 50
	Isohexanoate ($n = 6$)	2.3	1.1	2.1	131	306	388	4.1 ± 0.4	18 ± 1	140 ± 50
	Isoheptanoate ($n = 7$)	2.7	1.3	2.1	102	322	386	4.6 ± 0.2	22 ± 2	220 ± 10
	Pivalate ($n = 5$)	1.8	0.9	2.1	202	307	376	22.3 ± 0.6	7 ± 2	430 ± 70
	Tertiary butyl acetate ($n = 6$)	1.9	0.9	2.1	168	348	382	—[a]	—[a]	—[a]

a) A cast film is too brittle to undergo a tensile test.

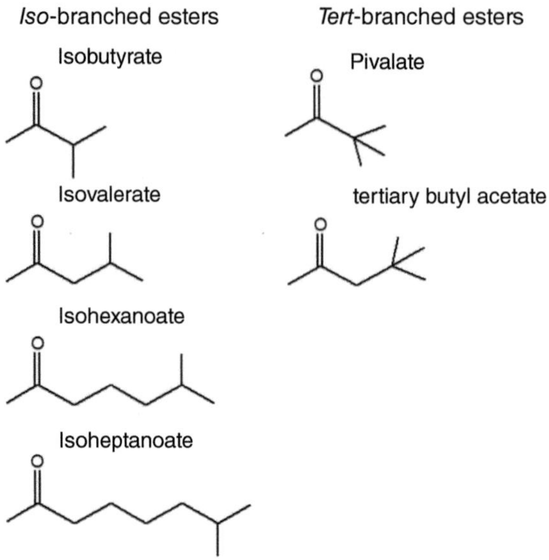

Figure 9.11 The structures of terminally branched ester groups introduced into polysaccharides.

with acyl carbon numbers of 4–7 are crystalline polymers, similar to the cellulose linear esters. Although the *iso*-branched esters with acyl carbon numbers of 5–7 have T_m and T_g values similar to those of the corresponding linear esters, the *iso*-branched esters with an acyl carbon number of 4, cellulose isobutyrate, showed a higher T_m (248 °C) and T_g (115 °C) than linear cellulose butyrate (n = 4, T_m = 182 °C, and T_g = 87 °C). Therefore, shorter *iso*-branched ester side chains resulted in higher thermal phase transition temperature, strength, and stiffness.

Furthermore, all cellulose *iso*-branched esters form colorless and transparent cast films. With increasing side-chain carbon number, the cast films become softer in the same manner as linear esters. Furthermore, the cast films of branched esters are harder (higher Young's modulus) and more brittle (lower elongation at break) than linear esters with the same acyl carbon number.

Regarding the crystalline structure, an explanation for the higher thermal phase transition temperatures and rigid mechanical properties of the branched esters can be proposed. Specifically, cellulose isobutyrate has threefold helix crystalline parameters similar to those of cellulose propionate. This suggests that isobutyric substituents are closely packed in the helix molecular chain conformation, resulting in unique physical properties compared with the linear esters.

9.6.2 β-1,3-Glucan (Curdlan) Branched Esters [15]

For curdlan, *tert*-branched acyl (pivalic substituent) and *iso*-branched acyl groups can be introduced using the same method as for cellulose-branched esters. Curdlan *iso*-branched esters with acyl carbon numbers of 4–7 and *tert*-branched esters with an acyl carbon number of 5 are all crystalline. The *iso*-branched esters have significantly higher T_m and T_g values than those of linear esters with the same acyl carbon number. Above all, curdlan isobutyrate has an extremely high T_m of 339 °C owing

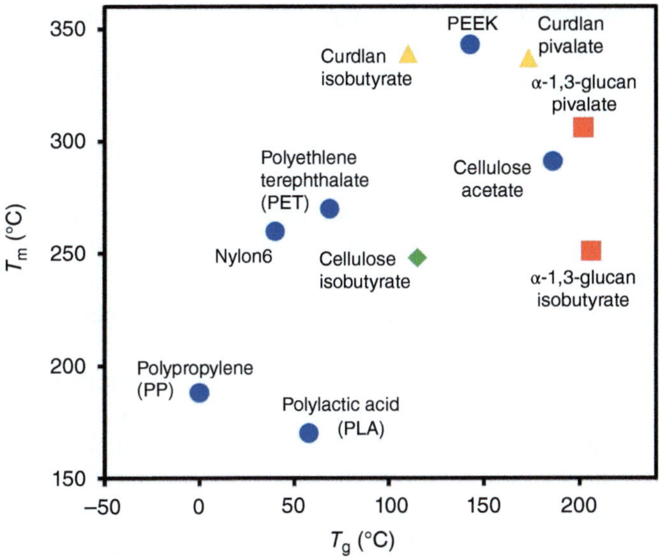

Figure 9.12 T_m and T_g values of polysaccharide-branched esters and conventional plastics.

to the closely packed crystalline structure, which cannot be realized by other curdlan esters and cellulose esters. Furthermore, *tert*-branched ester derivative, curdlan pivalate, has even higher T_m and T_g values than those of linear and *iso*-branched esters with the same carbon number. Remarkably, the T_g of curdlan pivalate (173 °C) is the highest of all the curdlan esters. As the T_m of curdlan pivalate (337 °C) is almost equal to that of curdlan isobutyrate, curdlan pivalate is the most thermally stable of all the curdlan esters, leading to the conclusion that introducing *tert*-branched ester groups is highly effective for improving thermal stability (Figure 9.12).

As observed for other polysaccharide esters, curdlan-branched esters form colorless and transparent cast films, which become softer with increasing side-chain length. Furthermore, cast films of the branched esters have higher tensile strength, higher Young's modulus, and lower elongation at break compared with linear esters with the same acyl carbon number. Regardless of the terminal shape (linear, *iso*-branched, or *tert*-branched), the tensile strength and Young's modulus of curdlan propionate, curdlan isobutyrate, and curdlan pivalate are similar. In contrast, the elongation at break values of curdlan isobutyrate and curdlan pivalate are about 5 and 10 times higher than that of curdlan propionate, respectively. Therefore, the introduction of branched ester groups also enhanced film toughness.

9.6.3 α-1,3-Glucan-Branched Esters [16]

For α-1,3-glucan, *iso*- and *tert*-branched esters can be synthesized in the same manner as those of cellulose and curdlan. These derivatives show crystallinity, although α-1,3-glucan pivalate is almost amorphous. The T_m of α-1,3-glucan-branched esters is proportional to the acyl carbon number, unlike all other linear or branched polysaccharide esters. The change in T_m with acyl carbon number is, as usual, inversely proportional owing to the internal plasticizing effects of side chains. This usual feature led to a conspicuous T_m of 348 °C for the α-1,3-glucan *tert*-butyl acetate, which is the highest T_m among all fully substituted polysaccharide esters reported to date. Furthermore, the T_g values of α-1,3-glucan-branched esters, which decrease with increasing acyl carbon number, are slightly higher than those of the linear esters. Among them, α-1,3-glucan isobutyrate and pivalate have T_g values of over 200 °C (Figure 9.12), which are highest among all fully substituted polysaccharide esters reported to date, and comparable to, or higher than those of conventional super engineering plastics, as represented by polyether ether ketone (T_g = 142 °C) [49].

Furthermore, α-1,3-glucan isobutyrate and pivalate show thermoplasticity at temperatures too low for thermal degradation to occur. Specifically, α-1,3-glucan isobutyrate shows thermofluidity at around 270 °C, which is slightly higher than T_m (251 °C), while α-1,3-glucan pivalate shows thermofluidity at around 250 °C, which is between the T_g (202 °C) and T_m (307 °C), owing to its low crystallinity. The presence of both high-dimensional stability and thermoplasticity is uncommon and highly attractive for industrial applications and mass production by injection molding. Furthermore, even commercial oil-based super engineering plastics with comparable T_g values, which generally have aromatic rings in the polymer backbone, seldom realize thermoplasticity alone. Accordingly, α-1,3-glucan isobutyrate and pivalate are promising thermoplastic resins with valuable properties utilizing a bio-based polysaccharide main chain and eventually fossil-based ester side chains.

9.7 Enzymatic Esterification of Polysaccharides

Enzymes have attracted much attention as biocatalysts in chemical reactions owing to their numerous advantages, including high selectivity, ability to operate under mild conditions, biocompatibility, and catalyst recyclability. Compared with reactions using chemical catalysts, the use of enzymes in the synthesis of polysaccharide esters has several characteristic differences. The main advantages are not only an environmentally benign process under mild conditions but also having high substrate specificity, allowing for the synthesis of polysaccharide esters with well-designed structures and functionalities. As mentioned in Section 9.1.2, polysaccharide esters are commercially manufactured using strong acid catalysts under severe conditions. These chemical processes usually lead to the production of highly substituted polysaccharide esters presenting significantly increased thermoplasticity and hydrophobicity. However, this chemical route has several disadvantages in the final product, such as toxicity from the residual reagent, undesired by-products, and partial polymer degradation. Furthermore, this chemical route has limitations when producing amphiphilic polysaccharide esters with adjusted hydrophilic–hydrophobic balance.

As an alternative approach to addressing these problems, the esterification of polysaccharides using enzymes has received much attention. This renewable and

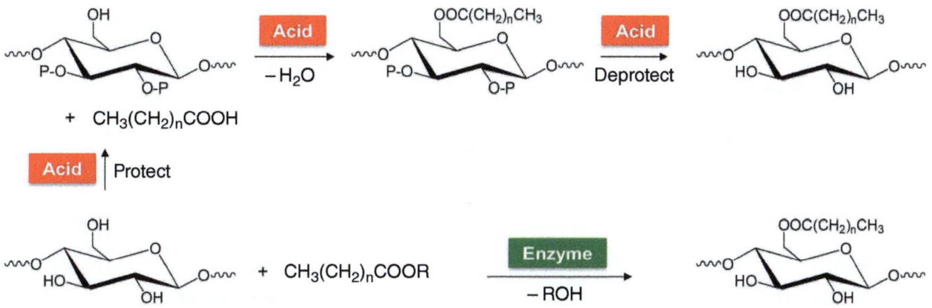

Figure 9.13 Representative routes of chemical and enzymatic esterification in the synthesis of regioselectively substituted polysaccharide ester.

eco-friendly process emphasizes environmental efficiency as an example of green chemistry. The synthesis of polysaccharide esters using enzymes is conducted under mild conditions at a relatively low temperature, which avoids polymer degradation. Enzyme esterification is also substrate specific and regioselective. These features facilitate the regioselective synthesis of polysaccharide esters with amphiphilic properties. Although regioselectivity substituted polysaccharide esters can be obtained through chemical esterification, this process usually requires sequential protection and deprotection reactions. These additional steps can cause unexpected side reactions and result in an expensive and uneconomical purification. Accordingly, as a one-step reaction for synthesizing regioselectivity substituted polysaccharide esters without the need for protection and deprotection reactions, enzymatic esterification has industrial advantages. Figure 9.13 shows the representative routes of chemical and enzymatic esterification for the synthesis of regioselectivity substituted polysaccharide esters.

9.7.1 Enzymes as Biocatalysts

As polysaccharide esterification with an enzyme is based on the original features of the enzyme, identifying the nature of enzymes is important. Enzymes are biological molecules, typically proteins comprising amino acids linked together via peptide bonds. Within its structure, the enzyme has an active site that can bind to the reactive groups (such as acetamido, amino, carboxyl, and/or hydroxyl groups) of polysaccharides. In other words, enzymes have unique interactions with specific substrates that can fit into their active site. Therefore, the selection of enzymes and polysaccharides to obtain substrate specificity is the most important factor for a successful reaction. In a prior study, many enzyme series were investigated in terms of their specificity for monosaccharides and oligosaccharides [50, 51]. Enzyme specificity related to various polysaccharides, such as cellulose, starch, chitosan, and dextran, has also been reported [52].

Among many enzymes, lipase (EC 3.1.1.3, triacylglycerol hydrolases), esterase (EC 3.1.1.1 carboxyl ester hydrolases), proteases (EC 3.4.21.x, peptidase or proteinase), and cutinases (EC 3.1.1.74, cutin hydrolases) have gained particular attention as efficient biocatalysts in the synthesis of polysaccharide esters. These enzymes are

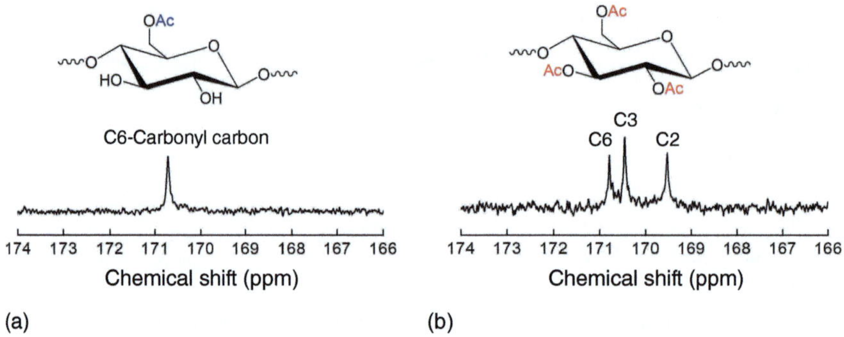

Figure 9.14 ^{13}C-NMR of dextrin acetate by enzymatic (a) and chemical esterification (b) related to the region of carbonyl carbon.

intrinsically catalysts for ester hydrolysis owing to the presence of serine in their catalytic active site but catalyze the reverse esterification reaction in the absence of water (namely, in an organic solvent). As this reaction is in the thermodynamic equilibrium state, water produced by the reaction must be controlled continuously to increase the conversion. An efficient strategy is transesterification using a vinyl ester because the alcohol by-products are easier to remove compared with water [53]. Figure 9.14 shows ^{13}C-NMR spectra of a regioselectively substituted polysaccharide acetate (from dextrin produced by partial starch hydrolysis) [17]. The spectrum of dextrin acetate catalyzed by the enzyme indicates a completely regioselective reaction, showing a single peak corresponding to a carbonyl carbon at the C6 position. To synthesize polysaccharide esters with a well-designed structure, optimum conditions in which various factors are taken into account are required.

9.7.2 Reaction Mechanism

Enzymatic esterification has an ordered mechanism including acylation and deacylation. First, the enzyme active site is adjusted to an accurate position by interacting with the substrate. The polysaccharide ester is then released from the enzyme surface, and the enzyme is regenerated for cycling of another reaction. Among the most important enzymes exhibiting high catalytic efficiency is lipase, which can esterify the primary hydroxyl group (C6-OH) of glucose units [54]. Figure 9.15 shows the catalytic mechanism of lipase for synthesizing polysaccharide ester. Lipases include a catalytic triad, comprising aspartic acid, serine nucleophile, and histidine base, as the active site. The reaction occurs at this site, with the first step proceeding to form a primary tetrahedral intermediate with acyl donor and serine. The enzyme then forms a link with the acyl group by electron transfer. The covalent acyl–enzyme intermediate finally derives a polysaccharide ester via a substitution reaction with a reactive group in the polysaccharide and is converted back to the original structure [55].

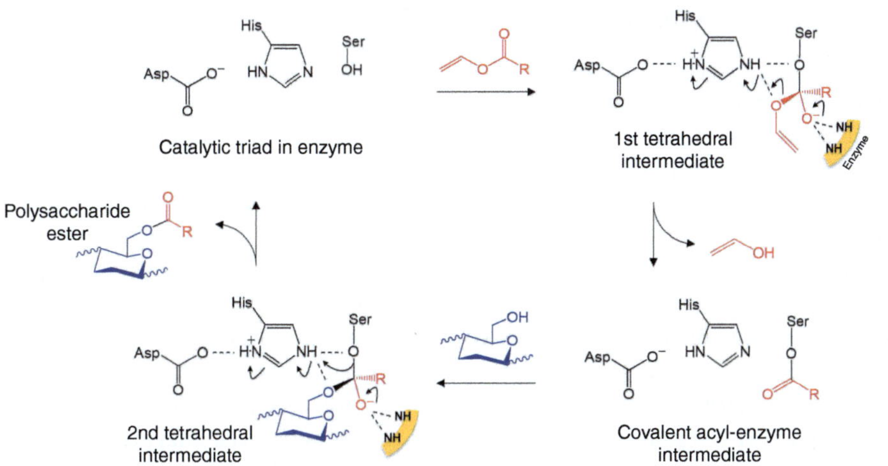

Figure 9.15 Reaction mechanism of polysaccharide ester using lipase in the organic solvent.

9.7.3 Factors Influencing Enzyme Activity

Enzymatic esterification is extremely sensitive and susceptible to various factors. Therefore, reaction under appropriate conditions is required for enzyme activation. The rate and degree of enzyme activation is basically affected by the frequency of collision with the substrate, which depends on reaction temperature, enzyme amount, and substrate concentration.

First, the temperature directly affects the structure and activity of the enzyme protein, strongly influencing the enzyme activity. At high temperatures (usually more than 50 °C), enzyme denaturation can occur, resulting in low activity. Enzymes have activity only within a limited temperature range, with many enzymes having specific temperature for catalytic activity.

Second, the amount of enzyme used in enzymatic esterification is also a factor directly influencing the degree of reaction. If the reaction rate at which an enzyme catalyzes is proportional to the enzyme amount, the reaction system with doubling the amount of enzyme should allow twice as much substrate to be catalyzed. However, the presence of excess enzyme in the reaction system can lead to a lower degree of reactivity. The lower degree of reactivity caused by excess enzyme is attributed to enzyme movement to neighboring hydroxyl groups being interrupted by other enzymes in a sequential manner [56].

Lastly, substrate concentration plays an important role according to the collision theory of chemical reaction. As molecules must collide with each other to react, increasing the substrate concentration increases the collision frequency, resulting in higher degree of reactivity. However, a substrate concentration that is too high might increase the overall viscosity of the response system, thereby limiting enzyme movement, which causes the reaction efficiency to deteriorate.

9.7.4 Strategies for Efficient Biocatalyst Processes

Enzymes are water-soluble proteins. An enzyme is generally dissolved in solution when performing a reaction. Therefore, the enzyme activity in organic solvents is different to that in aqueous solution. Generally, most enzymes degenerate easily in organic solvents, losing their activity. To mitigate this loss of function, two representative strategies can be used, which are enzyme immobilization and the use of ionic liquids.

Immobilization is a technical process in which enzymes are fixed to a support carrier, creating a heterogeneous immobilized enzyme system. This immobilized enzyme maintains its original catalytic activity and substrate specificity with enhanced thermal and solvent stability. The use of immobilized enzymes in the enzymatic esterification of polysaccharides has advantages, including improved yield and purity of polysaccharide esters [57, 58]. Immobilized enzymes are easily separated from the reaction product and can be reused owing to their catalytic activity being maintained. There are several techniques used to immobilize enzymes, including adsorption, covalent-bonding, cross-linking, entrapment, and encapsulation. For the immobilized enzyme to exhibit its catalytic activity, the active site must be prevented from changing during immobilization. When selecting the immobilization method, the purpose of use must be considered in accordance with the original enzyme properties.

Ionic liquids are ionic compounds (salts) that exist as liquids at ambient temperatures. Recently, ionic liquids have been effectively used as green solvents owing to their unique characteristics, such as low volatility, high chemical and thermal stability, and tunable solvent properties. Ionic liquids are advantageous as alternatives to organic solvents in enzymatic reaction systems. First, the ionic liquid reaction system is more stable and eco-friendly than the organic solvent system because it avoids the high volatility, flammability, and environmentally harmful characteristics of organic solvents. Second, solvent characteristics, such as polarity, acidity, and hydrophobicity, can be adjusted according to the combination of positive and negative ions. Therefore, enzymatic reactions can be conducted over a wider range of reaction conditions without denaturation of the enzyme protein in contrast to polar organic solvents. The use of well-controlled ionic liquids in the enzymatic reaction of polysaccharides can react regioselectively with improved activity [59]. Furthermore, ionic liquids have the advantage of dissolving polysaccharides. As most polysaccharides form strong hydrogen bonds between molecules, they are generally insoluble in ordinary organic solvents. In contrast, ionic liquids are effective solvents for polysaccharides. For example, an ionic liquid consisting of 1-butyl-3-methylimidazolium (C4mim) and chloride ions shows excellent cellulose solubility [60].

9.7.5 Development Trend and Prospects

In the last few years, many studies on the enzymatic synthesis of various carbohydrate molecules (monosaccharides, oligosaccharides, and polysaccharides) have been reported for their applications with environmental potential, including biodegradable emulsifiers, plasticizers, compatibilizers, and detergents [61, 62]. Based on the production of monosaccharide and/or oligosaccharide esters using an enzyme, some studies have been conducted on the synthesis of polysaccharide esters, such as cellulose, chitosan, starch, and dextran. Esterified polysaccharides with an appropriate acyl chain length can act as internal plasticizers, exhibiting improved thermoplasticity. Accordingly, cellulose and its derivatives have received particular research attention. Gremos et al. have reported the enzymatic esterification of cellulose with a series of vinyl esters (12–18 carbon atoms), which present a low degree of esterification of 0.9–1.3% [63]. This low enzyme activity is due to restricted enzyme penetration, which is caused by strong molecular interactions in the crystalline polymer. Comparatively, starch achieves a high conversion in the synthesis of starch esters using enzymes. Many studies have reported the synthesis of starch esters with a wide range of degrees of esterification of up to 90% [57, 58, 64, 65]. However, these reactions exhibit a lack of regioselectively owing to the specific conditions used to amplify the enzyme catalytic activity (substrate pretreatment, microwave treatment, and the addition of ionic liquids). A completely regioselective reaction has been reported using a branched polysaccharide derived from starch (namely, dextrin) [17]. Generally, branched polysaccharides show better efficiency than linear polysaccharides in enzymatic esterification, owing to good solubility in the organic solvent (a consequence of the extra degrees of freedom provided by rotation at the branched position). Dextran is another branched polysaccharide whose derivatives produced by enzyme catalysis have been studied in terms of potential for biomedical applications in particular [66, 67]. Dordick and coworkers have prepared dextran-based hydrogels using an enzymatic esterification reaction between dextran and divinyladipate [68]. This material exhibited superior mechanical properties (higher elastic modulus for a given swelling ratio) compared with chemically prepared dextran-based hydrogels.

Although much research has been conducted on the enzymatic esterification of polysaccharides, their potential has yet to be fully explored owing to the problem of low solubility resulting from the abundance of hydroxyl groups in polysaccharides. Furthermore, in terms of enzyme activation, predicting substrate specificity between enzymes and polysaccharides is difficult because polysaccharides have different complex structures and, therefore, exhibit unique properties depending on the constituents and conformations of repeating units (α- or β-glucosidically linked structures based on about 40 different monosaccharides). Therefore, no commercial process is yet available for the production of polysaccharide esters using enzymes. Nevertheless, many studies on enzymatic reactions of polysaccharides are being conducted for sustainable and environmentally friendly social development. Compared to chemical esterification of polysaccharides, enzymatic reactions do not show satisfactory results in terms of final properties, but this issue can be solved through continuous research. Among many research fields, the development of protein engineering and solvent engineering is important for improving catalytic efficiency, specificity, and stability in the enzymatic synthesis of polysaccharide esters. In the future study of this topic, the optimization of enzymatic processes should be focused on improving the process efficiency. The development of innovative enzymatic reactions as green chemistry is expected to meet the needs of various industries and, therefore, make sustainable contributions to all of the society.

9.8 Biodegradation of Polysaccharide Ester

The conventional research in the biodegradability of polysaccharide esters is mainly for cellulose acetate [69]. The biodegradation rate depends on the DS of acetate in cellulose hydroxyl groups. On the condition that the DS of acetate is lower than 2.5, cellulose acetate biodegrades at the almost same rate as virgin cellulose. However, with the increase of DS in a range of over 2.5, the biodegradation rate decreases, and fully substituted cellulose acetate (DS = 3) hardly biodegrades. As the problem of plastic waste in the ocean garners increased attention, in 2020, Daicel Co., a Japanese chemical company, developed highly biodegradable cellulose acetate named as CAFBLO™ [70].

As for other polysaccharide esters, the biodegradability has not been investigated yet, and an intensive study is strongly needed. From 2019, our group is investigating the years-long biodegradability of various kinds of polysaccharide esters and other biodegradable and common plastics on the sea floor at a depth of over 850 m near Hatsushima island, Shizuoka prefecture, Japan, and 5000 m near Minami-Tori-Shima island in the northwestern Pacific Ocean, located 1800 km southeast of Tokyo.

9.9 Summary

As an alternative for petroleum-based plastics, polysaccharides have great potential as a bio-based backbone for high-performance plastics. In the last few years, we are committed to develop polysaccharide esters having a wide range of adjustable physical properties by choosing different kinds of polysaccharides, as well as adjusting length and branch of ester groups. Through these efforts, we have succeeded in producing a considerable variety of plastics with excellent properties such as high thermal stability, outstanding transparency, and adhesive property. There are still many methods for expanding the performance, for instance, using the mix esterification technique, changing functional groups or achieving regioselectivity of ester groups with controllable DS using enzymatic esterification. Above all, we believe that we can build a sustainable material production system and protect the global environment for future generations.

References

1 Iwata, T. (2015). Biodegradable and bio-based polymers: future prospects of eco-friendly plastics. *Angewandte Chemie International Edition* 54 (11): 3210–3215.

2 Södergård, A. and Stolt, M. (2002). Properties of lactic acid based polymers and their correlation with composition. *Progress in Polymer Science* 27 (6): 1123–1163.

3 Lu, J., Tappel, R.C., and Nomura, C.T. (2009). Mini-review: biosynthesis of poly (hydroxyalkanoates). *Journal of Macromolecular Science, Part C: Polymer Reviews* 49 (3): 226–248.

4 Edgar, K.J., Buchanan, C.M., Debenham, J.S. et al. (2001). Advances in cellulose ester performance and application. *Progress in Polymer Science* 26 (9): 1605–1688.

5 Fordyce, C.R. and Meyer, I.W.A. (1940). Plasticizers for cellulose acetate and cellulose acetate butyrate. *Industrial and Engineering Chemistry* 32 (8): 1053–1060.

6 Enomoto-Rogers, Y., Iio, N., Takemura, A., and Iwata, T. (2015). Synthesis and characterization of pullulan alkyl esters. *European Polymer Journal* 66: 470–477.

7 Danjo, T., Enomoto, Y., Shimada, H. et al. (2017). Zero birefringence films of pullulan ester derivatives. *Scientific Reports* 7: 46342.

8 Togo, A., Enomoto, Y., Takemura, A., and Iwata, T. (2019). Synthesis and characterization of dextran ester derivatives and their adhesive properties. *Journal of Wood Science* 65 (1): 1–8.

9 Marubayashi, H., Yukinaka, K., Enomoto-Rogers, Y. et al. (2014). Curdlan ester derivatives: synthesis, structure, and properties. *Carbohydrate Polymers* 103: 427–433.

10 Gan, H., Enomoto, Y., Kabe, T. et al. (2017). Synthesis, properties and molecular conformation of paramylon ester derivatives. *Polymer Degradation and Stability* 145: 142–149.

11 Gan, H., Kabe, T., and Iwata, T. (2020). Manufacture, characterization, and structure analysis of melt-spun fibers derived from paramylon esters. *Journal of Fiber Science and Technology* 76 (5): 151–160.

12 Puanglek, S., Kimura, S., Enomoto-Rogers, Y. et al. (2016). *In vitro* synthesis of linear α-1,3-glucan and chemical modification to ester derivatives exhibiting outstanding thermal properties. *Scientific Reports* 6 (1): 1–8.

13 Puanglek, S., Kimura, S., and Iwata, T. (2017). Thermal and mechanical properties of tailor-made unbranched α-1,3-glucan esters with various carboxylic acid chain length. *Carbohydrate Polymers* 169: 245–254.

14 Danjo, T. and Iwata, T. (2018). Syntheses of cellulose branched ester derivatives and their properties and structure analyses. *Polymer* 137: 358–363.

15 Zhai, W., Danjo, T., and Iwata, T. (2018). Synthesis and physical properties of curdlan branched ester derivatives. *Journal of Polymer Research* 25 (3): 181.

16 Fukata, Y., Kimura, S., and Iwata, T. (2020). Synthesis of α-1,3-glucan branched ester derivatives with excellent thermal stability and thermoplasticity. *Polymer Degradation and Stability* 177: 109130.

17 Lee, H.Y., Kimura, S., and Iwata, T. (2018). Lipase-catalyzed regioselective synthesis of dextrin esters. *Biomacromolecules* 20 (2): 705–711.

18 Leathers, T.D. (2003). Biotechnological production and applications of pullulan. *Applied Microbiology and Biotechnology* 62 (5, 6): 468–473.

19 Singh, R.S., Saini, G.K., and Kennedy, J.F. (2008). Pullulan: microbial sources, production and applications. *Carbohydrate Polymers* 73 (4): 515–531.

20 Farris, S., Unalan, I.U., Introzzi, L. et al. (2014). Pullulan-based films and coatings for food packaging: present applications, emerging opportunities, and future challenges. *Journal of Applied Polymer Science* 131 (13).

21 Shibata, M., Asahina, M., Teramoto, N., and Yosomiya, R. (2001). Chemical modification of pullulan by isocyanate compounds. *Polymer* 42 (1): 59–64.

22 Teramoto, N. and Shibata, M. (2006). Synthesis and properties of pullulan acetate. Thermal properties, biodegradability, and a semi-clear gel formation in organic solvents. *Carbohydrate Polymers* 63 (4): 476–481.

23 Shibata, M., Nozawa, R., Teramoto, N., and Yosomiya, R. (2002). Synthesis and properties of etherified pullulans. *European Polymer Journal* 38 (3): 497–501.

24 Heinze, T., Liebert, T., Heublein, B., and Hornig, S. (2006). Functional polymers based on dextran. In: *Polysaccharides Ii* (ed. D. Klemm), 199–291. Berlin, Heidelberg: Springer.

25 Vandamme, E., Bruggeman, G., Debaets, S., and Vanhooren, P. (1996). Useful polymers of microbial origin. *Agro Food Industry Hi-Tech* 7 (5): 21–25.

26 De Groot, C.J., Van Luyn, M.J., Van Dijk-Wolthuis, W.N. et al. (2001). *In vitro* biocompatibility of biodegradable dextran-based hydrogels tested with human fibroblasts. *Biomaterials* 22 (11): 1197–1203.

27 Cadée, J.A., Van Luyn, M.J.A., Brouwer, L.A. et al. (2000). *In vivo* biocompatibility of dextran-based hydrogels. *Journal of Biomedical Materials Research* 50 (3): 397–404.

28 Berry, C.C., Wells, S., Charles, S., and Curtis, A.S. (2003). Dextran and albumin derivatised iron oxide nanoparticles: influence on fibroblasts in vitro. *Biomaterials* 24 (25): 4551–4557.

29 Togo, A., Enomoto, Y., Takemura, A., and Iwata, T. (2019). Synthesis of biomass-based adhesives derived from dextran ester derivatives. *Journal of the Adhesion Society of Japan* 55 (9): 315–322.

30 Harada, T., Misaki, A., and Saito, H. (1968). Curdlan: a bacterial gel-forming β-1,3-glucan. *Archives of Biochemistry and Biophysics* 124: 292–298.

31 Clarke, A.E. and Stone, B.A. (1960). Structure of the paramylon from *Euglena gracilis. Biochimica et Biophysica Acta* 44: 161–163.

32 Marchessault, R.H., Deslandes, Y., Ogawa, K., and Sundararajan, P.R. (1977). X-ray diffraction data for β(1→3)-D-glucan. *Canadian Journal of Chemistry* 55 (2): 300–303.

33 Takeda, H., Yasuoka, N., Kasai, N., and Harada, T. (1978). X-ray structural studies of (1→3)-β-D-glucan (curdlan). *Canadian Journal of Chemistry* 10: 365–368.

34 Marchessault, R.H. and Deslandes, Y. (1979). Fine structure of (1→3)-β-D-glucans: curdlan and paramylon. *Carbohydrate Research* 75: 231–242.

35 Deslandes, Y., Marchessault, R.H., and Sarko, A. (1980). Triple-helical structure of (1→3)-β-D-glucan. *Macromolecules* 13 (6): 1466–1471.

36 Shibakami, M., Tsubouchi, G., and Hayashi, M. (2014). Thermoplasticization of euglenoid β-1,3-glucans by mixed esterification. *Carbohydrate Polymers* 105: 90–96.

37 Shibakami, M., Tsubouchi, G., Sohma, M., and Hayashi, M. (2015). One-pot synthesis of thermoplastic mixed paramylon esters using trifluoroacetic anhydride. *Carbohydrate Polymers* 119: 1–7.

38 Shibakami, M. and Sohma, M. (2017). Synthesis and thermal properties of paramylon mixed esters and optical, mechanical, and crystal properties of their hot-pressed films. *Carbohydrate Polymers* 155: 416–424.

39 Johnston, I.R. (1965). The composition of the cell wall of *Aspergillus niger*. *Biochemical Journal* 96 (3): 651.

40 Leal, J.A., Guerrero, C., Gómez-Miranda, B. et al. (1992). Chemical and structural similarities in wall polysaccharides of some *Penicillium, Penicillium* and *Aspergillus* species. *FEMS Microbiology Letters* 90 (2): 165–168.

41 Wang, T., Deng, L., Li, S., and Tan, T. (2007). Structural characterization of a water-insoluble (1→3)-α-D-glucan isolated from the *Penicillium chrysogenum*. *Carbohydrate Polymers* 67 (1): 133–137.

42 Choma, A., Wiater, A., Komaniecka, I. et al. (2013). Chemical characterization of a water insoluble (1→3)-α-D-glucan from an alkaline extract of *Aspergillus wentii*. *Carbohydrate Polymers* 91 (2): 603–608.

43 Loesche, W.J. (1986). Role of *Streptococcus mutans* in human dental decay. *Microbiological Reviews* 50 (4): 353.

44 Simpson, C.L., Cheetham, N.W., and Jacques, N.A. (1995). Four glucosyltransferases, GtfJ, GtfK, GtfL and GtfM, from *Streptococcus salivarius* ATCC 25975. *Microbiology* 141 (6): 1451–1460.

45 Simpson, C.L., Giffard, P.M., and Jacques, N.A. (1995). *Streptococcus salivarius* ATCC 25975 possesses at least two genes coding for primer-independent glucosyltransferases. *Infection and Immunity* 63 (2): 609–621.

46 Giffard, P.M., Simpson, C.L., Milward, C.P., and Jacques, N.A. (1991). Molecular characterization of a cluster of at least two glucosyltransferase genes in *Streptococcus salivarius* ATCC 25975. *Microbiology* 137 (11): 2577–2593.

47 Banas, J.A., Simon, D., Williams, L.K. et al. (1994). Analysis of a primer-independent GTF-I from *Streptococcus salivarius*. *FEMS Microbiology Letters* 123 (3): 349–354.

48 Giffard, P.M., Allen, D.M., Milward, C.P. et al. (1993). Sequence of the *gtfK* gene of *Streptococcus salivarius* ATCC 25975 and evolution of the *gtf* genes of oral streptococci. *Microbiology* 139 (7): 1511–1522.

49 Mark, J.E. (ed.) (2009). *Polymer Data Handbook*, 2e. New York: Oxford University Press.

50 Gumel, A.M., Annuar, M.S.M., Heidelberg, T., and Chisti, Y. (2011). Lipase mediated synthesis of sugar fatty acid esters. *Process Biochemistry* 46 (11): 2079–2090.

51 Chang, S.W. and Shaw, J.F. (2009). Biocatalysis for the production of carbohydrate esters. *New Biotechnology* 26 (3–4): 109–116.

52 Karaki, N., Aljawish, A., Humeau, C. et al. (2016). Enzymatic modification of polysaccharides: mechanisms, properties, and potential applications: a review. *Enzyme and Microbial Technology* 90: 1–18.

53 Kumar, A., Dhar, K., Kanwar, S.S., and Arora, P.K. (2016). Lipase catalysis in organic solvents: advantages and applications. *Biological Procedures Online* 18 (1): 2.

54 Kobayashi, S. (2010). Lipase-catalyzed polyester synthesis–a green polymer chemistry. *Proceedings of the Japan Academy, Series B* 86 (4): 338–365.

55 Puskas, J.E., Sen, M.Y., and Seo, K.S. (2009). Green polymer chemistry using nature's catalysts, enzymes. *Journal of Polymer Science Part A: Polymer Chemistry.* 47 (12): 2959–2976.

56 Satoh, T., Imai, T., Ishihara, H. et al. (2005). Synthesis, branched structure, and solution property of hyperbranched D-glucan and D-galactan. *Macromolecules* 38 (10): 4202–4210.

57 Horchani, H., Chaâbouni, M., Gargouri, Y., and Sayari, A. (2010). Solvent-free lipase-catalyzed synthesis of long-chain starch esters using microwave heating: optimization by response surface methodology. *Carbohydrate Polymers* 79 (2): 466–474.

58 Chakraborty, S., Sahoo, B., Teraoka, I. et al. (2005). Enzyme-catalyzed regioselective modification of starch nanoparticles. *Macromolecules* 38 (1): 61–68.

59 Wang, F., Chen, Z.G., and Zhu, H.J. (2013). An efficient enzymatic modification of lily polysaccharide in ionic liquid under ultrasonic irradiation. *Biochemical Engineering Journal* 79: 25–28.

60 Xu, A., Wang, J., and Wang, H. (2010). Effects of anionic structure and lithium salts addition on the dissolution of cellulose in 1-butyl-3-methylimidazolium-based ionic liquid solvent systems. *Green Chemistry* 12 (2): 268–275.

61 Janssen, A.E.M., Lefferts, A.G., and Van't Riet, K. (1990). Enzymatic synthesis of carbohydrate esters in aqueous media. *Biotechnology Letters* 12 (10): 711–716.

62 van den Broek, L.A. and Boeriu, C.G. (2013). Enzymatic synthesis of oligo-and polysaccharide fatty acid esters. *Carbohydrate polymers* 93 (1): 65–72.

63 Gremos, S., Zarafeta, D., Kekos, D., and Kolisis, F. (2011). Direct enzymatic acylation of cellulose pretreated in BMIMCl ionic liquid. *Bioresource Technology* 102 (2): 1378–1382.

64 Junistia, L., Sugih, A.K., Manurung, R. et al. (2008). Synthesis of higher fatty acid starch esters using vinyl laurate and stearate as reactants. *Starch-Stärke* 60 (12): 667–675.

65 Adak, S. and Banerjee, R. (2016). A green approach for starch modification: esterification by lipase and novel imidazolium surfactant. *Carbohydrate Polymers* 150: 359–368.

66 Bi, Y., Yu, M., Zhou, H. et al. (2016). Biosynthesis of oleyl oleate in solvent-free system by *Candida rugosa* Lipase (CRL) immobilized in macroporous resin with cross-linking of aldehyde-dextran. *Journal of Molecular Catalysis B: Enzymatic.* 133: 1–5.

67 Ghugare, S.V., Chiessi, E., Cerroni, B. et al. (2012). Biodegradable dextran based microgels: a study on network associated water diffusion and enzymatic degradation. *Soft Matter* 8 (8): 2494–2502.

68 Ferreira, L., Gil, M.H., Cabrita, A.M., and Dordick, J.S. (2005). Biocatalytic synthesis of highly ordered degradable dextran-based hydrogels. *Biomaterials* 26 (23): 4707–4716.

69 Komarek, R.J., Gardner, R.M., Buchanan, C.M., and Gedon, S. (1993). Biodegradation of radiolabeled cellulose acetate and cellulose propionate. *Journal of Applied Polymer Science* 50 (10): 1739–1746.

70 DAICEL Corporation (2020). Highly biodegradable cellulose acetate. https://www.daicel.com/cell_ac/en/cellulose/marine_biodegradable.html (accessed 10 October 2021).

10

Biodegradable Lignin-Based Plastics

Yi-ru Chen and Simo Sarkanen

University of Minnesota, Department of Bioproducts and Biosystems Engineering, 2004 Folwell Ave., Saint Paul, MN 55108, USA

10.1 Lignocellulose Biorefineries

Outlook. *Over 2×10^{11} t of lignocellulose is biosynthesized annually in the form of plant biomass, ~80% of which is embodied in trees. From 2008 onward, huge investments were regularly allocated to converting a sizable share of these renewable materials to biofuels and chemicals. By the year 2020, however, these ventures had not achieved profitability because lignin valorization had failed to amass sufficient economic gain.*

Biomass collectively represents the most abundant renewable source of materials from which liquid biofuels (such as bioethanol) and valuable chemicals can be produced. The total annual plant growth (namely, primary biomass production) has been estimated [1] to be 220 billion oven-dry tonnes, 80% of which is manifested in trees [2]. The prospect of partially exploiting this vast potential has attracted enduring investments in Europe, Asia, and the United States. In 2006, the European Union published a forecast declaring that, by 2030, a quarter of road transport fuel needs could be derived from biofuels [3]. Correspondingly, the United States projected that, by 2030, enough biomass could be sustainably harvested for producing a quantity of biofuels equivalent to 25% of the nation's 2015 energy consumption – without encroaching on demands for food, feed, and fiber [4].

The single most abundant renewable polymer is cellulose, which consists of (unbranched) 1-*O*-4′ linked β-D-glucopyranose chains. Some of it occurs in a rather pure form, such as cotton plant seed hairs, for example, but most appears in combination with lignin and other polysaccharides (i.e. hemicelluloses) in the cell walls of vascular plants, including trees [5]. The lignins in these composite plant materials are aromatic; they represent the second most abundant group of biopolymers on the Earth. Annually, around 2×10^{10} t of new lignins have been independently estimated to form among the metabolic products of photosynthesis [6].

It is clearly understood that the high cost of converting the polysaccharides in plant cell walls to biofuels arises from the recalcitrance of lignocellulosic biomass

Biodegradable Polymers in the Circular Plastics Economy, First Edition.
Edited by Michiel Dusselier and Jean-Paul Lange.

toward chemical and/or enzymatic deconstruction [7]. Indeed, a 2013 technoeconomic analysis of cellulosic diesel range fatty acid blend stock production from corn stover uncovered an alarming situation. The report concluded that the corn stover lignin fraction must acquire an economic value of equal importance to the fermentable carbohydrates if the 2022 target selling price of US$0.80 per gasoline equivalent liter was going to be attained for the renewable fuel component [8]. In 2017, the prospect of profitable cellulosic ethanol sales had been weakened by comparable fears. The unsubsidized production cost of this liquid biofuel was not expected to fall below US$1.20 per gasoline equivalent liter, even by 2025 [9].

Various means have been contemplated as plausible routes for valorizing cellulose and hemicelluloses more profitably than converting them to biofuels. These have been succinctly organized [10] by way of consecutive acid-catalyzed transformations to successive platforms based on monosaccharides (from polysaccharide cleavage), furan derivatives (formed through monosaccharide dehydration), and finally levulinic acid, followed by a reductive step to γ-valerolactone (GVL). This appealing blueprint has fostered the further expectation that profitable lignin valorization will ultimately depend on efficient depolymerization to non-macromolecular compounds [10].

Lignin depolymerization may be achieved either pyrolytically, enzymatically, or catalytically. Both pyrolysis and biological depolymerization require, for intrinsically different reasons, extensive purification schemes to obtain products in profitable yields, while the costs of catalysts for the third alternative must be kept down to achieve economic viability [10]. Thus, formulations for plastics with very high (~90 wt%) lignin contents should also be evaluated thoroughly: success could provide inexpensive solutions to the formidable task of biorefining lignocellulose profitably, if adequate performance indices for such new materials can be secured.

Lignocellulose composition varies quite substantially between the cell walls of different vascular plants [11a]. Among (monocotyledonous) grasses and other non-woods, (eudicotyledonous) hardwoods, and (gymnospermous) softwoods or conifers, cellulose contents extend from 30 to 50 wt%, while hemicelluloses (lower molecular weight polysaccharides composed of pentoses and hexoses) encompass a 15–35 wt% range, and lignins contribute 15–35 wt% to the dry plant matter [11a, 12].

It has been generally believed that covalent bonds between lignin and hemicelluloses are responsible for the difficulties encountered in separating carbohydrates completely from lignin, when fractionating milled wood or other plant materials by physicochemical means [13]. However, solid-state NMR analyses of ^{13}C-labeled maize stems have revealed that the cell wall lignin and polysaccharide domains are dynamically distinct. Lignin exhibited the shortest ^{1}H-$T_{1\rho}$ and longest ^{13}C-T_1 relaxation times, whereas the reverse was observed in regard to the polysaccharides. Numerous electrostatic interactions were envisaged to occur between hydrophobic lignin nanodomains and the hemicellulose, xylan, but any effects from covalent linkages between the separate phases fell below the detection limits of the solid-state NMR techniques employed [14].

10.2 Macromolecular Lignin Configuration

Debated revision smoldering for almost 60 years. *Between four and six linkages interconnect the hydroxyphenylpropanoid units that form lignin chains. In 1960, lignin macromolecules were considered to be cross-linked biopolymers. By 2019, branching and cross-linking had tended to fade from native macromolecular lignin structures. This drastic revision could reinforce prospects for lignin valorization, but strong disagreement persisted in the lignin field.*

Lignins are assembled by the dehydrogenative polymerization of one, two, or (occasionally) three monolignols that are usually derived from phenylalanine biosynthesized in the plastid, where the aromatic amino acid precursor is formed by means of the shikimate pathway from phosphoenol pyruvate and erythrose 4-phosphate [15]. After being shuttled to the cytosol, phenylalanine is deaminated by phenylalanine ammonia lyase to cinnamate, which is converted through the phenylpropanoid pathway to *p*-coumaryl alcohol, coniferyl alcohol, and sinapyl alcohol (Figure 10.1). Following enzyme-catalyzed single-electron oxidation to the corresponding radicals, these traditional monolignols undergo coupling to form *p*-hydroxyphenyl (H), guaiacyl (G), and syringyl (S) units, respectively, in lignin macromolecules [16]. It is worth mentioning that commelinid monocotyledonous plants such as maize [17] or *Brachypodium distachyon* [18] may biosynthesize significant portions of their lignins through deamination of tyrosine directly to *p*-coumarate by the action of a bifunctional Phe-and-Tyr ammonia lyase.

The proportions of *p*-coumaryl, coniferyl, and sinapyl alcohols that are incorporated into lignins vary characteristically between gymnospermous softwoods, angiosperm eudicotyledonous hardwoods, and monocotyledonous grasses. Softwood lignins are composed almost exclusively of guaiacyl (G) units interconnected by six distinctive linkages [19]. Interestingly, spruce lignin has been separated (by Gellerstedt and coworkers) into two readily distinguishable fractions that preferentially interact (covalently or non-covalently) with the two pre-eminent cell wall hemicelluloses (Figure 10.2). The lignin chains forming complexes with xylans contain only β-*O*-4′ ether structural linkages between guaiacyl units, while those participating in complexes with glucomannans embody all six linkages in appropriately adjusted proportions [19].

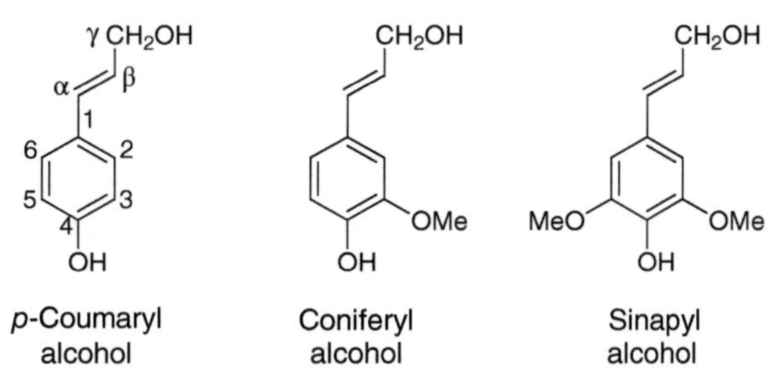

Figure 10.1 Traditional monolignols that respectively form hydroxyphenyl (H), guaiacyl (G), and syringyl (S) units in lignin macromolecules.

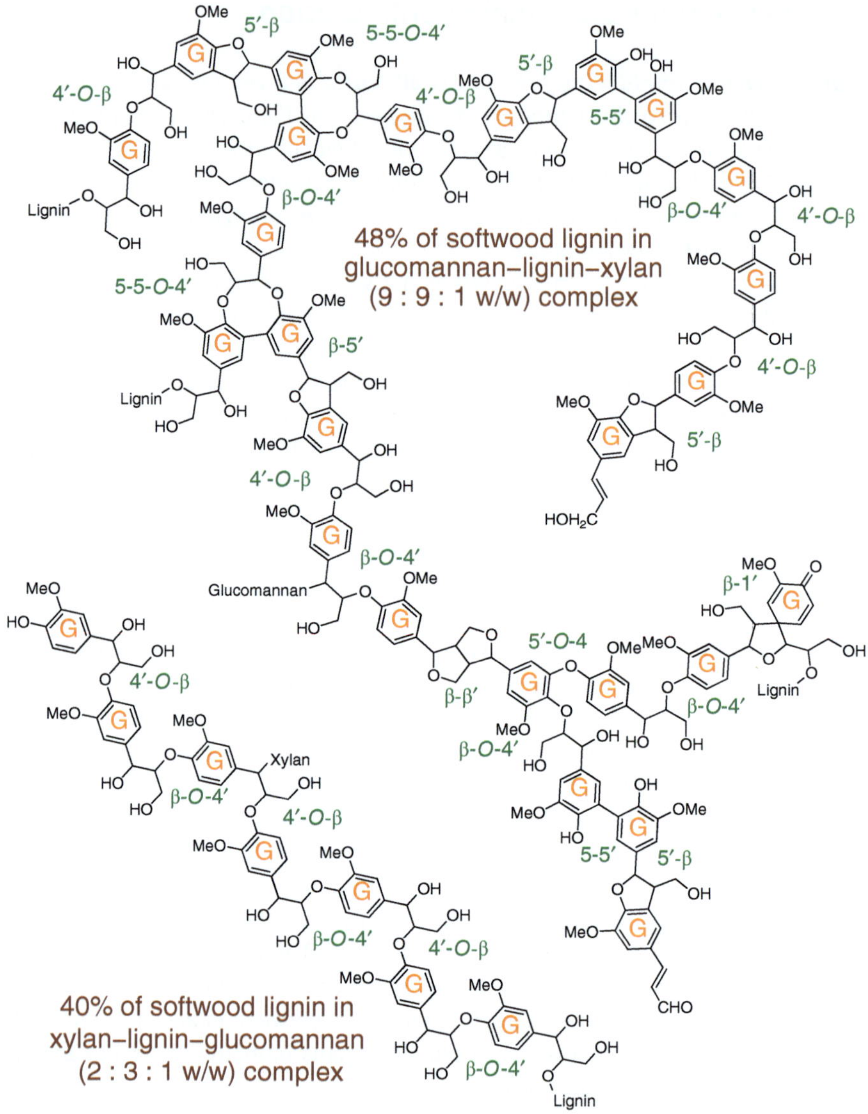

Figure 10.2 Structures suggested for softwood lignin–carbohydrate complexes isolated by successive fractionation of endoglucanase-treated spruce wood with aqueous 8 M urea, alkaline borate, and/or barium hydroxide. Source: Adapted from Göran Gellerstedt (2007) cited in Chen and Sarkanen [19] / with permission of Elsevier.

On the other hand, hardwood lignins are composed of syringyl (S) and guaiacyl (G) units in similar proportions interconnected primarily through four different linkages (Figure 10.3), among which β-*O*-4′ ethers are the most frequent [20]. In contrast, a representative 20-unit chain segment in a monocotyledonous lignin [16] may be composed of 13 syringyl (S) and 5 guaiacyl (G) residues with a *p*-hydroxyphenyl (H) unit at the phenolic end and a 5-linked ferulate ester (F) at the non-phenolic terminus (Figure 10.4). Macromolecular chain assembly involves formation of four

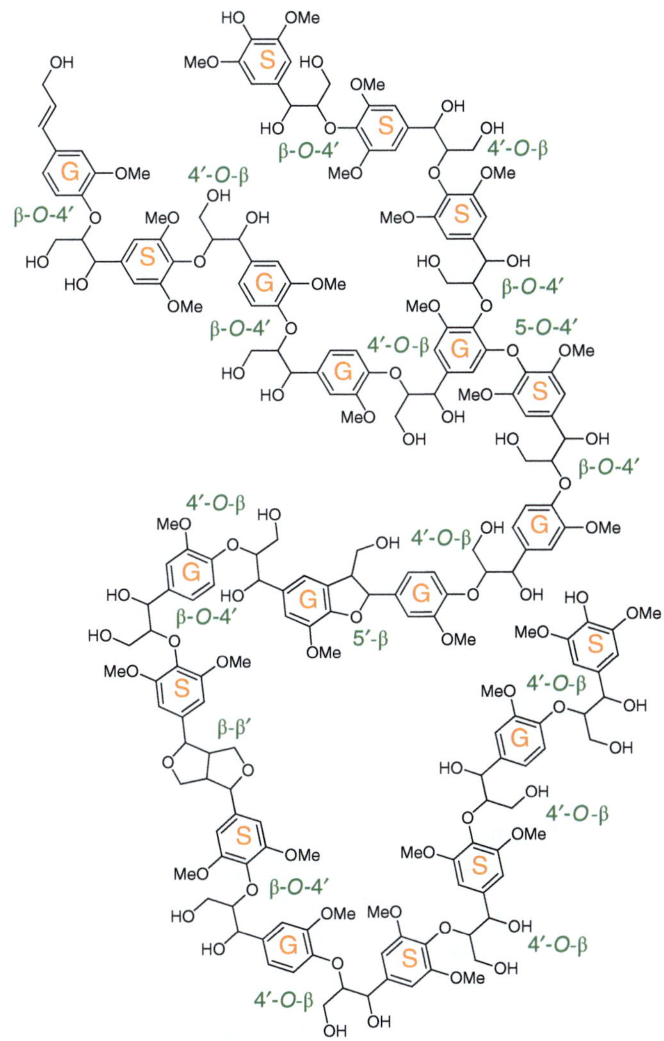

Figure 10.3 Twenty-unit model section of a hardwood (poplar) lignin chain. Source: Ralph et al. [20] / with permission of Springer Nature.

inter-unit linkages, among which β-*O*-4′ ethers are again predominant. A coumarate ester group is attached to the γ-carbon atom in four of the monomer residues, while an acetate ester occupies the γ-position in two other units (Figure 10.4).

A particularly noticeable aspect of this monocotyledonous lignin chain is the absence of branching. It has been claimed that, as of 2019, no structural evidence for branching in lignins had come to light [16]. Consequently, the schematic representations of softwood and hardwood lignin chains [19, 20] that were previously published (Figures 10.2 and 10.3, respectively) could be thought of in their revised forms [16], but the matter remains unresolved. A high degree of branching and/or cross-linking in lignin macromolecules has been a widely held belief since

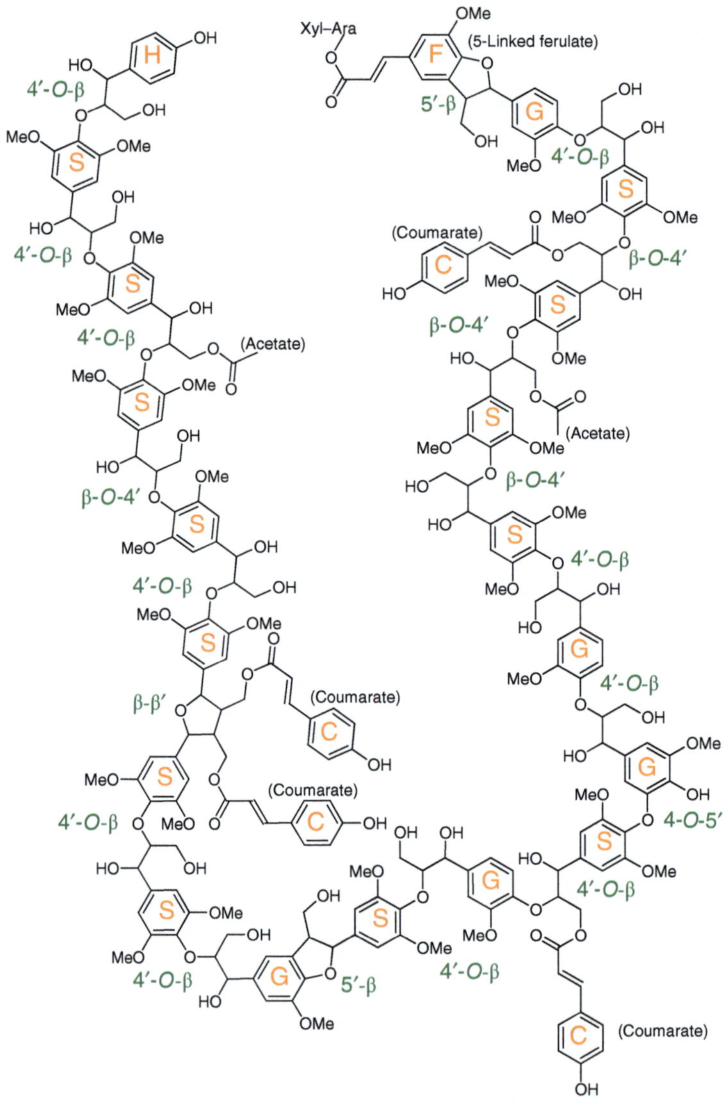

Figure 10.4 Twenty-unit model section of a monocotyledonous lignin chain. Source: Ralph et al. [16] / with permission of Elsevier.

1960 [21]; it will require further analyses to establish unanimous agreement about this crucial question [22]. The answer profoundly affects the perceived likelihood of creating functional thermoplastics with high lignin contents. The inevitable rigidity of highly branched [22] or cross-linked [21] lignin chains would lessen the possibility of overcoming the adverse impact of interstices between neighboring molecular entities in such materials.

Regardless of the case, there is a strong correlation between the configuration of a softwood lignin macromolecule and the adjoining hemicellulose chains (Figure 10.2) in a plant cell wall domain. Moreover, the non-covalent forces between

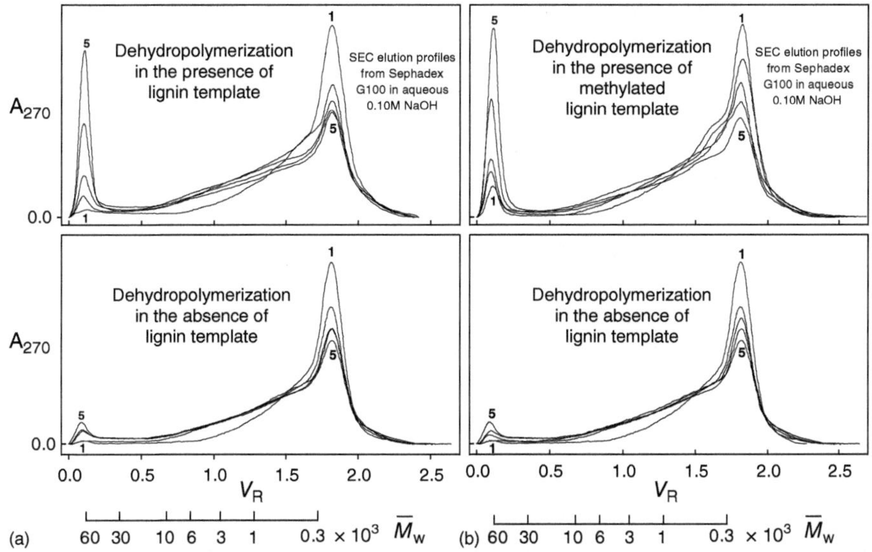

Figure 10.5 (a) Molecular weight distributions of dehydropolymerizates successively formed from coniferyl alcohol in the presence (upper profiles) and absence (lower profiles) of unmethylated macromolecular lignin template (weight-average molecular weight $\overline{M}_w = 206\,000$, 1.0×10^{-8} M initial concentration). Monolignol and H_2O_2 (1 : 1.05 ratio) were added gradually to aqueous 20% dioxane solution (pH 7.3) containing peroxidase (2.6 unit/l); total concentration of monomer units: 3.08×10^{-5} M (1) after 20 hours, 6.29×10^{-5} M (2) after 50 hours, 8.18×10^{-5} M (3) after 75 hours, 8.34×10^{-5} M (4) after 77.5 hours, and 8.50×10^{-5} M (5) after 80 hours. (b) Molecular weight distributions of dehydropolymerizates successively formed from coniferyl alcohol in the presence (upper profiles) and absence (lower profiles) of methylated macromolecular lignin template (weight-average molecular weight $\overline{M}_w = 15\,400$, 2.7×10^{-8} M initial concentration). Monolignol and H_2O_2 (1 : 1.05 ratio) were added gradually to aqueous 20% dioxane solution (pH 7.3) containing peroxidase (2.6 unit/l); total concentration of monomer units: 3.1×10^{-5} M (1) after 20 hours, 6.3×10^{-5} M (2) after 50 hours, 7.9×10^{-5} M (3) after 70 hours, 8.2×10^{-5} M (4) after 75 hours, and 8.5×10^{-5} M (5) after 80 hours. Source: Guan et al. [23] / with permission of Elsevier.

cofacially disposed aromatic rings in macromolecular lignin structures are thought to be stronger than those governing GC/CG base pair interactions in DNA double helices [19]. Thus, pre-existing lignin macromolecules in lignifying plant cell walls are also poised to influence the configurations of new lignin chains as they are formed during biosynthesis.

Such an effect can be induced in an open solution by small quantities of macromolecular lignin components (Figure 10.5) during the peroxidase-catalyzed dehydrogenative polymerization of coniferyl alcohol (Figure 10.1) by H_2O_2 [23]. When the intermediate monolignol radical concentration remained low, dramatic increases occurred in the populations of the largest dehydropolymerizate species that were formed (Figure 10.5a). The result was not caused by preferential monolignol radical coupling with any unpaired electron that may have appeared in the pre-existing lignin macromolecules: prior methylation of the aromatic hydroxyl groups in the latter did not suppress this striking outcome (Figure 10.5b).

Thus, under these conditions, pre-existing lignin macromolecules may act, by non-covalent means, as templates for the chains of lignin substructures that are formed through the dehydrogenative polymerization of coniferyl alcohol. No information came to light about the replicative degeneracy of the process, but the matter is clearly worthy of further investigation.

10.3 Industrial Availability of Lignins

Reliable quantities of lignin derivatives. *Traditional kraft pulping of wood chips has dominated production of cellulosic fibers from lignocellulose. Two industrial processes have been devised for purifying the co-product kraft lignins.*

Recognition of plant biomass as a potential inventory of renewable starting materials for producing liquid biofuels and biodegradable plastics has prompted development of new approaches to fractionating lignocellulose. For example, compared to dilute aqueous acid alone, 25 minutes of pretreatment in 1 : 1 (v/v) THF/water with 0.5 wt% sulfuric acid at 150 °C engenders marked improvements in sugar yields resulting from enzyme-catalyzed saccharification of corn stover [24]. Alternatively, dissolution of corn stover, hardwood, or softwood in homogeneous mixtures of GVL and water at elevated temperatures enables fractionation of lignocellulose into cellulose along with sugars from hemicelluloses and lignin in which structures close to the native biopolymer are initially preserved [25]. Thus, mildly acidic 4 : 1 (v/v) GVL/water at 120 °C can thermocatalytically saccharify maple wood lignocellulose as the carbohydrates and lignins in the biomass dissolve completely within 30 minutes [26].

Interesting developments such as these could form the basis of "second-generation" lignocellulose biorefineries, but their unsubsidized profitability has not yet been demonstrated. Therefore, it may be premature to invest substantially in valorizing lignins formed as co-products of such fractionation procedures, promising although they may be. Rather, it would be more productive to focus on fully optimized "first-generation" processes such as kraft pulping, which has been used since 1890 to convert wood chips into cellulosic fibers for making paper. In 2015, the kraft process was generating about 80% of virgin pulp worldwide [27]. The annual production capacity of co-product kraft lignin was estimated to be 265 000 t in 2018 [28], but an optimistic forecast for the year 2025 predicted an almost 10-fold increase in commercial kraft lignin production [28].

Delignification during kraft pulping takes place in aqueous solution at pHs above 12 in the presence of ~0.5 M (bi)sulfide over one to three hours at a maximum temperature of 160–180 °C [29]. Two commercial processes have been developed for partially purifying the kraft lignin that can be isolated from the so-called "black liquor" into which it dissolves during pulping in the digester. The LignoBoost process (developed by Innventia and licensed to Valmet) uses CO_2 to bring the black liquor from the digester to pH ~10, whereupon the precipitated kraft lignin is washed with dilute aqueous sulfuric acid (pH 2.0–3.5) to remove impurities [30]. On the other hand, the LignoForce System (developed by FPInnovations and NORAM) uses O_2

to detoxify sulfur compounds in the digester black liquor at 80 °C before the kraft lignin is separated from solution by lowering the pH to 9.5 with CO_2; the resulting precipitate is then washed with dilute sulfuric acid and water [11b, 31].

The dissolution of wood cell wall lignin during kraft pulping is a complicated process involving macromolecular depolymerization (primarily through β-*O*-4′ alkyl aryl ether cleavage), conversion of native structural features to styryl ethers and stilbenes, demethylation of methoxyl groups, and repolymerization through the formation of new inter-unit linkages involving aromatic ring C-atoms [32, 33]. Detailed analyses of acetone-soluble and acetone-insoluble softwood kraft lignin fractions led to the structural schemes depicted in Figure 10.6 [33]. These kraft lignin components have clearly undergone extensive chemical transformations in relation to the native biopolymer (Figure 10.2).

10.4 Compelling Traits in Physicochemical Behavior of Kraft Lignin Species

Governing structural features. *Despite some pronounced chemical changes, kraft lignins preserve vestiges of native lignin structures that restore continuity within subsets of interacting molecular components in solution.*

There are underlying traits among the physicochemical properties of kraft lignin components that do not emerge from the structural complexity portrayed in Figure 10.6. One distinctive feature has appeared in semilogarithmic absolute molecular weight calibration curves for the size exclusion chromatographic elution of softwood and hardwood kraft lignins in aqueous 0.10 M NaOH through Sephadex G100: these curves are parallel to the corresponding plot for poly(styrene sulfonate) fractions [34]. Thus, the hydrodynamic behavior of kraft lignin components is consistent with what would be expected for appreciably expanded unbranched random coil chain conformations. In agreement, as far as the ratio of z-average to weight-average molecular weight ($\overline{M}_z/\overline{M}_w$) is concerned, the polydispersity of paucidisperse softwood kraft lignin fractions in aqueous 0.10 M NaOH does not increase with their molecular weight [35].

The justification for choosing kraft lignins as a foundation for developing biodegradable plastics lies in the pronounced tendency of the individual molecular components to associate extensively with one another. The associative/dissociative processes are well defined under a broad range of solution conditions. Under aqueous alkaline conditions at pHs within a fairly narrow range above 11.5, the molecular components and complexes in >100 g/l kraft lignin solutions undergo reversible association. On the other hand, the associated complexes dissociate reversibly in aqueous 0.10 M NaOH solutions containing <1 g/l kraft lignin [36]. The apparent molecular weight distributions of kraft lignin preparations that have been thus modulated are illustrated in Figure 10.7 (after having been desalted). The elution profiles intersect through a common (isosbestic) point, indicating that the members of the set of kinetic rate processes are somehow coupled to the same rate-controlling step.

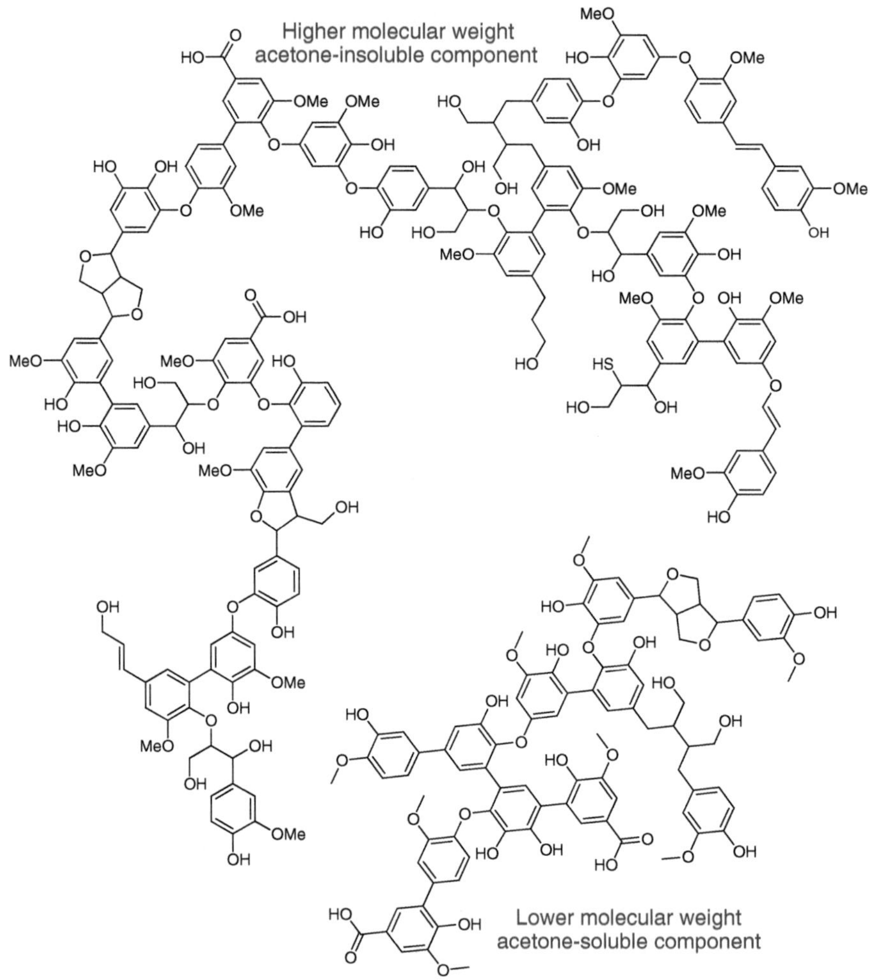

Figure 10.6 Structures proposed for acetone-insoluble higher molecular weight fragment and one of four acetone-soluble lower molecular weight components in softwood kraft lignin. Source: Crestini et al. [33] / with permission of the Royal Society of Chemistry.

Throughout the overall process, the number-average product of the molecular weights of interacting kraft lignin complexes and components, $\langle m_c m_l \rangle$, remains constant despite the marked change in overall degree of association. A plot of weight-average ($\overline{M}_w$) vs. reciprocal of number-average ($1/\overline{M}_n$) molecular weight [36] generates a curve with slope $-2\langle m_c m_l \rangle$, which, in this and other limiting cases, is invariant during the course of the associative/dissociative changes that result in kraft lignin preparations altered solely by incubation in aqueous alkaline solution (Figure 10.7). Evidently, associated kraft lignin complexes behave as well-defined entities that can only interact productively with individual kraft lignin components in a selective manner.

The highest degrees of association that can be attained in aqueous alkaline solutions are limited, inter alia, by the negative charge densities on the polyphenolic

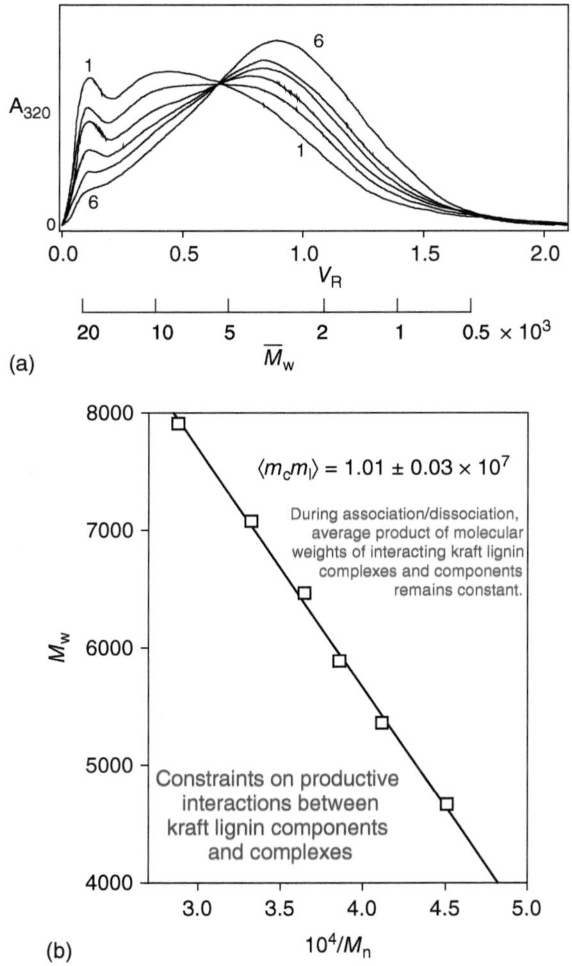

Figure 10.7 (a) Apparent molecular weight distributions and (b) relationship between $\overline{M}_w$ and $\overline{M}_n$ for a homologous series of kraft lignin samples obtained by desalting (through Sephadex LH20/aqueous 35% dioxane) after association and dissociation in aqueous alkaline solutions for (1) 300 hours, (2) 144 hours, and (3) 48 hours at 170 g/l in 1.0 M ionic strength aqueous 0.40 M NaOH; (4) 0 hours, (5) 144 hours, and (6) 644 hours at 0.50 g/l in aqueous 0.10 M NaOH (Sephadex G100/aqueous 0.10 M NaOH elution profiles). Source: Dutta et al. [36] / with permission of American Chemical Society.

kraft lignin complexes. Non-aqueous polar solvents, however, can tolerate a far greater upper limit to the size of soluble complexes that are formed through spontaneous self-assembly of individual kraft lignin components. For example, the formal weight-average molecular weight of an underivatized kraft lignin preparation can surge from approximately 7000 g/mol in aqueous 0.10 M NaOH to around 1.1×10^8 g/mol in DMF (Figure 10.8).

Such effects are quite dramatic; fortunately, the underlying non-covalent interactions are not restricted to hydrogen bonding alone. After acetylation (Ac_2O/pyridine) and methylation (CH_2N_2), the unimodal size exclusion chromatographic elution profiles of the underivatized kraft lignin preparations in DMF are replaced with multimodal distributions of supramacromolecular complexes [37] extending to very high molecular weights (again around 1.0×10^8 g/mol). In each acetylated methylated sample, the relative proportions of species with apparent molecular weight above 2×10^6 g/mol are related to the degree of association of the corresponding underivatized preparation in aqueous alkaline solution (Figure 10.8). These findings

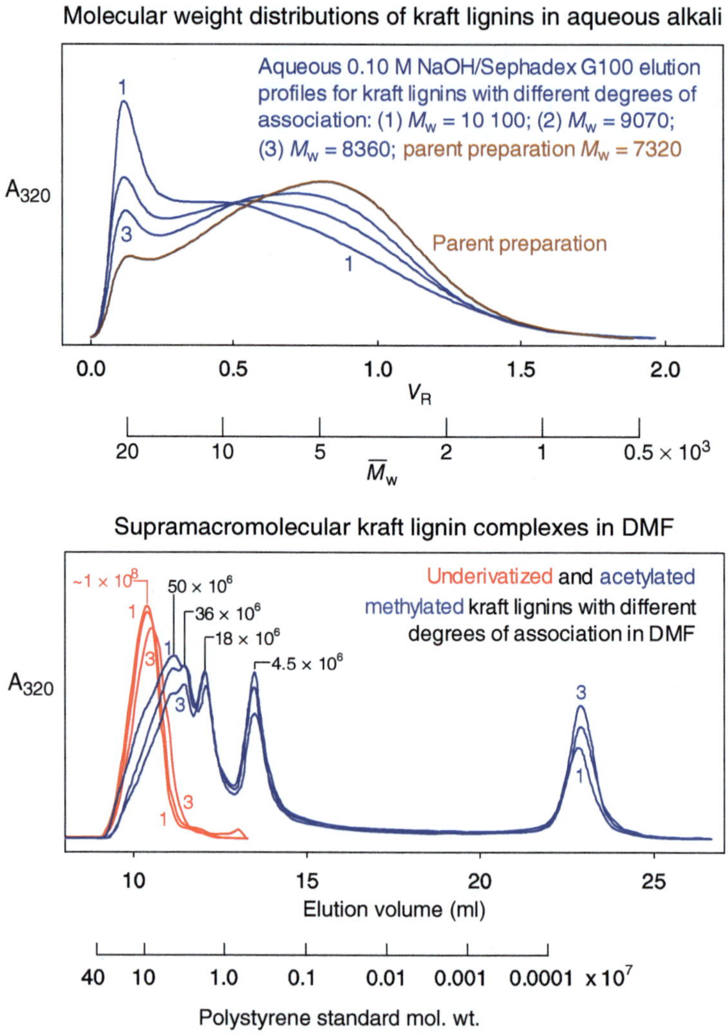

Figure 10.8 Apparent molecular weight distributions during elution in aqueous 0.10 M NaOH through Sephadex G100 (blue) of kraft lignin samples subjected to Sephadex LH20/aqueous 35% dioxane fractionation after association for (1) 6740 hours, (2) 3910 hours, and (3) 1630 hours at 195 g/l in 1.0 M ionic strength aqueous 0.40 M NaOH, compared with the same samples, following their acetylation and methylation, during elution in DMF through a 10^7 Å pore size poly(styrene-divinylbenzene) column; relationship to apparent molecular weight distributions in DMF (red) of underivatized kraft lignin samples after Sephadex LH20/aqueous 35% dioxane fractionation following association for (1) 300 hours (M_w = 7900), (2) 144 hours (M_w = 7100), and (3) 0 hours (M_w = 5900) at 170 g/l in aqueous 0.40 M NaOH (made up to 1.0 M ionic strength with 0.6 M NaCl). Source: Adapted from Refs. [36, 37].

suggest that the process of association advances through discrete stages. At a particular point in its development, every assembly of associating kraft lignin components will interact preferentially with a matching complex rather than a nascent molecular entity. This would ultimately lead to the formation of continuous material domains devoid of prominent interstices that could initiate crack propagation in our targeted kraft lignin-based plastics, when under mechanical stress.

10.5 Kraft Lignin-Based Plastics

Molecular configuration. *Plastics with high kraft lignin contents are composed of associated macromolecular complexes rather than individual kraft lignin components. Nevertheless, blends of kraft lignin with miscible eco-friendly components are fully biodegradable.*

The operation of a traditional kraft pulp mill is organized in such a way that the rate-limiting step in the overall process is housed in the recovery boiler (where "black liquor," concentrated from the digester, is burned to maximize chemical and heat recovery). Consequently, an increase in cellulosic pulp production would require removal of some of the kraft lignin in the black-liquor stream. Ordinarily, 15–30% of the kraft lignin can be thus separated from a pulp mill without the need for other modifications [32, 38]. As a result, there are strong incentives to gain value from the surplus kraft lignin in black liquor.

The most obvious way of valorizing co-product lignins from biorefineries of any kind would be to create biodegradable plastics with high lignin contents. Since 1960, however, arguments in favor of this simple goal have been received with profound skepticism: descriptions of lignin macromolecules in terms of highly branched or cross-linked chains [21, 22] have raised concerns that a search for such formulations might not succeed.

Nevertheless, the first polymeric materials with industrial softwood kraft lignin contents of 85 wt% revealed that, as expected from their size exclusion chromatographic behavior [37], the constituent species were associated macromolecular complexes rather than individual kraft lignin components (Figure 10.8). These plastics were produced by solution casting so as to take the form of heterogeneous blends with 12.6% poly(vinyl acetate), 1.6% diethyleneglycol dibenzoate, and 0.8% indene [39]. The formulations were based on a series of kraft lignin preparations in which the degree of association between individual components had been altered by incubation at different concentrations in aqueous alkaline solution (Figure 10.9). Although none of these materials surpassed polyethylene in their mechanical properties, their tensile strengths and Young's moduli depended linearly on the weight-average molecular weight ($\overline{M}_{w}$) of the kraft lignin preparation used. Thus, the individuality of the associated complexes was preserved as material continuity in the kraft lignin domains gradually developed during the solution casting process.

Homogeneous kraft lignin blends can be far more effective for developing plastics formulations than biphasic blends. This is exemplified (Figure 10.10) in the tensile behavior of 95–98 wt% industrial softwood kraft lignin-based plastics containing small quantities of anthraquinone, phenanthrenequinone, *p*-nitrophenol, melamine, poly(vinyl alcohol), and poly(vinyl acetate). After solution-casting from dimethyl sulfoxide (DMSO), the tensile strengths of these plastics may approach values 1.5-fold greater than polystyrene as elongations at break become almost 7-fold

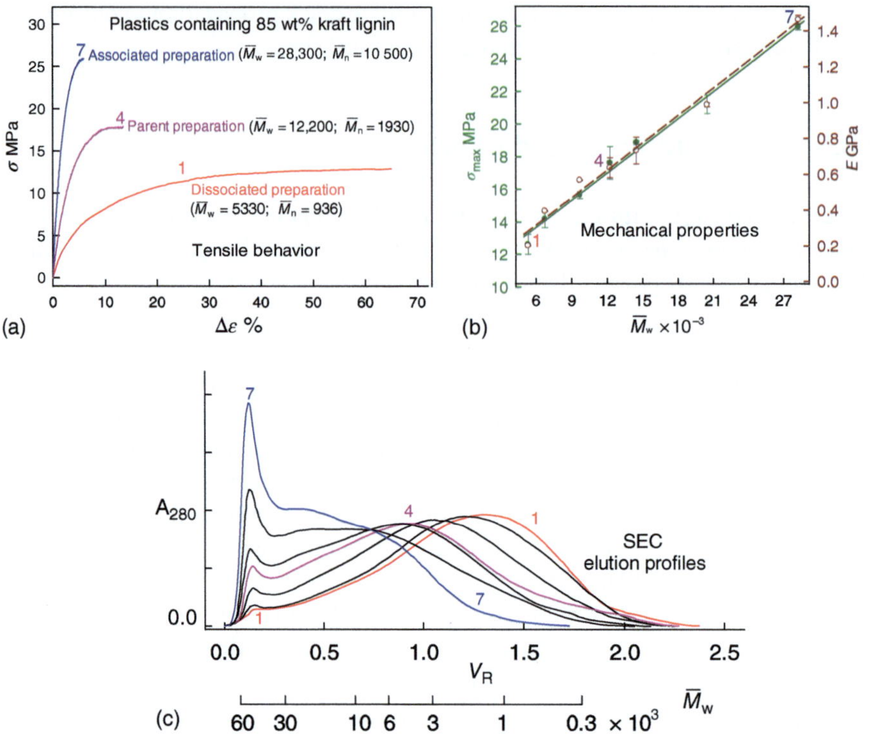

Figure 10.9 (a) Tensile behavior of thermoplastics containing biphasic blends of 85 wt% kraft lignin preparations that differ only in the degree of intermolecular association, with 12.6% poly(vinyl acetate), 1.6% diethyleneglycol dibenzoate, and 0.8% indene. (b) Dependence of (•) tensile strength (σ_{max}) and (○) Young's modulus (E, determined from the initial slope of stress–strain curve) upon weight-average molecular weight of kraft lignin preparations incorporated at 85 wt% levels in thermoplastics with the same blend composition. (c) Apparent molecular weight distributions (during elution in aqueous 0.10 M NaOH through Sephadex G100) of kraft lignin samples after dissociation for (1) 1220 hours, (2) 626 hours, and (3) 93 hours at 0.10 g/l in 0.10 M aqueous NaOH solution, (4) parent preparation, and after association for (5) 92.5 hours, (6) 618 hours, and (7) 1220 hours at 160 g/l in 1.0 M ionic strength aqueous 0.40 NaOH (Sephadex G100/aqueous 0.10 M NaOH elution profiles). Source: Li et al. [39] / with permission of John Wiley & Sons.

larger. The molecular weight distribution of the kraft lignin preparation can have a strong effect on the mechanical behavior of the formulations produced from it. The tensile strength of the plastic containing kraft lignin on its own (Figure 10.10) is reduced threefold when the DMSO solution is filtered through a 4.0–5.5 μm pore size fritted disc before casting [34].

The biodegradation of lignins is brought about by white-rot fungi through a process that requires some extraneous carbohydrate. The cleavage of native lignin macromolecules under these circumstances is thought, albeit implausibly, to be catalyzed primarily by extracellular fungal peroxidases and/or laccases [40]. Coincidentally, the final step in lignin biosynthesis involves catalysis by comparable (although not identical) plant peroxidases and laccases. Reliance on the same kinds

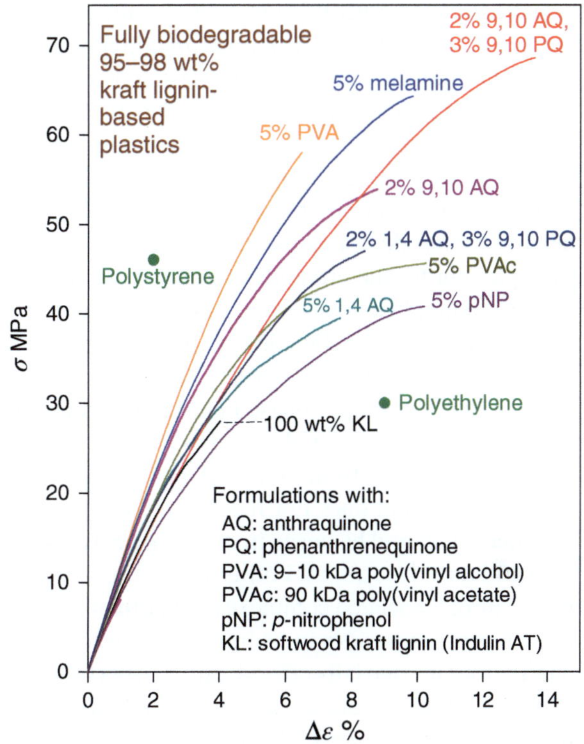

Figure 10.10 Tensile behavior of polymeric materials composed of industrial softwood kraft lignin alone and in blends with 2–5% biodegradable non-lignin components. (Test pieces filed to dog bone shape.)

of enzyme activities, even if modulated, for constructing and deconstructing lignin macromolecules *in vivo* has been called "the lignin enigma" [41]. Nevertheless, the relevant catalytic activities, regardless of their fungal enzyme origins, do extend to softwood kraft lignins in neutral homogeneous culture solutions inoculated, for example, with *Trametes cingulata* mycelium. The primary vegetative white-rot fungal growth is accompanied by marked polymerization of the kraft lignin, as solution conditions become moderately acidic, but then a subsequent metabolic phase witnesses complete degradation of the dissolved high molecular weight kraft lignin species [42]. On the other hand, undissolved softwood kraft lignin may also undergo extensive degradation when immersed in (pH ~4.5) solutions containing *Phanerochaete chrysosporium* mycelial culture [43]. A later section (Section 10.11) more fully describes the biodegradation of all blend components in the kraft lignin-based plastics of Figure 10.10.

10.6 Tuning Strength and Production Cost of Plastics with High Kraft Lignin Contents

Adequacy of miscible blend components in small quantities. *Plastics with high kraft lignin contents are primarily composed of associated macromolecular complexes that interact with one another through their peripheral domains. Miscible monomeric* *or polymeric components blend preferentially with these peripheral regions, and thus quite small proportions may have dramatic effects on mechanical behavior.*

The mechanical properties of plastics with kraft lignin contents near 95 wt% (Figure 10.10) are exquisitely sensitive to blend composition. Just 5 wt% of a miscible blend component such as melamine can increase the tensile strength by ~100–150%. Moreover, even though the non-lignin blend component is present in very low concentration, its configuration strongly affects its impact on mechanical behavior. This does not usually happen in homogeneous polymer blends. For example, the tensile strength of polystyrene increases by only 7% when blended with 5% poly(phenylene oxide) [44] with which it is miscible, while it decreases by 20% or 25% when blended with eco-friendly plasticizers such as dicinnamyl succinate or sebacate at 5% levels [45].

The 95 wt% kraft lignin-based plastics could be regarded as exhibiting antiplasticization characteristics in their behavior (Figure 10.10) because of the simultaneous increase in tensile strength and elongation at break [46]. Conceptually, however, such a description does not recognize the underlying mechanism responsible for the effect. It should be borne in mind that plastics containing very high proportions of kraft lignin [37, 39] and other lignin preparations generally [47] are composed primarily of associated complexes rather than individual components. The integrity of each complex resides in its inner domain where the aromatic rings interact with one another in cofacially offset arrangements [47]. On the other hand, the peripheral domain of this macromolecular entity embodies a higher frequency of edge-on aromatic ring arrangements [47]. The interactions between the neighboring associated complexes take place through interdigitation between the aromatic rings in these peripheral domains as material continuity develops during casting [34]. Non-lignin blend components interact preferentially with the peripheral domains [47]. In the process, they extend the motions of interacting aromatic rings in neighboring associated complexes in such a way as to enable more complete material continuity to be achieved in the biodegradable plastic being cast.

Even though the 95–98 wt% kraft lignin-based plastics illustrated in Figure 10.10 have been produced by solution casting, it is possible to estimate what an upper limit would have been in the year 2020 to the production cost (in US$) of compounding such biodegradable materials. Thus, ~1.4 kg of industrial kraft lignin (US$0.83 cost of commercial kraft lignin [48] that yields 0.95 kg purified starting material) is thoroughly washed with water (US$0.55/kg [30]), air-dried, and blended with 5 wt% miscible blend component (US$0.05 for 50 g melamine per kg blend [49]). The resulting mixture is then compounded (US$0.21/kg [50]) to create a plastic blend with an expected production cost of about US$1.64/kg. Even though this estimate does not take economies of scale into account, it compares favorably with the 2021 polystyrene trading price (~US$2.10/kg) [51] which was expected to rise.

In the quest to improve formulations for plastics with high kraft lignin contents, it is important to bear in mind that both hydrogen bonding and other non-bonded orbital interactions between aromatic rings contribute to the irrepressible inclination of lignin components to associate extensively with one another. Alkylation can eliminate the impact of hydrogen bonding, but it will not annul the predisposition for intermolecular association [37].

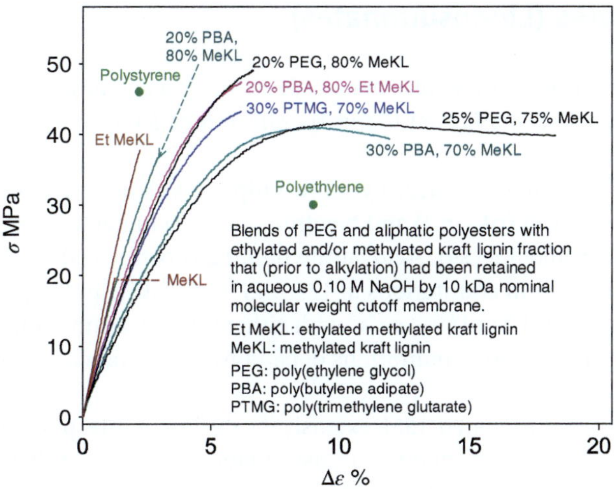

Figure 10.11 Tensile behavior of ethylated and/or methylated kraft lignin-based polymeric materials in blends with poly(ethylene glycol) and aliphatic polyesters (test pieces filed to dog bone shape). Source: Data from Refs. [52, 53].

As might be expected, then, the tensile strength of a methylated softwood kraft lignin-based plastic on its own (20 MPa, 1.2% elongation at break) is appreciably lower (Figure 10.11) than that (Figure 10.10) of its unmethylated counterpart (28 MPa, 4.0% elongation at break). Nevertheless, plastics formed from miscible blends composed of methylated and unmethylated kraft lignin preparations (separately) at high levels both exhibit better mechanical behavior than the unblended materials. Thus, methylated kraft lignin blends with 20 wt% poly(butylene adipate) and 20 wt% poly(ethylene glycol) (PEG) possess (Figure 10.11) tensile strengths of 52 MPa (4.3% elongation at break) and 49 MPa (6.7% elongation at break), respectively. These surpass the tensile strength of polystyrene (46 MPa, 2% elongation at break) and are not very far below the corresponding parameter (Figure 10.10) for the unmethylated kraft lignin blend with 5 wt% poly(vinyl alcohol) (58 MPa, 6.4% elongation at break). Moreover, methylated kraft lignin blends with 30 wt% poly(butylene adipate) and 25 wt% PEG exhibit (Figure 10.11) the tensile strengths of 42 MPa (~9% elongation at break) and 43 MPa (~10% elongation at break), respectively. These exceed the tensile strength of polyethylene (30 MPa, 9% elongation at break) by a substantial margin.

The findings in Figures 10.10 and 10.11 are not sufficiently detailed in scope to identify immediate improvements for plastic formulations with very high kraft lignin contents, but they strongly suggest that a broad range of novel compositions could be profitably examined. Apparently though, blends of hardwood kraft lignin with poly(ethylene oxide) do not follow the same pattern as those involving the softwood kraft lignin (Figure 10.10) and methylated softwood kraft lignin (Figure 10.11). Hardwood kraft lignin on its own and when blended with 5 wt% poly(ethylene oxide) forms brittle plastics with tensile strengths around 24 MPa (0.6% elongation at break), and blends with poly(ethylene oxide) levels of 25–50 wt% suffer from pronounced reductions in that modest initial strength [54].

10.7 Ligninsulfonates (Lignosulfonates)

***Sulfite pulping processes**, in various versions, are less frequently used; they produce polyanionic lignin derivatives that are harder to incorporate at high levels into functional plastic formulations.*

Less than 10% of virgin pulp is now produced by sulfite pulping, which treats wood chips with aqueous sulfurous acid (p$K_a \approx 2$) and bisulfite (p$K_a \approx 7$) under a variety of conditions between pH 1.5 and 4.5 [29]. Acid sulfite pulping is performed at pH ~1.5–2.0 in the presence of calcium counterions at 125–145 °C. Bisulfite pulping is carried out in a pH range of ~3.0–4.5 at 160–180 °C with either sodium, ammonium, (sometimes) potassium, or magnesium counterions depending on the process conditions employed [29].

Lignin macromolecules are sulfonated [55] extensively during sulfite pulping (Figure 10.12), but cleavage and other chemical transformations (such as stilbene formation) occur to a much smaller extent than the analogous changes that occur under kraft pulping conditions (Figure 10.6). For example, absolute molecular weight determinations for an industrial spruce ligninsulfonate (LS) sample revealed a weight-average molecular weight of 64 000 with a very high polydispersity index ($\overline{M}_w/\overline{M}_n$) of 8.8 [56]. Paucidisperse fractions from this parent preparation were isolated by progressively dissolving the corresponding precipitated barium ligninsulfonate from a cellulose column into ethanol–water in steps facilitated by successive reductions in the proportion of ethanol. Six fractions were isolated spanning a 4.6–400 kg/mol range in molecular weight.

Multi-angle light scattering analyses of the components in these ligninsulfonate fractions during their serial elution through a suitable size exclusion chromatographic column enabled the respective z-average ($\overline{M}_z$), weight-average ($\overline{M}_w$), and number-average ($\overline{M}_n$) molecular weights to be reliably determined. The contiguous semilogarithmic column–calibration curve was linear between 2 and 700 kg/mol. The two polydispersity indices, $\overline{M}_z/\overline{M}_w$ and $\overline{M}_w/\overline{M}_n$, for each fraction were, within experimental error, equal to one another; their ratio (0.93 ± 0.07) displayed no systematic trend within the overall molecular weight range. Thus, all six fractions [56] possessed a distinct logarithmic normal distribution [57] of ligninsulfonate components characterized by the same column–calibration curve. These results imply that long-chain branching and cross-linking are absent from ligninsulfonate components. Such a conclusion is reminiscent of a previous finding that $\overline{M}_z/\overline{M}_w$ for a series of paucidisperse softwood kraft lignin fractions in aqueous 0.10 M NaOH remains constant throughout the span of their molecular weights [35].

The value of pursuing functional polymeric materials with a high polyanionic lignin derivative content has been investigated using a ligninsulfonate formed by calcium bisulfite pulping of softwood [58]. This ligninsulfonate preparation possessed an uncommonly low $\overline{M}_w$ of 7100 g/mol with polydispersity index $\overline{M}_w/\overline{M}_n = 3.8$. Phenolic hydroxyl-group methylation was carried out with dimethyl sulfate in aqueous dioxane. As shown in Figure 10.13, both the underivatized and methylated ligninsulfonate exhibited quite modest tensile behavior, reaching ~21 MPa in strength with ~3% elongation at break [58]. Nevertheless, simply as

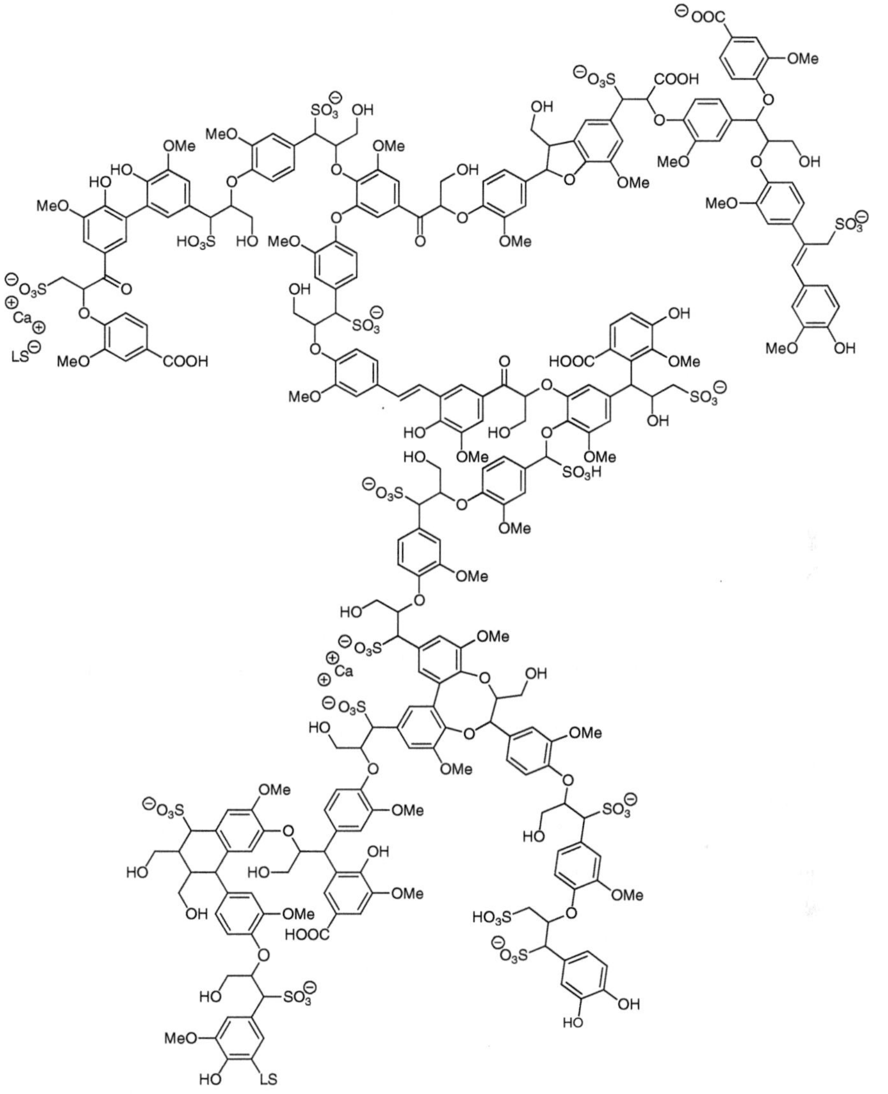

Figure 10.12 Model of 28 units for softwood ligninsulfonate fragment. Source: Adapted from Lund [55]. (LS denotes ligninsulfonate.)

a result of blending with 15 wt% levels of various miscible aliphatic polyesters, a cluster of stress–strain curves manifested themselves, attaining tensile strengths between 30 and 47 MPa with elongations at break from 4% to 7% in the mechanical testing of an entirely new series of 85 wt% ligninsulfonate-based materials (Figure 10.13). When blended with 15 wt% poly(trimethylene glutarate) (PTMG), on the other hand, the resulting 85 wt% methylated ligninsulfonate-based material exhibited a tensile strength of 46 MPa (equal to that of polystyrene) with ~11% elongation at break (Figure 10.13).

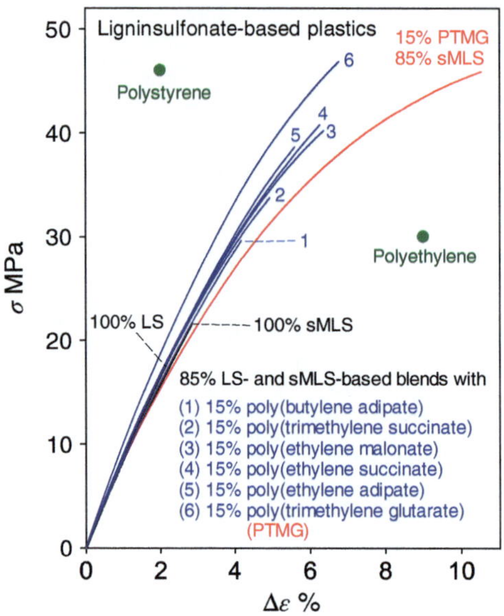

Figure 10.13 Tensile behavior of plastics composed of unmethylated and methylated ligninsulfonates (LS and sMLS) blended with aliphatic polyesters. Source: Data from Sarkanen et al. [58] and from S. Sarkanen & Y.-Y. Wang, unpublished results (2015).

Despite their polyanionic characteristic, immersion of ligninsulfonate- and methylated ligninsulfonate-based test pieces in distilled water for 120 hours under ambient conditions did not engender any component dissolution or reduction in tensile strength. However, the molecular weight distribution of the softwood ligninsulfonate can have a substantial impact on the mechanical properties of plastics formed from it alone, from its simple derivatives, or nearby blends [58, 59].

Ligninsulfonate biodegradability. The white-rot fungus *P. chrysosporium* is quite capable of depolymerizing ligninsulfonate preparations in aqueous solution through a two-phase process where polymerization precedes macromolecular cleavage. However, dissolved SO_2 in industrial calcium bisulfite liquors tends to inhibit the enzyme-catalyzed degradation of polymerized ligninsulfonate [60]. The intervention of ligninsulfonate polymerization before the onset of depolymerization exemplifies the inevitability of "the lignin enigma" [41] that would arise if the activities of peroxidases and/or laccases were actually responsible for white-rot fungal cleavage of lignin macromolecules during lignin biodegradation [40]. Almost identical kinds of enzyme activity engender mono- and oligolignol dehydropolymerization during lignin biosynthesis in plant cell walls; their catalytic effects assuredly contribute to the dehydrogenative polymerization of any phenolic lignin components that may be present in fungal culture solutions.

10.8 Laboratory Ball-Milled Lignins

Native structural characteristics *endow underivatized and methylated ball-milled lignin-based plastics with promising mechanical properties in a range of formulations with monomeric and polymeric blend components.*

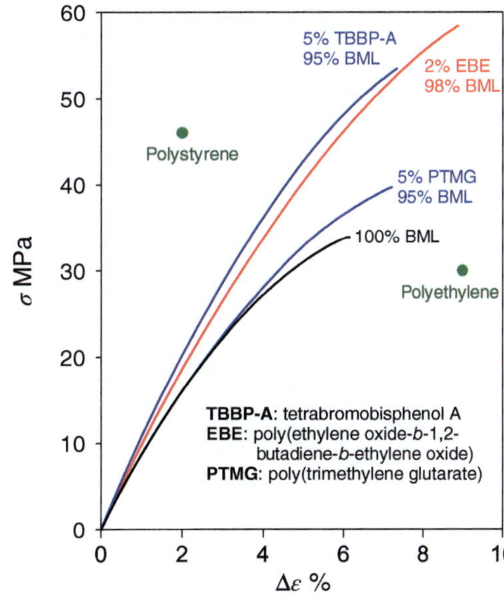

Figure 10.14 Tensile behavior of ball-milled lignin (BML)-based polymeric materials. Source: Chen et al. [34] / MDPI / CC BY 4.0.

Ball-milled lignin (or milled-wood lignin) is thought to be structurally close to the native biopolymer, except that some homolytic cleavage would have occurred during the first step in its isolation. The procedure is straightforward: according to the currently accepted protocol, ball-milling of wood or other plant material is followed by 24-hour extraction with aqueous 96% dioxane [61]. Plastics solution-cast from softwood ball-milled lignin alone (Figure 10.14) in their tensile strength (34 MPa, 6% elongation at break) surpass polyethylene [34]. Small quantities of non-lignin blend components such as 5 wt% PTMG, 5 wt% 3,3′,5,5′-tetrabromobisphenol A, and 2 wt% poly(ethylene oxide-*b*-1,2-butadiene-*b*-ethylene oxide) can result [34] in increasing tensile strengths of 40, 54, and 58 MPa with 7%, 7.5%, and 9% elongations at break, respectively (Figure 10.14). Although the two sets of blend compositions differ, such examples of tensile behavior are comparable (but not identical) to those of plastics with 95–98 wt% proportions of softwood industrial kraft lignin (Figure 10.10). It is conceivable that the underlying skeletal structures of these two dissimilar lignin derivatives (Figures 10.2 and 10.6) may be quite similar.

The tendency of ball-milled lignin components to form, when solution-cast, reasonably strong solid domains (by associating with one another) of course is not governed by hydrogen bonding alone. Methylation of softwood ball-milled lignin can result, through solution-casting, in a material that on its own (Figure 10.15a) surpasses polystyrene in its tensile behavior (~50 MPa strength with 7–8% elongation at break) [47]. Interestingly, the tensile strength of 100% methylated ball-milled lignin-based plastics can be adversely affected during casting by the loss of methylated lignin oligomers (Figure 10.15a). Removal of lower molecular weight lignin components by ultrafiltering the parent ball-milled lignin before methylation brings about analogous effects in plastics formed from blends with poly(ethylene oxide-*b*-1,2-butadiene-*b*-ethylene oxide), as shown in Figure 10.15c [47].

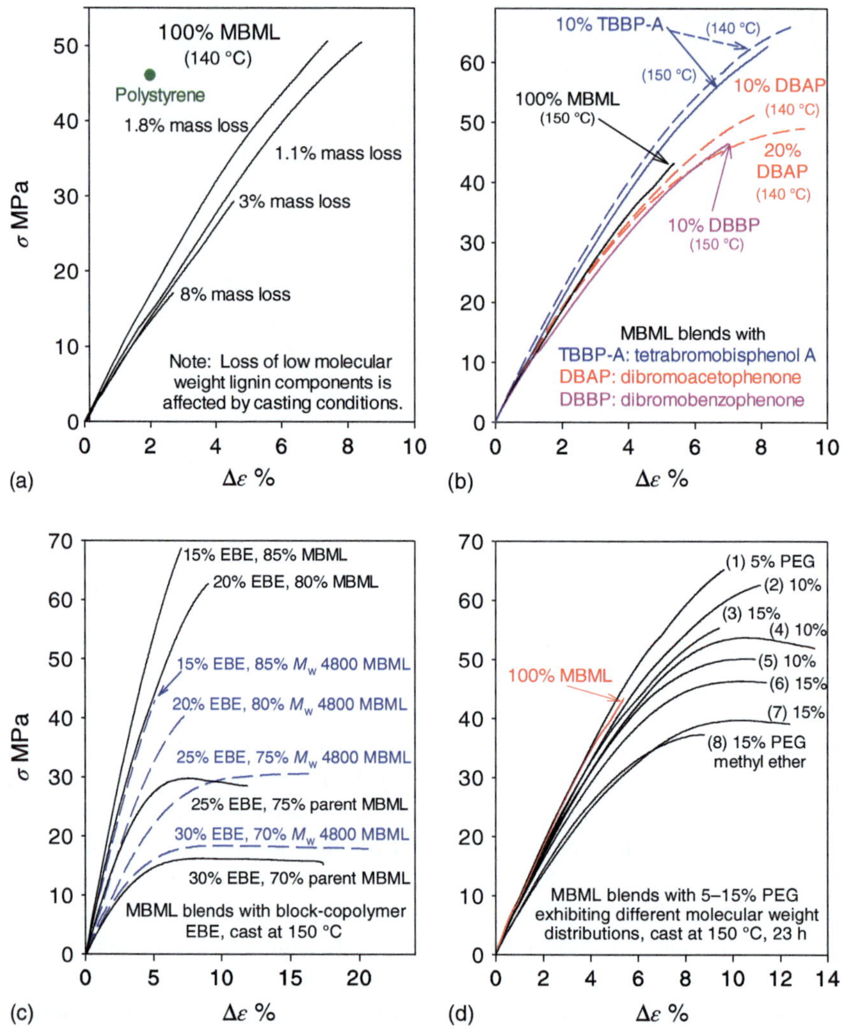

Figure 10.15 Tensile behavior of methylated ball-milled lignin (MBML)-based polymeric materials. (a) 100% MBML-based plastics prepared by solution casting at 140 °C. (b) Blends of 80–90 wt% MBML with bromo-substituted aromatic compounds cast at 140 or 150 °C. (c) Progressive plasticization of parent and higher molecular weight MBML-based materials with triblock poly(ethylene oxide-*b*-1,2-butadiene-*b*-ethylene oxide) (EBE). (d) Plasticization effects of 5–15 wt% of poly(ethylene glycol)s (PEGs) with different molecular weight distributions on MBML-based blends: 100% MBML; (1) 5% PEG, M_n 400 Da; (2) 5% PEG, M_n 400 Da and 5% PEG, M_n 10 kDa; (3) 15% PEG, M_n 4.6 kDa; (4) 5% PEG, M_n 400 Da and 5% PEG M_n 4.6 kDa; (5) 10% PEG, M_n 400 Da; (6) 15% PEG, M_n 2 kDa; (7) 15% PEG, M_n 10 kDa; and (8) 15% poly(ethylene glycol) methyl ether, M_n 5 kDa. Source: From Wang et al. [47].

Sizable electron-withdrawing atomic substituents in miscible non-lignin blend components had been expected to improve the tensile strength and/or ductility of methylated ball-milled lignin-based plastics. Thus, 10 wt% proportions of 3,3′,5,5′-tetrabromobisphenol A (a flame retardant) can engender a ~50% increase in tensile strength (Figure 10.15b), but a similar result was not obtained with compounds such as 3′,5′-dibromoacetophenone or 4,4′-dibromobenzophenone [47].

The triblock copolymer, poly(ethylene oxide-*b*-1,2-butadiene-*b*-ethylene oxide), brought about a tensile strength of almost 70 MPa with 7% elongation at break at 15 wt% levels when blended in a miscible way with solution-cast methylated ball-milled lignin (Figure 10.15c), but larger (20, 25, and 30 wt%) proportions of the block copolymer resulted in progressive reductions in tensile strength with accompanying increases in elongation at break [47]. Curiously, blends of a higher molecular weight ($\overline{M}_w$ 4800, $\overline{M}_w/\overline{M}_n$ 1.4) polydisperse methylated ball-milled lignin fraction with 15 and 20 wt% poly(ethylene oxide-*b*-1,2-butadiene-*b*-ethylene oxide) manifested considerably worse tensile behavior than the corresponding plastics based on the parent lignin preparation (Figure 10.15c). Low molecular weight methylated ball-milled lignin components evidently possess significant plasticizing capacity in these miscible blends [47].

Small proportions (5–15 wt%) of PEG in miscible blends with the parent methylated ball-milled lignin have unevenly dramatic effects. The tensile behavior is affected by both the quantity and the molecular weight distribution of PEG. For example, 5 wt% $\overline{M}_n$ = 400 PEG creates a material exhibiting a tensile strength of 65 MPa with ~9% elongation at break (Figure 10.15d). Otherwise, 5 wt% each of $\overline{M}_n$ = 400 PEG and $\overline{M}_n$ = 4600 PEG together result in a blend with 53 MPa tensile strength and ~13% elongation at break. However, a methylated ball-milled lignin-based blend containing 15 wt% $\overline{M}_n$ = 5000 PEG methyl ether possesses a tensile strength of only 37 MPa with an elongation at break below 9% [47].

10.9 Blend Configuration in Ball-Milled Lignin-Based Plastics Exemplifies the General Case

Thematic Preface. *X-ray powder diffraction patterns reveal how the* ***peripheral and inner domains of associated complexes*** *in methylated ball-milled lignin-based plastics vary with the proportion of blended PEG. Non-covalent rearrangements of aromatic rings at elevated temperatures in ball-milled lignin-based plastics affect glass transition temperature.*

Methylated ball-milled lignin-based plastics composed of blends with 0–15 wt% PEG exhibit X-ray powder diffraction patterns consisting of two overlapping Lorentzian peaks [47] with maxima at equivalent Bragg spacings (d) of 4.0–4.1 and 5.3–5.6 Å (Figure 10.16b–d). For semicrystalline polymers, chain segment packing in amorphous domains may follow a pattern that is characteristic of the corresponding crystal structure [62]. The approximately parallel aromatic rings in the crystal structures of dimeric [63, 64] and trimeric [65] lignin model compounds are separated by an average d-spacing of ~4.0 Å, while edge-on aromatic rings in the crystal structure of a (non-phenolic) β-*O*-4′ lignin-model dimer present d-spacings of ~5.3 Å [66]. These values are quite close to the two Lorentzian component peak maxima (4.1 and 5.5 Å) in the X-ray powder diffraction patterns of the methylated ball-milled lignin-based plastics (Figure 10.16b–d). Accordingly, the two Lorentzian

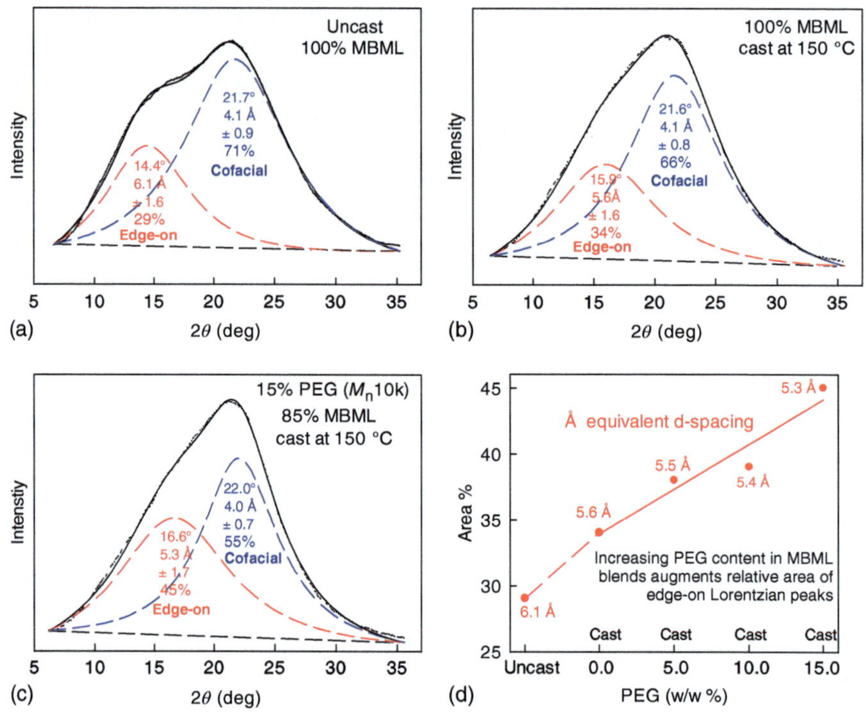

Figure 10.16 X-ray powder diffraction patterns of MBML-based materials. X-ray powder diffraction patterns of (a) uncast 100% MBML, (b) cast 100% MBML, and (c) MBML blend with 15 wt% poly(ethylene glycol) (PEG). The X-ray diffraction patterns of the amorphous polymer blends were analyzed by using two Lorentzian functions $I(x) = I(0)/(1 + x^2/\text{hwhm}^2)$, $x = 2\theta - 2\theta_k$, where $I(x)$ is the scattered intensity at x from the Bragg angle $2\theta_k$ for the peak, 2θ is the scattering angle, and hwhm is the half-width at the half-maximum of the peak. (d) Relative areas of lower-θ Lorentzian component peaks in X-ray powder diffraction patterns of methylated ball-milled lignin-based material blends with increasing PEG content (each point depicting the average area and equivalent d-spacing for blends with the same amount of PEG). Source: From Wang et al. [47].

peaks have been assigned [47], in part, to random distributions [67] of inner cofacially offset arrangements of aromatic rings within the associated lignin complexes (centered at $2\theta \approx 21.8°$, d ≈ 4.1 Å) and corresponding edge-on orientations of aromatic rings in the peripheral components of these macromolecular entities (centered at $2\theta \approx 16.2°$, d ≈ 5.5 Å).

The maximum of the Lorentzian peak circumscribing (in part) the edge-on aromatic rings shifts from d ≈ 6.1 Å to a smaller d-spacing (~5.6 Å) when the methylated ball-milled lignin powder is cast to form a plastic material (Figure 10.16a,b). The shift is accompanied by a significant increase in edge-on peak area because material continuity is achieved during casting through interdigitation between aromatic rings in the peripheral domains of adjacent associated lignin complexes [34]. When the methylated ball-milled lignin is blended with progressively greater proportions (5, 10, and 15 wt%) of PEG, the edge-on Lorentzian peak area undergoes further increases as the d-spacing at the peak maximum shifts to appreciably smaller values

(Figure 10.16c). Thus, the PEG interacts preferentially with the aromatic rings in the peripheral domains of the macromolecular complexes. Indeed, there is less of a change in the d-spacing of the maximum in the Lorentzian peak encompassing the inner cofacially offset aromatic rings in these associated complexes.

Plastics composed of unmethylated ball-milled lignin are also characterized by X-ray powder diffraction patterns that similarly consist of two overlapping Lorentzian peaks [58], as are those composed of methylated ligninsulfonates [59]. In marked contrast, however, materials composed of unmethylated ligninsulfonate components alone present X-ray powder diffraction patterns consisting of only one prominent Lorentzian peak ostensibly representing cofacially offset aromatic rings [59]. The evident absence of a distinct peripheral domain may afford some insight into the reasons for the lower tensile strengths encountered with some of these unblended lignin-based materials (Figure 10.13).

The integrity of the associated complexes, of which plastics with high lignin contents are primarily composed, resides in their inner domains through cofacially offset arrangements of aromatic rings. The development of mechanical strength during casting occurs by establishing continuity between the peripheral domains of adjoining associated complexes. This is illustrated in reference to the contours on a methylated ball-milled lignin-based plastic surface, where the interacting associated complexes are clearly visible (Figure 10.17). Concomitant changes in the distributions of aromatic ring arrangements are reflected in the extent to which new contributions to the two Lorentzian peaks alter the overall X-ray powder diffraction

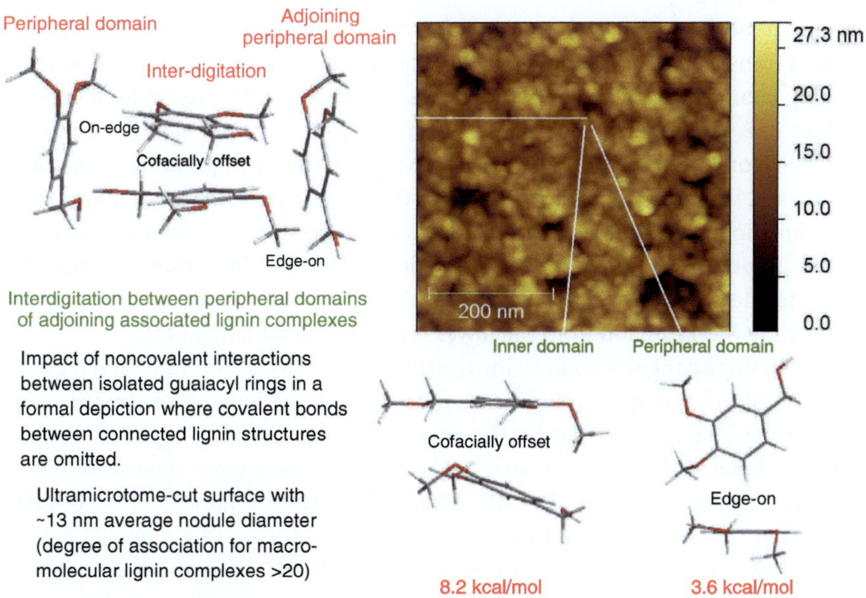

Figure 10.17 AFM height image of the associated lignin complexes on 100 wt% methylated ball-milled lignin-based plastic surface. Material continuity between adjoining complexes is established by inter-digitation between aromatic rings in interacting peripheral domains. Source: Chen et al. [34] / MDPI / CC BY 4.0.

pattern. The resulting populations of aromatic ring arrangements are determined by how readily cofacially offset orientations can be organized by inter-digitation between "edge-on" and "on-edge" aromatic rings on the interacting peripheral domain surfaces of neighboring lignin complexes (Figure 10.17). Thus, the manifest characteristic of the peripheral domains may be augmented or diminished within the bounds of the lignin-based plastic that is formed through casting [34].

Differential scanning calorimetric (DSC) thermograms of ball-milled softwood lignin, its methylated derivative, and higher molecular weight fractions thereof provide critically important insight into the physicochemical behavior of lignin preparations as they are cast into plastic test pieces. Thermograms of the uncast solvent-free parent methylated ball-milled lignin are systematically displaced toward higher temperatures during successive thermal cycles (Figure 10.18a). In the process, the apparent glass transition temperature (T_g) rises in gradually decreasing increments. A comparable change occurs in the glass transition temperature of an uncast higher molecular weight methylated ball-milled lignin fraction as it similarly responds to successive thermal cycles, but the temperature range over which T_g increases is about 40 °C above that found for the parent methylated preparation (Figure 10.18b). On the other hand, thermal cycling of the unmethylated parent ball-milled lignin resulted in a greater change in T_g over a temperature range, only part of which overlapped with that of the higher molecular weight methylated fraction (Figure 10.18c). Accordingly, the three related but contrasting lignin preparations exhibited comparable changes in T_g over different ranges of temperature [47].

A persistent rise in T_g resulting from thermal cycling of a lignin preparation past the glass transition region may occur for more than one reason. A reduction in the enthalpy change during traversal of the transition could be caused by increasing covalent cross-link density [67, 68]. Alternatively, the greater stability conferred by more favorable arrangements of non-covalently interacting lignin components would result in a much smaller reduction in the enthalpy change accompanying the glass transition to the rubbery plateau. This enthalpy change was calculated from the area under the thermogram between the glass transition onset and end with respect to the tangent extending from the heat flow–temperature curve segment approaching the T_g region from the lower temperature direction [47].

The enthalpy change characterizing the traversal across the T_g transition region in the 10 °C/min thermograms (Figure 10.18d) shows little dependence on the extent to which the T_g rises, whether for the parent ball-milled lignin (Figure 10.18c), or for its methylated derivative (Figure 10.18a), or for the higher molecular weight methylated fraction (Figure 10.18b). These observations indicate that covalent cross-linking is not occurring to a significant extent during DSC thermal cycling. Presumably, the same conclusion can be drawn about the configurations of the macromolecular species in corresponding plastics produced by solution casting from DMSO. Thus, the rise in T_g during thermal cycling may be ascribed to the increasing occurrence of more stable aromatic ring arrangements between interacting lignin components that are facilitated by the heightened thermal motions at temperatures above the glass transition.

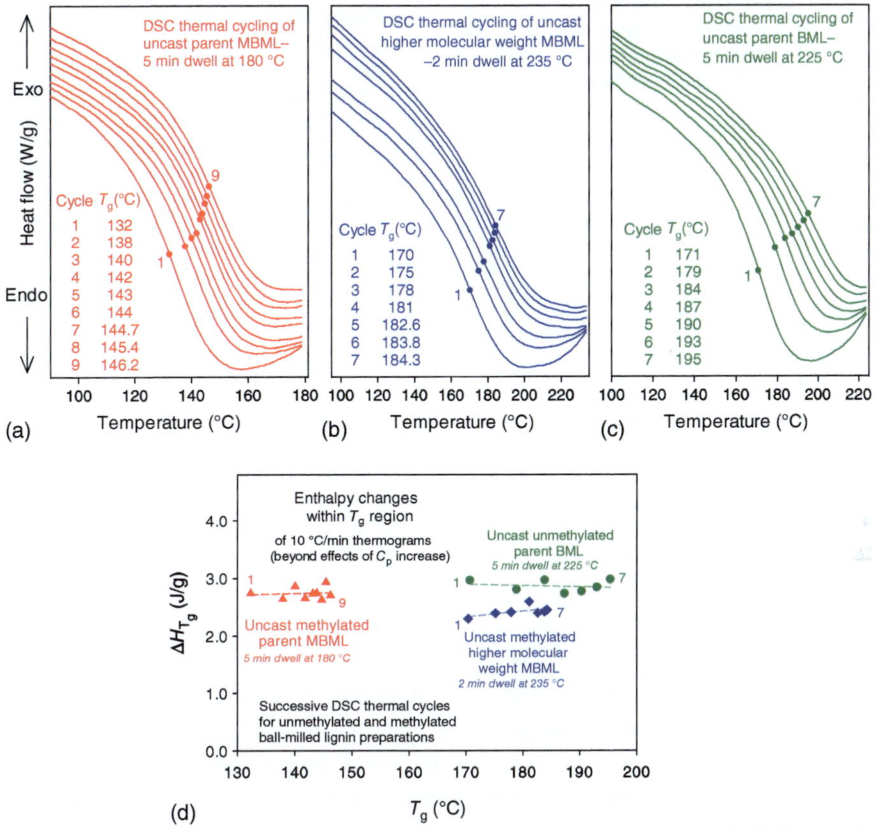

Figure 10.18 DSC thermograms characterizing T_g variations of methylated and unmethylated ball-milled lignin-based polymeric materials. Successive thermal cycles of (a) uncast methylated parent ball-milled lignin (MBML), (b) uncast higher molecular weight M_w 4800 MBML, and (c) uncast unmethylated parent ball-milled lignin (BML). (d) Enthalpy changes (beyond requirement for heat capacity [C_p] increase) accompanying passage through T_g region in 10 °C/min thermograms that reflect DSC thermal cycling of uncast methylated parent ball-milled lignin (MBML), uncast higher molecular weight MBML, and uncast unmethylated parent ball-milled lignin (BML). Source: Wang et al. [47].

10.10 Lignin–Lignin Blends

Extrinsic lignin preparations can impart functionality to incipient lignin-based plastics *composed of interacting associated complexes with peripheral domains that, on their own, are insufficient for thorough interpenetration.*

It is important to bear in mind that the apparent T_g values of plastics with very high lignin contents are determined largely by thermal motions in the peripheral domains of the associated complexes of which these materials are made. For example, for a representative plastic composed solely of a softwood (Douglas fir) ligninsulfonate ($\overline{M}_w = 9700, \overline{M}_w/\overline{M}_n = 5.0$), the X-ray powder diffraction pattern consisted of just a single Lorentzian peak reflecting the cofacially offset

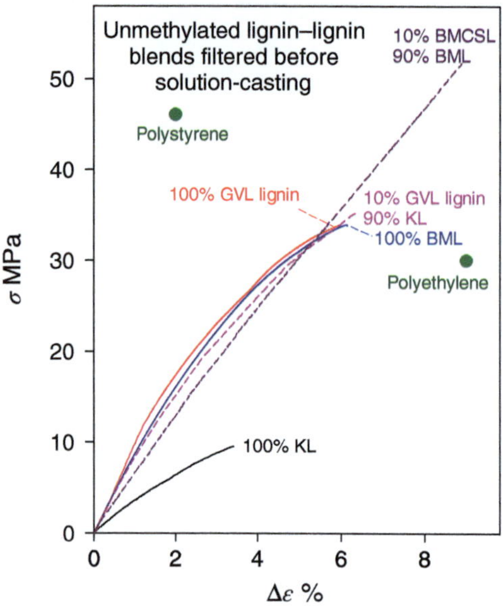

Figure 10.19 Enhancement of tensile behavior in lignin–lignin blends: ball-milled softwood lignin (BML) and ball-milled corn stover lignin (BMCSL); industrial softwood kraft lignin (KL) and maple γ-valerolactone (GVL) lignin. Source: Chen et al. [34] / MDPI / CC BY 4.0.

aromatic ring arrangements in the inner domains of the constituent associated complexes [59], while the corresponding DSC thermograms manifested no discernible T_g. Thus, it would be expected that small proportions of extrinsic lignin preparations could improve the performance of lignin-based polymeric materials with inadequate mechanical properties. It is actually possible to achieve such an outcome with lignin–lignin blends.

Potentially restorative lignin–lignin blends [34] have been found for lignin preparations as diverse as industrial softwood kraft lignin and laboratory ball-milled softwood lignin (Figure 10.19). In regard to creating the kraft lignin-based plastics in these studies, a DMSO solution of the industrial sample was filtered through a 4.0–5.5 μM pore size fritted disc before casting. This intervening step led to a plastic (Figure 10.19) with an approximately threefold lower tensile strength (9 MPa) than obtained without filtration (cf. Figure 10.10). However, the plastic formed from a blend of the kraft lignin with just 10 wt% maple GVL lignin [26] exhibited a tensile strength (35 MPa) very similar to that of the GVL lignin on its own (Figure 10.19). A comparable effect has been observed with a ball-milled softwood lignin (Figure 10.14) that had a tensile strength of 34 MPa on its own. The plastic produced from a blend of this ball-milled lignin with just 10 wt% ball-milled corn stover lignin possessed a tensile strength of 52 MPa with 9% elongation at break, well beyond the corresponding parameters for polystyrene (Figure 10.19). Indeed, the stress–strain curve exhibited very little sign of plastic deformation [34]. Thus, lignin–lignin blends with small (as yet unexplored) quantities of non-lignin components may uncover new pathways to enhanced performance for lignin-based plastic formulations.

10.11 Biodegradation of Kraft Lignin-Based Plastics

Among the plastics with very high lignin contents described in this chapter, those based on industrial softwood kraft lignins are likely to be commercialized first. The pulp and paper industry has begun to deploy processes [30, 31] through which kraft lignins from pulp mills can be partially purified in routine ways. Moreover, the production cost of functional >90 wt% kraft lignin-based plastics (Figure 10.10) should not be prohibitive (vide supra).

As far as the overall life cycle assessment of kraft lignin-based plastics is concerned, controlled biodegradability will be a significant advantage. White-rot fungi are the pre-eminent microorganisms responsible for macromolecular lignin cleavage *in vivo* [40]; their biodegradative activity can be controlled by the requirement for some extraneous carbohydrate. However, a complete understanding of the relevant extracellular white-rot fungal enzymology has been hampered by "the lignin enigma" [41], a dilemma that seeks to address a potential conflict in assigning enzymes with comparable mechanistic capabilities (peroxidases and laccases) to both lignin biosynthesis and lignin biodegradation.

The extracellular white-rot fungal enzymes [40] generally assumed to be responsible for lignin depolymerization – laccase, manganese-dependent peroxidase, and lignin peroxidase – can all function as single-electron oxidants. Consequently, from phenolic or non-phenolic units in lignin macromolecules, all three can generate, directly or indirectly, phenoxy radicals that will undergo bimolecular coupling unless pre-empted from doing so by a competing transformation. To investigate the situation in regard to kraft lignin biodegradation, *T. cingulata* was selected because it does not express lignin peroxidase activity [42]; indeed, when Mn^{II} is absent from the culture solution, this white-rot fungus does not exhibit manganese-dependent peroxidase activity either.

In homogeneous culture solution, the kraft lignin substrate experienced conspicuous polymerization [42] during the vegetative growth of *T. cingulata* as the pH fell from 7.0 to ~4.0 (Figure 10.20a). Moreover, polymerization persisted (Figure 10.21a) after the removal of *T. cingulata* cells from the culture solution, in which laccase activity clearly remained [69]. Following a brief delay, the (still soluble) polymerized kraft lignin in the (original) *T. cingulata* culture solution underwent marked depolymerization [42] through a process that engendered complete preferential degradation of the high molecular weight species (Figure 10.20b). Upon adding FAD (flavin adenine dinucleotide) to the cell-free culture solution isolated during this secondary metabolic phase, the previously polymerized kraft lignin substrate was still degraded quite decisively (Figure 10.21b), *providing* that the solution had been treated with pepstatin A and phenylmethanesulfonyl fluoride [69].

These findings strongly suggested that the initial polymerization of the kraft lignin and its subsequent degradation were controlled by quite different enzymes. The most obvious candidates in white-rot fungal secretomes for the biochemical agents that could act as lignin depolymerases are flavin-dependent monooxygenases [70]. A fitting example would be salicylate hydroxylase, which exhibits broad substrate

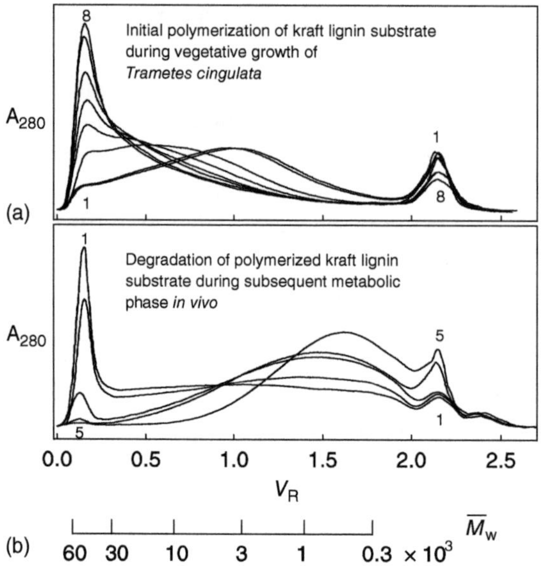

Figure 10.20 (a) Initial polymerization of softwood kraft lignin in *Trametes cingulata* liquid culture. Molecular weight distributions for species in extracellular solution (1) 24 hours, (2) 96 hours, (3) 154 hours, (4) 178 hours, (5) 202 hours, (6) 226 hours, (7) 274 hours, and (8) 322 hours after inoculation. (b) Subsequent degradation of polymerized kraft lignin in *T. cingulata* liquid culture. Molecular weight distributions for species in extracellular solution (1) 5 hours, (2) 149 hours, (3) 329 hours, (4) 353 hours, and (5) 509 hours beyond 39 days after inoculation (Sephadex G100/aqueous 0.10 M NaOH elution profiles). Source: Nutsubidze et al. [42] / with permission of Elsevier.

versatility [71] through a mechanism that can bring about lignin depolymerization without (re)polymerization.

In homogeneous aqueous solution at a pH below neutrality where a native polymeric (milled-wood) lignin preparation remains soluble, salicylate hydroxylase reduces the radius of gyration of the substrate rapidly [70] while the molecular weight increases more slowly due to non-productive enzyme-mediated association between the cleaved components. However, certain proteins (xylanase, for example) acting in an auxiliary capacity are capable of competing effectively with the interactions between the degraded lignin components and salicylate hydroxylase. Typically, the flavin-dependent monooxygenase requires NADH for its activity toward its substrate [71], but the native lignin itself can furnish the necessary reductive capacity when the natural cofactor is absent [70]. These considerations are important for the productive deployment of lignin depolymerases in future biorefineries.

As far as the non-lignin blend components in the (unalkylated) industrial kraft lignin-based plastics (Figure 10.10) are concerned, none will accumulate in the environment. Several soil bacteria and white-rot fungi are capable of degrading anthraquinone [72] and phenanthrenequinone [73, 74], and *p*-nitrophenol biodegradation may be carried out by a number of bacterial genera including *Pseudomonas* spp. [75]. Moreover, melamine is degraded by various soil bacteria

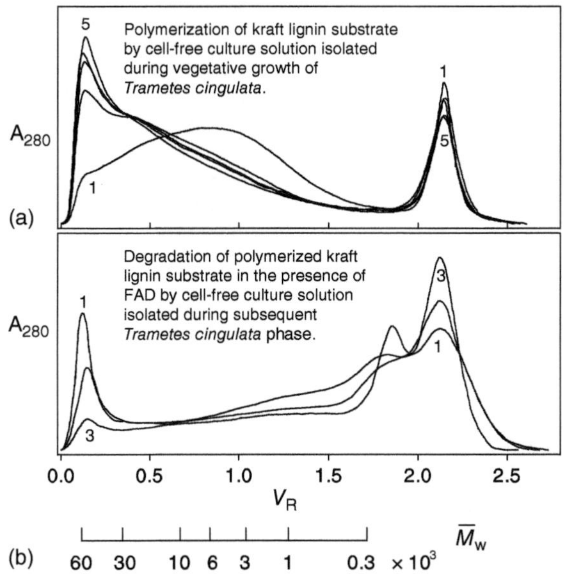

Figure 10.21 Polymerization and subsequent depolymerization of kraft lignin substrates by (unconcentrated) cell-free *Trametes cingulata* culture solutions: molecular weight distributions for species in extracellular solutions (a) alone (1) 5 hours, (2) 215 hours, (3) 385 hours, (4) 550 hours, and (5) 1220 hours beyond isolation nine days after inoculation and with (b) 1.6×10^{-4} M FAD (1) 25 hours, (2) 95 hours, and (3) 240 hours beyond isolation 22 days (530 hours) after inoculation. Molecular weight distributions are depicted in terms of Sephadex G100/aqueous 0.10 M NaOH elution profiles. Source: Adapted from Mlynár et al. [69].

such as *Arthrobacter* spp. [76], *Nocardioides* sp. [77], and *Rhodococcus* sp. [78]. In regard to bacterial strains that can degrade poly(vinyl alcohol), many species are found in the genera *Pseudomonas*, *Sphingomonas*, *Alcaligenes*, and *Bacillus* [79]. On the other hand, poly(vinyl acetate) is not itself water soluble, but it can swell when exposed to water; further abiotic hydrolysis of the acetate groups may facilitate entry into an oxidative pathway that leads to chain cleavage [79].

10.12 Alternative Formulations for Polymeric Materials Containing More than 50 wt% Lignin

In shedding light on the circumstances under which the foregoing biodegradable lignin-based plastics were developed, other approaches describing how polymeric materials with >50 wt% lignin contents have been formulated recently (after the year 2010) are germane. Whether or not these alternative polymeric materials would be biodegradable, they should be functional. A reasonable demarcation line for their tensile strengths should not lie far below 30 MPa, the value characteristic of polyethylene.

A series of polyurethanes formed from reactions of three underivatized kraft lignin (KL) fractions with 4,4′-methylenebis(phenyl isocyanate) (4,4′-MDI) in the

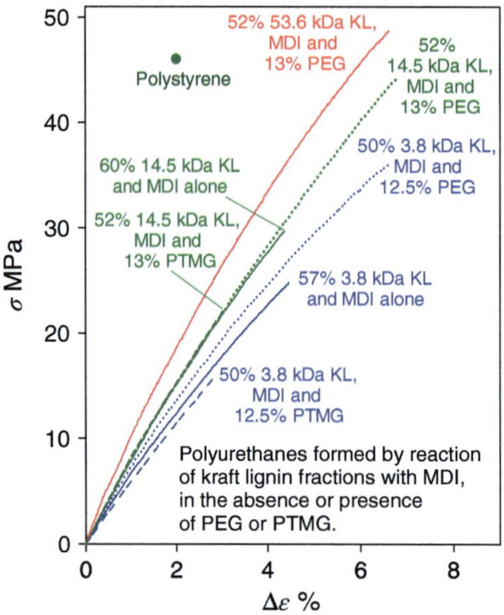

Figure 10.22 Tensile behavior of polyurethanes produced by the reactions of underivatized softwood kraft lignin fractions and MDI with or without 4.0 kg/mol PEG or 4.0 kg/mol PTMG as secondary polyols. Source: Wang et al. [80] / with permission of American Chemical Society.

absence or presence of ~13% PEG or PTMG were compared according to their tensile behavior [80]. The strengths of the polyurethanes produced by the reactions of MDI with 60 wt% 14.5 kg/mol KL alone (30 MPa) and 57 wt% 3.8 kg/mol KL alone (25 MPa) were within 20% of one another despite an approximately fourfold difference in KL molecular weight (Figure 10.22). This similarity was ascribed to the survival, in these two-component polyurethanes, of associated kraft lignin complexes that would interact differently with the PEG and PTMG used in corresponding three-part formulations. Thus, H-bonding of PEG at ~13% levels with the KL fractions improved polyurethane tensile strength approximately 1.5-fold, while less effective interactions involving ~13% PTMG resulted in a roughly 1.5-fold reduction in tensile strength (Figure 10.22). The polyurethane produced from 52 wt% of a polydisperse 53.6 kg/mol KL fraction, MDI, and 13 wt% PEG exhibited a tensile strength of 49 MPa, appreciably above that of polystyrene (Figure 10.22). However, the corresponding polyurethane formed in the absence of PEG did not furnish a point of comparison because it was not homogeneous [80].

On the other hand, epoxidized ethyl-acetate-soluble low molecular weight KL fractions, when cross-linked with a poly(propylene oxide) diamine ("Jeffamine D400"), have generated thermoset materials containing 66 wt% eucalyptus and spruce KL that exhibit tensile strengths between 56 and 66 MPa, respectively, with elongations at break of 10.5% and 8% [81]. These mechanical properties follow fairly closely the tensile behavior of 95 wt% unfractionated Southern pine KL-based plastic blends containing 5 wt% melamine or 2 wt% anthraquinone together with 3 wt% phenanthrenequinone (Figure 10.10). The 95 wt% unfractionated KL-based blends reach tensile strengths of 64 and 69 MPa, respectively, with 10% and 13.5% elongations at break (Figure 10.10). These blends contain greater proportions of the original kraft lignin and are cheaper to produce. Nevertheless, the similarities between the mechanical properties of the 95 wt% unfractionated KL-based blends (Figure 10.10) and the 66 wt% fractionated KL-based epoxy thermoset resins [81] have established reasonable expectations about what the tensile strengths should be of materials in which KL plays the more active role in determining the mechanical behavior.

Only in rare cases is an increasing proportion of polyanionic ligninsulfonate (LS) in a polymer blend accompanied by improvement in mechanical behavior. The more common outcome is exemplified by the results of mixing 10–70 vol% quantities of a ~1.9 kg/mol LS at elevated temperatures (180–220 °C) to create blends with polypropylene, polystyrene, polycarbonate, poly(methyl methacrylate), poly(lactic acid), and an ionomer composed of partially neutralized poly(ethylene-*co*-methacrylic acid) [82]. The LS was an industrial product with calcium counterions containing inorganic salts and sugars. Materials with 50 vol% LS spanned a range of tensile strengths between 10 and 35 MPa that, except for those of the ionomer-containing blends, fell on a downward slope as LS proportions were increasing.

Although all the blends were heterogeneous, none exhibited a T_g ascribable to the ligninsulfonate [82]. Moreover, none of these heterogeneous blends embodied LS contents that approached the 85 wt% levels in the homogeneous materials of Figure 10.13. Their tensile strengths would have fallen substantially below those of the homogeneous 85 wt% LS-based blends, had their formulations been able to preserve functionality with equivalent compositions. Actually, immersing 50–100 μm slices of the 10–70 vol% ~1.9 kg/mol LS-containing blends in water resulted in complete LS dissolution within 24 hours at ambient temperatures [82]. In contrast, immersion of the 85% 7.1 kg/mol LS-based test pieces (Figure 10.13) for 120 hours in water brought about no component dissolution nor reduction in tensile strength. The results with the heterogeneous LS blends [82] would have improved, had the industrial LS preparation been pre-purified.

A different approach to formulating polymeric materials with very high LS contents emerged from the laccase-catalyzed polymerization of a sulfonated lignin derivative produced in a magnesium bisulfite pulp mill [83]. Oxidative polymerization of the industrial LS hereby resulted in a ~ 15-fold increase in apparent molecular weight (from an initial value of 40 kg/mol) as brittle water-insoluble products were formed. Blending with 2.5–10 wt% levels of components such as sorbitol, glycerol, and xylitol or 0.6–20 kg/mol PEGs engendered striking improvements in tensile behavior (Figure 10.23).

The most promising formulations were those with 5 and 10 wt% contents of 0.6 and 1.0 kg/mol PEG [83]; the tensile strengths of these blends attained values as high as 74 MPa with elongations at break reaching 72% (Figure 10.23). Such results indicate that the skeletal structure in the polymerized LS chains allows preservation of material continuity as substantial material deformation takes place. It is unlikely that covalent cross-linking plays a central role in supporting the tensile behavior of the polymerized LS blends. Nevertheless, these findings are very encouraging, although a 20% average weight loss in the 90–97.5 wt% LS-based plastics occurred as a result of immersion in water. Thus, fundamental changes must be made if formulations like these [83] are going to produce useful materials.

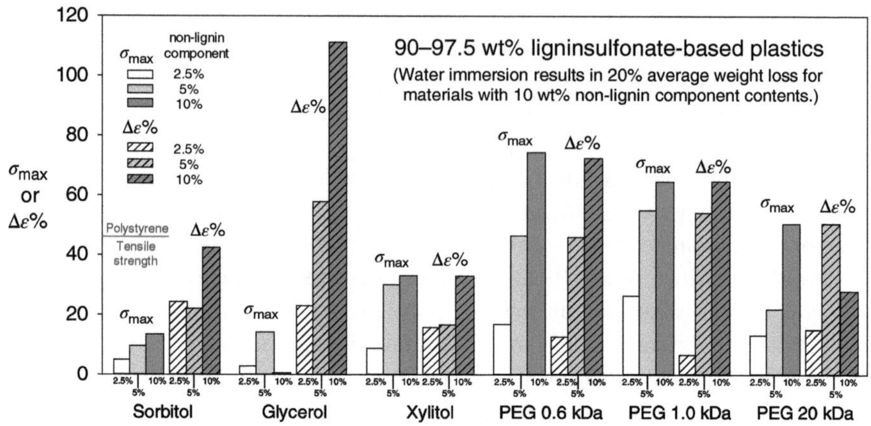

Figure 10.23 Tensile strengths (σ_{max}) and elongations at break ($\Delta\varepsilon$%) of partially water-soluble laccase-polymerized ligninsulfonate-based blends with 2.5–10 wt% sorbitol, glycerol, xylitol, and poly(ethylene glycol)s (PEGs). Source: Bartolome et al. [83] / with permission of Elsevier.

10.13 Concluding Remarks

The focus of this chapter has been confined to lignin-based plastics that are fully biodegradable. The limitation implies that the relevant thermoplastics or thermosets should embody lignin contents that are already above 50 wt% before blending or derivatization. Moreover, claims about overall biodegradability of these materials should be supported by reports about the susceptibility of all constituent components (whether covalently linked or not) to transformation by microorganisms under suitable conditions. Thus, numerous lignin-containing blends, composites, and copolymers that do not fit these criteria have been reviewed elsewhere [84], although "lessons from the past" [85] are unlikely to be productive because of entrenched views about macromolecular lignin cross-linking that were ultimately doubtful [16].

Blend alterations and transformations for compression and injection molding the industrial 95–98 wt% kraft lignin-based formulations (Figure 10.10) will enable materials to be manufactured that range from computer consoles and interior automobile components through doors and frames to manifold enclosures for which the dark color of these plastics would not be an issue. Later, identification and nullification of the dominant chromophores [86] will lead to many more applications for these renewable biodegradable plastics from lignins in plant materials and trees.

Acknowledgments

Work carried out at the University of Minnesota was supported by Agriculture and Food Research Initiative Grant no. 2011-67009-20062 from the U.S. Department of Agriculture (USDA) National Institute of Food and Agriculture (NIFA), a

Subaward (115808 G002979) from the "Northwest Advanced Renewables Alliance" led by Washington State University and supported by USDA Agriculture and Food Research Initiative Competitive Grant no. 2011-68005-30416, a subcontract (XFT-8-88522-01) from the U.S. Department of Energy BioEnergy Science Center, and McIntire–Stennis Funds from USDA NIFA project MIN-12-085. The authors gratefully acknowledge receipt of the maple GVL lignin provided by A.H. Motagamwala and G.W. Huber at the University of Wisconsin, and the corn stover ball-milled lignin provided by R. Katahira and D.K. Johnson at the National Renewable Energy Laboratory.

References

1 Rosillo-Calle, F. and Woods, J. (2015). Overview of bioenergy. In: *The Biomass Assessment Handbook*, 2e (ed. F. Rosillo-Calle, P. de Groot, S.L. Hemstock and J. Woods), 7–33. New York: Routledge.
2 Hall, D.O. and Rao, K.K. (1999). *Photosynthesis*, 6e. New York: Cambridge University Press, 214 pp.
3 Biofuels Research Advisory Council (2006). Biofuels in the European Union: A Vision for 2030 and Beyond. EUR 22066. 39 pp.
4 U.S. Department of Energy (2016). 2016 Billion-Ton Report: Advancing Domestic Resources for a Thriving Bioeconomy, Volume 1 (BT16): Economic Availability of Feedstocks. Langholtz, M.H., Stokes, B.J. and Eaton, L.M. (Leads). Oak Ridge National Laboratory, Oak Ridge, TN; 10.2172/1271651. http://energy.gov/eere/bioenergy/2016-billion-ton-report (accessed 17 December 2021).
5 Klemm, D., Heublein, B., Fink, H.-P., and Bohn, A. (2005). Cellulose: fascinating biopolymer and sustainable raw material. *Angewandte Chemie, International Edition* 44: 3358–3393.
6 Sandermann, H. Jr., Scheel, D., and Van der Trenck, T. (1983). Metabolism of environmental chemicals by plants – copolymerization into lignin. *Journal of Applied Polymer Science: Applied Polymer Symposium* 37: 407–420.
7 Himmel, M.E., Ding, S.-Y., Johnson, D.K. et al. (2007). Biomass recalcitrance: engineering plants and enzymes for biofuels production. *Science* 315: 804–807.
8 Davis, R., Tao, L., Tan, E.C.D. et al. (2013). Process Design and Economics for the Conversion of Lignocellulosic Biomass to Hydrocarbons. NREL/TP-5100-60223, National Renewable Energy Laboratory, Golden, CO. https://www.nrel.gov/docs/fy14osti/60223.pdf
9 Moriarty, K., Milbrandt, A., Lewis, J., and Schwab, A. (2020). 2017 Bioenergy Industry Status Report. NREL/TP-5400-75776, National Renewable Energy Laboratory, Golden, CO. https://www.nrel.gov/docs/fy20osti/75776.pdf, p. 17
10 O'Dea, R.M., Willie, J.A., and Epps, T.H. (2020). 100th Anniversary of macromolecular science viewpoint: polymers from lignocellulosic biomass. Current challenges and future opportunities. *ACS Macro Letters* 9: 476–493.
11 (a) Maki, K., Paleologou, M., Zhang, Y. et al. (2017). LignoForce kraft lignin extraction: process scale-up and product development. In: *Kraft Lignin Innovation Forum*. Iron Mountain, MI: Michigan Forest Biomaterials Institute,

visual 11. (b) Maki, K., Paleologou, M., Zhang, Y. et al. (2017). LignoForce kraft lignin extraction: process scale-up and product development. In: *Kraft Lignin Innovation Forum*. Iron Mountain, MI: Michigan Forest Biomaterials Institute, visual 14.

12 Galkin, M.V., Francesco, D.D., Edlund, U., and Samec, J.S.M. (2017). Sustainable sources need reliable standards. *Faraday Discussions* 202: 281–301.

13 Brunow, G. and Lundquist, K. (2010). Functional groups and bonding patterns in lignin (including the lignin-carbohydrate complexes). In: *Lignin and Lignans – Advances in Chemistry* (ed. C. Heitner, D. Dimmel and J.A. Schmidt), 267–299. Boca Raton, FL: CRC Press.

14 Kang, X., Kirui, A., Dickwella-Widanage, M.C. et al. (2019). Lignin–polysaccharide interactions in plant secondary cell walls revealed by solid-state NMR. *Nature Communications* 10: 347. https://doi.org/10.1038/s41467-018-08252-0.

15 Tohge, T., Watanabe, M., Hoefgen, R., and Fernie, A.R. (2013) Shikimate and phenylalanine biosynthesis in the green lineage. *Frontiers in Plant Science,* 4, 62. https://doi.org/10.3389/fpls.2013.00062.

16 Ralph, J., Lapierre, C., and Boerjan, W. (2019). Lignin structure and its engineering. *Current Opinion in Biotechnology* 56: 240–249.

17 Rosler, J., Krekel, F., Amrhein, N., and Schmid, J. (1997). Maize phenylalanine ammonia-lyase has tyrosine ammonia-lyase activity. *Plant Physiology* 113: 175–179.

18 Barros, J., Serrani-Yarce, J.C., Chen, F. et al. (2016). Role of bifunctional ammonia-lyase in grass cell wall biosynthesis. *Nature Plants* 2: 16050. https://doi.org/10.1038/nplants.2016.50.

19 Chen, Y.-r. and Sarkanen, S. (2010). Macromolecular replication during lignin biosynthesis. *Phytochemistry* 71: 453–462.

20 Ralph, J., Lundquist, K., Brunow, G. et al. (2004). Lignins: natural polymers from oxidative coupling of 4-hydroxyphenylpropanoids. *Phytochemistry Reviews* 3: 29–60.

21 Gupta, P.R. and Goring, D.A.I. (1960). Physicochemical studies of alkali lignins. III. Size and shape of the macromolecule. *Canadian Journal of Chemistry* 38: 270–279.

22 Balakshin, M., Capanema, E.A., Zhu, X. et al. (2020). Spruce milled wood lignin: linear, branched or cross-linked? *Green Chemistry* 22: 3985–4001.

23 Guan, S.-Y., Mlynár, J., and Sarkanen, S. (1997). Dehydrogenative polymerization of coniferyl alcohol on macromolecular lignin templates. *Phytochemistry* 45: 911–918.

24 Nguyen, T.Y., Cai, C.M., Kumar, R., and Wyman, C.E. (2015). Co-solvent pretreatment reduces costly enzyme requirements for high sugar and ethanol yields from lignocellulosic biomass. *ChemSusChem* 8: 1716–1725.

25 Luterbacher, J.S., Rand, J.M., Alonso, D.M. et al. (2014). Nonenzymatic sugar production from biomass using biomass-derived γ-valerolactone. *Science* 343: 277–280.

26 McClelland, D.J., Galebach, P.H., Motagamwala, A.H. et al. (2019). Supercritical methanol depolymerization and hydrodeoxygenation of lignin and biomass over reduced copper porous metal oxides. *Green Chemistry* 21: 2988–3005.

27 Suhr, M., Klein, G., Kourti, I. et al. (2015). Best Available Techniques (BAT) Reference Document for the Production of Pulp, Paper and Board, Industrial Emissions Directive. 2010/75/EU, Integrated Pollution Prevention and Control. European IPPC Bureau, Joint Research Centre, Seville, Spain. https://doi.org/10.2791/370629

28 Dessbesell, L., Paleologou, M., Leitch, M. et al. (2020). Global lignin supply overview and kraft lignin potential as an alternative for petroleum-based polymers. *Renewable and Sustainable Energy Reviews* 123: 109768.

29 Bajpai, P. (2018). *Biermann's Handbook of Pulp and Paper*, Raw Material and Pulp Making, 3e, vol. 1. Cambridge, MA: Elsevier.

30 Tomani, P. (2010). The LignoBoost process. *Cellulose Chemistry and Technology* 44 (1–3): 53–58.

31 Kouisni, L., Gagné, A., Maki, K. et al. (2016). LignoForce system for the recovery of lignin from black liquor: feedstock options, odor profile and product characterization. *ACS Sustainable Chemistry & Engineering* 4: 5152–5159.

32 Hu, Z., Du, X., Liu, J. et al. (2016). Structural characterization of pine kraft lignin: BioChoice lignin vs Indulin AT. *Journal of Wood Chemistry and Technology* 36: 432–436.

33 Crestini, C., Lange, H., Sette, M., and Argyropoulos, D.S. (2017). On the structure of softwood kraft lignin. *Green Chemistry* 19: 4104–4121.

34 Chen Y.-r., Sarkanen, S., and Wang, Y.-Y. (2019). Lignin-only polymeric materials based on unmethylated unfractionated kraft and ball-milled lignins surpass polyethylene and polystyrene in tensile strength. *Molecules* 24: 4611. https://doi.org/10.3390/molecules24244611.

35 Mlynár, J. and Sarkanen, S. (1996). Renaissance in ultracentrifugal sedimentation equilibrium calibrations of size-exclusion chromatographic elution profiles. In: *Strategies in Size Exclusion Chromatography*, ACS Symposium Series No. 635 (ed. M. Potschka and P.L. Dubin), 379–400. Washington, DC: American Chemical Society.

36 Dutta, S., Garver, T.M. Jr., and Sarkanen, S. (1989). Modes of association between kraft lignin components. In: *Lignin – Properties and Materials*, ACS Symposium Series No. 397 (ed. W.G. Glasser and S. Sarkanen), 155–176. Washington, DC: American Chemical Society.

37 Li, Y. (1995). Noncovalent interactions between kraft lignin components. M.S. thesis. University of Minnesota.

38 Wallberg, O. and Jönsson, A.-S. (2006). Separation of lignin in kraft cooking liquor from a continuous digester by ultrafiltration at temperatures above 100 °C. *Desalination* 195: 187–200.

39 Li, Y., Mlynár, J., and Sarkanen, S. (1997). The first 85% kraft lignin-based thermoplastics. *Journal of Polymer Science Part B: Polymer Physics* 35: 1899–1910.

40 Chan, J.C., Paice, M., and Zhang, X. (2020). Enzymatic oxidation of lignin: challenges and barriers toward practical applications. *ChemCatChem* 12: 401–425.

41 Leisola, M., Pastinen, O., and Axe, D.D. (2012). Lignin – designed randomness. *BIO-Complexity* 3: 1–11.

42 Nutsubidze, N.N., Sarkanen, S., Schmidt, E.L., and Shashikanth, S. (1998). Consecutive polymerization and depolymerization of kraft lignin by *Trametes cingulata*. *Phytochemistry* 49: 1203–1212.

43 Ulmer, D.C., Leisola, M.S.A., Schmidt, B.H., and Fiechter, A. (1983). Rapid degradation of isolated lignins by *Phanerochaete chrysosporium*. *Applied and Environmental Microbiology* 45: 1795–1801.

44 Fekete, E., Földes, E., Damsits, F., and Pukánszky, B. (2000). Interaction-structure-property relationships in amorphous polymer blends. *Polymer Bulletin* 44: 363–370.

45 Worzakowska, M. (2015). Thermal and mechanical properties of polystyrene modified with esters derivatives of 3-phenylprop-2-en-1-ol. *Journal of Thermal Analysis and Calorimetry* 121: 235–243.

46 Wypych, G. (2017). Plasticizer motion and diffusion (Chapter 7, pages 165–186), Effect of plasticizers on properties of plasticized materials (Chapter 10, pages 209–332). In: *Handbook of Plasticizers*, 3e (ed. G. Wypych). Toronto: ChemTec Publishing.

47 Wang, Y.-Y., Chen Y.-r., and Sarkanen, S. (2017). Blend configuration in functional polymeric materials with a high lignin content. *Faraday Discussions* 202: 43–59.

48 Lindgren, K., Samuelsson, Å., and Kulander, I. (2017). Techno economic evaluation of lignin extraction in a dissolving pulp biorefinery. Nordic Wood Biorefinery Conference. 118–123.

49 www.alibaba.com/products/melamine_powder.html (accessed 28 October 2021).

50 Rosen, M. and Kiani, A. (2016). The role of plastics compounding for injection molding. *Plastics Engineering* 72: 24–28.

51 ChemAnalyst, Polystyrene Price Trend and Forecast, North America, December 2021. https://www.chemanalyst.com/Pricing-data/polystyrene-ps-11.

52 Li, Y. and Sarkanen, S. (2002). Alkylated kraft lignin-based thermoplastic blends with aliphatic polyesters. *Macromolecules* 35: 9707–9715.

53 Li, Y. and Sarkanen, S. (2005). Miscible blends of kraft lignin derivatives with low-T_g polymers. *Macromolecules* 38: 2296–2306.

54 Kadla, J.F. and Kubo, S. (2003). Miscibility and hydrogen bonding in blends of poly(ethylene oxide) and kraft lignin. *Macromolecules* 36: 7803–7811.

55 Lund, K. (2016). *Industrial Binders from Borregaard – Production and Applications*. Sarpsborg: Borregaard ASA, 25 pp.

56 Fredheim, G.E., Braaten, S.M., and Christensen, B.E. (2002). Molecular weight determination of lignosulfonates by size-exclusion chromatography and multi-angle laser light scattering. *Journal of Chromatography A* 942: 191–199.

57 Sarkanen, S., Teller, D.C., Abramowski, E., and McCarthy, J.L. (1982). Kraft lignin component conformation and associated complex configuration in aqueous alkaline solution. *Macromolecules* 15: 1098–1104, and references therein.

58 Sarkanen, S., Chen Y.-r., and Wang, Y.-Y. (2016). Journey to polymeric materials composed exclusively of simple lignin derivatives. *ACS Sustainable Chemistry & Engineering* 4: 5223–5229.

59 Wang, Y.-Y., Chen Y.-r., and Sarkanen, S. (2015). Path to plastics composed of ligninsulphonates (lignosulfonates). *Green Chemistry* 17: 5069–5078.

60 Ritter, D., Jaklin-Farcher, S., Messner, K., and Stachelberger, H. (1990). Polymerization and depolymerization of lignosulfonate by *Phanerochaete chrysosporum* immobilized on foam. *Journal of Biotechnology* 13: 229–241.

61 Björkman, A. (1956). Studies on finely divided wood. Part I. *Svensk Papperstidning* 59: 477–485.

62 Murthy, N.S., Minor, H., Bednarczyk, C., and Krimm, S. (1993). Structure of the amorphous phase in oriented polymers. *Macromolecules* 26: 1712–1721.

63 Stomberg, R. and Lundquist, K. (1987). The crystal structure of *trans*-2,3-dihydro-2-(4-hydroxy-3-methoxyphenyl)-3-hydroxymethyl-7-methoxybenzofuran. *Acta Chemica Scandinavica, Series B: Organic Chemistry and Biochemistry* 41: 304–309.

64 Lundquist, K. and Stomberg, R. (1988). On the occurrence of structural elements of the lignan type (β-β structures) in lignins. The crystal structures of (+)-pinoresinol and (±)-*trans*-3,4-divanillyltetrahydrofuran. *Holzforschung* 42: 375–384.

65 Roblin, J.-P., Duran, H., Duran, E. et al. (2000). X-ray structure of a trimeric 5,5′-biaryl/*erythro*-β-*O*-4-ether lignin model: evidence for through-space weak interactions. *Chemistry–A European Journal* 6: 1229–1235.

66 Stomberg, R. and Lundquist, K. (1986). Stereochemical assignment of the diastereomers of 1-(3,4-dimethoxyphenyl)-2-(2-methoxyphenoxy)-1,3-propanediol from X-ray analysis. *Acta Chemica Scandinavica, Series A: Physical and Inorganic Chemistry* 40: 705–710.

67 Shan, L., Robertson, C.G., Verghese, K.N.E. et al. (2001). Influence of vinyl ester/styrene network structure on thermal and mechanical behavior. *Journal of Applied Polymer Science* 80: 917–927.

68 Rico, M., López, J., Bouza, R., and Bellas, R. (2011). Thermal behavior of blends based on a thermoplastic-modified epoxy resin with a crosslinking density variation. *Journal of Thermal Analysis and Calorimetry* 105: 599–606.

69 Mlynár, J., Nutsubidze, N., Sarkanen, S., and Shashikanth, S. (1995). Enzymatic macromolecular lignin depolymerization in vitro without peroxidases. *Proc. 8th International Symposium on Wood and Pulping Chemistry*, **II**, pp. 279–284.

70 Chen, Y.-r., Sarkanen, S., and Wang, Y.-Y. (2017). Lignin-degrading methods. US Patent 9, 796, 993, issued October 24.

71 Suzuki, K., Gomi, T., Kaidoh, T., and Itagaki, E. (1991). Hydroxylation of *o*-halogenophenol and *o*-nitrophenol by salicylate hydroxylase. *Journal of Biochemistry* 109: 348–353.

72 Rodgers-Vieira, E.A., Zhang, Z., Adrion, A.C. et al. (2015). Identification of anthraquinone-degrading bacteria in soil contaminated with polycyclic aromatic hydrocarbons. *Applied and Environmental Microbiology* 81: 3775–3781.

73 Kanaly, R.A. and Hamamura, N. (2013). 9,10-Phenanthrenedione biodegradation by a soil bacterium and identification of transformation products by LC/ESI-MS/MS. *Chemosphere* 92: 1442–1449.

74 Torres-Farradá, G., Manzano-León, A.M., Rineau, F. et al. (2019). Biodegradation of polycyclic aromatic hydrocarbons by native *Ganoderma* sp. strains: identification of metabolites and proposed degradation pathways. *Applied Microbiology and Biotechnology* 103: 7203–7215.

75 Kowalczyk, A., Eyice, Ö., Schäfer, H. et al. (2015). Characterization of *para*-nitrophenol-degrading bacterial communities in river water by using functional markers and stable isotope probing. *Applied and Environmental Microbiology* 81: 6890–6900.

76 Hatakeyama, T. and Takagi, K. (2016). Bacterial biodegradation of melamine-contaminated aged soil: influence of different pre-culture media or addition of activation material. *Environmental Science and Pollution Research* 23: 14997–15002.

77 Takagi, K., Fujii, K., Yamazaki K.-i. et al. (2012). Biodegradation of melamine and its hydroxy derivatives by a bacterial consortium containing a novel *Nocardiodes* species. *Applied Microbiology and Biotechnology* 94: 1647–1656.

78 Dodge, A.G., Wackett, L.P., and Sadowsky, M.J. (2012). Plasmid localization and organization of melamine degradation genes in *Rhodococcus* sp. strain Mel. *Applied and Environmental Microbiology* 78: 1397–1403.

79 Amann, M. and Minge, O. (2012). Biodegradability of poly(vinyl acetate) and related polymers. In: *Synthetic Biodegradable Polymers, Advances in Polymer Science*, vol. 245 (ed. B. Rieger, A. Künkel, G.W. Coates, et al.), 137–172. Heidelberg: Springer-Verlag.

80 Wang, Y.-Y., Wyman, C.E., Cai, C.M., and Ragauskas, A.J. (2019). Lignin-based polyurethanes from unmodified kraft lignin fractionated by sequential precipitation. *ACS Applied Polymer Materials* 1: 1672–1679.

81 Gioia, C., Colonna, M., Tagami, A. et al. (2020). Lignin-based epoxy resins: unravelling the relationship between structure and material properties. *Biomacromolecules* 21: 1920–1928.

82 Romhányi, V., Kun, D., and Pukánszky, B. (2018). Correlations among miscibility, structure, and properties in thermoplastic polymer/lignin blends. *ACS Sustainable Chemistry & Engineering* 6: 14323–14331.

83 Bartolome, M.J., Bischof, S., Pellis, A. et al. (2020). Enzymatic synthesis and tailoring lignin properties: a systematic study on the effects of plasticizers. *Polymer* 202: 122725. https://doi.org/10.1016/j.polymer.2020.122725.

84 Chung, H. and Washburn, N.R. (2012). Chemistry of lignin-based materials. *Green Materials* 1: 137–160.

85 Glasser, W.G. (2019). About making lignin great again – some lessons from the past. *Frontiers in Chemistry* 7: 565. https://doi.org/10.3389/fchem.2019.00565.

86 Ajao, O., Jeaidi, J., Benali, M. et al. (2018). Quantification and variability analysis of lignin optical properties for colour-dependent industrial applications. *Molecules* 23: 377. https://doi.org/10.3390/molecules23020377.

11

Design of Recyclable Thermosets

Bryn D. Monnery, Apostolos Karanastasis, and Louis M. Pitet

Hasselt University, Institute for Materials Research (IMO), Advanced Functional Polymers Laboratory, Department of Chemistry, Martelarenlaan 42, 3500 Hasselt, Belgium

11.1 Introduction

11.1.1 Polymers and Plastics

Plastics have become an indispensable class of materials, fueling modern civilization on every imaginable level – from packaging to healthcare, to construction, to space exploration, and everything in between. However, some of the very attributes (e.g. longevity/durability) that make plastics so appealing have also caused one of the world's most pressing crises – plastic waste. Worldwide plastics manufacturing has experienced exponential growth since its inception in the 1950s and continues growing unabated today [1]. It has been estimated that 7800 million metric tons (Mt) of plastics were manufactured between 1950 and 2015 [2]. A further 359 Mt of plastics were produced worldwide in 2018 [1]. This is the equivalent to c. 40 kg per person per year. Tragically, much of this has found its way into the environment [3]. Enormous quantities of solid waste persisting in our oceans continue to wreak havoc on marine environments (see Chapters 3 and 12) [4]. Manufacturing practices, the fossil-fuel origins of the monomers used to make most plastics, and the end-of-life incineration of plastics are exacerbating climate change. Plastic use unquestionably delivers unprecedented convenience, on which we have come to depend. However, this dependence is contributing overwhelmingly to a series of global crises. Thankfully, there is a concerted effort to remedy the current predicament with multidisciplinary exploration seeking new scientific breakthroughs. The very ingenuity responsible for creating the "original" plastic revolution will also enable the next generation of materials to address concerns related to resource origins, carbon output, and environmental impact. More responsible use of current plastic waste-streams (see Chapter 3) in conjunction with design of smarter chemical scaffolds are necessary steps along the path toward a sustainable, circular plastics economy. All parts of the value chain are being scrutinized in an effort to improve the conventional plastic life cycle.

Biodegradable Polymers in the Circular Plastics Economy, First Edition.
Edited by Michiel Dusselier and Jean-Paul Lange.

This chapter provides a brief overview of several strategies being employed in the last decades to re-design certain classes of plastics specifically for recyclability by purposefully incorporating degradable, reversible, or labile chemical bonds. This strategy enables reconfiguration of the chemical bonds responsible for imparting mechanical integrity in the original material, thereby rendering it degradable or reprocessable. In this manner, the reversibility in chemical bonding is utilized to rearrange the bonds, allowing the reshaping of polymer materials without substantial destruction of the underlying macromolecular scaffold. This strategy enables the original material to be directly transformed into an alternate product, thereby extending its useful lifetime substantially. This strategy has been applied predominantly to conventional thermoset-type constructs, which are typically considered un-recyclable and impossible to re-process. Ultimately, such a strategy may afford materials that are fully degradable at the end of their useful lifetime, or for materials that are disposed of irresponsibly (i.e. not collected for recycling).

11.1.2 Handling of Plastic Waste

Currently approximately 10% of global plastics are recycled, 14% incinerated, and 40% is disposed of in landfills (Figure 11.1) [5]. Landfilling is particularly concerning as it essentially represents a huge loss of potentially high-value chemical feedstocks and creates an enormous environmental burden, especially as most of this waste can essentially be considered non-degradable, except on a geological timescale. Worse, it has been estimated that 32% of polymers escape the waste collection process completely and pollute the general environment [2]. Plastics have been identified as efficient transporters of toxins for marine organisms as they travel through streams and rivers into the ocean [6].

Clearly, increasing the efficiency of waste collection will dramatically reduce the detrimental environmental impact of discarded plastics. Increased recycling must be pursued in conjunction. In part, recycling more plastic requires the intrinsic, structural redesign of certain plastic materials such that the recycling process can be conducted using manageable processes. The inherent recyclability of plastics can be influenced in several manners. The most convenient pre-requisite involves synthesizing polymers having functionality that is amenable to chemical transformations under mild conditions. On the other hand, volumes of recycled plastics can be increased by improving the processes that are already routinely used for recycling (see Chapter 12). This is often enabled by improved catalysis, which lowers the energy barriers responsible for the chemical transformations that occur during degradation [7]. In this chapter, we do not explicitly cover catalysis but instead focus on the design of polymer scaffolds and the chemical transformations responsible for polymerization/depolymerization and rearrangement in reconfigurable thermosets [5a, 8].

11.1.3 Chemical Nature of Plastics

Commercial plastics span an impressive range of physical properties, which is largely attributed to the chemical makeup (Chapter 04). While morphology,

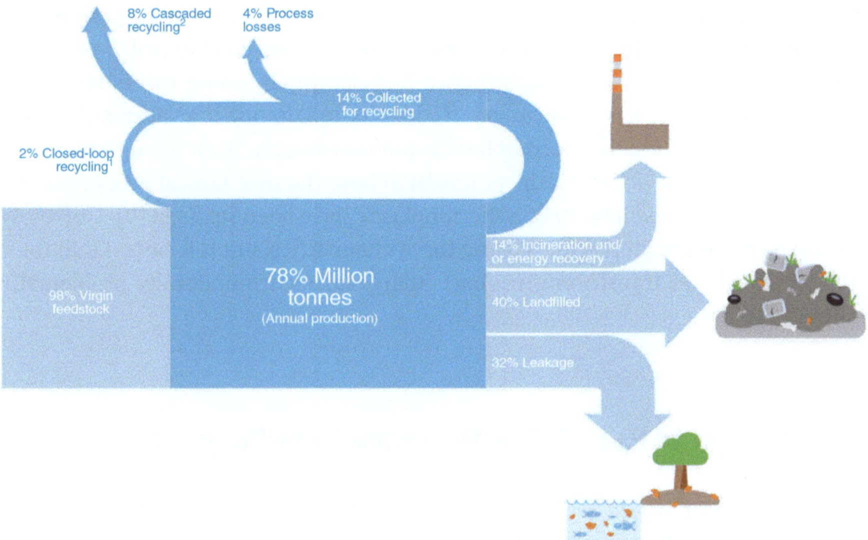

Figure 11.1 Estimated global flow of plastic packaging materials from 2013. 1: Closed-loop recycling: Recycling of plastics into the same or similar-quality applications; 2: Cascaded recycling: Recycling of plastics into other, lower-value applications. Source: Reproduced with permission from Ellen MacArthur Foundation [5b]. Copyright 2017 Ellen MacArthur Foundation.

architecture, and stereochemistry are all key determinants for physical and mechanical properties, the nature of the backbone chemistry has an enormous influence on reactivity, and is thus key for enabling efficient recycling. The term plastic, from the Greek word "plastikos" meaning "moldable," applies to the subset of polymers used in engineering applications. Plastics may be further subdivided into three types, based on the synthetic mechanism employed during buildup of molar mass.

1) Thermoplastics – strength provided by chain entanglement
2) Thermosets – strength provided by permanent covalent bonds between chains to generate a network
3) Vitrimers – a subset of thermosets wherein the covalent crosslinks are reversible, facilitating breaking and reforming of the network

Thermosets often exhibit superior mechanical properties and durability compared with thermoplastics due to their crosslinked network structure. The extensive interchain bonding provides improved durability and strength. The interconnected scaffold also enables swelling in solvents to make highly elastic gels. However, owing to the chemical bonding, thermosets essentially possess infinite molar mass and as such are not recyclable by conventional (i.e. mechanical or chemical) means. There are a few exceptions to this discussed below. End-of-life is typically landfill. In certain applications, the permanent structure afforded by crosslinking and the durability is welcome (e.g. tires). In that sense, thermosets fill several niche markets, in which thermoplastics are unsuitable.

The general reason for a material to be considered unrecyclable is the permanent, irreversible nature of the connections responsible for the mechanical integrity in thermoset polymers. An effort to introduce recyclability by using reversible chemistry has begun to bear fruit in a variety of contexts and employing various chemical moieties. This goes for both thermoplastics and thermosets. With the emergence of a subgroup known as vitrimers, the potential to have the mechanical properties of a thermoset, but the recyclability of a thermoplastic have been realized [9]. Currently there are no commercial products using these chemistries, but it is hoped that their introduction will be transformative, particularly for environmentally problematic segments like tires.

11.2 Design of Recyclable Thermosetting Polymers

Recycling has been an *ad hoc* matter; materials were developed and if they were found afterwards to be recyclable, then that was a bonus. Nowadays, we can do better; we can design thermoset materials that are specifically easily recoverable. This "closes the loop" allowing for reduction of the use of precious virgin, non-renewable resources. Furthermore, it is foreseen that by incorporating labile crosslinking into newly designed materials, the degradation can be promoted for materials that ultimately end up dispersed in the environment.

There are several complementary approaches to this, based primarily on differences in the nature of the crosslink (Figure 11.2). The crosslinks can be reversible (i.e. equilibrium reactions) and designed to be susceptible to various stimuli such as temperature, acids or bases, or light. These reversible or dynamic crosslinks are described under the collective term covalent adaptive networks (CANs), which can be further subdivided into two categories: dissociative CANs, where the dynamic crosslink concentration decreases above a critical threshold temperature, and associative CANs (aka vitrimers) where the bond concentration is constant and only the rate of exchange changes with changes in the external environment. The critical differences are shown in Figure 11.2 and elaborated further.

Along a parallel trajectory, recyclable thermosets can be achieved by incorporation of degradable (i.e. non-reversible dissociation) building blocks. This can be achieved in several manners, whereby every repeating unit is degradable (fully degradable in Figure 11.2), select units (i.e. comonomers) are degradable (partially degradable in Figure 11.2), or simply the crosslinking junctions are degradable (selectively degradable in Figure 11.2). Some select examples of these different classes of degradable thermosets will be discussed in the subsequent sections (Sections 11.2–11.4).

In essence, the premise behind CANs is to obtain a material with mechanical properties that are characteristic of traditional thermosets (e.g. creep resistance, durability) while retaining the appealing processing properties of thermoplastics (e.g. (re)moldability, recyclability). Marrying these two sets of properties leads to materials exhibiting the best of both worlds.

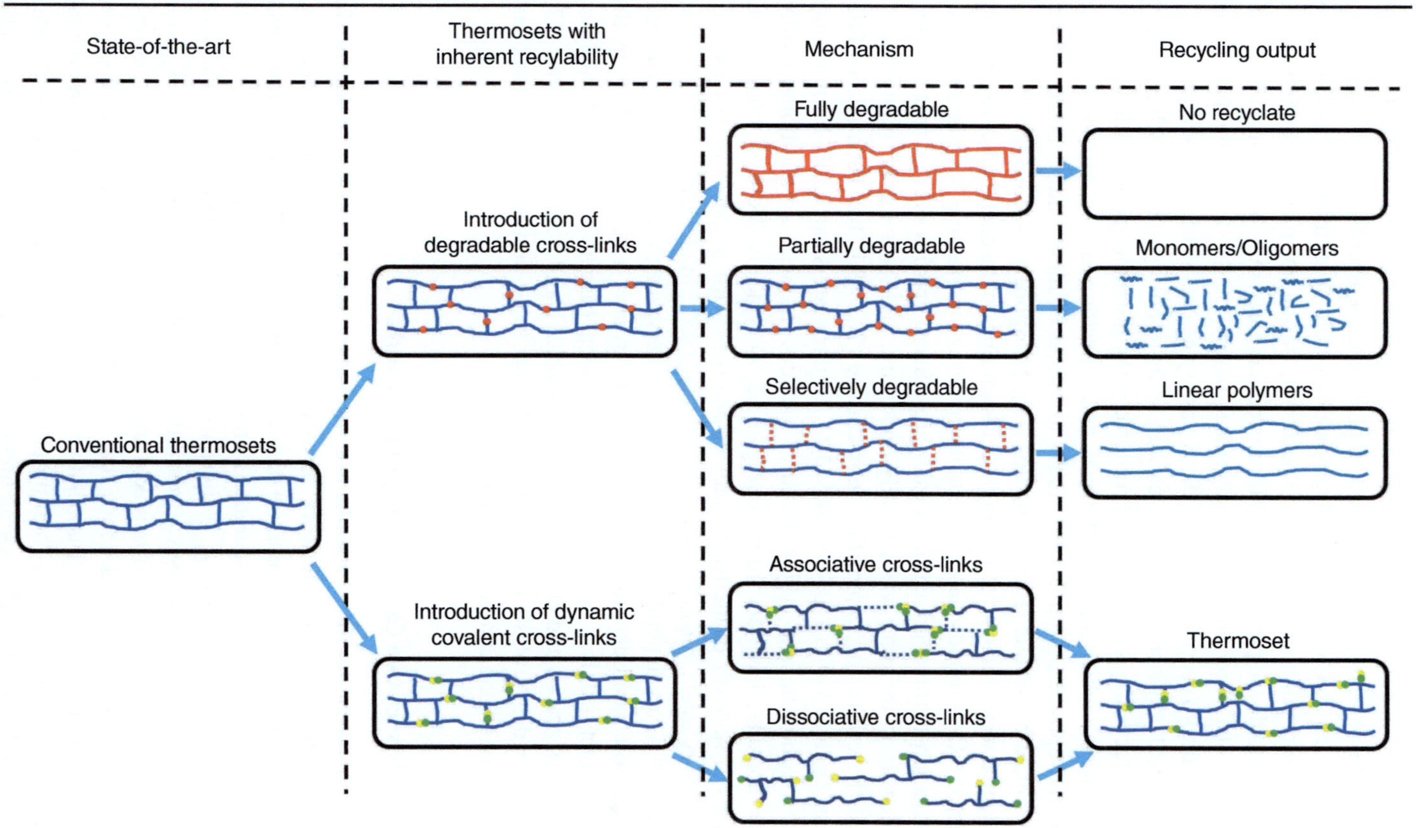

Figure 11.2 Strategies for introducing recyclability into thermosets. Top: two-stage trajectory via degradation and recovery of monomers/polymers; bottom: one-stage trajectory via reconfiguration of dynamic crosslinks. Source: Reproduced with permission from Post et al. [10]. Copyright 2019 Taylor Francis.

11.2.1 Recyclability by Triggered Degradation

Many conventional thermosets have a major issue with recycling due to the irreversibility of the crosslinks. Hence replacing these irreversible crosslinks with a crosslink capable of triggered degradation would in principle allow for easy recovery of the polymer. By focusing specifically on the reversibility of the network junctions and developing strategies for introducing dynamic behavior, in principle all thermosets, regardless of the chemical nature of the main chains, can be transformed into recyclable materials. A variety of chemistries are available [11], and chemical connections must be chosen so as to not open except when desired (i.e. at the recycling plant), to avoid compromising the material properties.

Chemistries, which rely on thermal decomposition of the linkers, result in thermally reprocessable polymers. Early examples of these materials were epoxy resins linked by acetals [12], carbonates [13], phosphoesters [13, 14], and sulfites [14a] which were stable up to 180–300 °C, depending on the chemistry and crosslink density. C—O—C bonds were more stable than those containing other heteroatoms. Investigation of the degradation of ester-crosslinked epoxies showed that the electron density in the ester was critically important, and there has been an effort to reduce the critical decomposition temperature (T_d). This has been successfully reduced from c. 340 to 120 °C (Table 11.1). Tertiary esters can easily form alkenes and the carboxylic acid [19]. α-Terpinyl epoxides (i.e. with an additional methyl on the epoxide) have activation energies (E_a's) typically 2 kJ/mol higher than epoxides [15]. With di(meth)acrylates crosslinked with a spacer of 1–4 carbons using tertiary esters, the networks were stable at 150 °C in all cases, but degraded at 180 °C with a rate dependent on the length of the spacer; increased spacer length leads to more rapid degradation [18]. There is an auto-acceleration effect, as the acidic products catalyze further reactions. Chemical triggers include acid-catalyzed hydrolysis of acetals and ketals [20], Schiff bases [21], and hexahydrotriazines [22], base-catalyzed hydrolysis of dilactones [23] and sugar-derived epoxies [24], and ozonolysis of alkene containing linkages [25].

11.2.2 Dissociative Covalent Adaptive Networks

In the CAN systems the crosslinks are reversible, leading to dissociation of the crosslinks responsible for network formation. This is often thermally induced. Above a certain critical temperature dissociation become predominant, and hence the network progressively undergoes a transition of material properties. Above a critical gelation temperature (T_{gel}), there are no crosslinks, and the material behaves as a conventional thermoplastic. Crosslinks re-form upon cooling below T_{gel} and the polymer behaves like a crosslinked material (i.e. thermoset), which contrasts with the materials discussed above that do not reform their crosslink when the external stimulus is stopped. Dissociative CANs are thereby potentially recyclable thermoset-like materials [26]. Their properties depend upon the crosslinking chemistry, crosslink density, and crosslink lifetime. Thus, they are rendered reprocessable and have in some cases been considered for design of recyclable thermosetting

Table 11.1 Degradable ester, ether, and carbonate crosslink functions used with associated degradation temperatures.

Linker chemistry	Degradation temperature	References
Epoxies		
Primary ester 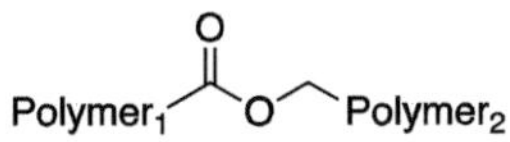	340 °C (202 °C with acid catalyst) (E_a = 186 kJ/mol)	[15c, 16]
Secondary ester 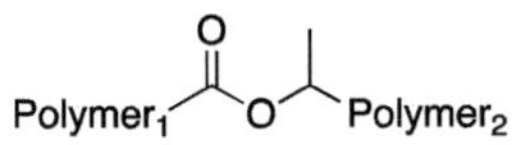	320 °C (E_a = 183 kJ/mol)	[15a]
Tertiary ester 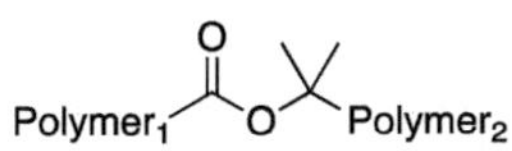	220 °C (reduced to 180–196 °C in presence of a strong acid) (E_a = 104 kJ/mol)	[15c, 16, 17]
p-cyclohexyl di-*tert*-ester 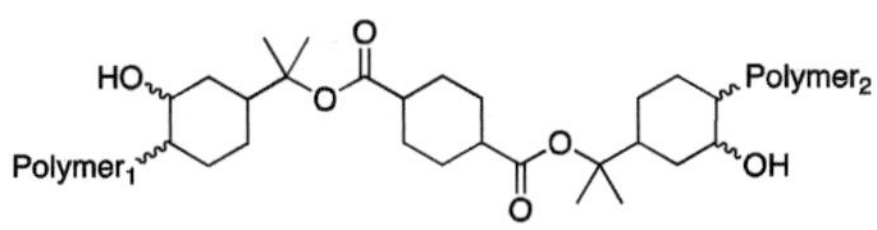	120 °C (110 °C with strong acid)	[15c, 16]
m-phenyl di-*sec*-ester 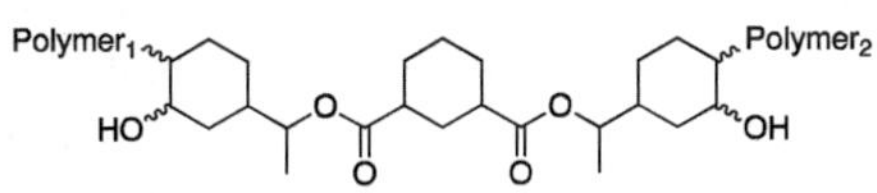	262 °C	[15c, 16]
m-phenyl di-*tert*-ester (terp) 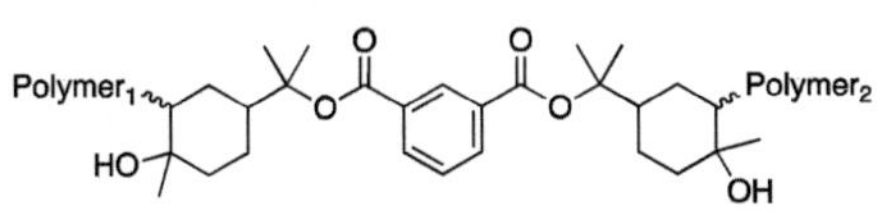	192 °C (165 °C with acid)	[15a, 16]
Cyclohexyl *tert*-ether (terp) 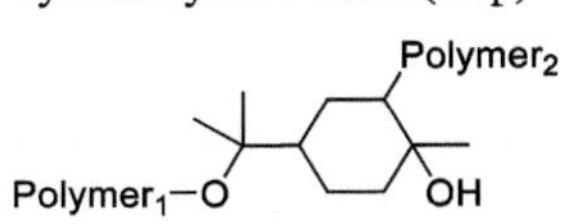	250 °C with acid	[17]
Cyclohexyl *tert*-ester (terp) 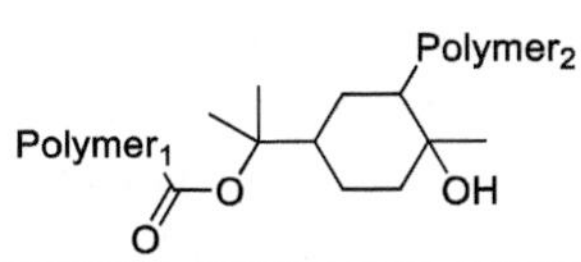	182 °C with acid	[17]

Table 11.1 (Continued)

Linker chemistry	Degradation temperature	References
Cyclohexyl *tert*-carbonate (terp) 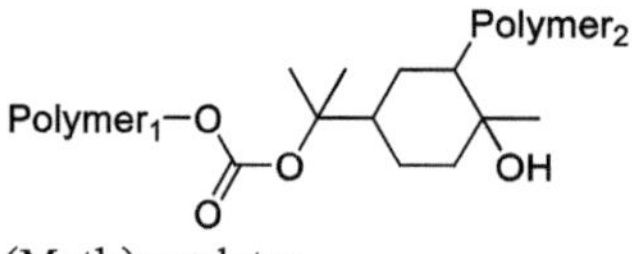	223 °C with acid	[17]
(Meth)acrylates		
(Meth)acrylyl *tert*-ester 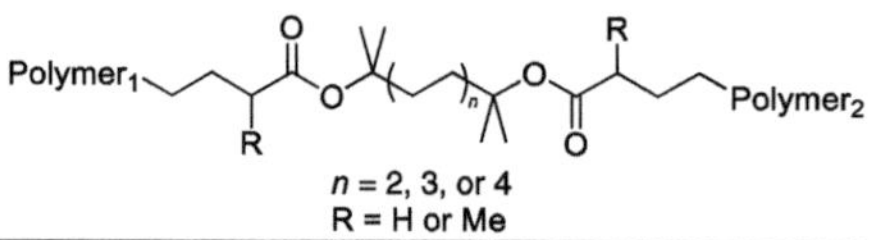	$180\,°C > T_{deg} > 150\,°C$	[18]

The thermally induced decomposition temperatures are consistent with breakdown to acid and olefins (and CO_2 in the case of carbonates), as observed by several analytical tools.

materials [27]. Various dissociative CAN systems are known, which have been retroactively dubbed "vitrimer-like" despite preceding vitrimers chronologically (see Section 11.2.3) [28].

Various crosslink systems have been explored in this context (Table 11.2). The use of photocatalysts also allows light to be used as a trigger instead of heat [38]. These dissociative CANs are generally researched as potential healable materials, but they have drawbacks compared to conventional thermosets. As their moduli do vary significantly upon even moderate heating, they have a tendency to creep, especially as the crosslinks are often of similar strength to hydrogen bonds (ΔG c. −20 to −30 kJ/mol, where ΔG is the free energy of bond-formation). There are however two reported systems that are significantly stronger than hydrogen bonds, both from the collaboration of Winne and Du Prez. The indole-triazolinedione system (TAD) is a strong but reversible reverse Diels–Alder adduct (Figure 11.3) [27, 32].

Du Prez and coworkers have also reported a system using thiol-yne Michael addition [37]. The initial product of the first Michael addition is weak (c. −20 kJ/mol), but the π-system can react again to form a stable crosslink giving essentially complete creep resistance up to 80 °C. Catalytic quantities of an organic base (TBD) are necessary to facilitate exchange reactions, and if the material if heated too strongly it is degraded. The crosslink strength of these two systems is such that they may be useful in the creation of reprocessable/healable materials that resist creep in normal use.

11.2.3 Vitrimers (Associative CANs)

Vitrimer networks are a variety of covalent adaptable networks in which the crosslink density remains constant [39]; no free reactive groups exist on a long

Table 11.2 Functional groups utilized as crosslinkers in dissociative CANs.

System	T_{gel}	ΔG	Notes	References
Furan-maleimide 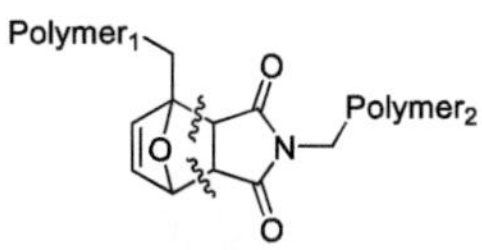	c. 90 °C	−8 to −15 kJ/mol	Requires heating to c. 40 °C to form network	[29]
Vulvene-activated acrylate 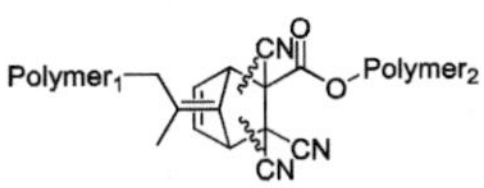	c. 50 °C	−8 to −15 kJ/mol		[30]
Cyclopentadiene-thioester 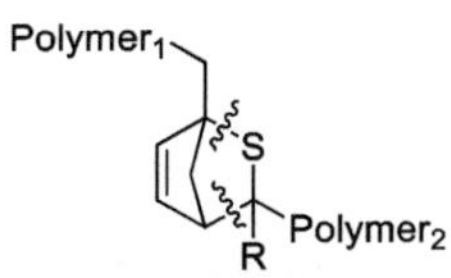	c. 80 °C	−35 to −45 kJ/mol		[31]
Indole-triazolinedione 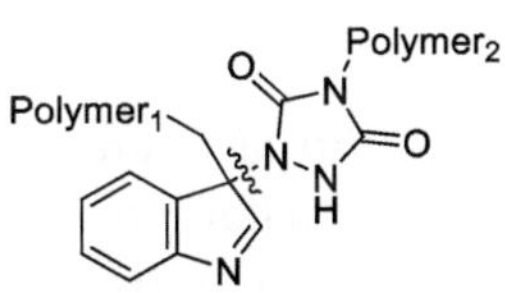	c. 90 °C	−60 to −70 kJ/mol		[27, 32]
Disulfides 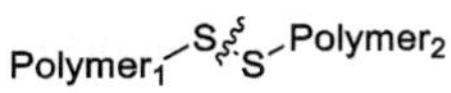	c. 60 °C			[33]
Alkoxyamines 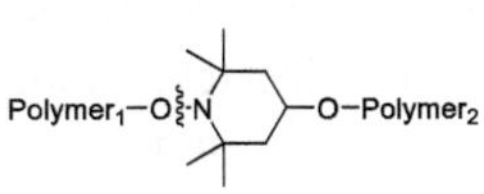				[34]
Activated urea 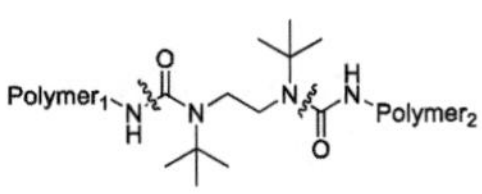	c. 90 °C	−10 to −50 kJ/mol		[35]
Thiol-ene Michael 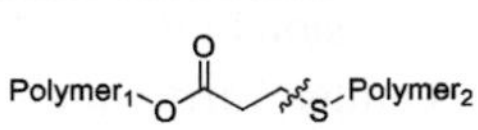				[36]
Thiol-yne double Michael 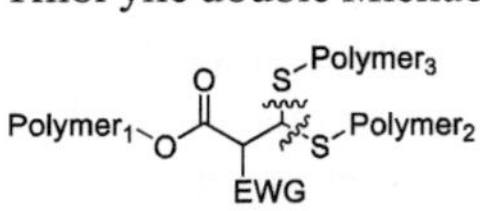		−160 to −252 kJ/mol (double reacted) (E_a = 120 to 200 kJ/mol)	Requires base (TBD) as catalyst. Creep resistant to 80 °C	[37]

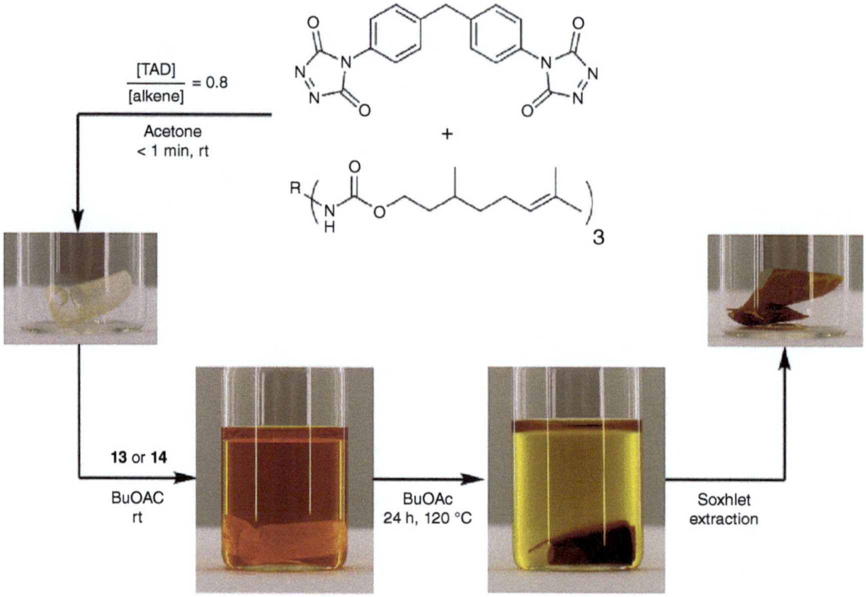

Figure 11.3 Reversible TAD dissociation used to transfer a dye to an unsaturated polymer. Source: Reproduced with permission from Houck et al. [32]. Copyright 2017 Royal Society of Chemistry.

molecular timescale. Hence vitrimers are associative CANs. Vitrimers are polymers utilizing an alternate crosslinking to thermosets, in an effort to create thermoset-like materials that are "infinitely" recyclable by secondary recycling. Several types of dynamic crosslinks have been employed in this context. A mechanistic/thermodynamic comparison illustrating the contrast between associative and dissociative CANs provides some insight (Figure 11.4).

The mechanically distinct feature of these associative systems (i.e. vitrimers) is that the storage and loss moduli are practically invariant up to a critical temperature, which prevents creep [27]. Viscous flow is observed above a certain temperature threshold. This contrasts with the previously considered dissociative systems as a consequence of the different network topology.

There are two important transition temperatures in typical vitrimers. The first is the glass transition temperature (T_g). The second is the vitrification transition temperature (T_v) – that is the temperature at which the topology of the network becomes unfrozen and transitions from a viscoelastic solid to a liquid. Diffusion limits the displacement of bonds between the T_g and T_v and the Williams–Landel–Ferry equation determines the viscosity; hence elastomeric behavior is observed [41]. Above the T_v, Arrhenial behavior is observed with the material being a viscoelastic liquid.

The T_v is conceptually similar to T_m in semicrystalline thermoplastic systems, but the vitrimer does not become a free-flowing liquid as the number of crosslinks is essentially constant. They are simply exchanging on a rapid timescale, enabling reconfiguration of the network topology across the entire material and rendering the system processable. These two numbers are governed by the following factors:

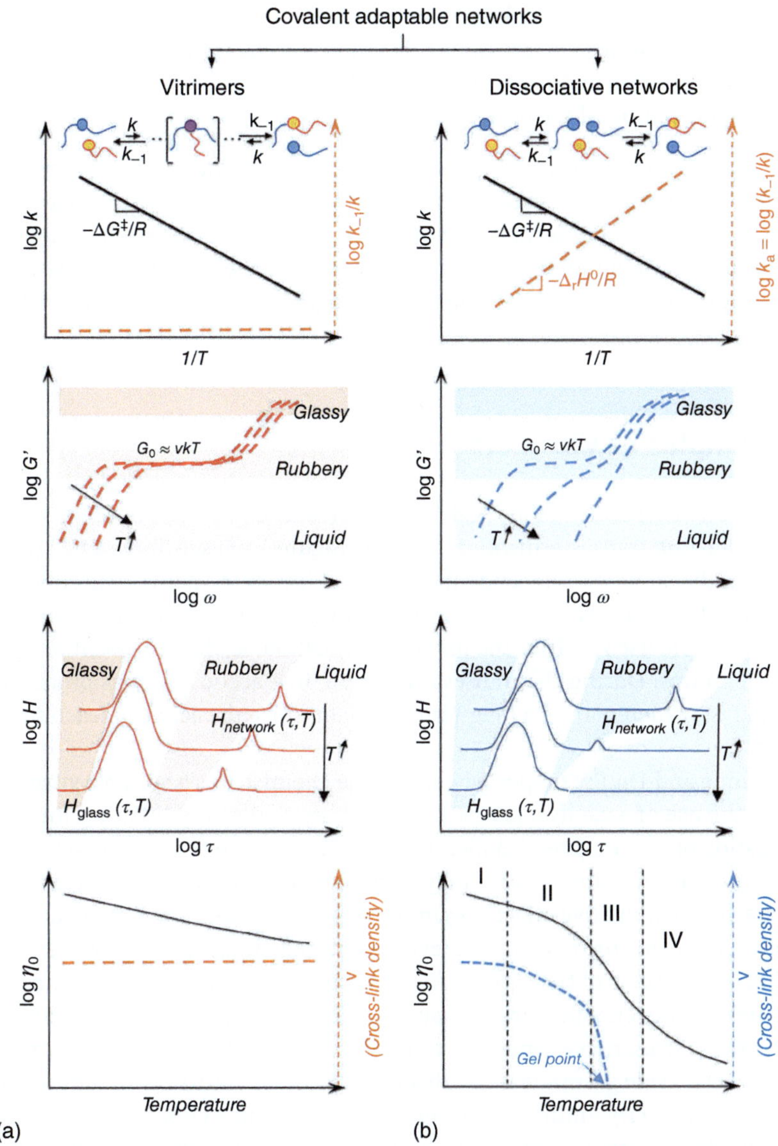

Figure 11.4 Difference between (a) associative CANs (vitrimers) and (b) dissociative CANs. (First line) The rate of exchange (k, black line) varies in the same manner for both, but the association equilibrium constant (K_a, orange line) decreases with increased temperature for associative CANs, while it is constant for vitrimers. (Second line) Pseudo-3D plot showing the evolution of storage moduli (G') against frequency (ω) and temperatures. (Third line) Pseudo-3D plot of relaxation time spectra showing temperature differences in transition to a viscoelastic liquid. (Fourth line) Evolution in zero sheer viscosity (black line) and crosslink density (orange/blue line) against temperature. Source: Reproduced with permission from Jourdain et al. [40]. 2020 American Chemical Society.

1) Chemical nature of the crosslinking moieties
2) Density of the crosslinks and crosslinkable groups (if one is in excess)
3) Catalyst concentration and nature (if applicable)
4) The underlying nature of the base polymer.

All these factors must be considered together for the properties of the final material. The crosslink density and nature of the base polymer is a matter of design. In addition to simple heating, thermal radiation can be used if an efficient acceptor is incorporated. If carbon nanotubes or graphene is embedded in the network as a filler, they heat the network when illuminated, bringing the temperature above the T_v [42].

11.3 Examples of Vitrimers

Vitrimers were conceptually pioneered by Liebler and coworkers utilizing ester crosslinks and a zinc transesterification catalyst (Figure 11.5; see Table 11.3) [43]. The strength of the network is dependent upon the ability of the network to conduct these transesterifications, with higher catalyst loading leading to more extensibility [43b]. Epoxides were used to drive the initial esterification, leading to essentially no free hydroxyl groups. During the transesterification, free acetic acid was produced from the catalyst, which may be a long-term problem, since the acid can leach from the vitrimer.

In 2014 Winne and Du Prez reported enaminone chemistry as a suitable vitrimer that avoided the use of a catalyst [27, 50]. In this system, the crosslinks are formed by the amidation of a conjugated 1,3-dicarbonyl with an amine to form a vinylogous urethane; that is the crosslink is essentially a urethane with a π-system inserted into it to resonate via the inductive effect (Figure 11.6a,b). These crosslinks are hydrolytically stable, and indeed the reaction can even be carried out in water. However, the system relies on an excess of free amines in the system, which are oxidatively unstable. Nevertheless, this chemical transformation has been successfully exploited to produce various vitrimeric systems. For example, vinyl-acrylate-based copolymers with beta-keto ester side groups was effectively transformed into a vitrimer by introducing branched primary amines in various amounts (Figure 11.6c) [52]. The proportion of functional groups was shown to be critical, as in nearly all vitrimers. Exchange kinetics can be fine-tuned by adjusting the relative building block concentrations (i.e. branched monomers vs. comonomer in the acrylate backbone). The diketoenamine moiety has recently also been incorporated into mixed plastic waste feeds, essentially converting plastic waste into useful thermosets, which are in turn themselves recyclable [53].

Since vitrimers are "silica like" the obvious chemistry to exploit was that of silica itself. The first report of such a system was by Guan and coworkers [46]. They used classical chemistry for the formation of silyl ethers; the reaction of either an alkyl silyl chloride or an activated aminoalkyl silyl methoxide with a hydroxyl group. They found that the alkyl crosslinks were very slow in their exchange, with a window

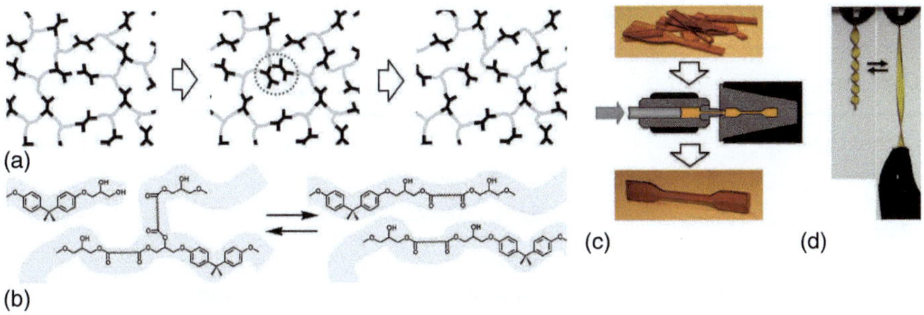

Figure 11.5 First report of a vitrimer exploiting thermally activated esters as the interchangeable groups (a and b). This leads to reprocessable thermoset materials (c) with shape memory behavior (d). Source: Reproduced with permission from Montarnal et al. [43a]. Copyright 2011 Science.

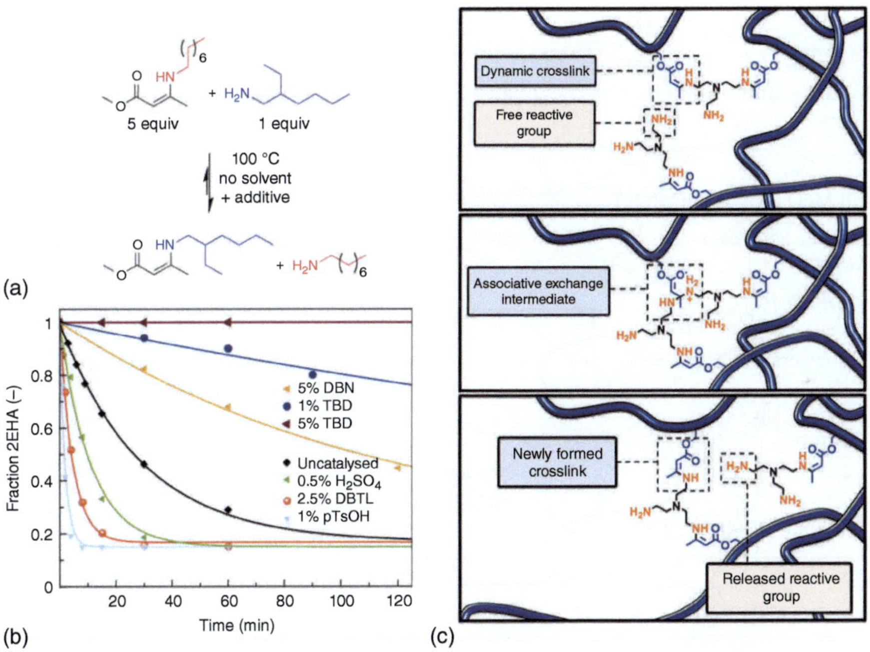

Figure 11.6 (a) Chemical exchange responsible for vinylogous urethane vitrimers, showing (b) the kinetic dependence on catalyst type (DBN = 1,5-Diazabicyclo[4.3.0]non-5-ene; DBTL = dibutyl tin laurate; pTsOH = toluene sulfonic acid). (c) schematic illustration of the reconfiguration of the junctions in a polymer network. Source: (a and b) Denissen et al. [51] / Springer Nature / CC BY 4.0. Reproduced with permission from Lessard et al. [52]. Copyright 2019 American Chemical Society.

between the T_v and T_g of only 6 °C and an activation energy of 174 kJ/mol. However, the amino-activated silyl crosslinks were much more dynamic, lowering the T_v to 47 °C (see Table 11.3).

Guan and coworkers purported that boronic acid esters would make excellent vitrimeric crosslinkers based on small-molecule studies [24]. However, it was

Table 11.3 Example chemistries used as vitrimeric linkers.

Chemistry	T_v	E_a	References
Carboxylate 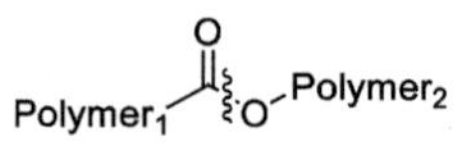	68 °C (1% Zn^{2+}) 57 °C (5% Zn^{2+}) 53 °C (10% Zn^{2+})	68 kJ/mol (7.5% Zn^{2+})	[43a, 44]
Enaminone 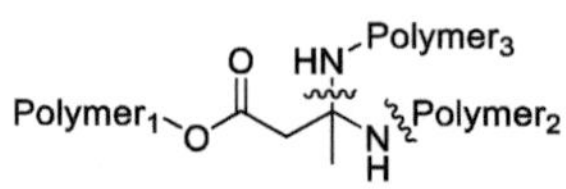	27 °C uncatalyzed, −1 °C with 0.5% TsOH or −63 °C with 0.95% DBTL	81 kJ/mol uncatalyzed (70 kJ/mol catalyzed with 0.5% TsOH or 30 kJ/mol with 0.95% DBTL)	[45]
Siloxane alkyl ether 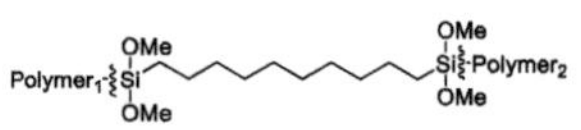	117 °C	174 kJ/mol	[46]
Siloxane aminoalkyl ether 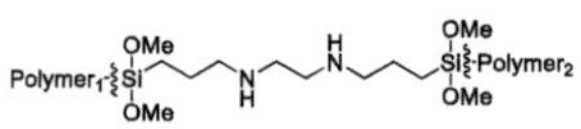 (all MeO groups are labile)	47 °C	81 kJ/mol	[46]
Boronic acid ester 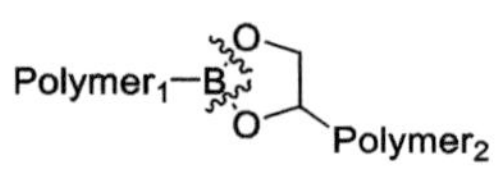	60 °C	28 kJ/mol	[47, 48]
Thioesters 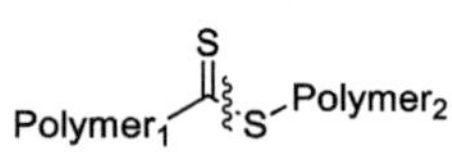		$\Delta^{\ddagger}H = 143$ kJ/mol (uncatalyzed)	[49]
1,2,3-Triazolium 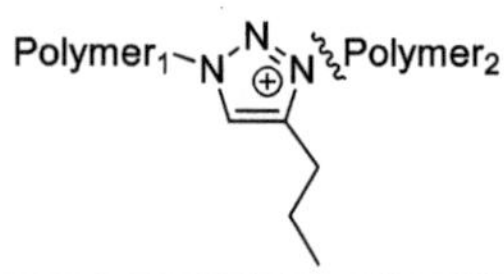	Not a true vitrimer		[40]

Leibler and coworkers who first reported the implementation of boronic acid esters as a crosslinking strategy [47]. This obviated the need for a catalyst with the T_v reduced to c. 60 °C. Since the boronic acid esters were incorporated into PMMA and high density polyethylene (HDPE), the properties of materials were similar to the mother polymer. At the T_v, PMMA was still glassy, and hence no change in properties occurred. However, the incorporation of boronic ester crosslinks improved the creep of HDPE significantly.

Bowman and coworkers reported the use of thioesters as vitrimeric crosslinks [49]. These required a basic catalyst, since the enthalpy of dissociation (ΔH^0) is estimated to be 143 kJ/mol without a catalyst, consistent with a very stable network.

In the presence of a base, which can deprotonate the thiol, the ΔH^0 is reduced to c. 17 kJ/mol, leading to a very dynamic system. The base can even be photoactivatable. However, doping the polymer with DBU or other strong bases was problematic. Bowman and coworkers have reported that in thiol-anhydride-based CANs with large excesses of thiol, there is some contribution of an associative thiol–thioester exchange [26].

Montarnal and Drockenmuller have reported 1,2,3-triazolium crosslinks produce a network which is mechanistically a dissociative CAN, but shows vitrimeric rheology [43a]. Essentially, the ΔH^0 for viscous flow is so high (232 kJ/mol) that the network can't enter that regime, and hence the material behaves as a vitrimer even though it is not one in the strictest sense.

11.4 Adaptable Cross-Linking of Conventional Polymers

While most of these chemistries are demonstrated with a "pure" vitrimer network, these materials are of very little use in engineering applications. Alternatively, the direct replacement of more conventional crosslinking chemistries employed in thermosets has also been considered. As seen above, the crosslinking of PMMA and HDPE with boron acid esters was very successful [47]. Caffy and Nicolaÿ reported an improvement wherein HDPE could be converted into a boronic-ester-crosslinked vitrimer using reactive extrusion at 170 °C that introduced the boronic esters using a TEMPO-based radical crosslinker [54]. The properties are essentially the same as HDPE vitrimer copolymerized from virgin feedstocks.

Unsaturated polyolefins are extensively used as precursors to thermosets and rubbers (i.e. polyisoprene), owing to the reactivity of the double bond, making it amenable to crosslinking. The cross-metathesis of double bonds is also widely used for introducing various functionalities. Therefore, metathesis offers an obvious opportunity in creating vitrimers. Guan and coworkers effectively used cross-metathesis to form reconfigurable thermosets, employing the Grubbs catalysts based on ruthenium (Figure 11.7) [55].

Essentially any polyester or polycarbonate can be converted into a vitrimer assuming it has crosslinking sites and is doped with an appropriate catalyst. Hillmyer and coworkers reported a vitrimer formed from star poly(L-lactide) (PLA) doped with dioctyl tin (which is also the polymerization catalyst) when crosslinked with diisocyanates (Figure 11.8) [56]. In the absence of catalyst the network was a conventional thermoset, and with it, it was a vitrimer for which the properties were tunable based on catalyst concentration. In another example of a polyester converted to a vitrimer by this strategy, PBT copolymerized with glycerol is vitrimeric with a zinc catalyst [22]. The T_m was slightly reduced, in line with the expected T_m decrease from the lowered M_w of the individual polymer chains, but above the T_m of the bulk polymer, vitrimeric activity was observed.

Ester interchange has also been applied to more complex architectures. Highly branched bottle-brush copolymers with pendant hydroxyl groups were combined with a bicyclic ester crosslinker [57]. The crosslinking reaction liberates additional

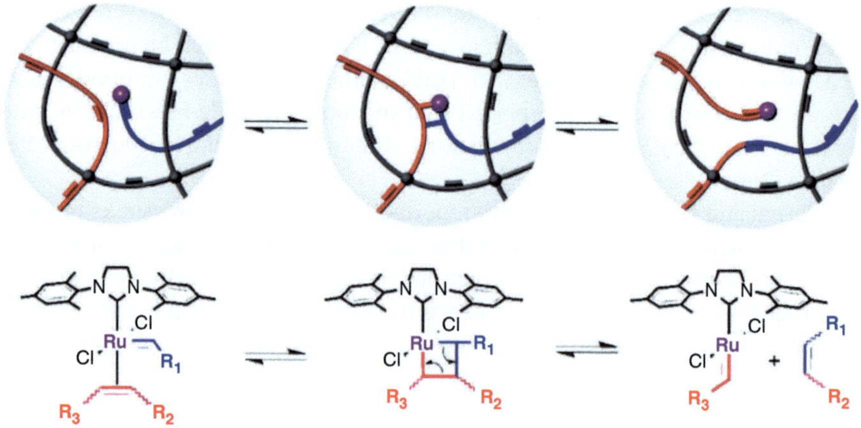

Figure 11.7 Reconfigurable unsaturated polyolefins using olefin metathesis. Source: Reproduced with permission from Lu et al. [55]. Copyright 2012 American Chemical Society.

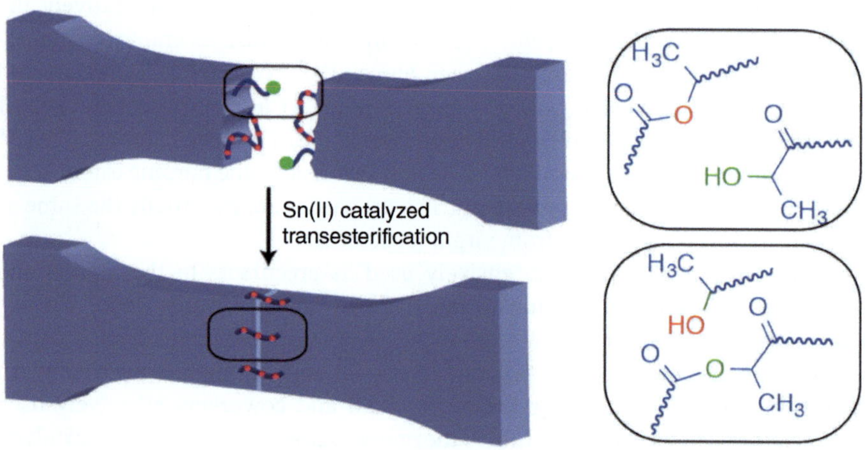

Figure 11.8 PLA vitrimers based on transesterification. Source: Reproduced with permission from Brutman et al. [56]. Copyright 2014 American Chemical Society.

hydroxyl groups that participate in the interchange at elevated temperatures. This feature, as with many vitrimers, is amenable to self-healing, which is otherwise prohibitive in conventional thermosets having irreversible bonding (Figure 11.9).

A similar approach utilizing carbonyl exchange was taken by using a polycarbonate with excess branched polyol building blocks, whereby carbonate exchange was responsible for the transfiguration upon heating [58]. The same research group took this concept one step further and applied interchangeable groups to crosslinked polyurethane foams [59]. In a twin-screw extruder, the polyurethane materials were heated in the presence of added catalyst to promote rapid exchange of urethanes, thereby rendering the materials reprocessable (Figure 11.10).

To date, research into vitrimers has concentrated on finding various temperature-sensitive chemical transformations. Several innovative strategies have exquisitely

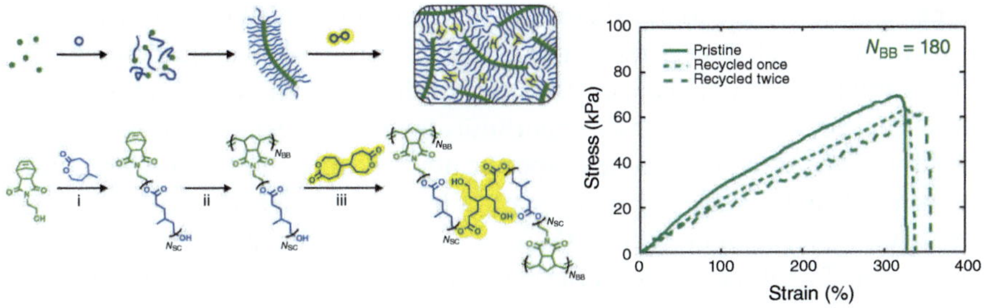

Figure 11.9 Crosslinked bottle-brush copolymers with ester-interchanged responsible for the vitrimeric behavior (N_{BB} = Backbone degree of polymerization). Source: Reproduced with permission from Self et al. [57]. Copyright 2020 American Chemical Society.

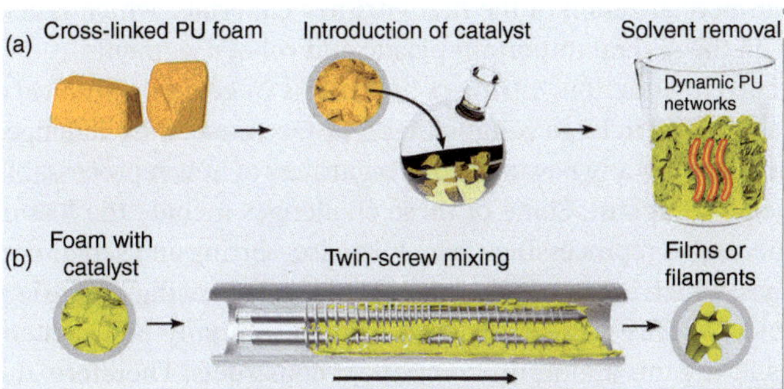

Figure 11.10 Polyurethane-based vitrimers shown to be reconfigurable in a twin-screw extruder. Source: Reproduced with permission from Sheppard et al. [59]. Copyright 2020 American Chemical Society.

been employed to various polymer backbones, effectively transforming thermosets into reprocessable materials to various extents. To move forward, these chemistries have to be incorporated in polymers with real-world applications.

11.5 Outlook and Summary

Thermoset polymers have historically been considered unrecyclable due to the permanent nature of the crosslinks. Initial efforts into degradable crosslinks may ultimately allow for some recyclability, but dynamic crosslinks offer a potentially appealing alternative, whereby the properties and reconfiguration conditions can be finely tuned via the molecular makeup. Dynamic, reversible crosslinking also provide other opportunities in addition to recycling/reprocessing. For example, dissociative CANs can self-heal upon heating. On the other hand, the crosslink strength and thermal properties have to date not proven strong enough to compare with traditional thermosets, and as such they have failed to provide the performance of true thermosets. Such disadvantages are a topic of active research, and this remains an area with some promise. Associative CANs or vitrimers retain a constant crosslink density and so their material properties fall off in a far more gradual manner and hence are useful thermoset replacements which can in principle be mechanically recycled by remolding. Translating this to real applications and real markets still has many challenges. This remains firmly in an exploratory stage.

Nevertheless, vitrimers offer the opportunity for designing new materials from the ground-up, where the end-of-life is specifically built into the material constructs. This concept makes recycling a key feature in newly designed materials, instead of merely an afterthought, such as the case with conventional commodity plastics. The concept of vitrimers, whereby the materials comprise the remarkable properties associated with thermosets, offers new avenues for reconfiguration and reprocessing of plastics that have traditionally been exclusively reserved for a select subset of thermoplastics. Rapid progress has been made in the decade since the inception of vitrimers, and the coming decades surely offer many new exciting developments. Most exciting is the commercialization of the first vitrimer materials, which is certain to have an impact on the several important plastic and rubber segments.

With respect to the current situation, vitrimers and CANs in general still have a long way to go to gain a foothold in large volume markets. Of course, the challenges associated with recycling apply to any new material, regardless of how reprocessable it is from a technological standpoint. Some of these challenges include the loss of properties with each successive reprocessing cycle. Likewise, sorting and separation offer logistical challenges, which are naturally exacerbated by increasing the variety of chemical building blocks. This is currently seen in the essentially non-existent recycling of mixed plastic streams and multi-component constructs. Therefore, the development of recyclable thermosets will certainly encounter the same roadblocks that currently impede the widespread recycling of commodity thermoplastics. In the near future, at least, the reprocessing and reuse of reconfigurable thermosets will be limited to clean sources that are well separated. Furthermore, chemical recycling methods that rely on specific decomposition mechanisms (e.g. ester exchange for poly(ethylene terephthalate) (PET)) are challenging when a multitude of chemical moieties must be dealt with.

Parallel to opportunities in recycling technology lies the (bio)degradation pathways. If collecting, sorting, and chemical recycling is impractical or deemed unsustainable, ideally natural degradation pathways should be built into the material, leading to benign byproducts after breakdown. The use of hydrolysable links in the main polymer chain and their cross-link is a significant step toward that goal. Of course, biodegradation should be tuned/timed to start well after the expected usability/durability of the product. This very feature has made widespread use of bio-based plastics in engineering applications (see Chapter 7 on PLA). To date, the express focus on biodegradable vitrimers has received minimal attention. However, this remains an area that holds many opportunities for further investigation.

The next generation of polymer design provides great opportunities, wherein the end-of-life can specifically be taken into account along with the properties for various applications. Adaptive networks and vitrimers offer a particularly appealing avenue for attaining the superior physical/mechanical properties and durability often observed in thermosets, even if it is particularly suited for niche applications. At least the concept takes into consideration the need for rationally designed systems, where the full life cycle is accounted for. This chapter offers some insight into several strategies that have emerged with this circularity in mind, applied specifically to crosslinked networks. The research outlined may provide viable pathways toward making environmentally troublesome waste streams from ever escaping. Among the products ripe for investigation are of course rubber tires, polyurethanes, urea-formaldehyde, and others. There is great promise for designing new plastics that even contribute to sustainability without sacrificing the material after its (initial) useful lifetime.

References

1 PlasticsEurope (2019). Plastics – the Facts 2019. https://plasticseurope.org/knowledge-hub/plastics-the-facts-2019/ (accessed 01 February 2021).

2 Geyer, R., Jambeck, J.R., and Law, K.L. (2017). *Science Advances* 3: e1700782.

3 Jambeck, J.R., Geyer, R., Wilcox, C. et al. (2015). *Science* 347: 768–771.

4 (a) Lebreton, L., Slat, B., Ferrari, F. et al. (2018). *Scientific Reports* 8: 4666. (b) Eriksen, M., Lebreton, L.C., Carson, H.S. et al. (2014). *PLoS One* 9: e111913.

5 (a) Okan, M., Aydin, H.M., and Barsbay, M. (2019). *Journal of Chemical Technology Biotechnology* 94: 8–21. (b) Ellen MacArthur Foundation 2017. http://www.ellenmacarthurfoundation.org/publications (accessed 28 December 2021).

6 (a) Galloway, T.S., Cole, M., and Lewis, C. (2017). *Nature Ecology & Evolution* 1: 116. (b) Lebreton, L.C.M., van der Zwet, J., Damsteeg, J.W. et al. (2017). *Nature Communications* 8: 15611.

7 (a) Bossion, A., Heifferon, K.V., Meabe, L. et al. (2019). *Progress in Polymer Science* 90: 164–210. (b) Zhang, X., Fevre, M., Jones, G.O., and Waymouth, R.M. (2018). *Chemical Reviews* 118: 839–885.

8 Sohn, Y.J., Kim, H.T., Baritugo, K.A. et al. (2020). *Biotechnology Journal* 15: e1900489.

9 Yue, L., Bonab, V.S., Yuan, D. et al. (2019). *Global Challenges* 3: 1800076.

10 Post, W., Susa, A., Blaauw, R. et al. (2019). *Polymer Reviews* 60: 359–388.

11 Akato, K.M., Nguyen, N.A., Bonnesen, P.V. et al. (2018). *ACS Omega* 3: 10709–10715.

12 (a) Wang, L. and Wong, C.P. (1999). *Journal of Polymer Science Part A: Polymer Chemistry* 37: 2991–3001. (b) Zhang, X., Chen, G.-C., Collins, A. et al. (2009). *Journal of Polymer Science Part A: Polymer Chemistry* 47: 1073–1084.

13 Liu, W., Wang, Z., Chen, Z., and Zhao, L. (2012). *Polymer Degradation and Stability* 97: 810–815.

14 (a) Chen, Z., Zhao, L., and Wang, Z. (2013). *Polymer* 54: 5182–5187. (b) Liu, W., Wang, Z., Xiong, L., and Zhao, L. (2010). *Polymer* 51: 4776–4783.

15 (a) Chen, J.-S., Ober, C.K., and Poliks, M.D. (2002). *Polymer* 43: 131–139. (b) Chen, J.-S., Ober, C.K., Poliks, M.D. et al. (2004). *Polymer* 45: 1939–1950. (c) Yang, S., Chen, J.-S., Körner, H. et al. (1998). *Chemistry of Materials* 10: 1475–1482.

16 Okamura, H., Toda, S., Tsunooka, M., and Shirai, M. (2002). *Journal of Polymer Science Part A: Polymer Chemistry* 40: 3055–3062.

17 Okamura, H., Yamauchi, E., and Shirai, M. (2011). *Reactive and Functional Polymers* 71: 480–488.

18 Ogino, K., Chen, J.-S., and Ober, C.K. (1998). *Chemistry of Materials* 10: 3833–3838.

19 Ito, H. and Ueda, M. (1988). *Macromolecules* 21: 1475–1482.

20 (a) Buchwalter, S.L. and Kosbar, L.L. (1996). *Journal of Polymer Science Part A: Polymer Chemistry* 34: 249–260. (b) Bulmus, V., Chan, Y., Nguyen, Q., and Tran, H.L. (2007). *Macromolecular Bioscience* 7: 446–455. (c) Hashimoto, T., Meiji, H., Urushisaki, M. et al. (2012). *Journal of Polymer Science Part A: Polymer Chemistry* 50: 3674–3681. (d) Xiong, M., Schneiderman, D.K., Bates, F.S. et al. (2014). *Proceedings of the National Academy of Sciences of the United States of America* 111: 8357–8362.

21 Harada, M., Ando, J., Yamaki, M., and Ochi, M. (2015). *Journal of Applied Polymer Science* 132: 41296.

22 Yuan, Y., Sun, Y., Yan, S. et al. (2017). *Nature Communications* 8: 14657.

23 Gallagher, J.J., Hillmyer, M.A., and Reineke, T.M. (2014). *Macromolecules* 47: 498–505.

24 Cromwell, O.R., Chung, J., and Guan, Z. (2015). *Journal of the American Chemical Society* 137: 6492–6495.

25 (a) Lou, L., Okamura, H., and Matsumoto, A. (2017). *Journal of Applied Polymer Science* 134: 44229. (b) Nomura, K., Tsujii, A., and Matsumoto, A. (2017). *Macromolecular Chemistry and Physics* 218: 1700156. (c) Tsujii, A., Namba, M., Okamura, H., and Matsumoto, A. (2014). *Macromolecules* 47: 6619–6626.

26 McBride, M.K., Worrell, B.T., Brown, T. et al. (2019). *Annual Review of Chemical and Biomolecular Engineering* 10: 175–198.

27 Billiet, S., De Bruycker, K., Driessen, F. et al. (2014). *Nature Chemistry* 6: 815–821.

28 Denissen, W., Winne, J.M., and Du Prez, F.E. (2016). *Chemical Science* 7: 30–38.

29 Boutelle, R.C. and Northrop, B.H. (2011). *The Journal of Organic Chemistry* 76: 7994–8002.

30 (a) Boul, P.J., Reutenauer, P., and Lehn, J.M. (2005). *Organic Letters* 7: 15–18. (b) Reutenauer, P., Buhler, E., Boul, P.J. et al. (2009). *Chemistry* 15: 1893–1900.

31 Inglis, A.J., Nebhani, L., Altintas, O. et al. (2010). *Macromolecules* 43: 5515–5520.

32 Houck, H.A., De Bruycker, K., Billiet, S. et al. (2017). *Chemical Science* 8: 3098–3108.

33 Michal, B.T., Jaye, C.A., Spencer, E.J., and Rowan, S.J. (2013). *ACS Macro Letters* 2: 694–699.

34 Higaki, Y., Otsuka, H., and Takahara, A. (2006). *Macromolecules* 39: 2121–2125.

35 Ying, H., Zhang, Y., and Cheng, J. (2014). *Nature Communications* 5: 3218.

36 Zhang, B., Digby, Z.A., Flum, J.A. et al. (2016). *Macromolecules* 49: 6871–6878.

37 Van Herck, N., Maes, D., Unal, K. et al. (2020). *Angewandte Chemie International Edition* 59: 3609–3617.

38 (a) Defize, T., Thomassin, J.M., Ottevaere, H. et al. (2019). *Macromolecules* 52: 444. (b) Lendlein, A., Jiang, H., Junger, O., and Langer, R. (2005). *Nature* 434: 879–882. (c) Zheng, Y., Micic, M., Mello, S.V. et al. (2002). *Macromolecules* 35: 5228.

39 Kloxin, C.J. and Bowman, C.N. (2013). *Chemical Society Reviews* 42: 7161–7173.

40 Jourdain, A., Asbai, R., Anaya, O. et al. (2020). *Macromolecules* 53: 1884–1900.

41 Williams, M.L., Landel, R.F., and Ferry, J.D. (1955). *Journal of the American Chemical Society* 77: 3701–3707.

42 (a) Yang, Y., Pei, Z., Zhang, X. et al. (2014). *Chemical Science* 5: 3486–3492. (b) Yang, Z., Wang, Q., and Wang, T. (2016). *ACS Applied Materials & Interfaces* 8: 21691–21699.

43 (a) Montarnal, D., Capelot, M., Tournilhac, F., and Leibler, L. (2011). *Science* 334: 965–968. (b) Capelot, M., Montarnal, D., Tournilhac, F., and Leibler, L. (2012). *Journal of the American Chemical Society* 134: 7664–7667.

44 Snijkers, F., Pasquino, R., and Maffezzoli, A. (2017). *Soft Matter* 13: 258–268.

45 Denissen, W. (2016). PhD Thesis. Ghent University.

46 Nishimura, Y., Chung, J., Muradyan, H., and Guan, Z. (2017). *Journal of the American Chemical Society* 139: 14881–14884.

47 Rottger, M., Domenech, T., van der Weegen, R. et al. (2017). *Science* 356: 62–65.

48 Roettger, M. (2016). Université Pierre et Marie Curie.

49 Worrell, B.T., Mavila, S., Wang, C. et al. (2018). *Polymer Chemistry* 9: 4523–4534.

50 Denissen, W., Rivero, G., Nicolaÿ, R. et al. (2015). *Advanced Functional Materials* 25: 2451.

51 Denissen, W., Droesbeke, M., Nicolay, R. et al. (2017). *Nature Communications* 8: 14857.

52 Lessard, J.J., Garcia, L.F., Easterling, C.P. et al. (2019). *Macromolecules* 52: 2105–2111.

53 Christensen, P.R., Scheuermann, A.M., Loeffler, K.E., and Helms, B.A. (2019). *Nature Chemistry* 11: 442–448.

54 Caffy, F. and Nicolaÿ, R. (2019). *Polymer Chemistry* 10: 3107–3115.

55 Lu, Y.-X., Tournilhac, F., Leibler, L., and Guan, Z. (2012). *Journal of the American Chemical Society* 134: 8424–8427.

56 Brutman, J.P., Delgado, P.A., and Hillmyer, M.A. (2014). *ACS Macro Letters* 3: 607–610.

57 Self, J.L., Sample, C.S., Levi, A.E. et al. (2020). *Journal of the American Chemical Society* 142: 7567–7573.

58 Snyder, R.L., Fortman, D.J., De Hoe, G.X. et al. (2018). *Macromolecules* 51: 389–397.

59 Sheppard, D.T., Jin, K.L., Hamachi, L.S. et al. (2020). *ACS Central Science* 6: 921–927.

12

Managing Plastic Wastes

Jean-Paul Lange[1,2]

[1] *University of Twente, Faculty of Science and Technology, Sustainable Process Technology, P.O. Box 217, 7500 AE Enschede, Netherlands*
[2] *Shell Global Solutions International B.V., Projects & Technology, Grasweg 31, 1031 HW Amsterdam, Netherlands*

12.1 Introduction

The petrochemical industry is a major field that upgrades about 10% of fossil hydrocarbons into about 350 Mt of polymers annually. The majority of them (40% in Europe) are used in packaging [1, 2]. However, polymers are also used in construction (20%), automotive (9%), electrical/electronic (6%), and many other applications. Eventually, polymers touch all facets of today's life, from housing to health, clothing, sport, transport, food, water, and many more. Often, however, the polymers eventually end up in the environment after use, being in the form of litter on land and in water or as CO_2, soot, and other gaseous contaminants when incinerated [3]. Beyond being unacceptable for our planet and all its living species, such (mis)management is also a waste of resources because spent plastics could in fact be used as valuable feedstock for new materials. This chapter discusses the issues of plastic waste and reviews the various options we have in hand to valorize end-of-life (EoL) polymers (Figure 12.1) in a circular economy. This chapter briefly addresses the challenges and opportunities in mechanical recycling, offers some in-depth details about the options for chemical recycling to polymer constituents or polymer feedstock, and finally addresses waste destruction in the form of energy valorization and biodegradation. Eventually, this chapter addresses life cycle analyses (LCAs) to compare the impact of various EoL options. Obviously, this chapter does not stand alone but establish earlier broad and insightful books and reviews [5–8].

12.2 Plastic Waste

To develop a circular economy, the petrochemical industry needs to complete the carbon cycle by collecting and recycling spent carbon. A minor fraction of spent

Biodegradable Polymers in the Circular Plastics Economy, First Edition.
Edited by Michiel Dusselier and Jean-Paul Lange.

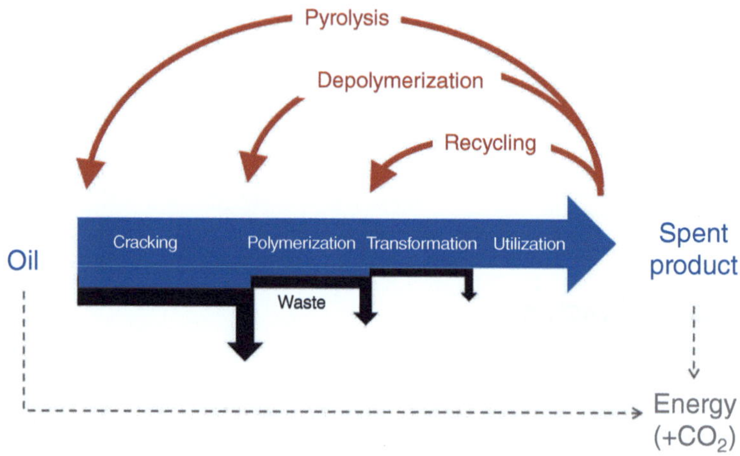

Figure 12.1 Options for recycling plastic waste for retransformation (mechanical recycling), repolymerization (chemical recycling to monomer), or recracking (chemical recycling to feedstock). Source: Adapted from Lange [4].

plastic is collected as a clean and well-sorted stream, e.g. post-industrial or car scrap wastes. However, the majority is collected in municipal solid waste (MSW), together with food scraps, yard trimmings, textile, paper, and other inorganic waste [9–11]. The world is producing about 1.1 Gt/yr MSW, with a production rate varying from <1 to >2 kg/capita/d, depending on the average income [12]. Approximately 10 wt% of plastic present in MSW mainly consists (~60 wt%) of polyolefins (i.e. high/low/linear-low density polyethylene HDPE, LDPE, LLDPE, and polypropylene (PP); the remaining fraction consists of polyethylene terephthalate (PET), polyvinyl chloride (PVC), polystyrene (PS), and other minor polymers [10, 11].

An efficient recycling of spent polymers not only ensures an efficient recycling of carbon but also aims at minimizing the consumption of energy and the production of waste over the life cycle of the product. This generally implies to operate through the shortest possible recycle loop. Depending on the quality and purity of waste, priority should therefore be given to reuse, reprocessing (mechanical recycling), depolymerization, conversion to hydrocarbon feedstock, and, as a last resort, energy recovery, as illustrated in Figure 12.1. This priority list is often presented as the priority pyramid in waste management.

Globally, about 12% of the spent plastic is recycled mechanically, while <1% is recycled back to its monomer [13]. The rest is either incinerated (25%) or ends up in the environment in landfill or unmanaged dumps.

With its ambitious "Green Deal," the European Union (EU) aspires to become climate neutral by 2050, have developed a circular economy, have restored biodiversity, and have cut pollution [14]. Of specific interest for our discussion are the EU's ambitions on circular economy [15] and, more specifically, on circular plastics, packaging, and textiles: the EU aims indeed at reducing waste, stimulating reuse (e.g. by banning single-use products whenever possible), and stimulating recycling. These ambitions further support the EU's Bioeconomy Action Plan, which explicitly aims, among other things, at reducing its dependence on non-renewable, unsustainable resources, whether sourced domestically or abroad [16].

The circular economy is further supported by numerous brand owners such as Coca Cola, Unilever, Henkel, P&G, and many others that have pledged (i) to make their packaging reusable and/or recyclable, (ii) to reduce the use of virgin resin, and (iii) to stimulate the use of recycled resins and/or plant-based materials. Specific information on these pledges can be found on the individual brand owner's websites. In these advertisements, however, the brand owners remain silent about the additional costs of these pledges and the unpopular fact that the consumer will eventually get the bill.

12.3 Mechanical Recycling

So far, plastic recycling mainly includes mechanical recycling, with the focus on three dominant packaging polymers polyethylene (PE), PP, and PET. The recycling processes rely on cautious sorting of the cleanest and purest waste fraction, chipping the plastic into flakes, and washing them if necessary. The recycled flakes are then blended with a virgin polymer of the same family together with compatibilizers and additives to mitigate the shortcomings of the recycled material [6, 7, 17, 18]. Although very efficient and so far successful, mechanical recycling is bound to be limited to a few cycles using the minor fraction of the purest and cleanest waste stream. For instance, PET is generally the recycled once, from bottle to textile [19], while PP is claimed to technically support up to four recycles but is practically recycled once to textile and playground equipment [20]. In fact, mechanical recycling seems to mainly consist of downcycling. Hence, they need complementary recycling options, particularly chemical recycling. Therefore, let us review the limitations of mechanical recycling.

Firstly, the recycled materials, particularly post-consumer plastics, generally show lower performance than the virgin ones because of several reasons. Firstly, the sorted material may not come as a single-grade plastic but as the market-average grade. It may therefore not meet the requirement for high-end applications.

Secondly, plastic products may contain additives such as fillers, antioxidants, plasticizers, pigments, flame retardants, etc. Recycled materials will therefore contain the market average of these additives.

Thirdly, the sorted waste may not have the same purity as the virgin material. It may contain minor fractions of foreign polymers, for instance, traces of PET or PP in the sorted PE stream. As polymers are practically immiscible with one another, these polymer impurities tend to segregate into small foreign domains and, thereby, create weak spots into the recycled material. These shortcomings can be mitigated by the addition of small amounts of compatibilizers, e.g. short block copolymers with chain segments of the same nature as two polymers of the blend (e.g. PP–PE block compatibilizers) [6, 7, 17, 18]. Alternatively, the compatibilizers contain a main chain that resemble the target matrix and a reactive end group that can react with the functional group of the polymer impurity, e.g. one that can react with an alcohol group of PET or EVOH (ethylene-vinyl alcohol) polymers. The challenge of contamination may come in an even more difficult form, i.e. in multilayer films and in textiles, because these materials generally consist of multiple polymers that are interlayered or interwoven to reach the desired properties.

Fourth, the recycled polymer chains may be partly degraded, e.g. through oxidation or UV radiation upon use, or through thermal degradation upon repeated hot processing [6, 7, 17, 18]. Indeed, hydrocarbon polymers and PVC are reprocessed at 160–260 °C while performance polymers such as polyamide (PA) and PET are reprocessed at 220–320 °C [7]. As a result, the polymer may exhibit polar groups that need compatibilization. However, recycle polymers may also exhibit an increased M_w and higher viscosity. The resulting shortcomings can be mitigated by means of various additives [6, 7, 17, 18].

Fifth and final, some applications such as food packaging are forbidden to use materials that could be contaminated by traces of toxic impurities. Mechanical recycling to make plastics for food packaging is then not a viable option.

These shortcomings are particularly pronounced for post-consumer wastes. Some of them may be of lesser concern for well-defined post-industrial waste, which explains the pronounced interest of the recycling industry for these waste streams.

Obviously, the use of compatibilizers and additives is increasing the level of impurities in recycled materials. They will further increase upon multiple recycle loops to the point where it disqualifies the material for further mechanical recycling. More sophisticated recycling technologies are then required, e.g. dissolution/precipitation or chemical recycling, to be discussed in Sections (12.4 and 12.5).

Moreover, our discussion on mechanical recycling has so far assumed that the material to recycle is a thermoplastic: it can be melted and reshaped at will upon heating. This obviously excludes thermoset materials such as polyurethane mattresses, vulcanized rubber tires, and cross-linked unsaturated polyester composites to name but a few. For such materials, which represent about one-third of the polymeric materials [21], spent products can be chipped and "downcycled" as a filler for new products. This applies to rubber [22], epoxy blends [23], and polyurethane [24]. These materials are not prone to dissolution either. Hence, recycling is bound to proceed via deep deconstruction, i.e. via chemical recycling.

12.4 Dissolution/Precipitation

An approach to remove contaminants from spent polymers, being additives or foreign polymers, is dissolution/reprecipitation [8, 25]. Accordingly, the spent polymer is dissolved in an appropriate solvent, separated from insoluble impurities and additives and reprecipitated upon addition of an antisolvent. Obviously, the solvent and the antisolvent need to be separated for reuse at the dissolution or precipitation stage of the process. This step consumes much energy but maybe less than the alternatives: recovering the polymer by evaporating the solvent instead of precipitating it with the antisolvent. Dissolution processes have been developed for recycling PS [17, 26], PVC [24], PA [27], polymethylmetacrylate (PMMA) [24, 28],

PE/PP [24], and multilayer films [24]. Most of these processes seem to be at pilot or demonstration stage [8], with the exception of APK's dissolution of multilayer films [29].

12.5 Chemical Recycling

Generally, the waste plastic stream does not meet the stringent quality requirements of mechanical recycling. For instance, the plastic has suffered minor, although significant, degradation upon use and/or reprocessing at elevated temperature (e.g. oxidation and/or minor depolymerization), or the sorted and cleaned plastic still contains traces of toxic components for use in food packaging. Chemical recycling is then necessary. Some polymers can and should be depolymerized back to their monomers. Other can only be converted back to a general feedstock. Still others cannot fit in either of these loops in an attractive manner. However, how to decide what is promising?

The basis for such a selection is illustrated in Figure 12.2 [4, 30]. The horizontal axis helps identifying the polymers that are easily depolymerized back to their monomer (heat of depolymerization dH < 70 kJ/mol of broken bonds, typically condensation polymers) and those that are only cracked to a general hydrocarbon through more severe pyrolysis (dH > 70 kJ/mol, typically addition polymers). The *y*-axis represents the cumulative amounts of resources that are wasted when producing the polymer. It, thereby, represents the "incentive" to recover the monomer rather than degrading it back to hydrocarbon. This simple mapping readily recommends cracking polyolefins back to general feedstocks (lower right quadrant) but depolymerizing PET and polyamides (PA) back to their monomers

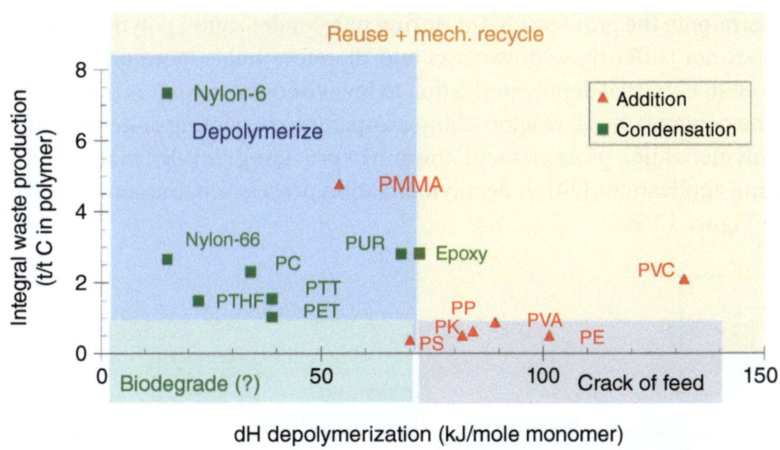

Figure 12.2 Options for plastic recycling (PC: polycarbonate, PTHF: polytetrahydrofuran, PTT: polytrimethylene terephthalate, PET: polyethylene terephthalate, PMMA: polymethylmetacrylate, PUR: polyurethane, PS: polystyrene, PP: polypropylene, PK: polyketone, PVA: polyvinylalcohol, PE: polyethylene, and PVC: polyvinylchloride). Source: Adapted from Lange [28].

(upper left quadrant). Polymers falling in the upper right quadrants (e.g. PVC) are demanding to make and are impossible to depolymerize back to the monomer. Society could consider abandoning them. Finally, although empty, the lower left quadrant would contain the ideal polymers that are easy to make and easy to depolymerize. These polymers would be promising candidates for biodegradation as they would consume only limited energy and chemistry invested in making them. Although not shown here, today's biodegradable plastics (e.g. polyhydroxyalkanoate (PHA) and polylactone) fall in the upper left quadrant as they are easily depolymerized but are fairly demanding to produce. However, let us now discuss how such depolymerization is carried out for the various polymers.

12.5.1 Depolymerization of Condensation Polymers

Condensation polymers are formed by nucleophilic substitution reactions that link the monomers through polar bridges, most commonly through ester linkage (–C(O)O–), amide linkage (–C(O)NH–), or urethane/carbamate linkage (–C(NH)O–). Most of them are prone to opening through hydrolysis, transesterification, or transamidation. Let us illustrate the approach for PET, PA, and polyurethane (PUR), which are well advanced [8, 17, 26].

The polyester PET is most commonly depolymerized by alcoholysis, i.e. methanolysis to dimethyl terephthalate of glycolysis (with ethylene glycol) to dihydroxyethyl terephthalate [31, 32]. The reaction is generally carried out at elevated temperature (~200 °C) in the presence of catalysts, traditionally a Lewis metal salt such as Zn acetate [33]. Methanolysis requires a complex product recovery and purification train, which becomes very challenging when depolymerizing copolymers, e.g. which contain some isophthalate (besides terephthalate) or some diethylene glycol (besides ethylene glycol). Deep solvent removal is indeed necessary since monoalcohols would terminate the growing chains during polycondensation polymerization. Glycolysis does not suffer these drawbacks and therefore holds more promises. It furthermore enables partial depolymerization to low-viscosity oligomers that can be fed back to the polymerization reactor. Many companies are piloting or demonstrating PET depolymerization processes with the aim of providing recycled materials for food packaging applications [34]. A depolymerization process scheme could look as presented in Figure 12.3.

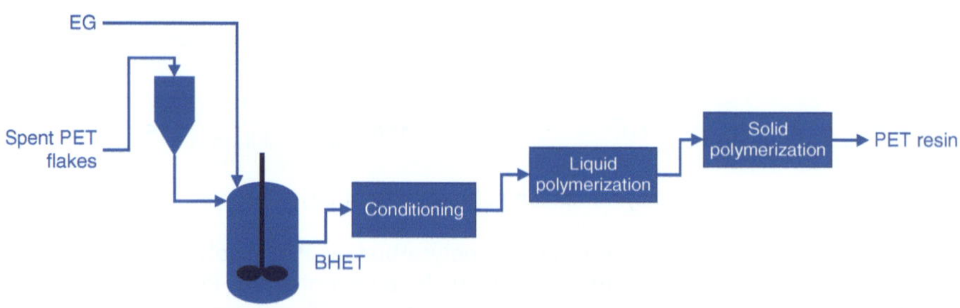

Figure 12.3 Process concept for PET chemical recycling (EG: ethylene glycol, BHET: bis-hydroxyethylene terephthalate). Source: Adapted from [35].

Polyamides (PA) are depolymerized in water at ~300 °C in the presence of inorganic acid [36]. However, this process seems limited to depolymerizing Nylon 6 to caprolactam for the depolymerization of Nylon 66 to adipic acid and hexamethylene diamine brings complications in product recovery and purification. Pyrolysis can also depolymerize Nylon 6 into caprolactam [37] but does not seem nearly as effective for nylon 66.

PUR can also be depolymerized by alcoholysis, glycolysis, and hydrolysis. However, it can also undergo aminolysis [38, 39]. In contrast to the previous cases, however, PUR depolymerization does not release the constituting monomers, i.e. propylene oxide and diisocyanates. Instead, it releases high-molecular weight products, namely, the oligomeric polyols and the aromatic, N-containing oligomers. The polyols can be recycled into new PUR, but the aromatic fraction is usually disposed of, as it cannot be properly upgraded to the original diisocyanates.

Although most elegant, selective depolymerization might still be economically challenging. The polymers that can be of interest typically represent a minor fraction of the total polymer waste. They are available in modest quantities and therefore require small-scale and costly logistics and reprocessing. This is surely the case when mixed with other plastic wastes. However, the small-scale logistics and reprocessing likely applies to well-sorted industrial waste as well.

Depolymerization of monomers will likely apply to the new and emerging bio-based polyesters such as polylactide (PLA), polyhydroxybutyrate (PHB), polybutylsuccinate (PBS), and polyethylenefuranoate (PEF). However, the small volume may challenge their recycling for a while.

12.5.2 Melt Pyrolysis of Polyolefins

Polyolefins cannot be depolymerized back to their monomeric constituents. Depolymerization requires harsh pyrolysis conditions and generally leads to a complex mixture of hydrocarbons, i.e. a general feedstock (lower right quadrant of Figure 12.2). Pyrolysis produces paraffinic/olefinic waxes under mild conditions, an aromatic product under more severe conditions, and gas and char at the highest severity [7, 26, 40, 41]. Such hydrocarbon products can be processed into a synthetic fuel. However, fractions of the aliphatic product produced under mild conditions can also be cracked into lower olefins, generally after removing the heaviest product and hydrotreating the desired distillate fraction. Steam cracking of plastic pyrolysis oil is expected to deliver olefins and aromatic base chemicals with ~65 wt% yield, with coproduction of ~10 wt% fuel gas, ~10 wt% of aromatic gasoline, and 15 wt% of aromatic fuel oil.

Pyrolysis is no new technology. Oil refineries are applying it at the large scale in various forms for decades for upgrading heavy oil fractions into gas and distillates. These technologies are then called thermal cracker, visbreaker, or coker [42]. Pyrolysis has also been explored for processing plastic waste by major chemical producers some 30 years. Although technically successful, these technologies were eventually abandoned because they could not compete with cheap crude oil. The rise in oil price in the early 2000 has encouraged small start-up companies to revisit plastic pyrolysis, which led to a plethora of technology providers today [17].

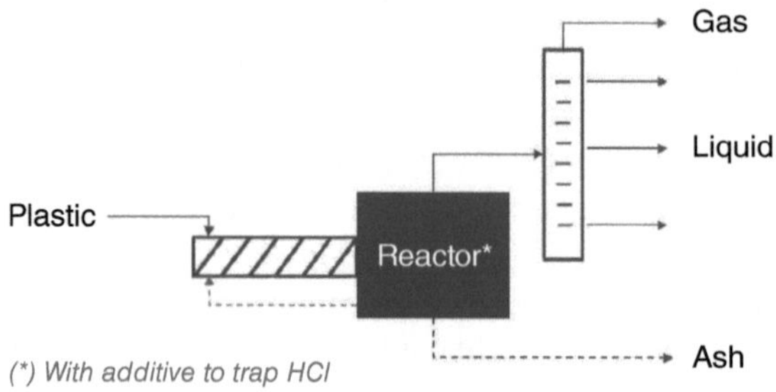

Figure 12.4 Process concept for melt pyrolysis of waste plastics.

The oil and chemical majors have also joined the effort, not so much with more pyrolysis technologies but rather with plans to process the resulting pyrolysis oil in their steam cracker. For instance, Shell announced in November 2019 to be processing a first truck of plastic pyrolysis oil in its cracker at Norco, USA, and announced its ambition to ramp up the volume of plastic recycling fed to the cracker to 1 Mt/yr by 2025 [43].

Polyolefins are generally pyrolyzed by the so-called melt-pyrolysis [35, 40]. A conceptual process scheme is provided in Figure 12.4. Accordingly, the plastic is fed and melted into an extruder before being fed to a large vessel that is heated to 450–500 °C and mechanically agitated. The cracked vapor is removed at the top of the vessel and, subsequently, condensed to liquid pyrolysis oil. The incondensable gases can be used for heating the reactor. The char is removed at the bottom of the vessel and disposed of. Tight temperature control and extensive agitation seem to be essential to minimize coke deposition and achieve high oil yields. These requirements seem to limit the scale of pyrolysis reactors at some 15–20 kt/yr, a scale that is very small when compared to the 3 Mt/yr liquid steam cracker that they are supposed to feed. The small scale is clearly harming the economic competitiveness of the process as it leads to high capital and operating costs. The pyrolysis technology is not fully omnivorous either. It is particularly suitable to process polyolefins, but it produces less oil and an oil that is more aromatic when the feed is contaminated with other polymers such as PS, PET, and PA. Small amounts of PVC in the feed are particularly nasty, as it liberates HCl that corrodes the equipment and make the oil unsuitable for further processing. One element of mitigation consists of heating the plastic waste in the feeding system and recovering the HCl-rich gas before feeding the plastic melt to the reactor. Another complementary approach is to feed caustic elements such as $CaCO_3$ to the reactor to trap and neutralize the remaining chloride. The resulting salt is then removed together with the coke.

12.5.3 Alternative Pyrolysis Processes

When run at the highest severity, the pyrolysis of polyolefins can deliver an olefin-rich gaseous stream as the main product. This could be considered as chemical recycling to monomers [35, 40]. However, the gaseous stream requires strict cryogenic fractionation to deliver polymer-grade ethylene, propylene, and butylenes, a purification that is uneconomical at the small scale of pyrolysis. The situation is more favorable when processing PS or PMMA waste, however. These addition polymers indeed crack back to their constituting monomer with reasonable selectivity upon moderate thermal treatment, as indicated by Figure 12.2. Moreover, the product can be condensed out of the gaseous by-products. Various processes are being developed and demonstrated for these streams [24, 26]. The small-scale pyrolysis might not be a critical limitation here since these waste streams are available in limited volume any way.

Beyond melt-pyrolysis, other technologies are also being developed to convert waste plastics into liquid hydrocarbons. For instance, the pyrolysis can be assisted by a catalyst, typically an acidic zeolite, to convert the pyrolysis vapors into cyclic and aromatic hydrocarbons [26]. Anellotech [44] and BioBTX [45] have developed such processes by modifying the process concept they developed earlier to upgrade lignocellulose to aromatic biofuels. The aromatic product holds promise for aromatic production and for fuel applications, but it is not suitable for cracking to olefins.

Hydrothermal liquefaction, as performed by Licella, is another process that has been developed for biomass and adapted for plastic waste [46]. Accordingly, the plastic waste is thermally degraded in near/super-critical water. This process is claimed to be more tolerant to the presence of engineering polymers in the feed. In fact, it can even digest the whole MSW, including its organic fraction, thereby producing an aromatic-rich oil that is partly of bio and plastic origin and could qualify as a low-carbon fuel (see Section 12.6).

Finally, the IH$^{2®}$ hydropyrolysis technology that has been developed by GTI and Shell runs the pyrolysis in the gas phase in the presence of hydrogen atmosphere and the hydrogenation catalyst (Figure 12.5) and is presently being demonstrated

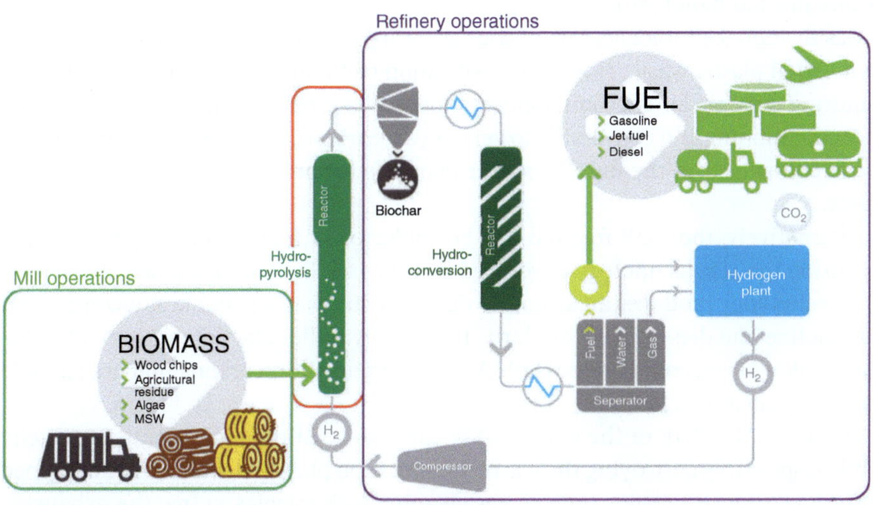

Figure 12.5 IH^2 Hydropyrolysis process, developed by GTI and Shell, converts MSW and RDF to hydrocarbon feedstock. Source: Based on Shell [47].

at the 2000 l/d scale [47]. It delivers a hydrocarbon stream that is rich in aromatics and free of heteroatoms. The technology is being developed for biomass but is compatible for biomass-rich waste fractions such as refuse-derived fuel (RDF) [48]. However, it remains unproven for pure waste plastic feedstock. Further discussion of this technology will therefore be moved to Section 12.6 that also considers gasification to valorize RDF.

Pyrolysis processes have also been explored for recycling thermoset resins such as PUR, rubber tires, or even epoxy composites [20–22]. While technically feasible, the product yield and quality does not appear very attractive so far.

12.6 Energy Recovery – Recycle Fuels and Incineration

The pyrolysis of plastic waste delivers a hydrocarbon fraction that can be made very suitable for transportation fuel. The crude pyrolysis oil may still need to undergo hydroprocessing to stabilize the fuel by hydrogenating the mono- and di-olefins and to remove residual oxygen and nitrogen. It may also need fractionation to select the desired boiling range and, possibly, dewaxing to avoid the gasoil-range product to crystallize at low temperature.

However, transportation fuels can also be produced from the more contaminated and cheaper stream such as the whole MSW or its sorting rejects, namely, the RDF or the solid-recovered fuel (SRF) than contains unsorted plastics still mixed with unsorted textile, paper/cardboard, and other organic fractions. These fractions can be upgraded to hydrocarbon fuels by means of gasification to synthesis gas (or syngas), a mixture of H_2 and CO, followed by syngas conditioning and conversion to fuel or chemicals [4]. For instance, Shell has partnered with other companies to help demonstrate Enerkem's gasification technology to convert MSW to syngas and, eventually, methanol [49].

Gasification technologies can, of course, be applied to mixed plastic waste or even well-sorted plastic waste. However, gasification technologies are expensive, thereby requiring large scale. They still deliver a low-value product, syngas, that needs further conditioning and conversion to get to hydrocarbons. It is therefore doubtful that it can compete with the much cheaper pyrolysis for processing well-sorted plastic waste.

Alternatively, the RDF fraction can be subjected to high-severity hydrotreating, e.g. using the IH^2 technology developed by GTI and Shell, as discussed above [46]. This technology produces an aromatic-rich distillate that is suitable as a component for gasoline and diesel fuels. Similarly, Licella's hydroliquefaction process (also discussed above) can convert the whole MSW to liquid hydrocarbons that are suitable for fuel applications [45].

Of course, the RDF or the whole MSW can also be burned to generate electricity while responsibly destroying the waste. Incineration plants should then be equipped with modern technologies such as gas cleaning technologies to free the exhausted gas from harmful components such as polyaromatic hydrocarbons, dioxins, NO_x, and SO_x before release to the atmosphere [50].

12.7 Waste Destruction – Biodegradation

Biodegradation, i.e. biological conversion to CO_2/H_2O/biomass, is truly wasting the energy and chemistry embedded in the polymer or is, at best, converted it to low-value compost. It is therefore not a recycle option but rather a waste destruction option. It can nevertheless be valuable in a few occasions. For instance, biodegradation is valuable for polymers that run a high risk to end up in the environment after use, e.g. for small-sized, single-use packaging. Biodegradation can also help not having to remove the polymer after use, as encountered in surgery, agricultural mulch films, or in plastic waste that is heavily spoiled with food rests (e.g. single-use food packaging and disposable cutlery).

Anaerobic biodigestion, i.e. biological conversion CO_2/CH_4/biomass, is more valuable when run in a well-controlled industrial setting as it delivers CH_4 as fuel. It thereby falls among the energy recovery options discussed above.

Both aerobic and anaerobic biodigestion processes are applicable to a limited set of polymers, those that are attacked by microorganisms. However, let us not dwell on this topic here as other chapters of this book will treat this topic in more detail. Chapter 2 (entitled "Fundamentals of Polymer Biodegradation Mechanisms").

12.8 Life Cycle Analyses

The literature and the web offer a wealth of LCAs that attempt to assess the environmental impact of plastics. Some aim at comparing plastics to other materials such as metal and glass; others aim at comparing bio-based to fossil-based plastics and still others aim at comparing various EoL scenarios such as landfill, incineration, conversion to fuel, or recycling to plastics. Some limit their analysis to the carbon footprint or green house gas (GHG) emissions, while others also consider other planetary boundaries such as land and water use, water and air pollution, biodiversity, etc. I personally find very delicate to draw solid conclusions from such studies because the overall conclusions heavily depend on too many factors, premises, boundaries, substitution scenarios, etc., which are often hidden in the small prints of the papers or in their supplementary information. I will nevertheless take the risk to extract a few important messages to be used in an indicative, at best semi-quantitative, manner. The reader is also referred to chapter 13 (chapter on LCA) for a deeper discussion.

Overall, the carbon intensity or GHG/CO_2 emissions of plastics seems to amount to 4–5 $t_{CO2}/t_{plastic}$ [51, 52], half of it being due to the production stage and the other half to the incineration at EoL [50]. This number is obviously an average over various plastics and is likely dominated by the most abundant plastics, i.e. the polyolefins. Indeed, the carbon footprint of the production stage varies largely with the type of polymers from 0.5 t_C/t_C (ton carbon wasted per ton carbon in the product) for polyolefins to 1—6 t_C/t_C for the major engineering plastics and likely more for high-performance polymers [28]. Recycling is generally seen as delivering significant GHG emission savings, when compared to incineration [51–54]. Obviously, the GHG savings vanish if one compares recycling with landfill, which could be

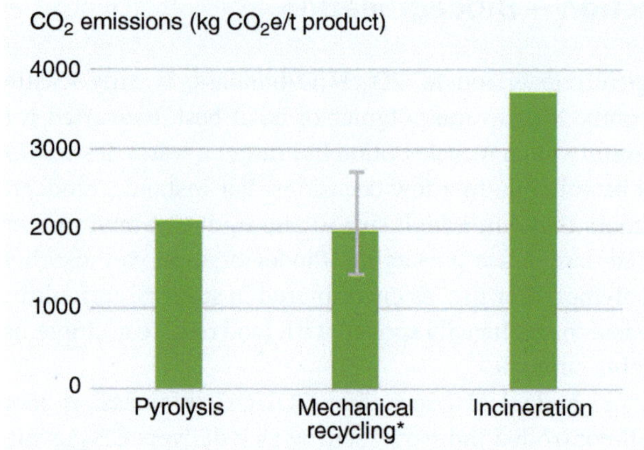

Figure 12.6 CO_2 emissions of various end-of-life scenarios for polyolefins [54] (The error bar reflects a different scenario of sorting quality and losses).

considered as a form of carbon sequestration if done responsibly, i.e. with long-term preservation of surrounding land, water, and air. However, numerous countries are discouraging landfill, and nine European countries have banned it [2, 11]. Considering the chemistry and energy requirement of the various recycling options, one is not surprised to see the CO_2 benefit decreasing in the order of mechanical recycling > chemical recycling by pyrolysis > chemical recycling by gasification > incineration for polyolefins [24]. According to BASF, the chemical recycling of polyolefin waste by pyrolysis would save 1 $t_{CO2}/t_{plastic}$, when compared to incineration of the waste, and does not underperform significantly over mechanical recycling, as illustrated in Figure 12.6 [55]. Vollmer et al. confirm the saving of 1 $t_{CO2}/t_{plastic}$ for pyrolysis of PE and report significant savings for chemical and mechanical recycling of acrylonitrile-butadiene-styrene (ABS), PA, and PET, when compared to incineration. They even find some savings when compared to landfill [8]. Intriguingly, they report the lowest C-footprint for the dissolution/precipitation approach.

Considering the topic of this book, a few words need to be devoted to the EoL by biodegradation. Qualitatively, we can expect biodegradation to deliver the highest C-footprint of all EoL options. Similar to incineration, biodegradation is converting the plastic to CO_2 and H_2O. Unlike incineration, however, it does not allow to recover the energy bound into the material and, thereby, does not save on other fuels. This conclusion is indeed confirmed by Posen et al. that report the CO_2 savings of PLA and PHB to decrease in the order of recycling > incineration > composting [56]. They also found CO_2 savings when substituting fossil polymers by PLA or PHB in the case the polymers are incinerated at EoL. No significant savings were found when the waste polymers were recycled.

12.9 Need for Fresh Carbon Input

However, recycle loops are inevitably accompanied with losses and will require to be replenished by a small fraction of fresh carbon. Fossil resources will keep supplying it for some decades ahead and the lost carbon will gradually need to be offset by CO_2 capture and sequestration, either artificially (CCS) or naturally by planting and preserving trees (nature-based solutions, NBS) to reduce the climate impact of the chemical industry. However, this will make the industry only partly circular. A fully circular economy will require fresh carbon input streams being fed by renewable carbon, i.e. by atmospheric CO_2.

Numerous research groups are exploring and developing technologies to capture CO_2 from the atmosphere and reduce it to fuels and chemicals with renewable electrons or renewable hydrogen [57]. Other groups are exploring biomass instead. Plants are indeed doing most of the work for us already; they capture the CO_2 from the atmosphere and reduce it from C(4+) carbon to C(0). Biomass appears indeed to be a much more attractive source of renewable carbon in the short to midterm. However, this will be discussed briefly in another Chapter 4 (on polymer tutorial).

12.10 Conclusion and Outlook

Over the years, the petrochemical industry has developed a plethora of polymers that are contributing to the comfort of modern societies. However, irresponsible disposal has also led to growing build-up of plastic litter, which is fouling the environment, harming wildlife, and wasting valuable resources. The industry has been struggling for a few decades to collect and (mechanically) recycle waste plastic. However, these efforts remain marginal in volume and are economically unsustainable. The litter problem has now become such that the society and industry have decided to join force to develop and deploy technologies to boost collection and recycling rate and to develop the regulation and financial schemes to support it. For instance, the chemical industry has partnered with industries along the whole plastic value chain from the manufacturer to the brand owners to waste processors and launched the *alliance to end plastic waste* to solve this problem through education, legislation, innovation, and environment cleaning [58].

Much work has already been accomplished in mechanical recycling of the highest quality materials. However, there are limits to the volume that is of suitable quality and to the number of times these streams can be recycled. Hence, the industry is developing or resuscitating complementary technologies that can recycle spent polymers of lesser quality, either to their monomers or to their general feedstock. The first technologies have now reached pilot or demonstration stage and offer the promise to recycle a much larger fraction than mechanically possible.

The largest potential is expected from the pyrolysis of polyolefins (PE and PP) because the technology is simple and robust, it can process the largest waste stream, it has modest purity requirements, and it delivers a synthetic hydrocarbon stream of reasonable value. Although well advanced, the pyrolysis technology still needs further improvements to reduce the cost and increase the value of the product. For instance, the technology should be made compatible for continuous operation, preferably in a fully automated manner and at the larger scale. When fed with mixed plastics, pyrolysis delivers an oil that is contaminated with aromatics as well as O- and N-containing components. These contaminants are not suitable for feeding to a steam cracker and need to be removed by post-treatment.

Gasification technologies are even more omnivorous than pyrolysis. However, they are also much more complex and costly. Moreover, they deliver a low-value product, i.e. synthesis gas that needs subsequent upgrading to deliver hydrocarbon streams. Hence, they will likely claim a role where no other recycling technologies can compete, e.g. for heavily heterogeneous and contaminated waste stream such as MSW or the RDF fraction left over after sorting the MSW.

Hydropyrolysis, e.g. IH^2, can also handle MSW and RDF. The production of higher value hydrocarbon might offset its lower overall yield and, thereby, deliver an advanced economy. Hence, we are looking forward to seeing it demonstrated and deployed.

Condensation polymers (PET, PUR, and PA) are minor components in the plastic waste and form, therefore, a less urgent issue. When properly sorted, these polymers should preferentially be recycled back to their constituting monomers and various technologies are being piloted for this purpose. However, the modest volume of these streams may eventually challenge the economics of sorting, logistic, and processing. The same will likely apply to the new and emerging bio-based polyesters such as PLA, PHB, PBS, and PEF. However, the small volume may challenge their recycling for a while.

LCA indicate significant CO_2 savings by recycling waste polymers instead of incinerating them. The savings seem to decrease in the following order:

$$\text{mechanical recycling} > \text{chemical recycling} > \text{incineration} > \text{biodegradation}$$

When combined, all these recycling technologies are promising to recycle a very large fraction of plastic waste. However, they will not be able to achieve full circularity because some waste stream will resist collections and all technologies will eventually deliver some reject stream that can, at best, end up as fuel or energy. These carbon losses will need to be compensated by virgin intake, initially from fossil resources with compensation by CCS/NBS, and at later stage from atmospheric CO_2 via artificial or natural photosynthesis. Bio-based plastics are very strong candidates to supply fresh carbon to the polymer cycles. The technology is ready for some polymers and in pilot stage for others. The conversion of atmospheric CO_2 and renewable electrons to polymers may come later, for the technology is in its infancy and threatens to remain prohibitive for many decades to come.

References

1 International Energy Agency (IEA) (2018). The Future of Petrochemicals, Towards a More Sustainable Chemical Industry, Technology Report — October 2018, https://www.iea.org/reports/the-future-of-petrochemicals.

2 PlasticEurope (2016). Plastics – The Facts 2016, An Analysis of the European Plastics Production, Demand and Waste Data. https://plasticseurope.org/knowledge-hub/plastics-the-facts-2016/.

3 ICCA/IEA/DECHEMA Technology Roadmap Catalysis. (2013). Energy and GHG Reductions in the Chemical Industry via Catalytic Processes. https://dechema.de/en/industrialcatalysis.html.

4 Lange, J.-P. (2002). *Green Chemistry* 4 (6): 546–550.

5 Francis, R. (2017). *Recycling of Polymers, Methods, Characterization and Applications* (ed. R. Francis). Weinheim: Wiley-VCH.

6 Ignatyev, I.A., Thielemans, W., and Vander Beke, B. (2014). *ChemSusChem* 7: 1579–1593.

7 Ragaert, K., Delva, L., and Van Geem, K. (2017). *Waste Management* 69: 24–58. http://dx.doi.org/10.1016/j.wasman.2017.07.044.

8 Vollmer, I., Jenks, M.J.F., Roelands, M.C.P. et al. (2020). *Angewandte Chemie International Edition* http://doi.org/10.1002/anie.201915651.

9 World Energy Council, World Energy Resources, Waste to Energy - 2016, https://smartnet.niua.org/sites/default/files/resources/weresources_waste_to_energy_2016.pdf.

10 Haig, S., Morris, L., Morton, R. et al. *Plastics to Oil Products*. Zero Waste Scotland report.

11 Villanueva, K.A. and Eder, P. (2014). End-of-Waste Criteria for Waste Plastic for Conversion. JRC Technical Report EUR 26843 EN. http://publications.jrc.ec.europa.eu/repository/bitstream/JRC91637/2014-jrc91637%20.pdf.

12 The world bank, WHAT A WASTE, A Global Review of Solid Waste Management (2012), https://documents1.worldbank.org/curated/en/302341468126264791/pdf/68135-REVISED-What-a-Waste-2012-Final-updated.pdf.

13 Hundertmark, T., Mayer, M., McNally, C. et al. (2018). *How Plastics-Waste Recycling Could Transform the Chemical Industry*. McKinsey on Chemicals http://www.mckinsey.com/industries/chemicals/our-insights/how-plastics-waste-recycling-could-transform-the-chemical-industry.

14 European Commission (2021), A European Green Deal - Striving to be the first climate-neutral continent, https://ec.europa.eu/info/strategy/priorities-2019-2024/european-green-deal_en.

15 European Commission (2020), Circular Economy Action Plan For a cleaner and more competitive Europe, https://ec.europa.eu/environment/pdf/circular-economy/new_circular_economy_action_plan.pdf.

16 European commission (2018), A sustainable Bioeconomy for Europe: Strengthening the connection between economy, society and the environment, https://eur-lex.europa.eu/legal-content/EN/TXT/?uri=CELEX%3A52018DC0673.

17 C&EN discovery report on the future of plastic. https://cen.acs.org/sections/discovery-reports/the-future-of-plastic.html.

18 Vishnu Sankar, S. and Anil Kumar, S. (2017). Common additives used in recycling of polymers. In: *Recycling of Polymers, Methods, Characterization and Applications* (ed. R. Francis), 11–53. Weinheim: Wiley-VCH.

19 Audrey Holmes (2021), How Many Times Can That Be Recycled?, https://earth911.com/business-policy/how-many-times-recycled/.

20 https://www.azocleantech.com/article.aspx?ArticleID=240#:~:text=This%20varies%20with%20the%20use,as%20clothes%20or%20playground%20equipment.

21 Energy Transition Commission (2018). Mission possible – reaching net-zero carbon emission from harder-to-abate sectors by mid-century; sectorial focus plastics. https://www.energy-transitions.org/publications/mission-possible/ (accessed 29 October 2021).

22 Parambath Swapna, V. and Stephen, R. (2017). Recycling of rubber. In: *Recycling of Polymers, Methods, Characterization and Applications* (ed. R. Francis), 141–161. Weinheim: Wiley-VCH.

23 Sunny, J.V. (2017). Recycling of polymer blends and composites (epoxy blends). In: *Recycling of Polymers, Methods, Characterization and Applications* (ed. R. Francis), 209–222. Weinheim: Wiley-VCH.

24 Sunny, J.V. (2017). Recycling of other layered mixed plastics or resins. In: *Recycling of Polymers, Methods, Characterization and Applications* (ed. R. Francis), 223–234. Weinheim: Wiley-VCH.

25 Zhao, Y.B., Lv, X.D., and Ni, H.G. (2018). *Chemosphere* 209: 707–720.

26 Simon, J.M. and Martin, S. (2019). *El Dorado of Chemical Recycling – State of Play and Policy Challenges*. Zero Waste Europe.

27 Griffith, A.T., Park, Y., and Roberts, C.B. (1999). Separation and recovery of nylon from carpet waste using a supercritical fluid antisolvent technique. In: *Polymer Plastic Technology and Engineering, Polymer and Fiber Recycling* (ed. M.K. Mishra), 411–431. New York: Marcel Dekker.

28 Sethi, B. (2017). Methods of recycling. In: *Recycling of Polymers, Methods, Characterization and Applications* (ed. R. Francis), 55–114. Weinheim: Wiley-VCH.

29 APK, Newcycling® - die innovative recyclingtechnologie der APK AG, https://www.apk-ag.de/.

30 Lange, J.-P. (2010). Resource efficiency of chemical manufacturing chains: present and future. In: *Sustainable Development in the Process Industry – Cases and Impacts* (ed. J. Harmsen and J.B. Powell), 23–38. Wiley.

31 Sangalang, A., Seok, S., and Kim, D.H. (2016). Practical design of green catalysts for PET recycling and energy conversion, Chapter 6. In: *Advanced Catalytic Materials – Photocatalysis and Other Current Topics* (ed. N. Luis), 141–166. IntechOpen.

32 Sinha, V., Patel, M.R., and Patel, J.V. (2010). *Journal of Polymers and the Environment* 18: 8–25.

33 Khoonkari, M., Haghighi, A.H., Sefidbakht, Y. et al. (2015). Chemical recycling of PET wastes with different catalysts. *International Journal of Polymer Science* 2015: 124524.

34 FDA (2020), Recycled Plastics in Food Packaging, https://www.fda.gov/food/packaging-food-contact-substances-fcs/recycled-plastics-food-packaging.

35 Buekens, A. (2018). *Kirk-Othmer Encyclopedia of Chemical Technology*. https://doi.org/10.1002/0471238961.1805032502151803.a01.pub3. Wiley.

36 Mustafa, N. (1993). *Plastic Waste Management, Disposal, Recycling and Reuse* (ed. N. Mustafa). New York: Marcel Dekker.

37 Bockhorn, H., Donner, S., Gernsbeck, M. et al. (2001). *Journal of Analytical and Applied Pyrolysis* 58, 59: 79–94.

38 Behrendt, G. and Naber, B.W. (2009). *Journal of the University of Chemical Technology and Metallurgy* 44: 3–23.
39 Yanga, W., Dong, Q., Liu, S. et al. (2012). *Procedia Environmental Sciences* 16: 167–175.
40 Lopez, G., Artetxe, M., Amutio, M. et al. (2017). *Renewable and Sustainable Energy Reviews* 73: 346.
41 Al-Salem, S.M., Antelava, A., Constantinou, A. et al. (2017). *Journal of Environmental Management* 197: 177–198.
42 Moulijn, J.A., Makkee, M., and van Diepen, A. (2001). *Chemical Process Technology*. Chichester: Wiley.
43 Shell (2019), Shell uses plastic waste to produce chemicals, https://www.shell.com/business-customers/chemicals/media-releases/2019-media-releases/shell-uses-plastic-waste-to-produce-chemicals.html.
44 https://www.anellotech.com.
45 https://biobtx.com.
46 www.licella.com.au/cat-htr.
47 Shell, iH^2 technology, https://www.shell.com/business-customers/catalysts-technologies/licensed-technologies/benefits-of-biofuels/ih2-technology.html (no date available).
48 A. Del Paggio (2018) iH^2 technology - Biomass Directly to Transportation Fuels, https://bioenergyinternational.com/app/uploads/2018/09/6.Alan_Del-Paggio.pdf.
49 Enerkem (2019), W2C Rotterdam project welcomes Shell as partner, https://enerkem.com/news-release/w2c-rotterdam-project-welcomes-shell-as-partner/.
50 Shell, Environmental catalysis, https://www.shell.com/business-customers/catalysts-technologies/catalysts/environmental-catalysts.html (no date available).
51 Energy Transition Commission (2018). Consultation paper "reaching zero carbon emissions from plastics".
52 Zheng, J. and Suh, S. (2019). *Nature Climate Change* 9: 374–378.
53 Lindgreen, E.R., Bergsma, G., and Jaspers, D. (2017) Innovatie afvalverwerkingstechnieken doorgelicht, CE Delft, publication number 17.2L47.160.
54 Gradus, R., van Koppen, R., Dijkgraaf, E., and Nillesen, P. (2016) A Cost-Effectiveness Analysis for Incineration or Recycling of Dutch Household Plastics, Tinbergen Institute Discussion Paper TI 2016-039/VI, https://research.vu.nl/en/publications/a-cost-effectiveness-analysis-for-incineration-or-recycling-of-du.
55 BASF (2020). ChemCyclingTM: environmental evaluation by life cycle assessment (LCA). https://www.basf.com/global/en/who-we-are/sustainability/we-drive-sustainable-solutions/circular-economy/mass-balance-approach/chemcycling/lca-for-chemcycling.html (accessed 29 October 2021).
56 Posen, I.D., Jaramillo, P., and Griffin, W.M. (2016). *Environmental Science and Technology* 50: 2846–2858.
57 Martens, J.A., Bogaerts, A., De Kimpe, N. et al. (2017). *ChemSusChem* 10: 1039–1055.
58 Alliance to end plastic waste, https://endplasticwaste.org/ (no date available).

13

Life Cycle Assessment of Bio-Based Plastics: Concepts, Findings, and Pitfalls

Li Shen

Utrecht University, Copernicus Institute of Sustainable Development, Group Energy and Resources, Princetonlaan 8a, 3584 CS Utrecht, The Netherlands

13.1 Introduction and Chapter Learning Objectives

The aim of this chapter is to introduce the environmental impacts of bio-based plastics. This chapter begins by a brief recap of the different groups of bio-based and biodegradable plastics (Section 13.2), followed by the principle of LCA (life cycle assessment), its history, international standards and norms commonly practiced, and the role of LCA in various decision-making contexts (Section 13.3). Next, seven LCA case studies of bio-based products are described in Section 13.4. These bio-based products are made from novel bio-based and/or biodegradable polymers. Their environmental impacts are then compared with the petrochemical competitors. The case studies provide the statue-of-the-art understanding of the environmental impacts of bio-based plastics as of 2019–2020. Once we learn about the case studies, I invite the readers to jointly reflect: are the bio-based plastics mentioned in the case studies a game changer?

After completing this chapter, you should understand the following concepts:

- *Recap*. Bio-based and biodegradable
- The initial knowledge of LCA with the focus on the assessment of bio-based innovation.
- The environmental impacts of bio-based plastics: the "knowns" and the "unknowns."

13.2 "Bioplastics" Is a Confusing Term

There are two dimensions when "bioplastics" are discussed: *biodegradability* and *bio-based content*. These two properties should not be confused. Bio-based polymers are not necessarily biodegradable; vice versa, biodegradable polymers are not necessarily made from biological resources. Yes, this is still confusing. Let us use these two properties to categorize polymers.

Biodegradable Polymers in the Circular Plastics Economy, First Edition.
Edited by Michiel Dusselier and Jean-Paul Lange.

In the first dimension, we use the narrow definition of "biodegradability" by the European Standard EN 13432:2000, i.e. if the polymer has passed the biodegradability test according to the standard, we call it "biodegradable" in this context. If a polymer fails to pass the test, we categorize it as "non-biodegradable." However, it needs to be kept in mind that in reality, biodegradability is not a simple "yes" or "no" property but rather a property depending on the duration of degrading, the shape and size of the material, and the ambient conditions.

In the second dimension, based on whether and to which extent the carbon in the polymer is originated from biological resources, we categorize polymers into three groups: fully bio-based, partially bio-based, and fully fossil fuel-based.

- *Fully bio-based polymers* in these polymers, all carbon (100%) in the polymer chain comes from biomass. For example, both PLA (polylactic acid or polylactide) and PHA (polyhydroxyalkanoates) are made from biochemical conversion of sugar, starch, or vegetable oils. For instance, bio-based PE (polyethylene) is commercially produced via dehydration of bio-ethanol [1]; bio-based PP (polypropylene) can be made from cracking bio-based naphtha-like oil made from hydrotreatment of used cooking oil [2]. In this category, we can further distinguish these polymers based on their biodegradability property into two groups: (i) *fully bio-based and biodegradable* polymers: for example, PLA, PHA, and starch plastics, and (ii) *fully bio-based and non-biodegradable* polymers: for example, bio-based PE and bio-based PP; they are identical polymers with petrochemical PE and PP and offer the same functionalities and applications. The only difference is the origin of carbon.
- For *partially bio-based polymers*, the origin of carbon is a mixture of bio-based and fossil fuel-based. Also, some of these polymers are biodegradable and some are not. This group can again be further categorized into two subgroups:
  (1) *Partially bio-based and fully biodegradable polymers*. For example, thermoplastic starch is often blended with biodegradable polyesters (PBAT, polybutylene adipate terephthalate, or PBS, polybutylene succinate) to create desired strength, which offers wide applications such as film extrusion and injection molding. Such a blend consists of partial bio-based carbon and is fully biodegradable.
  (2) *Partially bio-based but non-biodegradable polymers*. For example, partially bio-based PET (polyethylene terephthalate), widely used to make beverage bottles, is made from bio-based ethylene glycol co-polymerized with petrochemical terephthalic acid. About 30% of the mass in the polymer originated from biological resources and the remaining is fossil fuel-based. Partially, bio-based PET is an identical polymer as petrochemical PET; they have the same chemical and physical properties and offer the same applications.
- *Fossil fuel-based but biodegradable polymers*. These polymers are mostly polyesters such as PBAT and PBS. They are made from fossil fuels but are designed to biodegrade; the carbon has 100% fossil fuel origin and the polymers are able to degrade via biological processes.

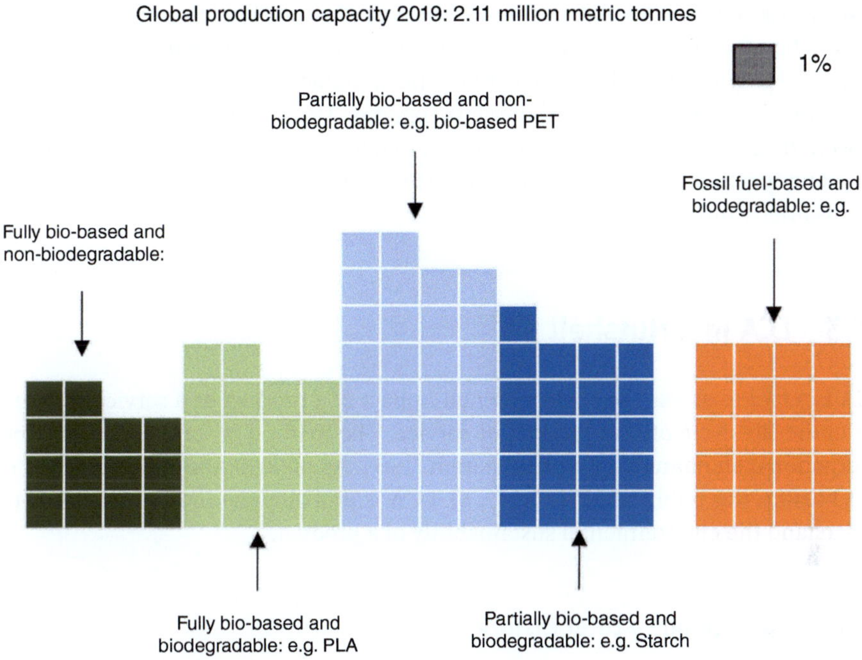

Figure 13.1 Share of global production capacity of bio-based and biodegradable plastics in 2019. Source: Compiled based on the data published by European Bioplastics [3].

In short, "bioplastics" refer to a mixed group of plastics made from biodegradable polymers *or* bio-based polymers, *or* sometimes both (see Figure 13.1). In 2019, the installed production capacity for these emerging groups of plastics was about 2.1 million metric tons (Mt) [3]. They fulfilled less than 1% of total polymers demanded presently in view of the annual global polymer production of 300–380 Mt. The share looks small, but it results from a very fast growing sector. In the past decades, the production volume has been increased drastically from 0.35 Mt in 2007 [4] to 2.1 Mt in 2019. This represented a 17% annual growth rate, whereas the global polymer market grew approximately 4% p.a. in the same period. The market demand is still growing strongly, primarily driven by two major factors: public concerns of poorly managed non-degradable plastic waste and the pressure from decarbonization of the material sector to combat climate change.

Many of these innovative bio-based plastics have the potential to offer GHG (greenhouse gas) emission savings, but they could also have other environmental trade-offs, and the savings can also depend on various factors, e.g. agricultural land management, energy and material efficiencies of production, responsible consumers, and effective waste management. To understand these potential trade-offs and to identify the key factors before making a decision, are more and more important for policy makers, company strategists and even for individuals. For this reason, the environmental impacts of the novel polymers need to be evaluated in a systematic way. LCA is a tool to carry out such evaluations.

Now, we know that "bioplastics" is not one type of plastic but has a myriad of polymers. When the environmental impacts are discussed, it is important to distinguish the type of polymer, their bio-based origin, the manufacturing processes, the logistic chains involved, and their end-of-life (EoL) options. These will be discussed in depth in Section 13.4 of this chapter. Before we go in depth into the case studies, we need to first understand the basic principles of LCA. In the following section, the method LCA is briefly introduced.

13.3 LCA in a Nutshell

LCA is a tool to assess the environmental impact of a product or a service through the entire life cycle of that product or service. The method is standardized by the international standardization organization (ISO). As of 2020, the valid standard is ISO 14040 series published in 2006 [5, 6]. LCA is probably the most applied tool to understand the environmental sustainability of a product.

13.3.1 Concept and a Brief History

The concept of LCA is straightforward. In LCA, the environmental impacts of a product are accounted for including all steps in the life cycle, i.e. raw material and resource extraction, chemical conversions, manufacturing steps, all logistic services in-between, storage, retailing, use phase, and EoL phase. In every step, energy, materials, and utilities are required, and emissions and waste are generated. In an LCA model, these data and information are gathered for all the steps through the life cycle. The resources were extracted from the environment and sent to the technosphere, and emissions released from the technosphere to the environment are quantified and then translated into environmental impacts (see Figure 13.2).

The earliest LCA-like research was probably the "Resource and Environmental Profile Analysis" carried out by the Midwest Research Institute for the Coca Cola company in 1969–1972 [7]. The energy crisis in the early 1970s strongly stimulated the research on energy balances and promoted system thinking, especially for the energy-intensive sectors. In 1977, the Swiss "Ecoscarcity" method was published; this was the first life cycle impact assessment method. In 1992, the first scientific guidance to conduct LCA, the Centrum voor milieuwetenschappen Leiden, in English: Institute of Environmental Sciences (CML) guide, was published by the University of Leiden. The world did not wait too long before the first international standards about environmental system management were made available. The ISO published the first 14 000 series in 1997. The current standards in use (ISO14040/44) were published in 2006 and are still valid to date. An important milestone in LCA was probably the International Reference Life Cycle Data System (ILCD) Handbook, published by the European Commission Joint Research Center in 2010 [8], where very detailed guide for conducting, reporting, and reviewing LCA is provided. This document is often considered by many practitioners as the "cook book."

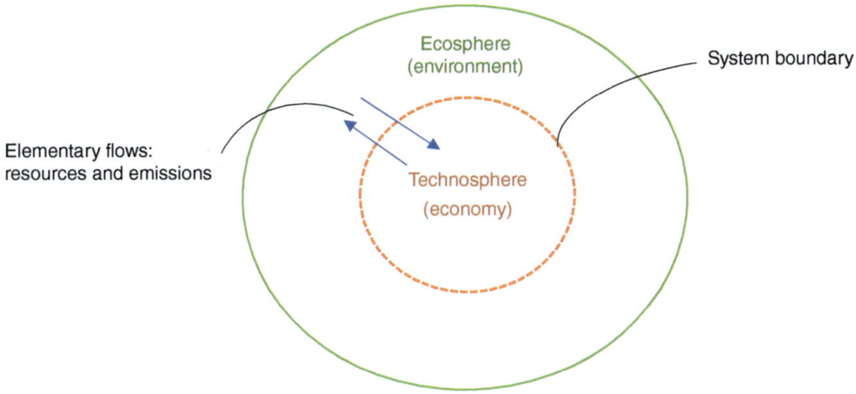

Figure 13.2 LCA is a tool to assess the environmental impact of the life cycle of a product by understanding the environmental interventions between the technosphere and the eco-sphere.

In 2013, the European Commission published the Product Environmental Footprint (PEF) method and the Organizational Environmental Footprint (OEF) method as one of the important actions of the EU initiative of Single Market for Green Products [9]. This document provides a common and consistent LCA methodological guidance for companies across different EU member states to market their products as environmentally friendly. For different markets and sectors, specific rules are made using the EF (environmental footprint) methods as the basis. These rules are called "Product Environmental Footprint Category Rules" (PEFCR) [10]. This is an important step forward to enable companies to use LCA to manage the supply chains and to appropriately market their products. In Section 13.4 of this chapter, we will introduce seven LCA case studies following some of the recommendations of the latest PEFCR.

In the past decades, LCA has been widely applied in many areas including but not limited to cooperate strategy, purchase decisions for individuals, corporates as well as public sectors, long-term policy recommendation, product and process eco-design, and supply chain management. The European Commission considers LCA as the "best framework for assessing the potential environmental impacts of products" [11].

13.3.2 Procedure, Jargons, and Sciences Behind

There are four steps to conduct an LCA, i.e. (i) *goal and scope definition*, (ii) *inventory analysis*, (iii) *impact assessment*, and (iv) *interpretation* [5]. In this section, the basic concepts and the most used terms are explained. It is important to understand some of those "jargons" because these are the terms used by professional practitioners and decision-makers all over the world to communicate LCA. It is also important to know what should be expected during each of these steps when we read, review, and interpret an LCA.

13.3.2.1 Goal and Scope Definition

Like any scientific research, the first step involves the research motivation. The ISO standard defines four requirements for the goal definition in an LCA (see Box 13.1). The motivation of the research is reflected by elements 2 and 3 (in Box 13.1). A well-reflected motivation is crucial for an LCA: the motivation of LCA determines, among many decisions during the research, the LCA decision context, the analytical scope, the types of inventory model, the choice of allocation method, tolerated data uncertainties, and the coverage of impact categories.

Box 13.1 The four elements in the goal definition of an LCA

1. *What* is the LCA about?
2. *Why* is this LCA needed?
3. *Who* are the audience of this LCA?
4. *By whom* is the study conducted?

An important concept in the goal and scope definition is *function unit*. A functional unit is the "quantified performance of a product system for use as a reference unit" [6]. It is a reference unit to be used to compare various options.

After the functional unit is defined, the scope of the LCA should be defined and described. This includes the identification of product systems, defining the geographical scope, temporal scope, technological scope, and the environmental impact categories (see also Section 13.3.2.3) [6].

13.3.2.2 Life Cycle Inventory Analysis (LCI)

In the second step of the LCA, data are gathered and verified and the inventory model is established based on the scope defined in goal and scope definition. The step of life cycle inventory (LCI) often represents one of the most time-consuming step for practitioners.

This step starts with the understanding how things are made. A flow sheet is often drawn in the first step to reflect the understanding. Figure 13.3 shows a simplified example of a flow sheet. A flow sheet consists of many processes in the life cycle of a product. The smallest process is called *unit process*. Unit processes are connected into a system that delivers the quantified function (i.e. functional unit) defined earlier. Such a system is called *product system* in LCA. *System boundary* is drawn to separate the economic activities (inside of the system boundary) from the environment (outside of the system boundary). The system boundaries can be adjusted to the question to answer (see Box 13.2). Substances that cross the system boundary (e.g. extraction of natural resources as arrows toward inside or emissions to the atmosphere as arrows toward outside) make *environmental intervention*. These substances crossing the system boundary are also called "*elementary flows*." A list of elementary flows that corresponding to the quantity defined by the functional unit forms the *life cycle inventory table*. This table is the aim of LCI. The inventory analysis has a massive

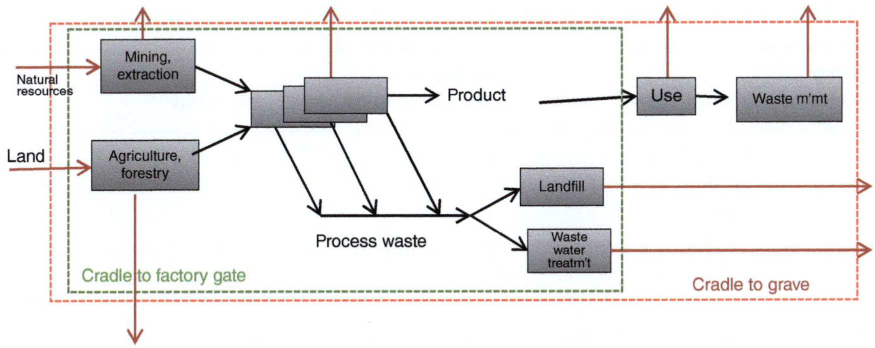

Figure 13.3 An example flow diagram in LCA with different system boundaries.

data processing requirement. The data processing is often carried out by database tools and LCA software.

Box 13.2 Cradle-to-factory gate and cradle-to-grave

A complete LCA should always be a cradle-to-grave LCA including the entire life cycle of the manufacturing, use, and post-consumer disposal phases (see Figure 13.3). However, in reality, many LCAs are conducted with the goal of assisting purchase decision in order to support supply chain management or to identify the significant environmental issues for an intermediate product instead of an end product. In those cases, a "cradle-to-factory gate" or a "cradle-to-user" systems is often analyzed, where the use phase and the EoL phase are not directly relevant for the goals of those LCAs.

13.3.2.3 Life Cycle Impact Assessment (LCIA)

The third step in LCA is life cycle impact assessment (LCIA). In this step, we move from the physical flow analysis in the technosphere to the environmental impact assessment. The aim of LCIA is to translate the environmental interventions (i.e. the result of inventory analysis) into environmental impacts. This is done in two steps: *classification* and *characterization*.

In the *classification* step, the inventory results are categorized into different environmental impact categories. For example, CO_2, N_2O, and CH_4 are GHGs and are classified into "Climate Change"; SO_2 and NO_x make the environment more acidic and are classified into "Acidification." In Appendix 13.A, the commonly used 16 impact categories are listed as recommended by the European Commission's EF's method [9]. Different *impact assessment methods* define different impact categories and their assessment frameworks. Table 13.1 shows a list of selected impact assessment methods commonly used by LCA practitioners.

In the characterization step, the classified inventory results are then quantified into environmental impacts. This step is determined by characterization models. The characterization models are cause–effect models. The outcome of these models is the so-called characterization factors, namely, a set of conversion factors to translate

Table 13.1 Commonly used impact assessment models and their references.

Name of the method	Developed by	Number of impact categories	Web links[a)]
EF	Joint Research Center, European Commission [12]	16 mid-point	https://eplca.jrc.ec.europa.eu/EFVersioning.html
ReCiPé	NL National Institute of Public Health and the Environment [13]	18 mid-point and 3 end point	http://www.rivm.nl/en/life-cycle-assessment-lca/recipe
CML	University of Leiden [14]	11 (baseline) mid-point	http://www.universiteitleiden.nl/en/research/research-output/science/cml-ia-characterisation-factors
TRACI	US EPA [15]	8 mid-point	http://www.epa.gov/chemical-research/tool-reduction-and-assessment-chemicals-and-other-environmental-impacts-traci
LIME	LCA Society of Japan [16]	7 end point	https://lca-forum.org/english/lime

a) Many LCIA methods constantly update characterization factors based on the latest knowledge in environmental impact assessment. Always check the latest available version when practicing LCA.

the inventory results into environmental impacts. In detail, an environmental impact is calculated by multiplying the inventory flows (emissions and resources) with the corresponding characterization factors of that impact category. The total environmental impact of that category is the sum of the impacts of all inventory flows that are classified to that impact category.

13.3.2.4 Interpretation

Like many scientific methods, LCA requires iteration. This means that after new information is obtained, the previous step(s) need to be revisited, and assumptions and data quality are reflected and adjusted wherever needed. The interpretation could occur in any steps in LCA. Most importantly, the interpretation of LCA results requires to answer the research questions and to evaluate and reflect the certainty and the quality of the answers. In this step, the consistency and completeness of the study are checked, and the analyses of the results are expected. Finally, like any scientific research, based on the results and interpretation, conclusions are drawn, limitations are pointed out, and recommendations are made.

13.4 LCA Case Studies of Seven Single-Use Plastic Items Made from Bio-Based Resources: Highlights and Lessons Learned

13.4.1 Background, Aim, and Scope of the BIO-SPRI Study

In this section, we will introduce seven LCA case studies of single-use plastic items made from bio-based resources and compare them with the petrochemical counterparts. The seven cases are summarized in Table 13.2. The seven cases cover three major commercialized innovative bio-based polymers, namely, bio-based PET (partially bio-based but not biodegradable), PLA (bio-based and biodegradable), and starch plastics (partially bio-based and biodegradable).

Table 13.2 Overview of the seven case studies, functional units, and product systems.

Case studies	Functional units	Bio-based product systems	Reference systems: petrochemical counterparts
Beverage bottles	Packaging of water in 100 bottles each 0.5 l providing a shelf life of at least 9 mo	Partially bio-based PET	PET
Single-use drinking cups	1000 single-use drinking vessels each for 200 ml of cold beverage	PLA from corn and sugarcane	PET and PP
Single-use cutlery	1000 sets of disposable cutleries each consisting of a knife, a fork, and a soup spoon		PS
Food packaging films	100 m^2 of transparent film packaging for fresh vegetables for 1 wk		PP
Horticultural clips	45 000 single-use clips used for horticultural purposes for 1 ha of land	Starch plastics[a)]	PP
Agricultural mulch films	Providing field mulching for 1 ha of land for 6 mo		LDPE
Single-use carrier bags	One single-use all-purpose lightweight plastic carrier bag with a volume of 20 l and 10 kg weight holding capacity		LDPE

a) Starch plastics is a generic term for a family of blends. For the three cases based on starch plastics, the compositions of these starch plastics are different.

The study was asked by the EU policy makers from the Commission's Directorate-General Research and Innovation (DG-RTD). The full project "Support to Research and Innovation Policy for Bio-based Products" (abbreviated BIO-SPRI) was conducted in 2017–2018. The policy makers asked for science-based evidence: are there environmental benefits if bio-based plastics are promoted? What are the critical environmental trade-offs? In the BIO-SPRI project, the seven cases were selected out of 20 candidate cases based on five criteria: market potential, promise for deployment, available LCA data, innovation, and potential sustainability benefits.

The aim of BIO-SPRI is to provide facts and evidence to support future policies about bio-based plastics, particularly in the context of supporting the implementation of the EU Plastic Strategy [17]. The goal of LCA was then further specified in twofolds: (i) to identify the key environmental hot spots of innovative bio-based plastics and (ii) to compare the environmental impacts with the petrochemical counterparts.

The research project, in a 745-page report, was published in February 2019 [18]. In this section, we will give a summary of this study and highlight the most important data and assumptions, results, and interpretation. The limitations of the study, the reflection of the methodology, and the lessons learned are summarized in Section 13.5.

The ISO standards [5, 6] were followed in general. Specifically, the study followed as much as possible the PEFCR guidance [10]. It needs to be kept in mind, even with the help of methodological guidance and protocols, in real LCA practice, there are often situations that are more complexed and require programmatic solutions. Some of these situations are highlighted in this section.

All cases were analyzed with the scope "**cradle to grave**" (see Figure 13.3) to align with the goal of LCA, i.e. EU-level policy support. The **geographical scope** includes the products sold, consumed, and disposed of in Europe. The supply chain of the product however, reflected the status quo reality, which is often global. For **technological scope**, only commercialized technologies that have been deployed by the industrial scale production were analyzed. Technologies under development were excluded from the scope. The **temporal scope** reflected the status quo production, use, and waste management, with a short-term vision into 2020–2025.

The LCAs were modeled in two software: SimaPro version 8 (simapro.com) from cradle to user and EASETECH (easetech.dk) for the EoL waste management. The background data were taken from the state-of-the-art LCI databases as of 2018, such as Ecoinvent (version 3.4), Agrifootprint (version 3), Gabi (version 2017), PlasticsEurope's Eco-profiles (2012–2017), and EASETECH. All detailed data and assumptions can be found in the report of the BIO-SPRI study [18].

The inventory models were built on a modular basis. For bio-based product systems, five modules were modeled separately for (i) biomass feedstock acquisition, (ii) polymer and material production from biomass, (iii) plastic conversion into end products, (iv) distribution to end user, and (v) the EoL waste management. The impact of the use phase of the packaging products was considered negligible.

Sixteen *environmental impact categories* recommended by the PEFCR (see Appendix 13.A) were analyzed. The study also included the *normalization* and *weighting* steps (see Box 13.3) in order to come to a single score of environmental impacts to simplify and to facilitate the communication. The normalization and weighting factors applied in this LCA can be found in Appendix 13.B.

Box 13.3 Normalization and weighting in LCA

Normalization and *weighting* are optional steps in LCA, according to the ISO [6]. Normalization is a step to normalize the LCA results (e.g. impacts at the mid-point) to the total environmental impacts of that impact category in a defined region and time. Appendix 13.B shows an example of normalization factors for per capita EU citizen using the EF impact assessment method developed by the European Commission in 2020. Normalization step can be done by dividing the mid-point LCA results by the normalization factors. The normalized results show the relative contribution of the environmental burden associated with the functional unit investigated to the overall environmental burden in that region. Normalization does not involve subjective opinions.

Weighting, on the contrary, involves subjective opinions about the relative importance among different impact categories. Sometimes in policy making and communication, it is difficult to come up with an overall conclusion if the environmental impact is higher in some categories and lower in other categories. To determine the weighting factors, multi-criteria approaches with stakeholder engagements are often used. The weighting factors are influenced by public concerns as well as the concerns from specific stakeholders (i.e. it depends on from whom the opinions were collected). In Appendix 13.B, an example of the weighting factors based on per EU citizen is given using the EF impact assessment method [19]. In that example, the EU citizen considered climate change a more urgent issue than human toxicity.

13.4.2 Key Findings

In this section, we first discuss the key findings of the seven bio-based products. The comparison with their fossil fuel counterparts can be found in Section 13.4.3.

Figure 13.4 shows the *normalized* and *weighted* LCA results for seven bio-based products, breaking down into major life cycle stages. For all seven cases, the "polymer and/or material manufacturing" phases play a dominant role. The plastic conversion and distribution also have significant contributions to the overall impact. The contribution from the biomass feedstock acquisition is high for beverage bottles (partially bio-based PET) but less significant for the other six products. The impacts from the EoL waste management stage depend strongly on the type of waste management assumed, the type of application, and the properties of the polymers, which is further interpreted in Section 13.4.2.4.

For all seven bio-based products, the most important environmental impact categories identified are climate change, resources depletion (fossil fuels), and human

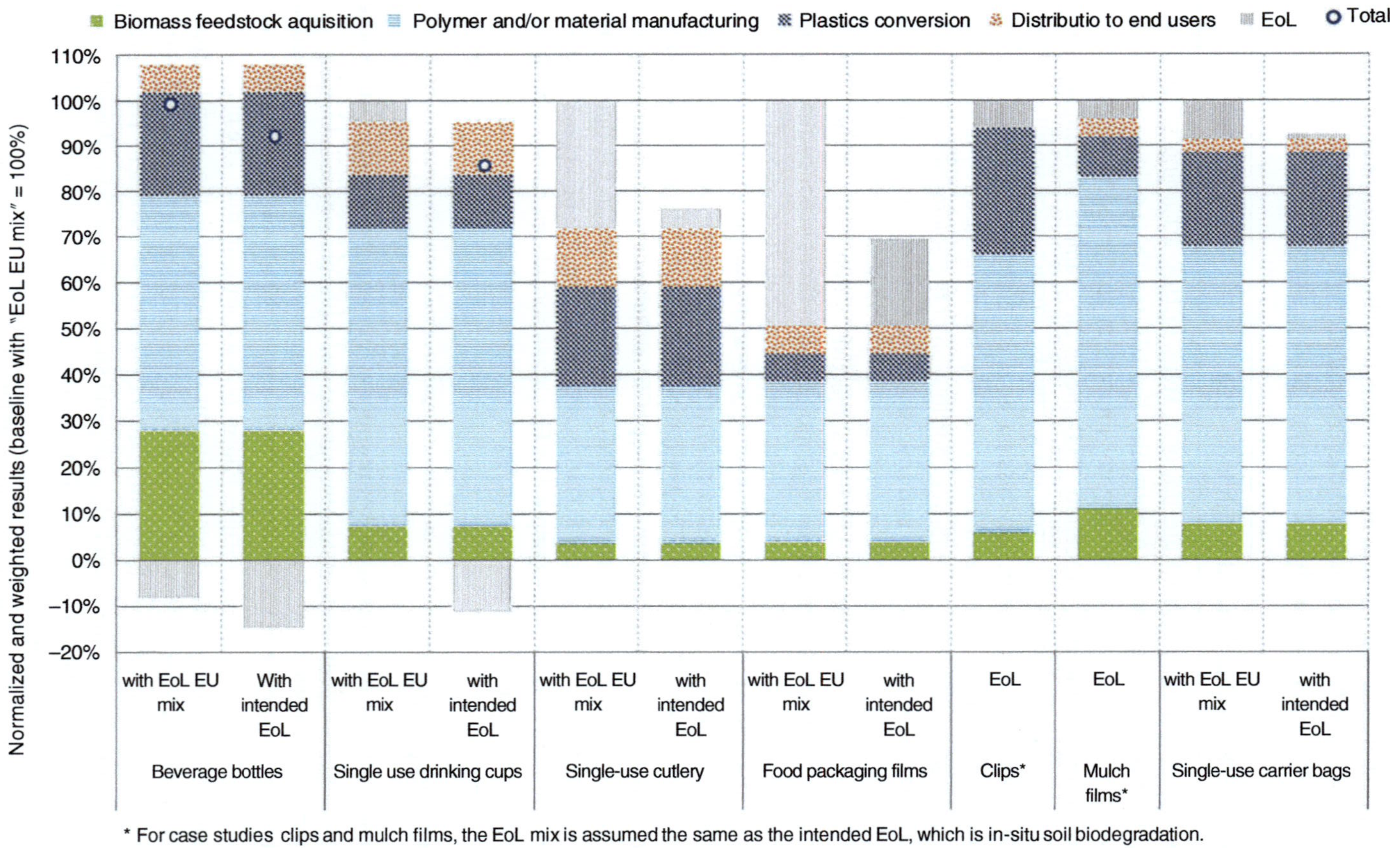

Figure 13.4 Cradle-to-grave environmental impacts of seven bio-based products, weighted results based on 16 PEFCR impact categories, breaking down into major life cycle stages. For each case, the baseline results (with EoL EU mix, see the detailed assumptions in Section 13.4.2.4) are then compared with the results based on intended end-of-life scenarios (intended EoL) to show the effects of different EoL waste managements. LCA results here are without taking into account the effect of direct or indirect land use changes. Source: COWI, DG RTD, and Utrecht University [18] / European Union / CC BY 4.0.

toxicity (see Figure 13.5). Together, these three impact categories account for approximately 30–60% of the total cradle-to-grave environmental impacts of the bio-based products. These three impact categories are highly associated with (predominantly fossil fuel) energy use in the manufacturing and distribution phases, as well as the direct air emissions from the EoL phase. We should also be aware that climate change is weighed the most urgent environmental issue based on the views provided by the EF method (see Appendix 13.B). In the following sections, the detailed impact breakdown per life cycle stages and the key data and assumptions are explained further.

13.4.2.1 Biomass Feedstock Acquisition

Sugarcane and corn are the two most important crops for the seven bio-based products studied (see Table 13.3). The LCA results presented in Figure 13.4 show that, for six out of seven bio-based products, the biomass production phase does not represent a major impact for the bio-based plastics in this study; in general, it accounts for less than 10% of the overall cradle-to-grave environmental impacts. The exception is the beverage bottles made from (partially) bio-based PET, where the high impact from biomass production (c. 28% of total cradle-to-grave impacts) is primarily contributed by the particulate matter, caused by the open-air cane residue burning in sugarcane harvesting in Brazil. It should be noted while this study was conducted in 2017–2018, Brazil is in the meantime phasing out the burning of cane trash by legislation. By 2031, the burning of cane trash should be fully phased out in the country (São Paulo State Law n. 11241/02) [20]. Thus, the impact of particulate matter is expected to be substantially reduced in the near future for sugarcane ethanol from Brazil.

Apart from the particular matter impact of sugarcane from Brazil, generally sugarcane is considered a more efficient feedstock compared to corn when fermentable sugar is required as the feedstock. For sugarcane, sucrose can be directly obtained by simple mechanical pressing and extraction, with a substantial energy benefit of using bagasse to fuel the milling and fermentation processes. For starch crops, glucose or dextrose needs to be first liberated from starch by wet or dry milling, which involves multiple-step conversions and separations, accompanied by many co- and by-products (e.g. proteins, fibers, and oils). Nevertheless, for the

Table 13.3 Biomass feedstocks used for the seven bio-based products.

Bio-based or partially bio-based polymers/materials	For the bio-based content: biomass feedstock and origin
Partially bio-based PET	Sugarcane from Brazil
PLA	Corn from the USA (67%) and sugarcane from Thailand (33%)[a)]
Starch plastics	Corn from the EU

a) The share represents the production capacities of the two major producers of PLA as of 2018.

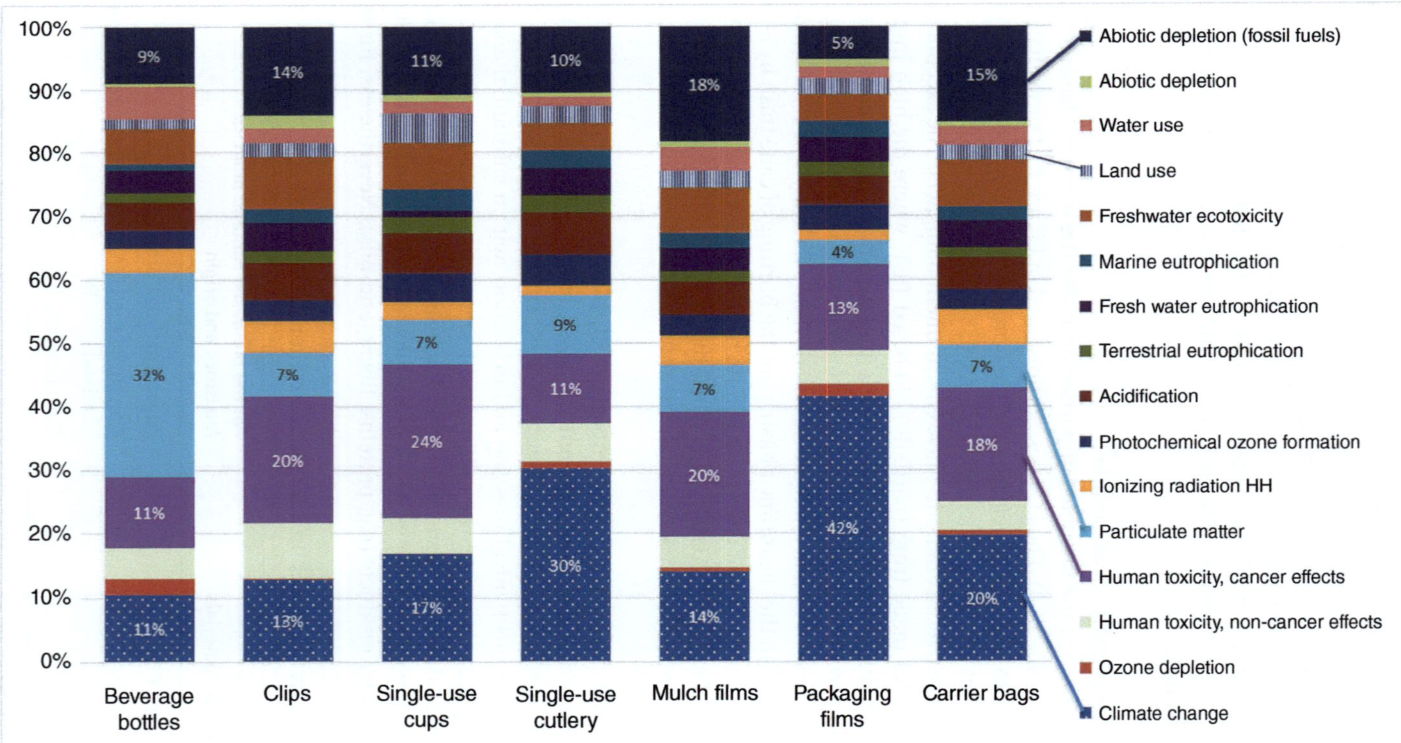

Figure 13.5 Cradle-to-grave environmental impacts of seven bio-based products, breaking down into impact categories. Normalized and weighted results assuming EoL EU mix. LCA results here are without taking into account the effect of direct or indirect land use changes. Source: COWI, DG RTD, and Utrecht University [18] / European Union / CC BY 4.0.

PLA products analyzed in the study, we observed rather negligible differences between sugarcane and corn, especially when the comparison is made from the entire life cycle perspective – i.e. the environmental impact of PLA is dominated by the monomer and polymer production and the material manufacturing phase, not by the impact from biomass acquisition (see Figure 13.4 for single-use drinking cups, single-use cutlery, and food packaging films).

Similarly, the impacts from the biomass acquisition phase for the three starch plastic products, namely, clips, mulch films, and carrier bags, also represent a small share of the overall impacts. For these starch plastic products, starch is utilized in its polysaccharide form – no sugar is required to be liberated. This leads to a relatively low impact from the biomass feedstock acquisition.

For both sugarcane and corn, the geographical differences do not represent as an influential factor because the industries source their feedstock based on the highest available yields in the regions, although bio-based products typically require (more) land than the extraction of fossil fuels. Land use and the associated impacts are often specifically investigated in the literature. In this study, we found that the impacts from land use and water use are rather limited compared to the same impacts from the manufacturing phase (Figure 13.5). This is partly explained by the high yield of the crops, partly explained by the dominance of the monomer/polymer and material production for these bio-based products.

For the impact category "Climate Change," it is important to note that the *biogenic carbon* removals were accounted for in the cradle-to-factory gate models (see Box 13.4), and it assigns a credit to the climate change impact (embedded carbon in the product is taken as a carbon sink: CO_2 is removed from the atmosphere by photosynthesis). At the EoL phase, the embedded biogenic carbon (in CO_2 equiv) was then modeled as biogenic CO_2 emission if fully oxidized or partially oxidized and form CO and/or CH_4 or stored in the compost (no emissions). This choice also results in a relatively low climate change impact for the biomass acquisition phases and a relatively high climate change impact for the EoL phases for all the bio-based products analyzed.

Box 13.4 Biogenic carbon in LCA

For bio-based products, the carbon embedded in the products is originally sequestered by biomass via photosynthesis in the "cradle" phase. This is different from the carbon embedded in the fossil fuel products, which was buried millions of years ago and intensified via geological formations. These two types of carbon need to be distinguished in the inventory analysis in LCA because the climate change impacts caused are different. The current ISO standards do not provide specific rules how to deal with biogenic carbon. Here, we follow the terms defined by a widely used method to calculate the carbon footprint of a product – PAS 2050 (Public Available Specification) (Figure 13.6) [21].

(Continued)

Box 13.4 (Continued)

When the biomass absorbs CO_2 from the atmosphere for photosynthesis, this flow of CO_2 enters the system boundary as an elementary flow (see Figure 13.5). This inbound flow is called biogenic *carbon removal*. When the product is burned at the end of its life and let us assume that all carbon is fully oxidized, CO_2 is released into the atmosphere. This outbound elementary flow is called biogenic *carbon emission*. If a part of the biogenic carbon stayed for a longer period of time in the product, e.g. biogenic carbon in a wooden table, this carbon is called biogenic *carbon storage*.

In inventory analysis, the principle to deal with biogenic carbon flows is to distinguish and report the flows separately [8–10, 21, 22]. The elementary flows of *carbon removals* and *carbon emissions* should be both assessed in the impact assessment for climate change. Note that *carbon storage* is not an elementary flow in LCA but it helps to create the correct overview of the carbon balance.

There are time-related issues about carbon removals (e.g. it takes time for forest to absorb CO_2) and carbon storage (e.g. delayed emissions could lead to reduced global warming effects). The time-related issues are relevant for assessing the climate change impact because climate change is often evaluated within a defined term, e.g. 100-year time frame. We will not expand the issue here. The current LCA norms, such as PAS 2050 and PEFCR, do not support the time-related assessment because the relevant research is still ongoing. The methodology and time-related dynamic databases that can be used for LCA are under development.

In addition, the impacts associated with *direct and indirect land use changes* (see concepts explained in Box 13.5) were investigated in this study. The study modeled the emissions from direct land use change (dLUC) as defined by PAS 2050 as well as the PEFCR guidance [10]. Indirect land use change (iLUC) was modeled using a deterministic model based on the statistical evidence observed in the past 10 years in the world for arable land converted via deforestation (expansion), e.g. using the Food and Agriculture Organization of the United Nations (FAO) statistics and literature knowledge, and to estimate the emissions due to the agricultural intensification, e.g. using the historical fertilizer statistics published by IFA (International Fertilizer Association). iLUC is not a compulsory step currently in LCA (see Box 13.5). It was modeled as an optional component in the study.

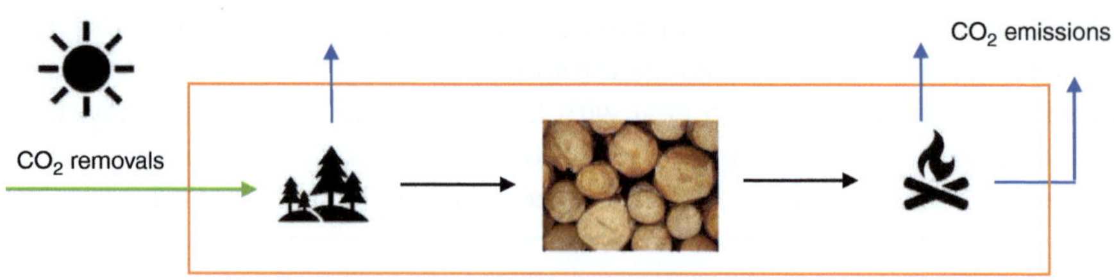

Figure 13.6 The biogenic carbon cycle: removals, emissions, and storage.

Box 13.5 Direct and indirect land use changes

Land use change is another important issue for bio-based production. Soil contains large amounts of organic carbon. When land use changes, the soil organic carbon content will also change. For example, if primary forest, which is organic carbon rich, is converted into arable land for crop production, the soil organic carbon content will decrease. The lost carbon is mostly emitted as CO_2. In LCA, this is called GHG arising from dLUC. The GHG emissions associated with land use changes was about 7% of global GHG emissions in 2017 [23].

Again, the current ISO standards do not provide specific rules about how to deal with emissions from land use changes. Both PAS 2050 and PEFCR recommend to account for GHG emissions related to *dLUC* if the land use change occurred on or after 1 January 1990. For the existing agricultural systems that are older than 30 years, the changes in soil organic carbon can be treated negligible [10, 21].

The concept of *iLUC* reflects the consequential thinking. The increase in demand of bio-based plastics creates pressure to the agricultural system, and some of the agricultural crops currently used for food or feed will be used for polymer production. In that case, the arable land is still an arable land; there is not dLUC attributed to bio-based plastics. However, the demand of food and feed remains the same. To meet the additional demand of bio-based plastics, additional land needs to be converted from primary forest. This is called *indirect land use change*. The carbon emitted from iLUC should be attributed to the bio-based plastic, which drives the additional demand of land. Alternatively, this additional demand could be met by intensifying agricultural outputs by, e.g. increasing yields. This is rather a limited response because yields cannot be increased infinitely. According to FAO statistics [24] from 2000 to 2010, about 85% of the additional demand of land was met by expansion, which leads to land use changes, and 15% of the additional demand was met by agricultural intensification.

In LCA, assessment of iLUC is not required by the current norms and specifications because it requires complexed global demand and supply models and databases. This is an important research area. Many high-resolution databases and methods are currently under development.

For the seven bio-based cases, an average iLUC factor of 4.0 t CO_2 equiv/ha/yr (with a range of 1.2–5.2 t CO_2 equiv/ha/yr) was found. The main impacts affected by the inclusion of iLUC were climate change and photochemical ozone formation. When the effects of land use changes are accounted for, the impacts of biomass production do increase. However, while the share of biomass production in the overall impacts was not dominant in the first place, the additional land use changes share to the overall cradle-to-grave impacts is relatively marginal: on average, 14% (4–17%) for climate change, 10% (4–30%) for photochemical ozone formation, and between 0.01% and 2.4% for all other impact categories.

13.4.2.2 Manufacturing Phase: From Biomass to Polymers, Materials, and End Products

The impacts of the manufacturing phase consist of the impacts from three sub-stages: the production of bio-based polymers (i.e. converting biomass into polymers), the production of (often) fossil fuel-based copolymers, and the plastic conversion step when the final products are formed. The first two sub-stages together are shown as "polymer and/or material manufacturing" in Figure 13.4.

The foreground data, especially for the processes involved in the conversion from biomass to platform chemicals and materials, compounding and plastic conversions, were the primary data collected from the key producers who have a dominant share of that product in the European market. The inventory data reflected the status quo as of 2018. If a product had more than one key producer, the weight average was formulated based on installed production capacities.

Figure 13.4 shows that the manufacturing phase has the highest contribution of all life cycle stage: it accounts for approximately 50% of the cradle-to-grave impacts of nearly all seven bio-based products studied. The contributions from the three sub-stages vary depending on the type of polymers or materials studied and summarized as follows.

- *Bio-based PET* is chemically identical with petrochemical PET. PET is a polyester made from polymerization of monoethylene glycol (MEG) and purified terephthalic acid (PTA). For the current bio-based PET available in the market, MEG is made from bio-based ethylene derived from sugarcane-ethanol, and PTA remains petrochemical based. The bio-based content is approximately 30% (by weight) in PET, representing the fraction of bio-MEG in the polymer. For (partially) bio-based PET, the manufacturing phase accounts for nearly three-quarter of the total impacts (see Figure 13.4), with the highest contributor of polymer production, which includes bio-MEG production, PTA production, and the polymerization process. These three production processes contribute to over 50% of the total cradle-to-grave impacts. The production of PTA, with a fossil fuel origin, has a dominant role. The second most important life cycle stage is the plastic conversion step (bottle blown molding), contributing to 23% of the total impacts. These impacts from the manufacturing phase are dominated by the process energy (heat and electricity) requirements.
- *PLA* (100% bio-based) is produced via polymerization of lactic acid or lactide. The monomers are products of sugar fermentation. For PLA, the conversion from sugarcane or corn into polymer PLA has the dominant impact, accounting for 40–60% of the total cradle-to-grave impacts of the three PLA case studies (i.e. single-use cups, single-use cutlery, and food packaging films). Process energy (electricity and heat) and chemical used in the lactic acid and PLA production are responsible for the major parts of the environmental impacts. The energy consumption is highly associated with the process water and its removal that needs to be dealt with in the biochemical conversion (fermentation of lactic acid). In the studied cases, a lion's share of the process energy (especially heat supply) is fossil fuel based, which contributes to the impacts of climate change and abiotic depletion (fossil fuels). The second most important contributor in the manufacturing phase of PLA is the

plastic conversion steps (e.g. thermal forming of cups, injection molding of cutlery, and film extrusion). Electricity is the key input of these conversion processes. Depending on the fuel mix of the electricity where the conversion step is located, the step could contribute to about 10% of the total cradle-to-grave impacts (e.g. single-use cups thermoformed in Europe) or up to 25% of the total impacts (e.g. cutlery molded in China where the electricity is more carbon intensive than the EU average).

- **Starch plastics** (partially bio-based) are starch thermoplastics (bio-based and biodegradable) blended with biodegradable petrochemical polymer(s). The bio-based content varies considerably depending on the formulation of the starch plastics for specific applications. For the three starch plastic applications studied, the bio-based contents range from 33% to 68%. The environmental impacts of studied starch plastics are dominated by the material manufacturing phase (c. 80% of the total cradle-to-grave impact), of which, the impacts from the fossil fuel-based co-copolymers are dominant (c. 50–60% of the total cradle-to-grave impacts). The impacts from plastic conversions (e.g. injection molding and film extrusion) account for 20–30% of the total impacts.

13.4.2.3 Distribution to End User: Impacts from Transportation

The impacts from transportation are in general not substantial (approximately 3–6% of the cradle-to-grave impact, see Figure 13.4). The exceptions are when the global supply chain is very long. For instance, based on the current supply chain of biomass, polymers, and end products, the bio-based carbon in the PLA cutlery travels nearly three quarters of the globe to come to the demand center – Europe. In this case, the impact of transportation is no longer negligible: about 13% of total impacts of the PLA cutlery are contributed by the transportation and distribution of the intermediates and the final product.

From cradle to user, the *energy requirements* along the chain, from process heat and electricity to transportation fuels, together play an important role. Energy consumption directly leads to the impact of fossil fuel depletion and climate change caused by CO_2 emitted from the combustion of fossil fuels (e.g. for heat and electricity). Combusting fossil fuels also lead to the impact of photochemical ozone formation (caused by nitrogen oxides emitted during fuel combustion) and particulate matter (dust emitted from coal combustion).

13.4.2.4 End-of-Life (EoL) Post-consumer Waste Management Scenarios

Modeling the EoL of niche products was challenging. The selected products are commercialized and have been produced on a large scale. However, compared to the petrochemical counterparts, the market share of these bio-based alternatives are still very small. European Bioplastics Association estimated about 1% of all plastics produced are bio-based (see Section 13.2). In the waste management stream, this amount has no statistical significance. Therefore, there was no so-called "status quo" primary data to model the EoL waste management of these bio-based products. The bio-based plastics waste could end up in the regular waste where they are intended for (e.g. recycling or composting) or where they are not expected (e.g. landfill or waste incineration). There is little empirical or statistical evidence to support either of these assumptions. In the BIO-SPRI project, to tackle this challenge, two EoL scenarios were distinguished:

(1) "EoL EU mix," where the plastics waste was assumed to end up in an average (mixture of) treatment that is most likely to happen to that packaging product. For example, for beverage bottles, both the bio-based and the petrochemical ones, the used bottles are likely to be disposed of for recycling, in a landfill or being incinerated in Europe (different member states have different waste collection schemes). The shares of the different waste treatments were determined using our exiting knowledge (from e.g. EuroStat) of the waste treatments of PET bottles in the EU.
(2) "Intended EoL," where the waste was assumed to end up in the ideal and desired waste treatments. For example, ideally, all PLA applications should all be either composted or recycled. Note that this is an extremely favorable ideal situation.

The reality is hard to predict. It highly depends on the waste management schemes in different member states as well as how well the consumer behaviors are steered. It is anticipated that the reality is probably somewhere in between the two scenarios.

An overview of the two EoL scenarios for the seven case studies are summarized in Table 13.4. The most foreground data were taken from the EU or country-level statistics and state-of-the-art literature data. The background data for the EoL modeling were taken from EASETECH databases and the modified Ecoinvent database. For the reference systems with petrochemical counterparts, in this book chapter, we only present the results of the average "EoL EU mix" as the benchmark (see Section 13.4.3).

It needs to be highlighted that for the two cases, "single-use cutleries" and "food packaging films" food residues were also assumed to be included in the product systems. Several studies [26–28] have reported that food leftovers often follow the packaging products and hence were often end up in the same EoL treatments as the packaging products. The impact of treating food residues was accordingly included in the system boundary.

The impacts of the *EoL* of the bio-based products studied are less dependent on the energy requirements of waste treatments but highly dependent on the type of waste treatment and the type of the plastic products. Except for PET bottles, the other six bio-based cases are all for biodegradable applications for which the materials are the newcomers for the existing waste management systems. For the six case studies where biodegradable materials are involved, the uncertainty of the EoL assessment is considered high.

- For *beverage bottles*, because PET is a currently highly recycled polymer in Europe (more than 60% collected for recycling in 2016 in Europe), the EoL waste treatment of PET bottles receives a net credit (see Figure 13.4) because they avoid the production of a virgin PET polymer. Based on the weighted LCA results for all 16 impact categories, the most favorable EoL technology of PET is recycling. By improving

Table 13.4 Key assumptions of the two scenarios of EoL waste management[a)].

Case studies (bio-based vs. reference)	Bio-based product systems: EoL EU mix	Bio-based product systems: Intended EoL	Petrochemical references, "EoL EU mix" as the benchmark EoL
Beverage bottles (bio-based vs. Pchem. PET)	Recycling 60% MSWI 20% Landfilling 20%	Recycling 100%	Recycling 60% MSWI 20% Landfilling 20%
Single-use drinking cups (PLA vs. PET, PP)	MSWI 39% Landfilling 31% Recycling 15%	Industrial composting 50% Recycling 50%	Recycling 30% MSWI 39% Landfilling 31%
Single-use cutlery (PLA vs. PS)	Industrial composting 15%	Industrial composting 100%	
Food packaging films (PLA vs. PP)		Industrial composting 100%	
Horticultural clips (starch plastics vs. PP)	*In situ* soil biodegradation 100%	*In situ* soil biodegradation 100%	MSWI 56% Landfilling 44%
Agricultural mulch films (starch plastics vs. LDPE)			Recycling 5% MSWI 53% Landfilling 42%
Single-use carrier bags (starch plastics vs. LDPE)	Industrial composting 30% MSWI 39% Landfilling 31%	Industrial composting 100%	Recycling 30% MSWI 39% Landfilling 31%

a) Data presented in this table were mostly based on European statistics [25]. In 2017, the average treatment for plastic waste within the EU member states was reported to be 39% MSWI (municipal solid waste incineration with and without energy recovery), 31% landfilling, and the remaining 30% was collected for mechanical recycling. We used these average shares to simplify the EoL modeling. The individual EoL technologies were modeled as an average European technology, e.g. indicating that incineration includes both with and without energy recovery and anaerobic digestion includes all combinations of dry/wet, thermophilic/mesophilic, and with/without post-maturation. It should be noted that there are differences between polymer types in terms of the actual recycling rates, which depend on the economic values of the recycled polymer and the recycling technology available for the polymer.

the 60% collection rate for recycling of today, to an idealistic 100% collection in the future, the overall impacts of the bio-based PET bottle could be further reduced by 7%. It should be noted that the difference between (partially) bio-based PET and petrochemical PET is the source of MEG. Both PET bottles do not differ in terms of EoL waste management.

- For *single-use cutleries and food packaging films*, where the reference flow includes food leftovers, the mixed EoL has a significant or even dominant contribution: this

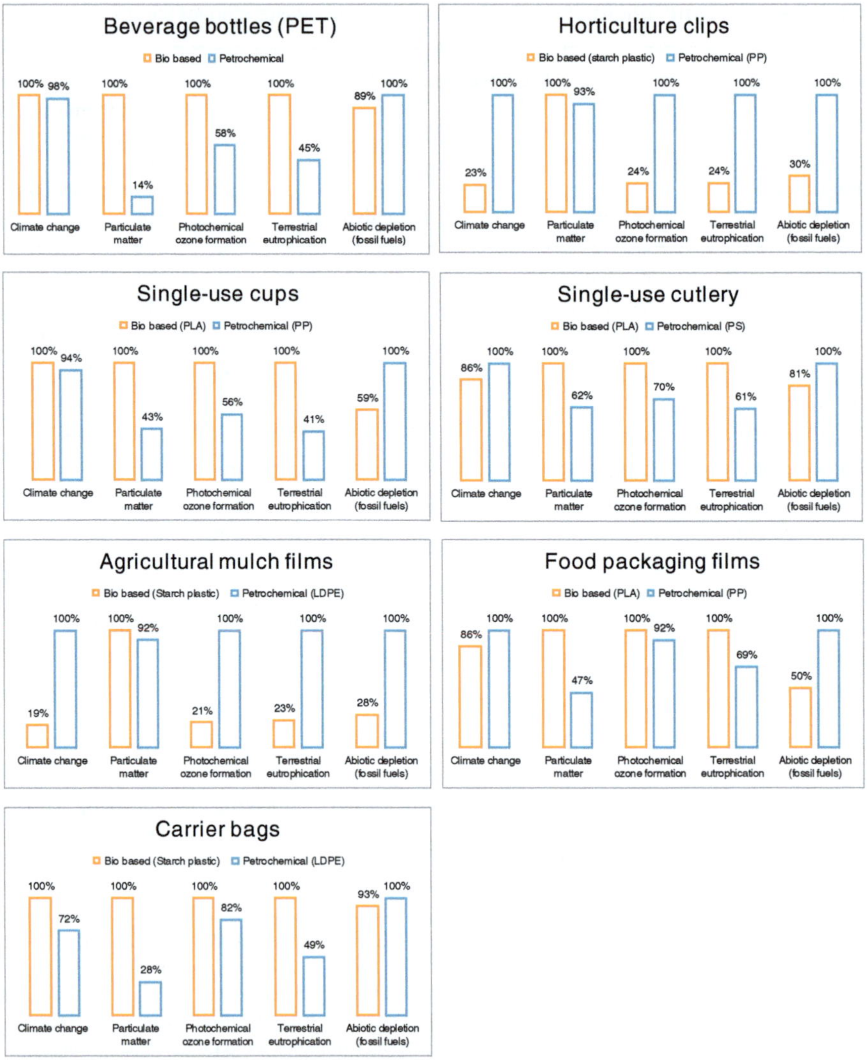

Figure 13.7 Comparing cradle-to-grave environmental impacts of the seven bio-based products with their petrochemical counterparts for five selected impact categories, results based on EoL EU mix only and without taking into account the effect of land use changes. The highest impact is normalized to 100%.

phase accounts for 30% and 50% of the cradle-to-grave environmental impacts for the two products, respectively. From Figure 13.7, it can be seen that if the "intended" EoL (industrial composting) is implemented, the overall impacts of PLA cutlery and food packaging films could decrease by 25% and 30%, respectively, largely because of the avoidance of methane emissions from landfilling. Therefore, the results of these two cases are highly sensitive to the assumed mix of waste management technologies as well as to the assumed organic waste ended up in the plastic waste streams.

- For *PLA cups and single-use carrier bags*, the differences between the "EoL EU mix" and "intended EoL" are less significant (see Figure 13.4). For starch plastics used for *mulch films* and *clips*, the impacts from EoL only contribute to 1% of the total cradle-to-grave impacts. Their EoL results are sensitive to the amount of soil assumed to be collected together with used plastics.

Overall, it can be concluded that recycling and industrial composting are more favorable than anaerobic digestion, incineration, and landfilling for the seven bio-based products studied.

13.4.3 Comparisons with Petrochemical Plastics

The types of petrochemical polymers were determined by industrial consultation and verified by market statistics. For petrochemical polymers, the average EU polymer production was assumed. The LCI data were obtained from the Eco-profiles published by PlasticsEurope [29]. The plastic conversion and final product manufacturing were modeled based on the status quo (as of 2018) plastic conversion technologies. Like the bio-based plastics, the use and disposal of the petrochemical plastics were assumed to take place in Europe. Note that because of the limited data availability, it is not possible to calculate the environmental impacts for all 16 impact categories for the comparison [18]. The Eco-profiles were compiled with industrial average data and were presented as "blackbox" data, which do not provide sufficient transparency subject to high-quality interpretation for all 16 impact categories [18]. The available transparency and data allowed us to compare only five impact categories, namely, climate change, particulate matter, photochemical ozone formation, terrestrial eutrophication, and abiotic depletion (fossil fuels), as shown in Figure 13.7.

From Figure 13.7, it can be seen that for all the seven cases, from cradle to grave, most of the bio-based products offer environmental benefits in two impact categories: *climate change* and *resource depletion fossil fuels*. All the seven bio-based products have higher impacts on *particulate matters* compared to the petrochemical counterparts. The higher impact of the particulate matter of bio-based products differs substantially from case to case. Bio-based PET has over six times higher impacts on particulate matter than the petrochemical PET because of the combustion of sugarcane trash in the biomass cultivation and harvesting phase. It should be noted that the burning cane trash will be soon phased out in Brazil. In the case of PLA food packaging films and cups, the particulate matter impacts are two times as much as the impact of their PP counterparts. PLA has a slightly higher energy requirement in the manufacturing phase and has much less environmental credits from the EoL phase compared to PP. For the remaining cases, the particulate matter is only marginally higher (3–8%) for the bio-based products than for their petrochemical counterparts.

For the remaining two impact categories, namely, photochemical *ozone formation* and *terrestrial eutrophication*, some bio-based products do not offer impact reductions, whereas for some other cases, substantial impact reductions are observed.

Table 13.5 Comparing cradle-to-grave environmental impacts of the seven bio-based products with their petrochemical counterparts and results indicated as environmental impact median savings of seven bio-based products (negative values stand for savings).

	Environmental impact categories	Cradle-to-grave environmental *savings* of bio-based products compared to petrochemical products, median values for seven cases	
		With EoL EU mix	With intended EoL
Lower	Climate change	−11% (−81% to 38%)	−61% (−87% to −14%)
	Resources depletion, fossil fuels	−45% (−72% to −7%)	−33% (−72% to −3%)
Higher	Particulate matter	94% (2% to 603%)	110% (6% to 597%)
Case specific	Photochemical ozone formation	23% (±79%)	19% (±79%)
	Terrestrial eutrophication	70% (−77% to 70%)	72% (−77% to 146%)

The comparisons between the bio-based product systems with *intended EoL* are listed in the second column. Savings are without taking into account the effect of iLUC.

Reasons are case-specific and no generic causes can be identified. It is advised to look into these impacts case by case for specific policy implications.

Table 13.5 shows that the choice of EoL has a strong influence on the overall impact savings of the bio-based products. The influence is especially significant for climate change impact. Treated with intended EoL, the bio-based products on average could offer more than 60% of the GHG emission savings compared to their petrochemical counterparts. This indicates a great potential of low carbon bio-based products if the EoL waste management is implemented appropriately.

13.5 Lessons Learned from the Case Studies and Looking Forward to a Circular Bio-Based Economy

From the seven case studies, we conclude that the bio-based products offer important environmental benefits to combat climate change and to reduce the limited fossil fuel resources, despite different assumptions about EoL waste management and alternative approaches of inventory modeling (e.g. to include the effect of iLUCs). This conclusion is in line with the findings from many previous LCAs [30–34].

However, like many LCA studies, the presented research struggled with limitation on data availability and LCA methodological issues. The uncertainties caused by these two issues should be taken into account when the results are interpreted and policy advices are given.

For the bio-based products, one of the key uncertainties is caused by unavailable data in the EoL waste management phase because there is little information about the new materials in the EoL waste streams. In the BIO-SRI project, this was dealt with by comparing average mix and an idealistic scenario (i.e. intended EoL). Even though many data were still absent, for example, assumptions needed to be made to anticipate the real compositions of different plastic items including unexpected additives and unknown food residues. The extremely heterogeneous waste collection, labeling, and management systems across different EU member states also made it challenging to come up with a representative average for a European level policy advice.

For the comparison with the reference cases based on petrochemical plastics, *data transparency* appeared to be the most crucial issue. The Eco-profiles published by PlasticsEurope are probably the most representative LCI data to assess the impacts of average EU polymer productions. However, the data were packed in a form of "black box" LCI because the information from real production facilities is often treated as commercial secrets that are carefully protected by the sector. The "packed" LCI data provided limited insights, for example, into allocation choices and data uncertainty interpretation. If the petrochemical polymers were to be used as the benchmarks for future policies of bio-based plastics, this benchmark needs to be more transparent to offer a prudent interpretation.

Next to the data availability and transparency issues, several *methodological issues* need to be tackled linked to the LCA of plastics. A few of those are highlighted below.

- The impact of land use changes remains a research hot spot in the field of LCA. In the case studies, attempt was made to assess the impacts from iLUC using a deterministic model. Well acknowledged data from the Intergovernmental Panel on Climate Change (IPCC) and the FAO were used as inputs. However, the method still relies on choices (e.g. share of intensification and expansion; type of intensification) that can greatly affect the results. The current LCA guidelines and norms do have the consensus on how iLUC should be modeled for bio-based production.
- *Littering* is nowadays one of the key environmental concerns both in the public and on the EU policy level, yet the exact amount (and relative share) of plastics that are littered is very uncertain and will largely differ among different countries, user behaviors, and end use applications. Also, the impacts of littering and microplastics are currently not properly captured in the toxicity-related impact categories (eco-toxicity and human toxicity, see Chapter 3) or impact categories related to eco-system services. There are methodological hurdles in impact assessment models. The fate, exposure, and effect models for macro- and microplastics (including additives) are still in the initial stage. Physical impacts (e.g. entanglement, ingestion of larger plastic particles, inhalation of micro- and nano-sized particles, and the effects), chemical impacts (e.g. because of microplastic formation, microplastics as carriers of other chemicals, and unknown impact of additives), and biological impacts (e.g. microplastics as carriers of germs/alien species) are largely unknown and are still not included in LCA. Therefore, for the time being, this is going to remain a major uncertainty in LCA of plastics. It thus will also remain a difficult trade-off to compare, e.g. the impacts of the feedstock production of biodegradable plastics (such as the bio-based starch film and clips) to the EoL impacts of their fossil counterparts. Having said all these, *bio-based and non-biodegradable plastics* will cause similar problems as their fossil counterparts in terms of the risk and the impact of littering.

- One general issue that is not taken into account in current LCA methodology is the *timing of emissions*, i.e. whether a GHG gas is emitted during or very soon after the use phase of whether this occurs only a long time after the production and use of the product. In the seven case studies investigated, average lifetimes are all very short (typically less than one year), but especially for more long-lived biomaterials and plastics (e.g. used in construction or automobile), this could lead to significant additional benefits of bio-based products. The issue of timing of emissions has been extensively discussed in the literature for bioenergy-related applications (especially, woody biomass for electricity and heat production, see e.g. the "carbon debt" described by Lamers and Junginger [35]) and should also be investigated in more detail for bio-based plastics.

Finally and, possibly, most importantly, the LCA analyses are typically bound to the present stage of technology and social/political choices (EoL). They, thereby, fail to show the potential of future technologies over the mature ones. The bio-based plastic industry is still in its initial stage. Most commercialized products are still "young" compared to the petrochemical counterparts. There are many ongoing efforts to optimize the manufacturing process of bio-based production. The efforts would likely lead to a lower environmental impact for the future bio-based products than today. The potential (decrease in) environmental impacts of bio-based plastics due to technological progress and the exogenously changing energy system should be assessed in the future. Ultimately, the assessment studies conducted today provide a starting point for the policy debate. Looking forward, the bio-based plastics will play an important role in reducing the environmental impacts, particularly in the battle of combating climate change. It is still not a silver bullet. Adopting bio-based plastics on a large scale in the future requires to understand the potential environmental trade-offs. LCA, which is still not perfect, does provide a good insight in the potential trade-offs. Toward a sustainable bio-based economy, both benefits and trade-offs of the innovative bio-based products need to be taken seriously in R&D decisions, individual consumer decisions, and ultimately policy decisions.

13.A General Structure of Classification and Characterization in LCIA, using the example of 16 Impact Categories Recommended by the EC EF (Environmental Footprint) Impact Assessment Methods

Name of mid-point impact category	Unit	Examples of inventory substances classified to this category	LCIA characterization models recommended by the EF method[a)]
Climate change	kg CO_2 equiv	CO_2, CH_4, N_2O, and F-gases, in total > 200 substances identified	IPCC fifth assessment report [36]
Ozone depletion	kg chlorofluorocarbon (CFC) – 11 equiv	CFCs, hydrofluorocarbons (HFCs), etc., in total 23 substances	World Metrological Organization (WMO) model [37]

Name of mid-point impact category	Unit	Examples of inventory substances classified to this category	LCIA characterization models recommended by the EF method[a)]
Human toxicity, non-cancer effects	CTUh	In total >15 k substances	USETox model
Human toxicity, cancer effects	CTUh	In total >9 k substances	USETox model
Particulate matter	kg PM2.5 equiv	Ammonia, SO_2, NO_2, NO_x, PM2.5, PM10, and regional specific impacts	RiskPoll model [38]
Ionizing radiation human health	kBq U235 equiv	Uranium 235, uranium 238	[39]
Photochemical ozone formation	kg non-methane volatile organic compounds (NMVOC) equiv	SO_2 (regional specific impacts), volatile organic compounds	CML model [40]
Acidification	Molc H^+ equiv	Ammonia, SO_2, NO_x, etc., in total 194 substances	CML model [41]
Terrestrial eutrophication	Molc N equiv	Ammonia, NO_x, regional specific impacts	CML model [41]
Freshwater eutrophication	kg P equiv	Phosphate, phosphoric acid, and phosphorous	ReciPe model [42]
Marine eutrophication	kg N equiv	Ammonia, NO_x, nitrates, nitrites, and regional specific impacts	ReCiPe model [42]
Freshwater eco-toxicity	CTUe	Organics, inorganics, and metals	USETox model
Land transformation	kg C deficit	Agriculture, forest, grass land occupations, and transformation in total 160 items	Soil organic matter model [43]
Water use	m^3 equiv	Regional specific impacts for surface water and groundwater	AWARE model [44]
Resource depletion, mineral	kg Sb equiv	Minerals and metals	CML model [45]
Resource depletion, fossil fuels	MJ equiv	Coal, crude oil, natural gas, and uranium	CML model [45]

a) This is an ongoing research area. The impact models are regularly updated (with the same impact categories). Always check the latest version of the models when performing an LCA.

13.B Normalization and Weighting Factors Recommended by the EF (Environmental Footprint) Method [12, 19, 46], Latest Update: May 2020

Impact categories	Unit for normalization factors	Normalization factors	Weighting factors (%)
Climate change	kg CO_2 equiv/person	8.10E+03	21.06
Particulate matter	Disease incidences/person	5.95E−04	8.96
Water use	m^3 water equiv of deprived water/person	1.15E+04	8.51
Resource use, fossils	MJ/person	6.50E+04	8.32
Land use	pt/person	8.19E+05	7.94
Resource use, minerals, and metals	kg Sb equiv/person	6.36E−02	7.55
Ozone depletion	kg CFC-11 equiv/person	5.36E−02	6.31
Acidification	mol H+ equiv/person	5.56E+01	6.20
Ionizing radiation, human health	kBq U-235 equiv/person	4.22E+03	5.01
Photochemical ozone formation – human health	kg NMVOC equiv/person	4.06E+01	4.78
Eutrophication, terrestrial	mol N eq./person	1.77E+02	3.71
Eutrophication, marine	kg N equiv/person	1.95E+01	2.96
Eutrophication, freshwater	kg P equiv/person	1.61E+00	2.80
Human toxicity, cancer	CTUh/person	1.69E−05	2.13
Eco-toxicity, freshwater	CTUe/person	4.27E+04	1.92
Human toxicity, non-cancer	CTUh/person	2.30E−04	1.84

References

1 Tsiropoulos, I., Faaij, A.P.C., Lundquist, L. et al. (2015). Life cycle impact assessment of bio-based plastics from sugarcane ethanol. *Journal of Cleaner Production* 90: 114–127.

2 Moretti, C., Junginger, M., and Shen, L. (2020). Environmental life cycle assessment of polypropylene made from used cooking oil. *Resources, Conservation and Recycling* 157: 104750.

3 European Bioplastics (2020). *Bioplastics: Facts and Figures for 2019*, p. 16.

4 Shen, L., Worrell, E., and Patel, M. (2010). Present and future development in plastics from biomass. *Biofuels, Bioproducts and Biorefining* 4 (1): 25–40.

5 ISO 14040 (2006). *Environmental management — Life Cycle Assessment — Principles and Framework. Technical Committee ISO/TC 207*. Environmental Management 3 (1): 28.

6 ISO 14044 (2006). *Life cycle assessment — Requirements and guidelines*. International Organization for Standardization, p. 14044.

7 Baumann, H. and Tillman, A.-M. (2004). *The Hitch Hiker's Guide to LCA*. Lund: Studentlitteratur AB.

8 European Commission (2010). *ILCD Handbook: General guide for Life Cycle Assessment – Provisions and Action Steps*. Ispra: European Commission, Joint Research Centre, Institute for Environment and Sustainability.

9 European Commission (2013). Recommendation 2013/179/EU on the use of common methods to measure and communicate the life cycle environmental performance of products and organisations. *Official Journal of the European Union* (L 124): 210.

10 European Commission (2018). *PEFCR Guidance document*. Guidance for the development of Product Environmental Footprint Category Rules (PEFCRs), version 6.3, December 14 2017.

11 European Commission (2003). Communication COM (2003:302) - Integrated Product Policy - Building on Environmental Life-Cycle Thinking. *Commission of the European Communities*, 30.

12 Fazio, S., Biganzioli, F., De Laurentiis, V. et al. (2018). Supporting information to the characterisation factors of recommended EF LCIA methods.

13 RIVM (2017). ReCiPe 2016 v1.1. *RIVM Rep. 2016-0104*.

14 Guinée, J.B., Gorrée, M., Heijungs, R. et al. (2002). *Handbook on Life Cycle Assessment. Operational Guide to the ISO Standards. I: LCA in Perspective. IIa: Guide. IIb: Operational Annex. III: Scientific Background*. Dordrecht: Kluwer Academic Publishers.

15 Bare, J. (2011). TRACI 2.0: The tool for the reduction and assessment of chemical and other environmental impacts 2.0. *Clean Technologies and Environmental Policy* 13 (5): 687–696.

16 Inaba, A. and Itsubo, N. (2018). Preface. *International Journal of Life Cycle Assessment* 23 (12): 2271–2275.

17 European Commission (2018). Communication from the Commission to the European Parliament, the Council, the European Economic and Social Committee and the Committee of the Regions: A European Strategy for Plastics in a Circular Economy. *COM(2018) 28 Final*, **SWD(2018)** (1), 1–18.

18 COWI, DG RTD, and Utrecht University (2019). Environmental impact assessments of innovative bio-based products.

19 Sala, S., Cerutti, A.K., and Pant, R. (2018). *Development of a Weighting Approach for Environmental Footprint*. Luxembourg: European Commission, Joint Research Centre, Publication Office of the European Union.

20 Carvalho, J.L.N., Nogueirol, R.C., Menandro, L.M.S. et al. (2017). Agronomic and environmental implications of sugarcane straw removal: a major review. *GCB Bioenergy* 9 (7): 1181–1195.

21 BSI (2011). Publicly Available Specification PAS 2050: 2011. Specification for the assessment of the life cycle greenhouse gas emissions of goods and services, pp. 1–45.

22 ISO 14067:2013 (2013). *Greenhouse gases — Carbon footprint of products — Requirements and guidelines for quantification and communication*. International Organization for Standardization, p. 64.

23 Navigant (2019). Total GHG emissions worldwide: 53.7 Gt CO2 eq (2017).

24 FAO (2010). Global Forest Resources Assessment 2010. Main report.

25 European Commission (2018). A European Strategy for Plastics in a Circular Economy. p. 18. https://eur-lex.europa.eu/legal-content/EN/TXT/?qid=1516265440535&uri=COM:2018:28:FIN.

26 Silvennoinen, K., Heikkilä, L., Katajajuuri, J.-M., and Reinikainen, A. (2015). Food waste volume and origin: case studies in the Finnish food service sector. *Waste Management* 46: 140–145.

27 Razza, F., Fieschi, M., Innocenti, F.D., and Bastioli, C. (2009). Compostable cutlery and waste management: an LCA approach. *Waste Management* 29 (4): 1424–1433.

28 Calderón, L.A., Iglesias, L., Laca, A. et al. (2010). The utility of life cycle assessment in the ready meal food industry. *Resources, Conservation and Recycling* 54 (12): 1196–1207.

29 PlasticsEurope (2010). Eco-profiles PlasticsEurope.

30 Weiss, M., Haufe, J., Carus, M. et al. (2012). A review of the environmental impacts of biobased materials. *Journal of Industrial Ecology* 16 (Suppl. 1): S169–S181.

31 Hottle, T.A., Bilec, M.M., and Landis, A.E. (2013). Sustainability assessments of bio-based polymers. *Polymer Degradation and Stability* 98 (9): 1898–1907.

32 Chen, G.Q. and Patel, M.K. (2012). Plastics derive d from biological sources: present and future. *Chemical Reviews* 112: 2082–2099.

33 Broeren, M.L.M., Molenveld, K., van den Oever, M.J.A. et al. (2016). Early-stage sustainability assessment to assist with material selection: a case study for biobased printer panels. *Journal of Cleaner Production* 135: 30–41.

34 van der Hilst, F., Hoefnagels, R., Junginger, M. et al. (2018). Biomass provision and use, sustainability aspects. In: *Encyclopedia of Sustainability Science and Technology* (ed. R.A. Meyers), 1487–1517. New York, NY: Springer.

35 Lamers, P. and Junginger, M. (2013). The 'debt' is in the detail: a synthesis of recent temporal forest carbon analyses on woody biomass for energy. *Biofuels, Bioproducts and Biorefining* 7 (4): 373–385.

36 Myhre, G., Shindell, D., Bréon, F.-M. et al. (2013). Anthropogenic and Natural Radiative Forcing. *Climate Change 2013: The Physical Science Basis The Working*

Group I contribution to the Fifth Assessment Report of the Intergovernmental Panel on Climate Change (IPCC), pp. 659–740.

37 WMO, NASA, NOAA, UNEP, and EC (1999). Scientific assessment of Ozone depletion.

38 Rabl, A., Spadaro, J.V., and Holland, M. (2014). Description of the RiskPoll software. *How Much Is Clean Air Worth?*

39 Frischknecht, R., Braunschweig, A., Hofstetter, P., and Suter, P. (2000). Human health damages due to ionising radiation in life cycle impact assessment. *Environmental Impact Assessment Review* 20: 159–189.

40 van Zelm, R., Huijbregts, M.A.J., den Hollander, H.A. et al. (2008). European characterization factors for human health damage of PM10 and ozone in life cycle impact assessment. *Atmospheric Environment* 42: 441–453.

41 Posch, M., Seppälä, J., Hettelingh, J.P. et al. (2008). The role of atmospheric dispersion models and ecosystem sensitivity in the determination of characterisation factors for acidifying and eutrophying emissions in LCIA. *International Journal of Life Cycle Assessment* 13: 477–486.

42 Struijs, J., Beusen, A., Zwart, D., and Huijbregts, M. (2010). Characterization factors for inland water eutrophication at the damage level in life cycle impact assessment. *International Journal of Life Cycle Assessment* 16: 59–64.

43 Milà i Canals, L., Bauer, C., Depestele, J. et al. (2007). Key elements in a framework for land use impact assessment within LCA (11 pp). *International Journal of Life Cycle Assessment* 12 (1): 5–15.

44 Boulay, A.-M., Bare, J., De Camillis, C. et al. (2015). Consensus building on the development of a stress-based indicator for LCA-based impact assessment of water consumption: outcome of the expert workshops. *International Journal of Life Cycle Assessment* 20 (5): 577–583.

45 van Oers, L., de Koning, A., Guinée, J.B., and Huppes, G. (2002). Abiotic resource depletion in LCA. Public Works and Water Management.

46 Saouter, E., Biganzoli, F., and Ceriani, L. (2013). Environmental Footprint: Update of Life Cycle Impact Assessment Methods-Ecotoxicity freshwater, human toxicity cancer, and non-cancer. JRC Tech. Rep. Eur. Comm.

14

How to Create "A New Plastics Economy"? Marketing Strategies and Hurdles – Finding Application Niches

Sil Nevejans and Stefaan De Wildeman

B4Plastics R&D, IQ-Parklaan 2A, Dilsen-Stokkem 3650, Belgium

14.1 Introduction

As was already thoroughly discussed in the previous chapters, the world anno 2021 seemingly has sufficient innovative ideas, (bio)chemistry researched, and novel materials prototyped which offer all the necessary bricks to start building the new plastics economy. However, the bottlenecks for bringing an idea or concept from the tech lab to a viable product in the market are manifold: after technicalities are tackled, it comes down to the wider business context such as production equipment (CAPEX), production capacity, economy of scale, marketing, intellectual property rights, and all of this in a rapidly changing legal framework. The social impact of new (bio)chemistries and material products cannot be overlooked. Debates about plastics and their use are appearing at all levels in social media. The awareness about old and new, about polluting and safe, about depleting and sustainable, is growing continuously – from expert to private consumer. And that is a positive trend: we have been tapping into the apparent richness and force of the fossil bubble, but many limits are coming up at an accelerated pace. The big words are out: biodiversity turns into mass extinction, a stable climate turns into uncontrolled tipping points, not to say that pandemic outbreaks are catalyzed by our ultra-global ways of acting in our business as well as private settings. The plastics industry is closely linked to many of these elements – in an increasingly holistic way.

This chapter discusses the business and socio-economic context of the various technical solutions discussed throughout this book: how to transfer viable products from concept to real products participating in tomorrow's new plastics economy? If we want to sustain the historically peaking living standards that many highly developed regions on the planet enjoy today, we need to transition to an industry that supplies the materials to support this situation in a sustainable way. Although there is common sense that this should happen, the road is paved with many challenges. As the attention for the plastic pollution grows, it seems these challenges will

Biodegradable Polymers in the Circular Plastics Economy, First Edition.
Edited by Michiel Dusselier and Jean-Paul Lange.

and are to be overcome by an expanding group of successful teams and companies around the globe offering sustainable solutions and finding their way to the market.

14.2 Stories from the Past

Starting from a breathtaking kick-off of the fossil plastics industry back in the 1930–1940s of the twentieth century, the industry considerably changed its rules over time. Some 80 years later, the industry has become massive. Investments have risen accordingly, and market leaders are confronted by ever-diminishing margins that can only sustain profitability via their often-depreciated mega plants. Exploration turned into maintenance and expansion: more of the same. "Incremental" has become the name of the game. Economy of scale has been ruling. Based on the past, it has become clear that more than a decade is needed to launch a new polymer – from discovery till full commodity. The driver for change has been decreasing over time because the fossil generation of plastic materials has been maturing and optimizing till unseen heights: in the framework of the fossil era they just work great.

However, the framework is changing nowadays. "Green" is rapidly shifting from *nice to have* to *need to have* and is becoming part of the business case. If big brands tend to follow their customers, and these customers want to live sustainably, the direction becomes very clear. As a consequence, the common sense about "green" is being sharpened at a fast pace: in life-cycle assessment (LCA) expertise all over the world, in legislation focusing on sustainability, and – not to underestimate – via social media and daily exposure to everyday products. In that sense, humanity is entering a new awareness era: "If we do not control our resources and waste, we cannot sustain our current wealth." Finding the materials and their production processes that match this new reality is one of the important challenges to overcome. The many examples in this book give us a positive technical outlook, albeit not as mature as other green developments such as solar or wind energy. Though the challenge for new sustainable plastics also goes beyond technology, the examples below prove that all parameters of the business context should switch to "green" too before a new biomaterial finds its way to the market.

14.2.1 Polyhydroxyalkanoates (PHAs)

Although massive cumulative R&D budgets have been spent on polyhydroxyalkanoates (PHAs) over the last three decades, their breakthrough has still not fully happened so far. Though renewed production promises have been made in the recent years, PHAs have long suffered from a few disadvantageous characteristics:

- They often come with complex purification – adding to cost
- They often have the characteristics to crystallize slowly upon processing
- Once they solidify fully, the natural polyhydroxy butyrate (PHB) variants tend to become brittle.

Many of these persistent technical flaws have been largely overcome in the developments of the last decade (Chapter 6 on PHAs). However, also other elements have blocked the introduction of PHAs for several decades. It is clear that the business factors have been interwoven with the technical flaws. Historically, the cost perspective of PHAs has been stuck in suboptimal growth efficiency of the PHA-producing microbial strains, and their harvesting from the microbial biomass. The most important hurdle for their entrance into the market might have been caused by their cumbersome processing: when conventional equipment and conditions are used, PHAs in their natural form tend to underperform vs. established fossil plastics such as polyethylene and polypropylene, while their cost outranges these polyolefins by far. In other words, the balance in cost – functionality – ecology has not been favorable enough to promote their market entrance. Over time, both technical improvements and business attractiveness for introducing bioplastics have increased the marketability of PHAs. Examples of companies producing commercial and specialty PHA materials are Bio-on, Danimer Scientific, Eggplant Srl, Kaneka, Pharadox, RWDC Industries, and TerraVerdae Bioworks (http://www.bio-on.it/index.php; https://danimerscientific.com; http://www.eggplant.it; www.kaneka.co.jp/en; https://pharadox.com; https://www.rwdc-industries.com; https://terraverdae.com).

14.2.2 Polylactic Acids (PLA)

Polylactic acids (PLA) have also built up a multidecade history in the bioplastics industry, starting in the late 1980s. They found their way to the first kilotons since 2010 after almost two decades of specialty applications. Nowadays, their low-cost profile combined with their food-contact certification allowed them to enter in food packaging markets, such as thin films. However, their intrinsic properties also caused popularity dips, such as deforming coffee cups when containing hot liquid (because of a low crystallinity degree combined with a glass transition temperature of about 50 °C – lower than hot coffee or tea) or unpleasant crisping of chips bags – even the sound of bioplastics has turned them at some moments against their swift market uptake. Not helping their reputation, often the "biodegradable" label had been claimed for PLA, already since the time biodegradability was not yet documented very well: often they have been found in leftovers of industrial composters. Indeed, their biodegradation is proven but is generally not fast enough to guarantee disintegration and full mineralization in an industrial composter. Since some leading companies have upscaled PLA production and the launching applications have been saturated over time, the more than three decades of business development with PLA materials have led to a mature business with strong growth figures. Commercial PLA materials are produced by Corbion, Floreon, NatureWorks, and Ricoh-PLAir (http://www.corbion.com; https://www.floreon.com; https://www.natureworksllc.com; https://industry.ricoh.com/en/plair). Mixtures of both PHA and PLA are also used in commercial materials, for example at Bioplas Plastic, GoodNaturedProducts, Nonoilen, and Vegware (www.bioplas.com.au; https://goodnaturedproducts.com; https://nonoilen.com/en; https://www.vegware.com/uk-en).

14.2.3 Polyethylenefuranoates (PEF)

As a final example, polyethylenefuranoates (PEF) as a plant-based, recyclable, and degradable alternative for the widely used polyethylene terephthalate (PET) bottles have been largely discussed in literature [1]. They have been picked up by several parties in the market. Starting from fructose-rich or fructose-yielding sugar streams, several pathways have been explored and piloted. Avantium has been marketing PEF as the outperforming sustainable solution to PET as they developed and patented the YXY technology platform, converting fructose into furandicarboxylic acid (FDCA), and have established partnerships with several big brands, e.g. with Coca-Cola, L'Oréal, and P&G under the Paper Bottle Project (https://www.avantium.com/technologies/yxy). Next to Avantium, Corbion has developed a proprietary process to produce 2,5-FDCA as plant-based monomer for PEF and works on implementing this building block into the value chain of partners (http://www.corbion.com/bioplastics/fdca). After the discovery of an efficient and fast method for PEF production at the ETH Zürich, Sulzer now works on this proprietary ring-opening polymerization (ROP) technology for implementation in industrial mass production [1] (https://www.sulzer.com/en/shared/applications/polyethylene-furanote-pef). Still, however, after a decade of efforts toward the commercialization and market introduction of PEF, no PEF bottles are yet commercially available on the market in 2021. This shows that even with a strong technology and with powerful partners, the swift introduction of a novel and long-awaited biopolymer into the market is not guaranteed. Still, all of the above companies and partnerships pave the way toward a new plastics economy.

14.3 Greenwashing vs. Growing Pains

Almost together with the birth of sustainable initiatives, the term "greenwashing" has appeared, and its resulting perceptions have been growing alongside the development of green alternatives. The phenomenon has been discussed by researchers from various sectors, e.g. business, communication, and sciences, and is commonly described as a deliberate corporate action with the presence of misleading elements, focused on the deception of stakeholders [2]. In the race toward zero-waste and zero-plastic commitments, companies have been forced to make promises in their sustainability campaigns that they could not keep up with in following sustainability reports [3]. Although it can be questioned if it is better to break promises than not to make any promises of efforts toward a more sustainable planet at all, it all comes down to the same result: companies do not meet the goals set which leads to consumer distrust. Often established plastic producers are stuck in conservative systems and invested production assets that do not allow fast implementation of sustainable visions and missions, let alone products and materials. To maintain a clean brand name, however, their marketing branch is pushed to promote their good intentions. Unfortunately, the rigid back offices, ruled by many structures, cannot live up to the goals set or claimed commitments and hence many companies are trapped into the greenwashing pitfall. Often, still, the story does not stop with their inability to deliver but continues with hidden corporate actions to stop legislation or pushing responsibilities to consumers.

Ways to overcome greenwashing and initiate global steps forward involve all stakeholders. Policymakers should introduce policies, bans, and taxes with extended producer responsibilities, while companies need to focus on supporting legislation, transparency, and reduction in the use of resources, and consumers have to stick to several *R-words* of waste management – refuse, reduce, reuse, repurpose, recycle – while consciously supporting legislation [3].

Despite this, one should also recognize that greenwashing accusations have followed on genuine intentions of innovation that could not be realized due to one of the many obstacles on the unpaved road toward the new plastics economy – even by very thought-out strategic teams in big corporates. This is counter to the argument on intentional greenwashing and can be attributed much more to "growth pains". After all, it is important to also ground these allegations scientifically and consciously analyze them accordingly. Misperceptions due to poor scientific backing or bad marketing can lead to false accusations of greenwashing and unjustified greenwashing complaints. The European Single Use Plastics (SUPs) Directive, for example, is a reasonable initiative to reduce the use of (sometimes non-essential) SUP articles. Still, the corresponding attention obtained can lead to misperceptions on the environmental impact of these articles compared to other LCA elements in other single-use products. For instance, the food inside a plastic packaging is usually accountable for a larger environmental impact than the packaging itself: food supply might here be a decisive element in the overall LCA [4].

Finally, it is easy to accuse a company of greenwashing, but it is then very hard for the accused company to lose that stamp. Companies, small medium enterprise (SMEs), scale-ups, and start-ups cannot be perfect from the start in their urge toward the new plastics economy. Before pointing a blaming finger, one should consciously analyze the efforts done and steps taken in the right direction – where proportion is key and should be balanced out. If a market leading company is only offering 5% of its effort toward sustainability, and it cannot fulfill 90% of the goals set, we have more reasons to criticize than blaming a start-up that is fully focused on the development of bio-based materials (i.e. 100% effort) and has been missing 5% of its commitments. Companies should act fast, cluster their competences, trust each other, and advance in the right direction but should also be given the liberty of making acceptable mistakes when doing so. It is still a lot better than doing nothing at all.

14.4 From Idea to Product: "Technical Readiness Levels"

14.4.1 Defining the Technical Readiness Levels

Since companies have started to "re-invent" themselves and have focused on renewing their customer base in view of a more sustainable future, the innovation process has received increasing attention. In this paragraph, the process to introduce novel products in the bio-based materials markets will be explained along the so called "Technical Readiness Levels" (TRLs). In the meantime, the language to describe the different levels has been standardized in subsidized innovation programs on a national up to continental level, such as the Horizon Europe Program in Europe and National Science Foundation in USA.

To indicate the status of innovation from idea to product, the following definitions described below apply for each TRL, unless otherwise specified. Additionally, to clarify its use, we apply the list to a *typical innovation pathway in bio-based materials* running from TRL1 to TRL9 according to proper expertise in leading innovation projects that are handed on a daily basis – as at the moment of auditing, both authors are participating in 11 national and European innovation projects developing novel biomaterials:

- TRL 1 – basic principles observed
  "The idea" (virtual = in minds or on paper)
- TRL 2 – technology concept formulated
  "Making the idea generally understandable and partially confirmed" by showing a proof-of-principle
- TRL 3 – experimental proof of concept
  Demonstrate minimal lab results that support the idea: mostly mg to g scale experimentation with analytical detection of target compounds or materials
- TRL 4 – technology validated in lab
  Develop the minimal results to a robust lab protocol that can be repeated and gives first sample material: mostly up to 1 kg
- TRL 5 – technology validated in (industrially – in the case of key enabling technologies) relevant environment
  Transfer the protocol out of the lab into a pilot plant that represents a miniaturized form of the characteristics of a full-scale production plant: mostly to produce the first 10–100 kg
- TRL 6 – technology demonstrated in (industrially – in the case of key enabling technologies) relevant environment
  Further upscale in similar miniaturized equipment up to 1000 kg, giving access to "demo sample material"
- TRL 7 – system prototype demonstration in operational environment
  The demo sample materials attain a "minimal viable product" label since it can be validated and certified on industrial processing equipment in-house or at partners: the sample material is converted and validated into downstream products
- TRL 8 – system complete and qualified
  The parameters of the demo production plant are fixed and ready for transfer to a full production plant
- TRL 9 – actual system proven in operational environment or in competitive manufacturing in the case of key enabling technologies
  Launchable production samples can be produced in a production plant

TRLs go hand in hand with time and budget consumption. If we compare three TRL clusters from TRL1 to TRL9, we can differentiate as follows:

- TRL1–4: lab scale up to 1 kg, often one to three years R&D, between 10% and 20% of the total R&D budget
- TRL5–6: pilot and demo scale between 10 and 1000 kg, often one to two years R&D, between 25% and 75% of the total R&D budget
- TRL7–9: validation and certifications of pre-production samples, often about one year R&D, between 25% and 50% of the total R&D budget.

Many innovative bioplastics companies make use of an objective stage gate process, to reach consensus about positioning innovative leads in the pipeline, and to control innovation costs and risks. Most often, cross-departmental evaluations are organized to guide the innovation process (R&D – marketing – sales – purchasing – production). The aim is to position each innovation lead in the pipeline of upcoming novel products. Criteria set by the different departments are very important to prioritize the innovation budget. This in its own indicates that bringing novel bioplastics to the market, goes far beyond technology alone. The balance between the anticipated market potential of the lead, its risk profile, and its uniqueness is a determining factor to shift innovation leads up or down in the pipeline, leading to higher or lower TRL levels.

14.4.2 Application of the TRLs

In this budgeting exercise and the used criteria to shift innovation leads up and down on the priority list, innovation opportunities can be categorized along different parameters. Generally big corporates tend to promote more de-risked opportunities that can have substantial impact by their high volume, in return for the required investment. At the other side of the range, start-ups accumulate high risks in their opportunities and search the impact more in the disruption of the market to change the rules of game. This might also require high investments, leading to a whole new investment world from *angel* to *venture* to *corporate* investors, besides increasing subsidies, awards, and governmental support in general. In that sense, disruptive start-ups and big corporates come ever more in a symbiosis with each other, handling other but complementary risk parameters to guide the innovation opportunities. Even more, many start-ups increase their credibility by linking their innovation to established companies in the field, while the latter are stimulated to scout for disruption at start-ups. They are searching and finding each other more and more.

Though the DNA of big corporates and start-ups shows a substantial psychological difference: most big brands state "to follow their customers," while start-ups choose rather "customers to follow them." No doubt the latter is by far the strongest guarantee for radically new products, but the chance of success of customers following an unknown brand is indeed much lower. This plays an important role, even nowadays, in the plastics industry since the experience on bioplastics stretches broader over the value chain. With block chain technology and increasing traceability methods throughout the chemical industry, people will gradually gain understanding about where their materials come from and what it is. These insights will help them to fulfill their unmighty feeling of how to contribute to a sustainable plastics industry by making responsible choices. The value linked to these new insights will further determine the popularity of certain bioplastics compared to others. The hidden times for chemicals and plastics will soon be over: consumers have the right to know where the materials come from, who has made them, and if they fulfill their needs and desires. In that sense, the plastics industry can follow the paths taken by the food industry, where traceability has reached higher standards.

Regarding budgeting innovation opportunities through the TRLs, there are more common cost differences between running or starting big and small businesses. If we spread our thoughts over the innovation funnel, we must reach out from TRL1 to TRL9. In the ideation and discovery phase, up to TRL3, costs are highest in human capital and lowest in CAPEX. Typically, in this phase, a highly innovative research team is active in state-of-the-art labs. Today, these research crews are surrounded with innovation partners, which require competences like project management, human resource management, and IP management to run the cooperation. Already in this phase, relations are built up with investors and banks, subsidy agencies, governments, and schools. In socially driven research teams creating breakthrough companies, society is involved in all its facets already from the beginning onwards. Human resources also play an increasingly important role: employees search much more nowadays than a solid salary package. In this stage, with the help of subsidy support and angel investors for disruptive innovation, new projects can root at the start of the innovation funnel. It is important to create sufficient supporting "family, friends, and (a lot of) fools" (FFF) to try out apparent crazy innovations.

Once reaching TRL4–5, CAPEX increases considerably to prove that the novel product can be viable in an industrial environment. Pilot equipment is designed, invested in and/or finetuned to confirm the novel material in a scale most often situated between 10 and 1000 kg. Here, the focus lies on safety measures, robustness, accurate monitoring of process parameters, and achieved purities of target materials. Angel investments switch more and more to venture and corporate private equity in a so-called round A, which corresponds nowadays with a single-digit million investment. During TRL6, ending the development phase, early calculations for pricing and LCA become more reliable because of the industrial conditions of the trials. It prepares for the last part of the innovation funnel: TRL7–9 (representing demonstration and deployment).

At the end of the innovation pipeline, production and market launch are further prepared, involving security of supply, certifications, batch repetitiveness studies, quality and control (Q&C), energy and raw materials balances, demo materials for marketing campaigns, and optimizing production protocols. Also maintenance schemes are being scheduled and installed. Depending on the equipment that can be used for future productions, this stage can be either relaxed in CAPEX (when equipment is installed with free capacity) or become very CAPEX intensive when unique equipment must be designed, tested, engineered, and newly constructed. Once reaching this last part of the innovation funnel, it is necessary to plan production and capitalize for it. This is the time for investment rounds B and C rising in a double- and sometimes even triple-digit million investment nowadays, respectively. Going hand in hand, the employee number is rising along in the double- and triple-digit volume. Worth to mention: from this point onwards, green field installation of new production units requires most often at least two years to become operational in the plastics industry, while the chemical industry is one of the most CAPEX-intensive market sectors throughout the industry.

14.4.3 Product(ion) Validation

Once we design and scale novel bioplastics, product validation and certification might pose a challenge. Often plastic granulate-processing equipment is not fit to accommodate novel plastic grades. Otherwise extra investments are necessary to tune the equipment to reach acceptable processing conditions. Also, a novel plastic must overcome the maturity level that is often obtained for established fossil plastic types. Specification sheets have been created decades ago for the established plastics: it remains a challenge for new plastic types to match these requirements and meet all expectations from the beginning onwards. Processors or brand owners should be intrinsically motivated to overcome these hurdles, being patient and trustful that all necessary parameters can be sufficiently tweaked till a satisfying level. Also analytics and polymerization (bio)chemistry should be engineered in such a way that they can pivot further toward the desired key requirements – if achievable at all. Besides validation, certification is often required for novel plastic types. From biodegradation to food contact to registration, evaluation, authorisation and restriction of chemicals (REACH) certifications: all take time and budget. This all adds up to the overall risk that should be considered. A certification can be fully derisked, but can mean that the innovation falls out of its optimal time window to enter the market. Candidate world-leading innovators in the biomaterials sector that request too much certainty on a technical or economical level early onwards are in that sense mostly not the pioneers and frontrunners in their field. Being a pioneer means that you are prepared to pay the price of taking calculated risks – which can fail. Only companies with this truly innovative DNA get a long-term competitive edge and are trendsetting. Sometimes, in this respect, there is a discrepancy however between receiving credits and accumulating cash by pioneership.

Many companies use stringent multidisciplinary protocols to shift innovation leads from TRL1–3 to TRL4–6 to TRL7–9: a lot of innovation budget can be saved by selecting the best risk-mitigated set of leads to promote them to next TRLs in the pipeline. "Fail fast" is a beneficial competence in innovative companies, since "fail slowly" simply costs much more innovation budget.

14.5 Five Innovation Rules to Create "A New Plastics Economy"

Companies and company cultures grown in the fossil era have several characteristics that are challenged by the current transition to a bioeconomy. That is certainly also the case for the transition within the field of biomaterials. In this paragraph, we will highlight five attitudes that are not evident in the often-rigid structures of big corporates, but that are proving increasingly successful to bring novel biomaterials from idea to innovative product in the market. Below, we list five rules for an optimal innovation culture at biomaterials disruptor companies nowadays – the antagonist behavior of standard material companies that have grown to market leaders in the fossil era.

14.5.1 Target Small-Volume, High-Value Applications to Open New Market Space

New materials can go big, but not from day one. Actually, every material starts from scratch, from an idea and a first gram. Often that evokes an extreme exercise in humility for scientists and investors that aim for big impacts – but have to start from the first grams and kilograms onwards. That is how polyethylene and polypropylene were created, the blockbusters of the (mostly) fossil plastics industry today. It is also how it is expected to proceed for the winning biomaterials of the new plastics economy. *If we would know who they are (these winning biomaterials), they would exist already at big scale.*

There is a general tendency to call something "big" when it is big in relation to the company where the idea originated and development takes place. For big corporates, that amounts to substantial business, whereas for scale-ups it can mean a novel business with rather limited revenues. Also here, "big" is a relative measure. For each product class, optima and thresholds can be calculated where margins are optimized and highest business profitability is obtained. This indicates for each market segment what should be called big or small. An example is found with the PET business that flourished worldwide in the 1950s and 1960s via batch-line processes of several *kT* plate capacity each. It allowed to produce in a flexible manner and develop specialties to reach higher margins. Though when massive PET production facilities were built as from 2000 onwards, these flexible plants lost their competitiveness because their lower scale brought non-competitive fixed costs. In other words, what was big enough to run in the 1950s became too small to survive in the same business less than 50 years later. This sweet spot has to be determined for each biomaterial – of course one has to rely on assumptions if the market is even still under construction.

An example for this is found in the massive trials of Avantium to launch FDCA and PEF on the market. To reach its goals, the company has received major investment capital to aim "big" for PEF production with its technology estimated at a market worth of over US$ 200 billion [5]. However, in this case, apart from finding the sweet spot about volume, the competition with conventional PET that is increasingly recycled, has been shown to be tough.

A further consideration about scale is given in the following thought. When expanding and debottlenecking chemical and plastics sites, specific fixed costs drop considerably. One could therefore argue why not to make one very big production site for every chemical or plastic that is existing – covering the world capacity for that chemical or plastic from one central point on the planet: "One production site per plastic type." The answer is simple: at a certain moment the transportation to bring raw materials to the site and to get the resulting plastic products away from the site outweighs the gain in specific fixed costs. That means that economies of scale will always be limited by logistics. In a bio-based plastics industry where most raw materials evolve into bio-feedstock, these limitations become more stringent and follow completely other rules. Even more, biomass is growing because "every square meter on earth" receives an equal amount of sunlight and CO_2, besides different nutrients and soil or aquatic conditions to determine the overall fertility package on each local spot. Biomass growth is therefore organized in an extremely decentralized way by nature since almost four billion years of evolution. Consequently, the most efficient bio-based plastics industry is also partially decentralized. This means that transitioning from a fossil to a bio-based industry comes with spread-out production sites following other rules in economy of scale.

And then there is the eagerness to (be able to) predict which material will go big. How many innovative pipelines have been managed from the risk-balanced "calculation" that *they would go big*? And how many times has reality confirmed this calculation? Predictions on how big novel materials can go in the market are seldomly proven accurate and therefore can rather be considered as hypotheses instead of predictions. Just as the target application only becomes clear after thorough test programs, the size of the market can only be estimated after sufficient validations and certifications – mostly an accumulation of many trials and adjustments. And that brings us to the second dogma.

14.5.2 Time Right Instead of Fast

Besides going big, most business plans want to take control of time. If there is one parameter that is hard to control, it is the exact timing to drop the innovation in the market. That is because many elements that open the opportunity window for launching a new biomaterial go far beyond the technical influence of a single company. Nowadays, the rhythm of biomaterial innovation is determined by multiple effects, some hampering, others stimulating the speed of development:

– the supply of sometimes seasonal- and climate-dependent renewable resources
– the production capacity to convert them into bio-based building blocks or materials
– the timing to design and construct novel production equipment for novel processes
– a lacking driver for change in application fields that match with the material properties
+ invested money that awaits return on investment
+ time-ticking patents and IP in general
+ media attention that creates unplanned momentum
+ consumer surveys and social media that push the transition.

Again, "fast" should be seen related to the context where the innovation has to happen: big corporates are generally impactful but slower, while start-ups and scale-ups start from scratch but can handle in a very agile way. When we talk about innovating with intrinsic novel materials containing new molecular backbones, there is again a common understanding about what is fast or slow. While the beginning of the fossil plastics industry has been characterized with swift developments, the innovation speed has been slowing down considerably after half a century. For instance, several historical leading European chemical sites, such as the BASF site in Ludwigshafen (Germany) and the Brightlands Chemelot site in Geleen (The Netherlands), have been built at the speed of one to two production plants per year shortly after World War II. By the end of the twentieth century, this speed went down by about one order of magnitude – till plants even got transferred in mothballing status, which is a controlled shut-down of a chemical plant to anticipate for its smooth restart later in time, or were even dismantled (https://www.basf.com/global/en/who-we-are/history/BASF-History-in-Figures.html; https://www.chemelot.nl/chemelot-en/history). Because the market was saturated with "those very mature and cost-efficient fossil plastics," it became very challenging to introduce any new materials in the market. The ideas and developments to overrule current plastics were scarce and risky, and the market had no driver for change. This led to a general statement in market leading corporates that "entering a novel plastic on the market takes at least 10 years," which also seemed to be confirmed by the market.

But then some existential questions were asked and start-ups were appearing in the chemical industry at the start of the twenty-first century – mirroring the developments in the pharma sector since the 1990s [6]:

- *Venture funds came alive in the chemical industry through the following question*: "Do we still invent novel plastics ourselves, or do we look for successful start-ups and acquire the technology once it is de-risked?"
- *But especially in the case of bio-based plastics*: "Do we have the time to wait for 10 years to develop novel bioplastics that present less ecological concerns?"

Currently, typical breakthrough bio-based plastics developers are underway to transfer their innovative materials from TRL1 to TRL9 in less than five years, reaching multiton scale to even kiloton scale in a timeframe of less than seven years. The tendency is that this timeframe will be further reduced and will be fueled by the sustainability revolution that is currently climbing to unseen heights.

14.5.3 Go Local

Since neoliberalism, our sense for global activities is overdeveloped. Fossil energy and carbon sources have contributed to this globalism since they have made transport costs very low. We take it for granted that our pen is made from a plastic type containing carbon atoms that are originating from fossil resources harvested at the other side of the planet. In the fossil industry, it is normalized that a pen will travel easily 20 000 km to find our hands to write with it. If this looks eye-opening today, it might be labeled *unforgivable* by our children.

Of course, there is a reason why we are not aware or conscious about these global journeys: plastics have become too cheap and too anonymous to steal our thoughts and our precious time. In the Dutch language, another word for plastics is, literally translated, "Art-Stuff" but they leave an impression far from anything artistic. They are too common for that – they slash our brains – even more: we hardly remark their presence. How many car owners know that there is more than 300 kg of plastic in their vehicle, let alone where it would come from? Even the salesman of the next-door car shop has nothing in his value proposition that comes close to anything like plastics. Plastics, apart from their claimed recycling percentage, are simply hardly or not at all mentioned in the catalog, apart maybe from a few articles that you can touch in the interior space, such as grass-filled panels or a recycled dashboard strip. Their absolute minority is meant to neutralize any guilt feeling – and surprisingly it still does. But luckily not for long anymore.

This global attitude – also counting for other materials like metals or glass – that dominates the current fossil plastics economy, has grown over the last centuries but demonstrates to have an irreversible effect on our planet [5]. In contrast, technical developments show a very promising outlook to disrupt this ultra-global non-conscious model: they show that the carbon in bioplastics now can be much more local and renewable. This evolution is even stimulated by logistics that are prone to hit new price records. In the case of PLA for instance, the LCA can be significantly improved when the resources for this material are harvested and used close to the respective market. Bio-based materials lose their effect when they travel half the planet to reach their market: their proximity between creation and use is just a differentiator vs. virgin fossil materials.

14.5.4 Take Risks

There is another dogma that is rooted deeply into the maturity of the current plastics industry: "Why should we take risks if our fossil industry is (still) doing so well?" Indeed, the plastics industry will nearly triple in the next three decades only, driven by demand, hitting close to one billion ton of plastics production per year by 2050 [7]. This evokes a dilemma: should we disrupt the plastics industry in the coming years with novel materials, or pay the climate bills of a *supermature fossil plastics industry* exhausting close to one billion tons of plastics to the planet every year?

So far, novel biomaterials have shown that their popularity or profitability cannot be predicted. To know their fate, one still must go through that dark tunnel of uncertainty: entrepreneurship in balance with good financial control is the best recipe – gradually stepping up till the products have been de-risked. It goes hand in hand with launching novel bioplastic specialties, that, upon successful market introduction, hit lower costs as they scale, opening more market space. The path flows from the top of the plastics pyramid to sink to levels that balance out pricing vs. economy of scale. Moderating the rights steps along this path, with the right timing and sufficient patience, is the best risk mitigator one can wish for creating a new plastics economy. For start-ups, this means that (back)loans, bank credits, angel investments, and subsidies span the financial strength required for each of the steps to take. And although the path is unknown, dividing it in overseeable steps helps to keep upright and move forward.

14.5.5 Go "Green"

Another challenge for plastic materials is the length of their lifetime. We created a lot of fossil plastics, but hardly laid-off some. They are landfilled. They are linear. Comparing plastics with the history of life on our planet, it seems the young existence of plastics is in no sense in line with the success factor of life in general: "Survival of the fittest," in which circularity rules. Ironically, almost all plastics that were ever made are somehow surviving so far – if not burned to black carbon dioxide – but not in a way that is compatible with nature. Now, as new ecological trends are kicking into our lives in numerous ways, also plastics should move on toward more sustainability by changing the rules for their existence. The examples in this book show that many fossil plastics can be, apart from proper recycling, substituted in their applications with alternatives that make up for a new ecological way of fulfilling the desired functionality. In that sense, plastics and their uses are starting their second generation – the new plastics economy – in search for their ecological siblings that fit a modern sustainable society. None of them should be invented or launched to grow forever. They should only last until a new optimized product has been developed that is more beneficial to sustain all life on this planet.

There is another trend that dominates the fossil plastics industry: plastics should be high performing. In many cases, "high" performance can be substituted by "out" performance. Many plastics are even "over" performing for their application. This is in contrast again with nature: all materials that have been accurately optimized to build fauna and flora, and human beings, have never been selected based on their "out-performance." They just did perform *enough* for what they were meant for. And that performance also contained an ecological performance: bio-based content, biodegradability, and unlimited recyclability by breakdown to biogas and biomass. If we start to suffer from CO_2 exhaust and from microplastics, why would we continue to outperform in a classical way? The classical way where plastic production creates high levels of CO_2 in their production and logistics, and in their toxic end-of-life? Would we not switch "out" for "enough" and get sustainability in return? Because we tend to forget that "enough" has been the key word of natural evolution on Earth, just as Darwin's theory showed that the "fittest" was the one that was "fit enough" to survive.

Talking about fit enough, one can argue how to define it for plastics. In that sense, "perfection" and "standardization" became credos of the fossil plastics industry. In fact, both are in contrast with nature again. Perfection and standardization are forbidden terms to guarantee a successful evolution promoting *the fittest*. How can you even have a fittest under perfect standardized conditions – making only copies and clones? *Cloning is nature's sin.* Nature worked for billions of years to perfection of DNA-based life on Earth: the one that allows a seldom *error* in the DNA code, besides recombining the DNA all the time, to exactly create what makes a living community strong in the end: *diversification, supporting biodiversity*. No two trees, no two people, no two creatures on this planet are exactly the same, since diversity created diverse survivors that were fit to varying circumstances. So why could we not tolerate small imperfections, losing maybe some cost efficiency in the production but allowing much more degrees of freedom in molecular and article design of novel bioplastics? (https://www.plasticpollutioncoalition.org/blog/2017/5/2/10-inspiring-works-of-art-about-plastic-pollution; www.hanaadahy.com). Let us not fight against proven models from nature – let's be inspired by them. Maybe this degree of freedom injects a lot of fresh oxygen to bright minds, legislation and producers, to create and establish the new plastics economy. We will not create it by endless restriction lists, but by shaping opportunities with new materials. Let the selection criteria for bioplastic survivors not be ruled by the fossil plastics industry, but let them be reflected by a world that has the choice between sustainability or human extinction.

14.6 Conclusion

The message of the five dogmas above is as simple as uncomfortable related to the plastics industry: the continuation of the wellbeing of developed societies lies in the hands of consumers, besides disruptive teams in corporates, SMEs, and start-ups, often flanked by academia, trying out new technologies. Open innovation at its best. Ethical radars switched on. It has the character of a gamble. But one can argue that the risk not to gamble is becoming higher than gambling itself – for novel bioplastics that bring functionality together with affordability and unprecedented ecology. Once new materials hit a new balance in their application and become subjected to upscaling dynamics driven by demand, they start to gain the sympathy of innovators, believers, partners, and finally end-consumers. Unluckily for the pioneers involved, there is no guarantee on big or fast successes. There is also a limited tendency to search at the other side of the world what can be found and harvested locally. Certainties are not existing – they have to be created. And the growth of the innovative biomaterial might be moderate or even stop at a certain scale. These pioneers will also have to teach the world that enough performance is good enough, and *imperfection* is the new perfection. Besides scientists and industrialists, they should become an inspiration that is followed – realizing the dream of a sustainable world. If soldiers liberate(d) us from wars, and medicinal forces saved us from a global pandemic, let bioplastics pioneers and technology pioneers in general help to save us from the most urgent threat of all: climate change. As people will also have *to help themselves* based on intrinsic motivation for a more sustainable planet, the transparency and inspiring elements should be safely conceived in novel plastic types. The point is that the origin of this climate change is very clear: it is not originating from a lab, but from the cumulative exhausts of the unreasonable consumption behavior of the civilized population that calls itself "well-being" [8]. And since you are reading this book: chances are big that you are one of them. This means in turn that there is very good news: the impact of your new ecological life is potentially extensive and can start today. Just redesign your daily life, leverage your awareness, choose with responsibility, and *be for plastics* described in this book that guide us to new sustainable horizons!

References

1 (a) Eerhart, A.J.J.E., Faaija, A.P.C., and Patela, M.K. (2012). Replacing fossil based PET with biobased PEF; process analysis, energy and GHG balance. *Energy and Environmental Science* 5: 6407–6422. (b) Rosenboom, J.-G., Hohl, D.K., Fleckenstein, P. et al. (2018). Bottle-grade polyethylene furanoate from ring-opening polymerisation of cyclic oligomers. *Nature Communications* 9: 2701.

2 de Freitas Netto, S.V., Sobral, M.F.F., Ribeiro, A.R.B., and da Luz Soares, R.G. (2020). Concepts and forms of greenwashing: a systematic review. *Environmental Sciences Europe* 32: 19, https://doi.org/10.1186/s12302-020-0300-3.

3 Changing markets foundation. Talking trash – The corporate playbook of false solutions to the plastic crisis. (2020), https://changingmarkets.org/portfolio/talking-trash/.

4 Miller, S.A. (2020). Five misperceptions surrounding the environmental impacts of single-use plastic. *Environmental Science and Technology* 54: 14143–14151.

5 Masson-Delmotte, V., Zhai, P., Pirani, A. et al. IPCC 2021: Climate Change 2021: The Physical Science Basis. Contribution of Working Group I to the Sixth Assessment Report of the Intergovernmental Panel on Climate Change. Cambridge University Press (2021).

6 Ali, S., Dream, R., Jornitz, M. et al. (2018). A 20 year retrospective: the pharmaceutical industry then and now. *American Pharmaceutical Review* https://www.americanpharmaceuticalreview.com/Featured-Articles/354573-A-20-Year-Retrospective-The-Pharmaceutical-Industry-Then-and-Now/.

7 Lange, J.-P. (2021). Towards circular carbo-chemicals – the metamorphosis of petrochemicals. *Energy and Environmental Science* 14: 4358–4376.

8 (a) Berman, J.D. and Ebisu, K. (2020). Changes in U.S. air pollution during the Covid-19 pandemic. *Science of the Total Environment* 739: 139864. (b) Fu, F., Purvis-Roberts, K.L., and Williams, B. (2020). Impact of the Covid-19 pandemic lockdown on air pollution in 20 major cities around the world. *Atmosphere* 11 (11): 1189. (c) Skirienė, A.F. and Stasiškienė, Z. (2021). Covid-19 and air pollution: measuring pandemic impact to air quality in five European countries. *Atmosphere* 12 (3): 290.

Index

Biodegradable Polymers in the Circular Plastics Economy, First Edition.
Edited by Michiel Dusselier and Jean-Paul Lange.